Fish Diseases

Fish Diseases

Prevention and Control Strategies

Edited by

Galina Jeney

Academic Press is an imprint of Elsevier
125 London Wall, London EC2Y 5AS, United Kingdom
525 B Street, Suite 1800, San Diego, CA 92101-4495, United States
50 Hampshire Street, 5th Floor, Cambridge, MA 02139, United States
The Boulevard, Langford Lane, Kidlington, Oxford OX5 1GB, United Kingdom

Library of Congress Cataloging-in-Publication Data
A catalog record for this book is available from the Library of Congress

British Library Cataloguing-in-Publication Data
A catalogue record for this book is available from the British Library

ISBN: 978-0-12-804564-0

For information on all Academic Press publications visit our website at
https://www.elsevier.com/books-and-journals

Publisher: Nikki Levy
Acquisition Editor: Patricia Osborn
Editorial Project Manager: Jaclyn Truesdell
Production Project Manager: Nicky Carter
Designer: Maria Ines Cruz

Typeset by TNQ Books and Journals

Dedication

The book is dedicated to my family—my husband, Zsigmond Jeney, and our lovely daughter, Annamaria Jeney—for their unconditional love and patience. The book is also dedicated to Doug P. Anderson, my teacher and friend.

Contents

List of Contributors

Claude E. Boyd Auburn University, Auburn, AL, United States

Ian Bricknell The University of Maine, Orono, ME, United States

Adelino V.M. Canário University of Algarve, Faro, Portugal

Mohamed Faisal Michigan State University, East Lansing, MI, United States

Trygve Gjedrem Nofima, Ås, Norway

Hana Kocour Kroupova University of South Bohemia in Ceske Budejovice FFPW CENAKVA, Vodnany, Czech Republic

Thomas P. Loch Michigan State University, East Lansing, MI, United States

Jana Machova University of South Bohemia in Ceske Budejovice FFPW CENAKVA, Vodnany, Czech Republic

Ana Patrícia Mateus University of Algarve, Faro, Portugal

Aaron A. McNevin World Wildlife Fund, Washington, DC, United States

R. Dinakaran Michael Vels Institute for Science, Technology and Advanced Studies, Chennai, India

Birgit Oidtmann Centre for Environment, Fisheries and Aquaculture Science (CEFAS), Weymouth, Dorset, United Kingdom

Deborah M. Power University of Algarve, Faro, Portugal

Edwige Quillet INRA, AgroParisTech, Université Paris-Saclay, France

Nicholas Andrew Robinson Nofima, Ås, Norway; The University of Melbourne, Parkville, VIC, Australia

Hamed Samaha Alexandria University, Edfina, Egypt

Ariadna Sitjà-Bobadilla Consejo Superior de Investigaciones Científicas (CSIC), Castellón, Spain

Parasuraman A. Subramani Vels Institute for Science, Technology and Advanced Studies, Chennai, India

Zdenka Svobodova University of South Bohemia in Ceske Budejovice FFPW CENAKVA, Vodnany, Czech Republic

Kim D. Thompson Moredun Research Institute, Midlothian, United Kingdom

Josef Velisek University of South Bohemia in Ceske Budejovice FFPW CENAKVA, Vodnany, Czech Republic

Preface

Fish production has more than doubled and become an increasingly significant source of healthy animal protein worldwide. Under intensive aquaculture conditions, fish are exposed to various stressors that are unavoidable components of the fish aquaculture environment. These stressful conditions caused an overall reduction in performance, including poor acclimation and growth performance, impaired reproduction, increased susceptibility to disease, and poor quality of the final product.

This intensive phase of development of aquaculture was accompanied with increasing numbers of disease occurrence and awareness about the quality of produced fish along with the water effluents from aquaculture facilities. Most of the important fish diseases have been intensively studied, with pathogens described and methods for treatments elaborated. However, most of these compounds were forbidden to use in aquaculture. For example, the use of antibiotics for controlling diseases was widely criticized because it was often very expensive and led to the selection of antibiotic-resistant bacterial strains, immunosuppression, environmental pollution, and accumulation of chemical residues in fish tissues that can be potentially harmful to public health. In the USA and Europe, governmental restrictions have limited the number of drugs that can be currently used in aquaculture and even in Asian countries a strict demand for fish products free of pollutants/antibiotics is increasing.

Due to the complex character of fish diseases, solutions are also complex, where all elements of the fish–environment–pathogens should be taken into account. Majority of the books on fish diseases are mainly focused on pathogens, diagnosis, and diseases treatment, and do not pay much attention to the early prevention, which leads to fish-diseases management falling behind. Good quality water, well-balanced food, and good farming practices are the basic preconditions of a good health status in fish culture. To reduce the losses due to diseases, several methods of prevention and treatment are applied by fish culturists.

This book provides general information about disease prevention in fish by different novel methods, including knowledge of maintaining environmental conditions, stress in fish, etc. Furthermore, it also concentrates on disease control strategies, such as integrated pathogen and area management, biosecurity, and novel disease control techniques. The contributing authors are the world's leading experts in the subjects, drawn from many countries where aquaculture is a significant and expanding part of the economy.

The book structure comes from the pathogen–host–environment triple interaction, developed by Snieszko and divided into three parts.

1. FISH

This part focuses on fish immunology and improvement of disease resistance by genetic methods. There is a need for understanding of factors in which disease resistance may emerge and how to improve immune responses.

Genetic methods described here are very detailed. For most species and diseases studied to date, a significant component of disease resistance is genetically determined, heritable, and affected by many genes. Novel methods of quantitative genetics and animal breeding, accounting for the unique biology and culture conditions for aquatic species, are described and can be applied to produce populations with greater resistance or tolerance to diseases.

2. PATHOGENS

This part of the book concentrates on disease transmission, prevention, and control, rather than just pathogens. How diseases are transmitted is a fascinating research area where environmental factors and the virulence of the pathogen often make the difference between a disease becoming established and the fish remaining fit and well. Understanding how these organisms complete their lifecycle is an important part of appreciating the complex world of pathobiology.

The various prophylactic and therapeutic methods available at present against diseases are also discussed. Essential research approaches, required in future, in the areas of immunoprophylactics and therapeutics are presented.

Integrated pathogen management, included in this part, is a holistic approach that gathers the best available preventive, treatment, and control strategies to minimize the impact of pathogens in fish production, while striving to increase sustainability. These strategies do not necessarily aim to eradicate pathogens; more likely, they are the methods to keep them at levels where they cause relatively little impact on fish welfare and overall farm production, and therefore minimize economic losses.

3. ENVIRONMENT

The major water-quality stressors are suboptimal levels or concentrations of the following: water temperature, salinity, and cation imbalance (usually site- or source water–related); pH, gas supersaturation, and toxic algae (both site- or aquaculture input-related and management-related); and dissolved oxygen, ammonia nitrogen, nitrite, carbon dioxide, and hydrogen sulfide (usually aquaculture input and management related). Adverse environmental conditions may decrease the ability of organisms to maintain an effective immunological response system, so that an increased susceptibility to different diseases might occur. This certainly happens in aquatic organisms, particularly fish, where acute and/or chronic pollution of surface waters can cause a reduction in the level of unspecific immunity to disease. Methods for controlling water quality are also discussed.

Chapter on effect of stress on fish concentrates on early conditioning to stress, which offers the potential to early reprogramming of behavior and metabolism to minimize the impact of stress later in life. The use of nonintrusive monitoring methods based on behavior and physiology offers a practical way to minimize or abolish the sources of stress.

The information of planning a fish health program is also provided in this part of the book. The development of a fish health management plan that capitalizes upon disease prevention and control strategies is a necessity not only at the aquaculture facility level, but also at the national and international levels. An effective fish health management plan serves three primary purposes: to ensure optimal rearing conditions, thereby avoiding an increased susceptibility to disease; to avoid pathogen introduction into the facility; and, to prevent pathogen spread throughout the facility and into the surrounding geographical area should an infectious disease outbreak take place.

At the end of the book, a chapter on aquatic animal health and the environmental impacts deals with the use of natural resources that can be inefficient in aquaculture operations if culture species do not reach harvest and market. The introduction of novel species can also bring novel disease-causing agents. Chemicals are used for a variety of aquatic animal health-related issues and have impacts on the surrounding environment. Aquaculture operations would be most efficient with chemical use and least impactful to the environment if these chemicals are not released before being taken up by target organisms in the farm or until they are adequately degraded.

ACKNOWLEDGMENT

I would like to thank all contributors of the book for their devoted and excellent work, and to the editors from ELSEVIER for their help and encouragement.

Galina Jeney

Part I

Fish

Chapter 1

Immunology: Improvement of Innate and Adaptive Immunity

Kim D. Thompson
Moredun Research Institute, Midlothian, United Kingdom

1.1 INTRODUCTION

One of the main factors responsible for production losses in aquaculture is disease, and effective disease management is therefore vital for maintaining a sustainable aquaculture industry. The fish's immune defenses are a crucial component of their resistance to disease, and studies related to this are one of the key areas of aquaculture-related research, especially with regard to economically important species. A deeper insight into the immune mechanisms and immune cells involved in host/pathogen interactions is needed to help support the current-day aquaculture practices and problems, such as emerging diseases, reducing the use of antibiotics, introduction of new aquaculture species, use of diploids *versus* triploids, understanding disease associated with larval development, sexual differences, climate change and pollution, and the effect of alterative feed ingredients on fish health. This information is essential not only for providing optimal culture conditions to prevent immuno-incompetence in intensively reared fish, but also for developing vaccination strategies, breeding programs for disease resistance, and the use of functional feeds, containing immunostimulants and probiotics, for example. As the external epithelial surfaces of fish are involved in pathogen entry, understanding the immune mechanisms and immune cell interactions associated with mucosal immunity are also important. Together, this information will allow us to understand how to manipulate and improve their innate and adaptive immune responses, which can then be incorporated into disease management strategies to help prevent the onset of disease outbreaks and reduce the losses associated with these.

1.2 TELEOST IMMUNITY

Fish have evolved effective immune responses to combat infections caused by bacterial, fungal, viral, or parasitic agents that cohabit the fish's aquatic environment and cause disease. The initial response of the host against invading pathogens is delivered by the innate immune response, acting as the first line of defense allowing time for the lymphocytes of the adaptive immune to respond, although the innate immune response alone can often stop infections from progressing further. Both of these responses are mediated by a variety of cells and humoral mediators. Proinflammatory cytokines such as tumor necrosis factor alpha (TNF-α) and interleukin 1 beta (IL-1β), produced by neutrophils and macrophages, induce the migration of phagocytic cells to the site of infection to kill and eliminate the pathogen. The phagocytes, in turn, present antigen to cells of the adaptive immune response, stimulating adaptive immunity (Secombes et al., 2001). Orthologue pattern recognition receptors (PRRs) on the surface of fish cells suggest that the innate immune system of teleost fish has similar pathogen recognition mechanisms and modes of action as those found in mammals. The presence of fish orthologues for antimicrobial peptides (AMPs), proinflammatory cytokines (TNF-α, IL-1β, IL-18, interferons (IFNs), chemokines, and IL-8), and complement has been the focus of the review by Plouffe et al. (2005), although our current knowledge on adaptive immunity in teleosts is still limited compared to mammals.

Skin, gill, and gut are important routes for pathogen entry because of the close contact that the fish has with its aquatic environment. The mucosa-associated lymphoid tissue (MALT) found in these tissues plays an important role in providing a first line of defense against the invading pathogen and subsequent infections, and is responsible for eliciting appropriate innate or adaptive immune responses through the activity of leukocytes (lymphocytes, macrophages, and granulocytes including eosinophilic granular cells) present in the MALT, and the cytokines produced by these cells. Mucosal tissues are covered with a thick mucus layer, which inhibits the invasion and proliferation of bacteria

Fish Diseases. http://dx.doi.org/10.1016/B978-0-12-804564-0.00001-6

and viruses through the presence of mucins and other biologically active molecules with biostatic and biocidal activities (e.g., complement, C-reactive proteins, proteases, lectins, lysozyme, hemolysins, agglutinin, proteolytic enzymes, AMPs, antibodies, immunoglobulins).

New research tools (e.g., complete genome sequences, in vivo and in vitro models, and monoclonal antibodies against cell markers and cytokines), and new techniques (e.g., reverse genetics, sensitive high-throughput analysis for gene expression) are being developed and used to help understand the intricacy of teleost immune response, which will undoubtedly lead to improved disease management and vaccination strategies for aquaculture.

1.2.1 Innate Immunity

The innate immune response is regarded as the first line of defense against both invading pathogens in plants and animals, and the subsequent infections that occur. It is also able to provide immediate defense against infection. The cells of the innate immune system recognize and respond to pathogens in a generic fashion, but do not confer long-lasting or specific immunity, unlike the adaptive immune response. In vertebrates, the innate response is delivered by both cellular and humoral components, and while there is considerable conservation between teleost fish and higher vertebrates in these responses, highlighted by the presence of orthologous PRRs and various cytokines, there also appears to be components and functions of the innate immune response, which are lacking in mammalian innate immunity (Aoki et al., 2008). Not only is the innate immune response involved in destroying pathogens and thus controlling infection through phagocytosis and cytokine production, it is also involved in the activation of adaptive immunity through antigen presentation.

The first innate defense the pathogen encounters is the physical barriers on the exterior of the fish that are directly exposed to the aquatic environment, such as skin, scales, and the epithelial cells of the gills, together with the mucus that covers these surfaces. These barriers are very effective at preventing pathogens from entering the body of the fish. If the pathogen manages to breach these barriers, however, it then encounters a variety of cellular and humoral components of the innate immune response.

The cells of the innate immune response are mainly myeloid in nature and comprise mononuclear and polymorphonuclear phagocytes. The mononuclear phagocytes include monocytes, which migrate from the bloodstream into the tissues, where they differentiate into resident macrophages and dendritic cells. The latter are very efficient at presenting antigen to T-cells, resulting in activation of the adaptive immune response, and as such act as a bridge between the innate and the adaptive immune systems. Further details about the activation of fish macrophages can be found in the review by Forlenza et al. (2011). Melanomacrophages are a unique phagocytic cell type found in fish that contain melanin within their cytoplasm. These cells are seen in the spleen and the head kidney, as well as at the sites of inflammation, and are believed to be involved in antigen presentation (Agius and Roberts, 2003; Koppang et al., 2003). The main polymorphonuclear phagocytes are neutrophilic granulocytes. These cells mount an early and potent antimicrobial response against invading pathogens, actively phagocytizing pathogens, utilizing toxic intracellular granules (Flerova and Balabanova, 2013), reactive oxygen species (ROS) (Filho, 2007; Katzenback and Belosevic, 2009; Rieger et al., 2012), and neutrophil extracellular traps (Palić et al., 2007; Pijanowski et al., 2013). They are the first leukocytes to migrate to the sites of infection and together with macrophages are involved in the inflammatory response in fish, recruited to sites of infection site by inflammatory cytokines (Havixbeck and Barreda, 2015; Havixbeck et al., 2015).

The cells help to clear the pathogen through the process of phagocytosis (Esteban et al., 2015). This involves the phagocyte detecting and recognizing the pathogen, and attaching to the pathogen through PRRs present on its surface. The phagocytes have a variety of soluble and cell-associated germline-encoded PRRs that recognize the pathogen through pathogen-associated molecular patterns (PAMPs). Most of the PRR families identified in mammals have also been found in fish (Jorgensen, 2014). These include toll-like receptors (TLRs) of which 17 have so far been described in fish (Palti, 2011), and although these are the best characterized PRRs in fish, C-type lectin receptors, RIG-1-like receptors, and NOD-like receptors have also been identified (Zhu et al., 2013). Viral nucleic acids, including single-strand RNA (ssRNA), double-strand RNA (dsRNA), and DNA, are detected by TLR7/8, TLR3, and TLR9, respectively (Gerlier and Lyles, 2011), and bacterial PAMPs are mainly recognized by TLR1, TLR2, TLR4, TLR5, TLR6, TLR7, and TLR9 (Kumar et al., 2011). The TLRs have a greater diversity compared to mammals with additional TLRs found in fish (Jørgensen, 2014), possibly due to gene duplication (Aoki et al., 2008). The binding of PRRs to the PAMPs, coated with opsonic molecules, leads to the activation of different cell signaling pathways. These in turn produce various transcription factors, including nuclear factor-κβ, which results in the production of inflammatory cytokines in dendritic cells and macrophages, and IFN regulatory responses as part of the JAK-STAT signaling pathway, involved in regulating transcription of IFNs (Kumar et al., 2011). The activated macrophages also produce chemokines, which summon other cells to the site of infection.

The binding of pathogen molecules to receptors on the surface of the phagocyte triggers the cell to engulf and internalize the pathogen within a phagosome. After pathogen engulfment, the phagosome containing the internalized pathogen fuses with a lysosome to form a phagolysosome. It is here that the internalized microbes are killed by various means, including superoxide radicals resulting from respiratory burst, inducible nitric oxide species, and AMPs, for example. The killed pathogen is then degraded by an assortment of hydrolytic enzymes contained within the lysosome. The digested material is subsequently presented by the macrophages and dendritic cells to lymphocytes of the adaptive immune response through the process of antigen presentation (Esteban et al., 2015).

Nonspecific cytotoxic cells (NCCs) and natural killer (NK)-like cells are two cell types associated with nonspecific cell-mediated cytotoxicity (CMC) in fish, although the relationship between these two cell types is unknown. Both cell types are able to kill xenogeneic and allogeneic cells. The NCCs of catfish are small agranular lymphocytes that express a novel type-III 34 kDa membrane protein (NCC receptor protein) (Evans et al., 1984), and are believed to be the evolutional precursors of mammalian NK cells (Jorgensen, 2014), while NK-like cells are large, granular cells (Shen et al., 2004). The NCCs are derived from lymphoid tissue, while NK-like cells are found in the peripheral blood. NCCs have been shown to spontaneously kill tumor cells, viral-infected cells, and protozoan parasites (Frøystad et al., 1998; Praveen et al., 2004), while the NK-like cells have been shown to kill via a perforin/granzyme-mediated apoptosis pathway (Secombes and Wang, 2012). The activities of these cells in channel catfish have been reviewed by Shen et al. (2002), although characterization of CMC in fish still continues (Jorgensen, 2014).

In addition to the cellular responses, there is a wide array of humoral responses involved in the innate response of fish that also offer protection against infections. These include a variety of proteins and glycoproteins that can kill or inhibit the growth of the pathogen, with many different factors involved such as antibacterial peptides, various lytic factors (lysozymes, cathepsin, chitinase), complement, agglutinins, precipitins, natural antibodies, cytokines including interleukin (IL)-1, IL-6, TNF-α, growth inhibitors, serum protease inhibitors (α2 macroglobulin, α1 antitrypsin), chemokines and acute phase proteins.

Over 90 fish AMPs have now been identified (Wang et al., 2009), which are characterized into β-defensins, cathelicidins, hepcidins, histone-derived peptides, and fish specific piscidins (Zhu et al., 2013; Masso-Silva and Diamond, 2014; Katzenback, 2015). In addition to their ability to kill bacteria, these peptides have been shown to also have a variety of immunomodulatory activities including activation of innate immunity, reviewed by Masso-Silva and Diamond (2014) and Katzenback (2015). AMPs are of particular interest, not only for their microbial activity, but also for their potential to be used as antimicrobial agents, vaccine adjuvants, inactivated vaccines, and antitumor agents.

The main functions of the complement system are to directly kill bacteria, promote inflammation, help clear damaged cells, and regulate the adaptive immune system (Nakao et al., 2011). The structure and activity of the teleost complement system is similar to that of mammals (Boshra et al., 2006). In mammals, it is composed of around 30 plasma and membrane-associated proteins, and it would seem that the teleost complement system is homologous to these. Several of these components have multiple paralogues in fish, due to genome or gene duplications (Nakao et al., 2006). Many of the proteins are produced by liver hepatocytes and secreted into the plasma, although a few are produced in extra-hepatic tissue (Jorgensen, 2014). Most of the proteins are normally found in an inactive form, but on recognition of the pathogen are sequentially activated as an enzyme cascade. As in mammals, this cascade can be activated via three different pathways (classical, alternative, or lectin dependent). The classical pathway is triggered by antibody–antigen complexes, while the alternative pathway is activated by the C3b protein binding directly to the pathogen or to foreign material and damaged cells. The mannose-binding lectin (MBL) pathway is activated by the binding of the MBL, or ficolin, to sugar residues present on the surface of pathogen. Ultimately, all pathways cause activation of the C3 convertase, which is split into C3b, which acts as an opsonin promoting phagocytosis, and C3a anaphylatoxin that aids to inflammation by acting as a chemotactic factor for macrophages and neutrophils, attracting them to the site of inflammation. Cleaving of the C3 convertase molecule then activates a lytic pathway, the end product of which (i.e., the membrane–attack complex), is able to kill the pathogen by damaging its plasma membrane. As well as the C3a anaphylatoxin, other byproducts of this cascade include the C4a and C5a anaphylatoxins.

Cytokines are small, secreted proteins (~5–20 kDa) that regulate the immune response though cell signaling, helping to control cell growth, differentiation, and activation, and are involved in many aspects of both innate and adaptive immune responses. They include chemokines, IFN, ILs, lymphokines, TNF, with homologues of mammalian cytokines found in teleost fish (Aoki et al., 2008; Secombes et al., 2011; Wang et al., 2011). The best characterized innate cytokines include the proinflammatory cytokines IL-1β and TNF-α and antiviral type-1 IFN. IL-1β is one of the earliest proinflammatory cytokines to be expressed which promotes a prompt response to infection by inducing a cascade of reactions leading to inflammation through up or downregulation of the expression of other cytokines and chemokines (Reyes-Cerpa et al., 2013). In mammals it is produced by wide variety of cells, in particular blood monocytes and tissue macrophages, as an inactive precursor that is cleaved by a cysteine protease, caspase 1, into a 17kDa bioactive protein. In fish, IL-1β has been shown to be involved in response to infection and to lipopolysaccharides (LPS) or polyinosinic-polycytidylic acid (poly (I:C)), while the use of recombinant IL-1β showed that it is involved in regulating immune genes, lymphocyte activation, migration of leukocytes, phagocytosis, and various bactericidal activities (Reyes-Cerpa et al., 2013). TNF-α is another important proinflammatory cytokine and is produced in mammals mainly by activated macrophages, although others cells such as $CD4^+$ lymphocytes, NK cells, neutrophils, and eosinophils also produce it. The cytokine has also been identified in several teleost species with multiple isoforms found (Zou et al., 2002; Saeij et al., 2003). In mammals it induces apoptosis, killing infected cells, inhibiting intracellular pathogen replication, and upregulating a diverse range of host response genes. Interestingly, in fish it is constitutively expressed in relatively high amounts in healthy fish and shows poor upregulation in response to immune challenge, and it seems to be mainly involved in recruiting leukocytes to the site of inflammation rather than their activation (Reyes-Cerpa et al., 2013).

IFNs are important defense molecules, which induce vertebrate cells to produce an antiviral response. There are three families of IFNs in mammals (type-I IFN, type-II IFN, and type-III IFN). Type-I IFNs include the IFN-α and IFN-β, and are induced by the presence of virus within cells, while type-II IFN, i.e., IFN-γ, is produced by NK cells and T-lymphocytes in response to IL-12, IL-18, mitogens, or antigens (Samuel, 2001). Type-I IFNs of fish show structural and functional similarities to those of mammals (Whyte, 2007; Robertsen, 2006) with multiple copies of the IFN1 gene are found (Zou and Secombes, 2011). Salmon have 11 different IFN1 genes divided into three subtypes (Sun et al., 2009), with distinct functions (Svingerud et al., 2012). They signal though the JAK/STAT pathway and upregulate the expression of IFN-stimulated genes (ISGs), which in turn deliver antiviral activity, e.g., Mx, ISG15, and protein kinase R (Jorgensen, 2014). The fish type-II IFN family consists of two members, IFN-γ with similar functions to mammalian IFN-γ and a teleost specific IFN-γ related (IFN-γrel) molecule, whose function is not fully elucidated (Zou and Secombes, 2011). IFN-γ induces the expression of many of the same ISGs induced by IFN1, although their expression is weaker by comparison and a relatively high dose of IFN-γ is needed to elicit an antiviral response (Sun et al., 2011).

1.2.2 Adaptive Immunity

Adaptive immunity elicits a specific response against a pathogen; it has a memory component that is able to quickly eliminate pathogens upon reencountering them. The adaptive immune system of fish is comparable to that of higher vertebrates, possessing T- and B-lymphocytes (cellular responses) and immunoglobulins (humoral response). It relies on B-cell and T-cell receptors (BCR and TCR), recombination activator genes (*RAG1* and *RAG2*), and the major histocompatibility complex (MHC-I and -II) to deliver this specificity. The lymphocytes circulate around the fish's body searching for pathogens, which they recognize and bind to through their cell-surface receptors. This in turn initiates clonal expansion of the cell, with progeny cells sharing the same antigen specificity.

1.2.2.1 Major Histocompatibility Complex

The MHC is a set of cell-surface molecules that the adaptive immune system uses to recognize as foreign molecules. The main function of these molecules is to bind to peptide fragments originating from the pathogen, and display these on the surface of the cell for recognition by appropriate T-cells (Janeway et al., 2001). The MHC proteins are highly polymorphic heterodimeric glycoproteins that are divided into three different groups (Klein et al., 1997). MHC class I molecules are present on all nucleated cells and mediate CMC against tumor cells or cells infected with intracellular pathogens. The antigen within the MHC class I molecule on the surface of the infected cell is presented to the CD8 receptor, in conjunction with TCR, on cytotoxic T-lymphocytes (CTLs), and this triggers the infected cell to undergo apoptosis (programmed cell death). MHC class II molecules are involved in antigen presentation. This involves the presentation of antigen fragments to T-cell lymphocytes, when bound to an MHC class II molecule located on specialized antigen-presenting cells (APCs), such as dendritic cells, macrophages and B-cells, whereby antigen enzymatically cleaved into smaller pieces, is displayed at the APC's surface in association with the MHC class II molecule. The peptide fragment of the pathogen originate from the digestion of the pathogen in the phagolysosome after phagocytosis, and it is loaded into an MHC class II molecule prior to its migration to the surface of the cell, where it binds with the CD4 receptor associated with the TCR on helper T-cells (Th cell), which has an affinity for the epitope displayed on the pathogen fragment, thus resulting in specific activation of the T-helper cell for that particular antigen. MHC class III are other immune proteins, outwith antigen processing and presentation, that comprised a group of genes involved in the inflammatory response (Deakin et al., 2006). The organization of the class III region in teleosts is split over several different chromosomes, which encode for components of the complement cascade, cytokines of immune signaling, and heat shock proteins protecting cells from stress.

1.2.2.2 B-Cells and Antibodies

The main function of B-lymphocytes is the production of antibodies. Until recently, IgM was thought to be the only antibody present in fish, but it has since been shown that teleosts also produce IgD and IgT. IgM is the most abundant circulatory antibody in fish, although it can be found in the skin, gut, gill mucus, and bile, as well as in the blood (Morrison and Nowak, 2002). The molecular structure of teleost IgM is different to that of mammals as the light and heavy chains are held together noncovalently instead of with disulphide bonds, as seen in mammalian IgM. This means that monomeric, dimeric, and tetrameric forms of the antibody can be present, possibly reflecting differences in their function. This noncovalent binding is believed to enhance the ability of tetramer forms to bind to different types of epitopes by allowing them to adjust their orientation (Solem and Stenvik, 2006). They are capable of opsonizing pathogens to enhance phagocytosis by macrophages (Secombes and Fletcher, 1992; Solem and Stenvik, 2006), are able to activate the classical complement pathway, and can effectively agglutinate pathogens. They also have relatively low intrinsic affinity and heterogenicity compared to mammalian antibodies (Ma et al., 2013; Solem and Stenvik, 2006), but they are still able to offer protection against bacterial and viral infections (Mutoloki et al., 2014).

The function of IgD on the other hand is still unclear. This immunoglobulin represents an ancient Ig isotype that is found in most vertebrate taxa (Edholm et al., 2011). Although it has been cloned in several different fish species, secreted IgD (sIgD) has only been characterized in catfish and trout, and its structure between these two species is very different (Parra et al., 2013). The sIgD of catfish does not possess Cμ1 or variable domains (Edholm et al., 2011), whereas these domains are present in trout IgD (Ramirez-Gomez et al., 2012). It is secreted as a monomer, and exists in at least two different isoforms in trout (Ramirez-Gomez et al., 2012). The fact that secreted IgD can bind to the IgD-binding receptor of granulocytes via their Fc receptor, suggests that it acts as a pattern recognition molecule (Edholm et al., 2011).

IgT has been associated with mucosal tissues, where multimeric forms of the molecule can be found, while it is found as a monomer in the serum of fish (Zhang et al., 2010). The ratio of IgT/IgM in plasma and in mucosal secretions was the first indicator that IgT plays a significant role in mucosal immunity, and the IgT/IgM ratio in rainbow trout gut, skin, and nasal mucus is much larger than in plasma in the absence of antigenic stimulation (Zhang et al., 2010; Xu et al., 2013; Tacchi et al., 2014; Salinas, 2015). The function of this isotype still remains to be fully elucidated.

B-cells are activated by signals from T-helper cells and develop into plasma cells producing antibody or memory B-cells. The plasma cells are characterized by large amounts of cytoplasm, where immunoglobulins are rapidly synthesized for secretion (Ellis, 1977). Memory B-cells in mammals are formed from activated B-cells with same antigenic-specific, and the BCR on the surface of the B-cell is specific for one particular antigen. In channel catfish three different subpopulations of B-cells have been observed, IgM^+/IgD^- secreting IgM, IgM^+/IgD^+, and IgM^-/IgD^+ secreting IgD. Only two subpopulations have been observed in rainbow trout, however, $IgM^+/IgD^+/IgT^-$ secreting IgM and $IgM^-/IgD^-/IgT^+$ secreting IgT. In all analyzed species except catfish, IgD is coexpressed with IgM on B-cells. It has been shown that rainbow trout IgM^+ B-cells and IgT^+ B have a potent phagocytic capacity with an ability to kill engulfed bacteria (Li et al., 2006; Zhang, 2010). Other teleost species have also been shown to have phagocytic B-cell populations (Li et al., 2006).

1.2.2.3 T-Cells

The T-lymphocytes populations in mammals include Th cells, CTLs, memory T-cells, regulatory T-cells (TREG cells), natural killer T-cells (NKT cells), and mucosal-associated invariant T-cells (MAITs).

1.2.2.3.1 T-Helper Cells

As explained earlier, the Th cells are activated by the presentation of peptide antigens to the TCR by APCs linked to MHC class II molecules, after which the activated Th cells rapidly divide and release cytokines, which in turn regulates the immune response. The Th cells can differentiate into Th1, Th2, and T17 cells among others (Mutoloki et al., 2014), and secrete different cytokines profiles. The cytokines produced by Th1 cells mediate CMC against intracellular pathogens, while the cytokines from Th2 cells promote humoral immunity against extracellular pathogens, and cytokines from Th17 cells enhance responses to extracellular bacteria. The effector functions of Th1 cells are exerted in part by production of IFN-γ, IL-2, and lymphotoxin-α (LTα), and those of Th2 cells by IL-4, IL-5, IL-10, and IL-13.

1.2.2.3.2 Cytotoxic T-cells

Cytotoxic T-cells, also referred to as $CD8^+$ T-cells because they express a CD8 glycoprotein on their surface, are the mediators of CMC, killing host cells infected with intracellular pathogens and tumor cells, using MHC class I restriction as explained in Section 1.2.2.1. They use three different mechanisms to destroy the target cell: (1) the granule exocytosis pathway, (2) the Fas pathway, and (3) the release of cytolytic cytokines (TNF-α, LTα, and IFN-γ). In the first method, the

CTLs release granules containing granzymes and perforin close to the target cell membrane. These are taken-up into the cell and cause apoptosis in the target cell. Granzymes also induce DNA fragmentation in the nucleus. With the second method, the Fas ligand on the surface of the activated CTL binds to the Fas receptor on the target cell, and this also inducing apoptosis. CMC was first demonstrated in channel catfish (Miller et al., 1986) and has since been observed in several other fish species as alloantigen- and virus-specific CMC (Somamoto et al., 2000; Stuge et al., 2000; Nakanishi et al., 2002; Fischer et al., 2006; Utke et al., 2007).

1.2.2.3.3 Regulatory T-cells

In mammals TREG cells, or suppressor T-cells, secrete soluble factors including, e.g., IL-10, TGF-β, fibrinogen-like protein-2 (FLG-2), granzyme, and adenosine which suppress excessive immune responses. Two cell types appear to exist: Thymus-derived naturally occurring TREG cells ($CD4^+$, $CD25^+$, $FOXP3^+$ TREG cells), and adaptive TREG cells or inducible TREG cells (Tr1 cells secrete IL-10; Th3 cells secrete TGF-β and IL-10; and $FoxP3^+$ TREG cells) (Peterson, 2012). Although Treg-like activity has been observed in cells displaying a CD4-2^+, CD25-$like^+$, and FOXP3-like phenotype in pufferfish (Wen et al., 2011), the presence of true Treg cells still remains to be confirmed.

1.2.3 Immune Tissues in Fish

Various immune response cells are produced by tissues and organs of the lymphoid system. Fish lack bone marrow, unlike mammals, and the primary lymphoid tissues where lymphocytes are initially produced in fish include the thymus and the head kidney, often referred to as the anterior kidney or pronephros. The lymphocytes migrate from these sites to the secondary (peripheral) lymphoid tissues and it is here where they are immunological active. The secondary lymphoid tissues in teleosts include the kidney, spleen, and MALT, composed of the gut-associated lymphoid tissue (GALT), skin-associated lymphoid tissue (SALT), gill-associated lymphoid tissue (GIALT), and the recently discovered nasopharynx-associated lymphoid tissue (NALT).

The thymus of fish appears to be responsible for the development of T-lymphocytes, as in other jawed vertebrates (Alvarez-Pellitero, 2008). It is located beneath the pharyngeal epithelium in the dorsolateral region of the gills, and its structure varies between fish species and life stages of the fish, but there does not appear to be any clear corticomedullary differentiation as seen in the mammalian thymus (Secombes and Wang, 2012). Much of the evidence for the thymus as a primary site of T-cell production has come from immunizing fish with T-dependent antigens (Ellsaesser et al., 1988), using monoclonal antibodies as cell-surface markers (Passer et al., 1996) or functional assays. Lymphocytes have been shown to migrate from the thymus to the spleen and kidney, suggesting that the teleost thymus is the main source of immunocompetent T-cells in rainbow trout (Tatner and Findlay, 1991).

The anterior kidney is the most important hematopoietic organ in fish (Zapata et al., 1996), where blood cell differentiation takes place and which is involved in early immune responses and early immune cell production. It has a critical role in a wide range of immune functions including phagocytosis (Danneving et al., 1994), antigen processing, and IgM production (Kaattari and Irwin, 1985; Brattgjerd and Evensen, 1996). The anterior kidney of most teleost fish contains melanomacrophagic centers (MMCs), which are aggregates of reticular cells, macrophages, lymphocytes, and plasma cells. These centers are thought to be involved in antigen trapping and have a role in immunologic memory (Secombes et al., 1982; Herraez and Zapata, 1986; Tsujii and Seno, 1990).

The spleen is a major secondary lymphoid organ in fish and appears to be involved in immune activity and blood cell formation and breakdown (Manning, 1994; Zapata et al., 1996; Galindo-Villegas and Hosokowa, 2004). Although the fish spleen contains red and white pulp, as seen in the mammalian spleen, the architecture of these structures is less defined. The red pulp consists of a supporting network of fibroblastic reticular cells containing lymphocytes of differing size and development, mature plasma cells, and macrophages. The white pulp is poorly developed and contains MMCs and ellipsoids. It is also believed that the MMCs are involved with antigenic stimulation, with the presence of immunoglobulin-positive cells evident after immunization (Herraez and Zapata, 1986), and which can retain antigens as immune complexes for long periods of time suggesting a possible role in the development of immunologic memory (Van Muiswinkel et al., 1991), as well as the kidney. Blood-borne material trapped within the ellipsoidal wall is subsequently phagocytosed by the macrophages surrounding these vessels which then migrate into the MMCs (Press and Evensen, 1999).

The liver of mammals is involved in both systemic and local innate immunity. It is the site of production for components of the complement cascade and acute phase proteins, and also contains cells of myeloid lineage (Kupffer cells and dendritic cells), and intrahepatic lymphocytes, including conventional T- and B-cells, NK cells, and nonconventional lymphoid cells (Nemeth et al., 2009). The relevance of the liver as an immune organ in fish is still unclear, but active expression of immune genes as a result of bacterial infection highlights that it is immunologically active (Martin et al., 2010; Millán et al., 2011).

The main MALT tissues of teleosts are located in the gut, skin, gill, and the recently discovered NALT. Teleost MALTs are composed of both innate and adaptive immune cells and molecules that work together to maintain homeostasis at the mucosa (Salinis, 2015). B- and T-cells are diffusely scattered throughout these tissues (D-MALT), and the phenotypes of these cells differ from their systemic counterparts and respond to both mucosal infection and vaccination (Salinis, 2015). Interbranchial lymphoid tissue (ILT), on the other hand, is a unique lymphocyte-rich tissue, consisting largely of T-cells embedded in a meshwork of epithelial cells (Aas et al., 2014; Koppang et al., 2010; Haugarvoll et al., 2008). IgT^+ B-cells are the preponderant B-cell subset in GALT, SALT, and NALT compared to the spleen or head kidney, where IgM^+ B-cells are the main subset (Zhang et al., 2010; Xu et al., 2013; Tacchi et al., 2014) and both IgT and IgM are present on the mucosal surfaces.

1.3 EFFECTORS OF THE IMMUNE RESPONSE

Many different factors are known to affect the physiology and immune response of farmed fish, especially those maintained under intensive rearing conditions. Some of these causes have recently been reviewed, in which the authors point out that the lack of well-established baseline data and natural variations in the fish's immune response makes interpretation of these causes difficult (Bowden, 2008; Makrinos and Bowden, 2016). Two of the most reported causes to affect the fish's immune response are temperature and stress (Pickering and Pottinger, 1989; Pulsford et al., 1994; Espelid et al., 1996; Wendelaar Bonga, 1997). We have reviewed the impact of seasonality on fish immunity (Bowden et al., 2007), of which photoperiod is an important component. Controlling the body clock is a normal husbandry practice to change the breeding cycle to produce eggs outwith the normal breeding period to allow more than one input into seawater each year.

1.4 IMPROVEMENT OF THE IMMUNE RESPONSE

A variety of approaches are used in aquaculture today to enhance the immune response, and thus the disease resistance of fish. Most importantly, improved farming practices have helped to optimize environmental conditions to reduce stress and improve the welfare of fish stocks. Selective breeding programs have also increased disease resistance of fish against certain diseases, such as infectious pancreatic necrosis and pancreas disease, as reviewed by Gjedrem (2015) in which he pointed out that the potential of these programs is not being fully utilized with only 8.2% of aquaculture production in 2010 based on family breeding programs. The use of fish derived from commercial breeding programs will undoubtedly increase in the future.

Substances that induce, enhance, or suppress the immune response are collectively called immunomodulators, and these have the potential to significantly reduce disease-related losses in aquaculture. There is a diverse range of substances (recombinant, synthetic, and natural) that offer an attractive alternative to antibiotics as they generally have fewer side effects than existing medicines, and it is less likely that the pathogen will develop resistance against them.

Immunostimulants are substances that stimulate the immune response of fish by inducing or increasing the fish's immune activity, either through antigenic-specific responses, such as vaccines, or nonspecifically, independent of antigenic recognition, such as adjuvants or nonspecific immunostimulators discussed in more detail in Section 1.4.2. Adjuvants are added to vaccines to help generate a stronger protective response to the antigens present in the vaccine, and to provide increased protection against the pathogen. Cytokines produced by the cellular immune system also act as immunostimulators and are able to enhance immune function (Secombes and Wang, 2012). Many vaccines and functional feeds, containing immunostimulants, nucleotides, and pre- and probiotics, are commercial available to help the farmer reduce mortalities associated with disease.

1.4.1 Vaccines

The most appropriate method of disease control, both on economical and ethical grounds, is vaccination. Vaccines are nonpathogenic preparations of the pathogen that induce an adaptive immune response in the host, enabling it to recognize and destroy the pathogen when it encounters it at a later date. Vaccines are considered to be a very environmentally friendly approach to disease control compared to antibiotics, and it is clear that as the use of vaccines has increased, the amount of antibiotics used by the industry has decreased, and that their use is directly related to the growth seen in fish farming in recent years (Evensen, 2009). The use of vaccines by the aquaculture industry has expanded rapidly in recent years, both with regard to the number of fish species and the number of microbial diseases addressed (Håstein et al., 2005; Evensen, 2009). Commercial fish vaccines are now available in more than 40 countries for more than 17 different species of fish, and although there are around 30 vaccines commercially available for aquaculture (Brudeseth et al., 2013), with many existing as multivalent products, there are still several diseases where no vaccine is available or little scope exists to improve the efficacy of the vaccine.

Most of the vaccines produced for aquaculture are inactivated whole cell bacterial products, and the use of recombinant vaccine technology is still limited. Injection vaccination is the most popular method of delivery, with automated vaccination machines being introduced to help mass vaccine fish. Vaccination by immersion is often used for smaller fish and oral administration in feed is also available for some commercial vaccine products. Salmonid fish tend to be immunized with multivalent vaccines by intraperitoneal (i.p.) injection, and although side effects are reported with injectable adjuvant vaccines (Midtlyng et al., 1996), they are used because of the long lasting and high level of protection they elicit compared to immersion or oral delivery. DNA vaccines are administered by intramuscular injection, such as with the commercial DNA vaccine against infectious hemorrhagic necrosis (IHN). Commercial immersion vaccines tend to be formalin inactivated bacterial suspension or live bacterial vaccines (used in USA), whereby the fish are dipped in a 1/10 dilution of a concentrated vaccine suspension usually for between 30 and 60s (Brudeseth et al., 2013). The efficacy of the vaccine is affected by the immunocompetence of the fry and fish need to be of a sufficient size to respond immunologically to the vaccine (Amend and Johnson, 1981). Vaccination regimes are designed to provide protection over the course of the production cycle. This may involve initially dip-vaccinating the fish, followed by an oral or immersion booster vaccination, or an injection vaccination.

An important area of research is establishing antigens/adjuvants vaccine systems that can be used for oral delivery of vaccines or delivery by other mucosal routes of administration. Understanding the role of mucosal immunity in protection against infection may offer a route to delivering new or improved vaccines. It is therefore important to establish a knowledge- and technology-base for the development of next generation mucosal vaccines for aquaculture.

In addition to formalin-killed vaccines, several other types of vaccines that have been investigated for use in aquaculture, some of which have been commercialized, include attenuated, subunit, and DNA vaccines. Attenuated vaccines are based on a mutant that is no longer pathogenic as a result of repeatedly subculturing the pathogen in nutrient medium resulting in random mutations that lead to attenuation, chemical processing or radiation, or genetic manipulation. Designing genetically modified organisms (GMOs) through targeted genetic engineering, on the other hand, allows better control of gene manipulation compared to random mutagenesis. The advantage of live attenuated vaccines is that they survive and replicate within the host, eliciting a strong cellular immune response as well as stimulating a humoral and mucosal immune response, conferring longer lasting immunity. There are safety concerns about using modified live vaccines because of the possibility that the modified organism might revert to virulence, which is an important consideration for its registration. The regulation for the release of live GMO vaccines into the aquatic environment is much stricter than the live vaccines developed conventionally through culture (Brudeseth et al., 2013). These concerns may explain why few live attenuated vaccines have been commercialized for aquaculture use.

There has been a great interest in DNA vaccines for use in aquaculture, especially for viruses and other intracellular pathogens, because of their ability to stimulate cell-mediated immunity (Evensen and Leong, 2013). DNA vaccination involves the injection of DNA plasmids, encoding the immunogenic antigens of interest into the muscle of the fish, and this technology has proven very effective for IHN in rainbow trout fry (Corbeil et al., 2000). The first DNA vaccine to be licensed for use in aquaculture was in Canada for IHN. However, the efficacy of DNA vaccines outwith the family Rhabdoviridae appears less convincing (Biering and Salonius, 2014). There are still concerns about the safety of DNA vaccines, in Europe for example, because of the risk that the DNA construct may integrate into the fish's chromosome or spread in the aquatic environment (Biering and Salonius, 2014).

Formalin inactivated vaccines are often weakly immunogenic, resulting in low levels of efficacy and duration of protection. This is also the case for recombinant antigens, which are generally less immunogenic than whole cell inactivated vaccines. The use of adjuvants in fish vaccines has been shown to improve protection for some antigens, although some injected adjuvants caused side effects and reduced growth in the short term (Mutoloki et al., 2004). There is therefore a need to develop new and improved adjuvants for use in aquaculture that can stimulate cellular immunity against intracellular bacteria and viruses, are suitable for use with mucosal and DNA vaccines, reduce the amount of antigen required in vaccines formulations, eliminate antigen interference in multivalent vaccines, and which can remove the need for booster vaccinations (Tafalla et al., 2013).

Adjuvants are added to vaccines to enhance the immune response to the target antigen in the vaccine by improving the level of protection elicited, and the onset and duration of immunity. The activity of adjuvants is classified into Type-I and Type-II facilitators (Tafalla et al., 2013). The action of Type-I facilitators improves the persistence or uptake of antigen in the tissues. Examples of these include mineral oils, nonmineral oils, or a mixture of both, which produce different types of emulsions that when delivered by i.p. injection result in the slow release of the antigen into the peritoneal cavity and increased antigen presentation and immunity. Encapsulation of the antigen in microparticles is a promising alternative to the use of oil emulsions. Particle-based adjuvants/delivery systems are manufactured using biodegradable, biocompatible particles composed of poly-lactide-co-glycolide, alginate, immune-stimulating complexes (ISCOMs) chitosan, liposomes, or virosomes. Type-II facilitators are compounds that directly enhance the immune response of the host by providing an enhanced stimulus, triggering enhanced antigen presentation and a heightened adaptive immune response. These traditionally include compounds such as aluminum hydroxide or bacterial toxin, but there has been great interest in trying to produce a more targeted response using molecules, such as TLRs, β-glucans, saponins, lipopeptides, flagellin (the structural protein of the flagella of Gram-negative bacteria), bacterial DNA and synthetic oligodeoxynucleotides expressing unmethylated CpG motifs, cytokines and polyinosinic-polycytidylic acid (poly I:C) (Tafalla et al., 2013).

1.4.2 Immunostimulants

Immunostimulants have been shown to improve fish's resistance to disease (Sakai, 1999; Bairwa et al., 2012; Meena et al., 2013) and enhance their immune response at times of stress (Ringø et al., 2012; Dong et al., 2015). Their use is now commonplace in disease control programs to help prevent infectious diseases in aquaculture, especially since they can be easily fed to fish. Immunostimulants are derived from both natural and synthetic sources. Although this list is not inclusive of the products tested in aquaculture, examples of immunostimulants include β-glucans, chitin, lactoferrin, levamisole, vitamins B and C, growth hormone, and prolactin (Sakai, 1999).

β-glucans are the most commonly used immunostimulants in aquaculture, especially β-glucan (β-1,3 and 1,6-glucans) derived from the cell wall of baker's yeast *Saccharomyces cerevisiae*, although other sources of β-glucan investigated include plants (e.g., wheat, rye, barley, oats, and *Echinacea* sp.), yeast (*Saccharomyces* sp.), seaweed (e.g., *Laminaria* sp.), mushrooms shiitake (*Lentinus edodes*), maitake (*Grifola frondosa*), reishi (*Ganoderma lucidum*), fungus (*Agaricus subrufesuns*; *Pneumocystis carini*; *Cryptococcus neoformans*, Schizophylan from *Schizophyllum commune*), and bacteria (e.g., *Rhizobiaceae* sp.) (Sirimanapong et al., 2015).

β-glucans are a heterogeneous group of glucose polymers, with a backbone of β-(1,3)-linked β-D-glucopyranosyl units and β-(1,6)-linked side chains. The distribution and length of these vary depending on the origin of the β-glucans. The β-glucans from oats and barley have linear β-(1,4) and β-(1,3) linkages, mushroom β-glucans have short β-(1,6)-linked branches with a β-(1,3) backbone, and yeast β-glucans have β-(1,3)-glycosidic-linked D-glucose subunits with irregular β-(1,6)-linked side chains of various lengths (Auinger et al., 2013; Meena et al., 2013). Differences in activity are attributed to the extraction of the β-glucans from their source material, which can affect their immunostimulatory properties (Sirimanapong et al., 2015). β-glucans increase the innate defense mechanisms of fish by enhancing phagocytic and bactericidal activities of macrophage (Ranjan et al., 2012), increased complement and lysozyme activity, and enhanced antibody responses (Sakai, 1999; Dong et al., 2015). There are numerous reports of immunostimulants enhancing fish's resistance to disease in the literature, recently reviewed by Newaj-Fyzul and Austin (2015).

There has also been increasing interest in the use of medicinal plant as modulators of the immune response of fish (Bulfon et al., 2015; Newaj-Fyzul and Austin, 2015; Hai, 2015), especially those linked with cultural practice to treat disease, such as herbs used in Chinese and Indian medicine (Jeney et al., 2009).

The increased efficacy of vaccines combined with immunostimulants was discussed earlier in Section 1.4.1 (Tafalla et al., 2013).

1.4.3 Probiotics, Prebiotics, and Synbiotics

The role of the gut flora in maintaining the health of the animal and preventing disease is well documented (Holzapfel and Schillinger, 2002). Probiotics are live microorganisms, derived from "normal" environmental or intestinal bacteria that have the potential to provide health benefits when administered to fish. They are defined as "beneficial live micro-organisms when administered to a host at an effective dose" (FAO/WHO, 2001). The success of probiotics and prebiotics in the livestock industry is now adopted by the aquaculture industry to improve the health and growth performance of fish. There seems to be a lot of interest in using them to improve the digestibility of nutrients, increase the fish's tolerance to stress, improve reproduction, and reduce disease outbreaks (see reviews Irianto and Austin, 2002; Balcázar et al., 2006; Cruz et al., 2012; Pandiyan et al., 2013; Priyadarshini et al., 2013 for more information related to this).

When choosing a new probiotic product for aquaculture, it is necessary to show that it is both effective and safe in vivo; that it is not harmful to the host; it is palatable to the host and has the potential to colonize the gut and replicate; and does not contain virulence resistance genes or antibiotic resistance genes (Kesarcodi-Watson et al., 2008). Selected bacteria within the species *Lactobacillus*, *Lactococcus*, *Leuconostoc*, *Enterococcus*, *Carnobacterium*, *Shewanella*, *Bacillus*, *Aeromonas*, *Vibrio*, *Enterobacter*, *Pseudomonas*, *Clostridium*, *Saccharomyces*, *Pediococcus*, and *Streptococcus* have been investigated as potential probiotics for aquaculture (Pandiyan et al., 2013; Mancuso, 2013).

The action of probiotics is based on their ability to stimulate the growth of specific microbes in the intestinal tract of the fish. They maintain the microbial equilibrium of the gut by competing with pathogenic bacteria for attachment sites on the mucosa of the gut and also by competing for nutrients. They have an antagonistic activity against the pathogen, as they produce a variety of antimicrobial substances (bactericidal or bacteriostatic) that prevent the replication and/or kill the pathogen, thus preventing the pathogen from colonizing the fish's gut. They also directly enhance the host's immune response against the pathogen (Irianto and Austin, 2002; Pandiyan et al., 2013).

The antagonistic properties of probiotics on fish pathogens were shown, in part, to be due to siderophores, produced by the probiotic grown under iron-depleted conditions within the host (Gram et al., 2001; Spanggaard et al., 2001; Brunt et al., 2007). Siderophores are high-affinity iron-acquisition molecules, when produced in vivo, giving the probiotic a competitive advantage for growth in iron-scarce environments of the host (Miethke and Marahiel, 2007). We recently demonstrated the inhibitory effects of *Pseudomonas* sp. M174 against *Flavobacterium psychrophilum*, the causal agent of rainbow trout fry syndrome (RTFS), both in vitro and in vivo (Korkea-aho et al., 2011), with the production of siderophores as an apparent mode of action for this inhibition (Korkea-aho et al., 2011; Ström-Bestor and Wiklund, 2011). Entericidin was shown to be involved in the probiotic activity of *Enterobacter* C6-6 against *F. psychrophium* (Schubiger et al., 2015), and the authors suggest that entericidin may present new opportunities for therapeutic and prophylactic treatments against similarly susceptible pathogens.

Probiotics are also thought to be able to modulate the host's gut immune defenses. Several studies have suggested that the attachment and colonization of probiotics in the fish's intestine may lead to stimulation of their innate immune responses (Balcázar et al., 2007; Panigrahi et al., 2005). A number of immunological parameters were found to be stimulated in fish fed with probiotics (for review see Nayak, 2010; Newaj-Fyzul and Austin, 2015), for example, in rainbow trout fed with *Bacillus subtilis* (Newaj-Fyzul et al., 2007), *Carnobacterium maltaromaticum* (Kim and Austin, 2006), *Kocuria* SM1 (Sharifuzzaman and Austin, 2010), *Enterococcus faecium* and *B. subtilis* (Panigrahi et al., 2007), *Pseudomonas* sp. (Korkea-aho et al., 2011, 2012) and *Enterobacter* sp. (Burbank et al., 2011). Increased total serum IgM levels were also noted (Nikoskelainen et al., 2003; Balcázar et al., 2007; Salinas et al., 2008). High mortalities in newly hatched fry due to disease outbreaks, e.g., as with RTFS, may reflect the lack of a developed adaptive immune response in these young fish, and the young fish rely on their innate immune response for protection (Mulero et al., 2007). Probiotics may help by stimulating their innate disease resistance before their adaptive immune systems become fully immunocompetent.

Prebiotics are indigestible carbohydrates that confer health benefits when fed to the host by stimulating the growth and/or the activity of selected bacteria in animal's gut (Roberfroid, 2007; Ringø et al., 2010). Fermentable carbohydrates are considered the most promising of these, exerting a positive effect on the composition and activity of indigenous microflora in the gut tract. There are several potential prebiotic carbohydrates that have been tested in aquaculture, including inulin, fructo-oligosaccharides (FOS), short-chain fructo-oligosaccharides (scFOS), mannan-oligosaccharides (MOS), galacto-oligosaccharides (GOS), xylooligo-saccharides (XOS), arabinoxylo-oligosaccharides (AXOS), isomalto-oligosaccharides (IMO), and GroBiotic, reviewed by Ringø et al. (2010). The prebiotics are metabolized in the gut of the host by bacteria such as *Lactobacillus* and *Bifidobacterium*, and these, in turn, produce metabolites such as short-chain fatty acids which are important for colon health. Butyrate, for example, is a primary energy source for colonic cells and also has anti-carcinogenic and antiinflammatory properties (Greer and O'Keefe, 2011). They also decrease the level of intestinal pathogens present in the gut (Ringø et al., 2010).

In addition to prebiotics being referred to as functional saccharides because of the energy they provide for gut microflora (Roberfroid, 1993), some prebiotics have also been referred to as immunosaccharides, because of their ability to directly stimulate the animal's innate immune response (Kocher, 2004). The immunomodulatory activity of prebiotics is a result of their ability to stimulate the innate immune system through the PRRs expressed on the surface of macrophages, such as β-glucan receptors and dectin-1 receptors (Yadav and Schorey, 2002) and this directly results in macrophage activation, neutrophil activation, activation of the alternative complement system, and increased lysozyme activity, or they can interact with microbe-associated molecular patterns (MAMPs) to activate innate immune cells (reviewed by Song et al., 2014). They can also enhance the growth of commensal microbiota in the fishes gut, and when probiotics and prebiotics are combined together and fed to fish they are referred to as synbiotic, with the prebiotic influencing the growth and activity of the probiotic. There is limited research on the use of synbiotics in aquaculture, although their application appears to have great potential as Ringø and Song (2016) pointed out in their recent review. They also make the point that research related to the use of synbiotics in aquaculture is very recent, with the first publication in 2009 (Rodriguez-Estrada et al., 2009). Their influence on fish growth performance, gut microbiota, gut histology, immune parameters, hematological and biochemical parameters, and improved disease resistance has subsequently been investigated, and some studies have used plant products or β-glucans in combination with probiotics (Ringø and Song, 2016). The influence of dietary supplements on the gut microbiota, gut morphology, and mucosal immune response of fish was recently reviewed (Ringø and Song, 2016; Ringø et al., 2016).

1.4.4 Nucleotide Diets

Dietary nucleotides are considered to be conditionally essential nutrients under certain physiological challenges. Nucleotides are low-molecular weight molecules comprising a nitrogenous base, a five-carbon sugar, and a phosphate group, and are necessary for many biological processes (Kenari et al., 2013). They are cofactors in cell signaling and metabolism and are precursors of DNA replication (Burrells et al., 2001; Peng et al., 2013). They are the core building blocks of nucleic acids and are therefore important in development and cell repair (Trichet, 2010). They have been shown to have an immunopotentiating effect on both the innate and adaptive immune response of fish, and can enhance resistance against viral, bacterial, and parasitic infections. It is, therefore, important that feed formulations are balanced to provide sufficient dietary nucleotides to meet physiological requirements. Dietary nucleotides have also been shown to have various effects on the gastrointestinal tract of higher vertebrates, influencing physiological, morphological, and microbiological parameters of the gut, and also the composition of their intestinal microflora. Although there have been few reports examining the role of exogenous nucleotides on the expression of immunity (Leonardi et al., 2003; Glencross and Rutherford, 2010; Tahmasebi-Kohyani et al., 2011; Ringø et al., 2012), diets supplemented with nucleotides appear to have beneficial effects in fish (Li and Gatlin, 2006) improving growth performance, enhancing intestinal morphology and function, aiding liver function as well as modulating innate and adaptive immune responses (Burrells et al., 2001; Sakai et al., 2001; Low et al., 2003; Malina et al., 2005; Li and Gatlin, 2006; Li et al., 2007; Jha et al., 2007; Lin et al., 2009; Tahmasebi-Kohyani et al., 2012; Peng et al., 2013). Further research is also needed to understand their effects on mucosal immunity in fish.

1.5 CONCLUDING REMARKS AND FUTURE PERSPECTIVES

Effective disease control is paramount for a sustainable aquaculture industry. This should include disease management strategies to prevent the onset of disease, and measures to reduce losses before and during the disease episodes. It is clear that the immune system of fish is very complex, and although tools are available to identify and measure changes in the expression of many immune genes at a molecular level, we still lack many of the tools necessary for characterizing immune cells and their products at protein level (Secombes and Wang, 2012). This is being addressed using newer, more sophisticated methods of analysis, such as next generation sequencing, genomics, proteomics, and the bioinformatics that accompany these techniques. This will allow us to identify correlates of disease resistance for selective breeding programs and help to minimize the number of fish used in experimentation. It will also inform on the development of next generation vaccines, based on novel adjuvants and delivery systems for administration via routes that stimulate mucosal immunity, for example by oral delivery. Mucosal surfaces of the fish are important routes for pathogen entry. Although there are many effective vaccines commercialized for use in aquaculture, they tend to be specific for one pathogen, or a few if multivalent vaccines are used. Thus, alternative methods of immune-enhancement are needed to complement the use of vaccines at times when fish are likely to be immunocomprised.

The fish's environment impacts on its health in many ways (Bowden, 2008). Routine husbandry practices and changing water conditions can lead to stress-related immunosuppression. Understanding the impact that these have on immune function and disease resistance can allow immunosuppressive events to be predicted and appropriate action taken to lessen their impact. Functional feeds have attracted considerable interest in terms of their ability to improve growth performance and increase the fish's resistance to disease, and there are many published studies describing the ability of probiotics, prebiotics, and immunostimulants to do this. In fact, a variety of these products are now commercially available for the fish farmer to use. They also offer some protection to young fish by stimulating their innate immune response before their adaptive response has developed full immunocompetence.

As these products are administered orally, it is important to understand the effects that they have on the MALT of the gut. The health benefits elicited by probiotics come from alternations to the intestinal microbial community of the host, which positively impact on growth, digestion, immunity, and disease resistance. Greater knowledge is needed about to the composition and activity of fish's intestinal tract microbiota and how it influences the development and function of the fish's immune response. Understanding this will allow us to manipulate the gut flora to improve fish health and performance. Synbiotics is a relatively new concept in aquaculture nutrition combining pre- and probiotics administration together to improve the survival, activity and efficiency of probiotics fed to fish.

Ultimately, future disease management programs should provide a holistic approach to address the challenge of disease in aquaculture systems, focusing on the overall wellbeing of the fish, and should include effective biosecurity, nutritionally balanced diets and feeding regimes containing functional ingredients, vaccines, and selective breeding programs.

REFERENCES

Aas, I.B., Austbø, L., König, M., Syed, M., Falk, K., Hordvik, I., Koppang, E.O., 2014. Transcriptional characterization of the T cell population within the salmonid interbranchial lymphoid tissue. Journal of Immunology 193, 3463–3469.

Agius, C., Roberts, R.J., 2003. Melano-macrophage centres and their role in fish pathology. Journal of Fish Diseases 26 (9), 499–509.

Alvarez-Pellitero, P., 2008. Fish immunity and parasite infections: from innate immunity to immunoprophylactic prospects. Veterinary Immunology and Immunopathology 126, 171–198.

Amend, D.F., Johnson, K.A., 1981. Current status and future needs of *Vibrio anguillarum* bacterins. In: International Symposium on Fish Biologics: Serodiagnostics and Vaccines, Developments in Biological Standardization, Leetown WV, USA,, vol. 49. S. Karger, Basel, pp. 403–417.

Aoki, T., Takano, T., Santos, M.D., Kondo, H., Hirono, I., 2008. Molecular innate immunity in teleost fish: review and future perspectives. In: Tsukamoto, K., Kawamura, T. (Eds.), Fisheries for Global Welfare and Environment, Memorial Book of the 5th World Fisheries Congress. Terrapub, Tokyo, Japan, pp. 263–276.

Auinger, A., Riede, L., Bothe, G., Busch, R., Gruenwald, J., 2013. Yeast (1,3)-(1, 6)-beta-glucan helps to maintain the body's defence against pathogens: a double-blind, randomized, placebo-controlled, multicentric study in healthy subjects. European Journal of Nutrition 52, 1913–1918.

Bairwa, M.K., Jakhar, J.K., Satyanarayana, Y., Reddy, A.D., 2012. Animal and plant originated immunostimulants used in aquaculture. Journal of Natural Product and Plant Resources 2, 397–400.

Balcázar, J.L., Blas, I.D., Ruiz-Zarzuela, I., Cunningham, D., Vendrell, D., Múzquiz, J.L., 2006. The role of probiotics in aquaculture. Veterinary Microbiology 114, 173–186.

Balcázar, J.L., de Blas, I., Ruiz-Zarzuela, I., Vendrell, D., Calvo, A.C., Marquez, I., Girones, O., Muzquiz, J.L., 2007. Changes in intestinal microbiota and humoral immune response following probiotic administration in brown trout (*Salmo Trutta*). British Journal of Nutrition 97, 522–527.

Biering, E., Salonius, K., 2014. DNA vaccines. In: Gudding, R., Lillehaug, A., Evensen, Ø. (Eds.), Fish Vaccination. John Wiley and Sons, Ltd, pp. 47–55. Chapter 5.

Boshra, H., Li, J., Sunyer, J.O., 2006. Recent advances on the complement system of teleost fish. Fish and Shellfish Immunology 20, 239–262.

Bowden, T.J., Thompson, K.D., Morgan, A.L., Gratacap, R.M., Nikoskelainen, S., 2007. Seasonal variation and the fish immune system: a fish perspective. Fish and Shellfish Immunology 19, 413–427.

Bowden, T.J., 2008. Modulation of the immune system of fish by their environment. Fish and Shellfish Immunology 25 (4), 373–383.

Brattgjerd, S., Evensen, O., 1996. A sequential light microscopic and ultrastructural study on the uptake and handling of *Vibrio salmonicida* in phagocytes of the head kidney in experimentally infected Atlantic salmon (*Salmo salar* L.). Veterinary Pathology 33 (1), 55–65.

Brudeseth, B.E., Wiulsrød, R., Fredriksen, B.N., Lindmo, K., Løkling, K.E., Bordevik, M., Steine, N., Klevan, A., Gravningen, K., 2013. Status and future perspectives of vaccines for industrialised fin-fish farming. Fish and Shellfish Immunology 35 (6), 1759–1768.

Brunt, J., Newaj-Fyzul, A., Austin, B., 2007. The development of probiotics for the control of multiple bacteria disease of rainbow trout, *Oncorhynchus mykiss* (Walbum). Journal of Fish Diseases 30 (10), 573–579.

Bulfon, C., Volpatti, D., Galeotti, M., 2015. Current research on the use of plant-derived products in farmed fish. Aquaculture Research 46, 513–551.

Burbank, D.R., Shah, D.H., LaPatra, S.E., Fornshell, G., Cain, K.D., 2011. Enhanced resistance to coldwater disease following feeding of probiotic bacterial strains to rainbow trout (*Oncorhynchus mykiss*). Aquaculture 321, 185–190.

Burrells, C., Williams, P.D., Southgate, P.J., Wadsworth, S.L., 2001. Dietary nucleotides: a novel supplement in fish feeds: 2. Effects on vaccination, salt water transfer, growth rates and physiology of Atlantic salmon (*Salmo salar* L.). Aquaculture 199 (1–2), 171–184.

Corbeil, S., LaPatra, S.E., Anderson, E.D., Kurath, G., 2000. Nanogram quantities of a DNA vaccine protect rainbow trout fry against heterologous strains of infectious hematopoietic necrosis virus. Vaccine 18, 2817–2824.

Cruz, P.M., Ibáñez, A.L., Monroy Hermosillo, O.A., Ramírez Saad, H.C., 2012. Use of probiotics in aquaculture. International Scholarly Research Notices: Microbiology. 2012:916845. http://dx.doi.org/10.5402/2012/916845.

Danneving, B.H., Lauve, A., Press, M.C., Landsverk, T., 1994. Receptor-mediated endocytosis and phagocytosis by rainbow trout head kidney sinusoidal cells. Fish and Shellfish Immunology 4, 3–18.

Deakin, J.E., Papenfuss, A.T., Belov, K., Cross, J.G.R., Coggill, P., Palmer, S., Sims, S., Speed, T.P., Beck, S., Marshall Graves, J.A., 2006. Evolution and comparative analysis of the MHC class III inflammatory region. BMC Genomics 7, 281.

Dong, X.-H., Geng, X., Tan, B.-P., Yang, Q.-H., Chi, S.-Y., Liu, H.-Y., Liu, X.-Q., 2015. Effects of dietary immunostimulant combination on the growth performance, non-specific immunity and disease resistance of cobia, *Rachycentron canadum* (Linnaeus). Aquaculture Research 46, 840–849.

Edholm, E.S., Bengten, E., Wilson, M., 2011. Insights into the function of IgD. Developmental and Comparative Immunology 35, 1309–1316.

Ellis, A., 1977. The leucocytes of fish: a review. Journal of Fish Biology 11, 453–491.

Ellsaesser, C.F., Bly, J.E., Clem, L.W., 1988. Phylogeny of lymphocyte heterogeneity. The thymus of the channel catfish. Developmental and Comparative Immunology 12, 787–799.

Espelid, S., Lokken, G.B., Steiro, K., Bogwald, J., 1996. Effects of cortisol and stress on the immune system in Atlantic salmon (*Salmo salar* L.). Fish and Shellfish Immunology 6, 95–100.

Esteban, M.A., Cuesta, A., Chaves-Pozo, E., Meseguer, J., 2015. Phagocytosis in teleosts. Implications of the new cells involved. Biology 4 (4), 907–922. http://dx.doi.org/10.3390/biology4040907.

Evans, D.L., Hogan, K.T., Graves, S.S., 1984. Nonspecific cytotoxic cells in fish (*Ictalurus punctatus*). III. Biophysical and biochemical properties affecting cytolysis. Developmental and Comparative Immunology 8, 599–610.

Evensen, Ø., 2009. Development in fish vaccinology with focus on delivery methodologies, adjuvants and formulations. In: Rogers, C., Basurco, B. (Eds.), The Use of Veterinary Drugs and Vaccines in Mediterranean Aquaculture. CIHEAM, Zaragoza, pp. 177–186 (Options Méditerranéennes: Série A. Séminaires Méditerranéens; n. 86).

Evensen, Ø., Leong, J.C., 2013. DNA vaccines against viral diseases of farmed fish. Fish and Shellfish Immunology 35, 1751–1758.

FAO/WHO, 2001. Health and Nutritional Properties of Probiotics in Food Including Powder Milk With Live Lactic Acid Bacteria. Report of Joint Food and Agriculture Organization of the United Nations/World Health (FAO/WHO) Expert Consultation on Evaluation of Health and Nutritional Properties of Probiotics in Food Including Powder Milk with Live Lactic Acid Bacteria, Cordoba, Argentina, October 1–4. , pp. 1–34.

Filho, D.W., 2007. Reactive oxygen species, antioxidants and fish mitochondria. Frontiers in Bioscience 12, 1229–1237.

Fischer, U., Utke, K., Somamoto, T., Kollner, B., Ototake, M., Nakanishi, T., 2006. Cytotoxic activities of fish leucocytes. Fish and Shellfish Immunology 20 (2), 209–226.

Flerova, E.A., Balabanova, L.V., 2013. Ultrastructure of granulocytes of teleost fish (Salmoniformes, Cypriniformes, Perciformes). Journal of Evolutionary Biochemistry and Physiology 49, 223–233.

Forlenza, M., Fink, I.R., Raes, G., Wiegertjes, G.F., 2011. Heterogeneity of macrophage activation in fish. Developmental and Comparative Immunology 35, 1246–1255.

Frøystad, M.K., Rode, M., Berg, G., Gjøen, T., 1998. A role for scavenger receptors in phagocytosis of protein-coated particles in rainbow trout head kidney macrophages. Developmental and Comparative Immunology 22, 533–549.

Galindo-Villegas, J., Hosokawa, H., 2004. Immunostimulants: towards temporary prevention of diseases in marine fish. In: Cruz Suárez, L.E., Ricque, D., Marie, M.G., Nieto López, M.G., Villarreal, D., Scholz, U., Gonzalez, M. (Eds.), Avances en Nutrición Acuícola VII. Memorias del VII Simposium Internacional de Nutrición Acuícola, November 16–19, Hermosillo, Sonara, México, pp. 279–319.

Gerlier, D., Lyles, D.S., 2011. Interplay between innate immunity and negative strand RNA viruses: towards a rational model. Microbiology and Molecular Biology Reviews 75, 468–490.

Gjedrem, T., 2015. Disease resistant fish and shellfish are within reach: a review. Journal of Marine Science and Engineering 3, 146–153.

Glencross, B., Rutherford, N., 2010. Dietary strategies to improve the growth and feed utilization of barramundi, *Lates calcarifer* under high water temperature conditions. Aquaculture Nutrition 16 (4), 343–350.

Gram, L., Loevold, T., Nielsen, J., Melchiorsen, J., Spanggaard, B., 2001. In vitro antagonism of the probiont *Pseudomonas fluorescens* strain AH2 against *Aeromonas salmonicida* does not confer protection of salmon against furunculosis. Aquaculture 199, 1–11.

Greer, J.B., O'Keefe, S.J., 2011. Microbial induction of immunity, inflammation, and cancer. Frontiers in Physiology 1, 168. http://dx.doi.org/10.3389/fphys.2010.00168.

Hai, N.V., 2015. The use of medicinal plants as immunostimulants in aquaculture: a review. Aquaculture 446 (1), 88–96.

Håstein, T., Gudding, R., Evensen, Ø., 2005. Bacterial vaccines for fish: an update of the current situation worldwide. In: Midtlyng, P.J. (Ed.). Midtlyng, P.J. (Ed.), Fish Vaccinology, vol. 121. Karger. Development of Biological Standards, Basel, pp. 54–75.

Haugarvoll, E., Bjerkas, I., Nowak, B.F., Hordvik, I., Koppang, E.O., 2008. Identification and characterization of a novel intraepithelial lymphoid tissue in the gills of Atlantic salmon. Journal of Anatomy 213, 202–209.

Havixbeck, J.J., Barreda, D.R., 2015. Neutrophil development, migration, and function in teleost fish. Biology 4 (4), 715–734.

Havixbeck, J.J., Rieger, A.M., Wong, M.E., Hodgkinson, J.W., Barreda, D.R., 2015. Neutrophil contributions to the induction and regulation of the acute inflammatory response in teleost fish. Journal of Leukocyte Biology 99 (2), 241–252.

Herraez, M.P., Zapata, A.G., 1986. Structure and function of the melano-macrophage centres of the goldfish *Carassius auratus*. Veterinary Immunology and Immunopathology 12 (1–4), 117–126.

Holzapfel, W.H., Schillinger, U., 2002. Introduction to pre and probiotics. Food Research International 35, 109–116.

Irianto, A., Austin, B., 2002. Probiotics in aquaculture. Journal of Fish Diseases 25, 633–642.

Janeway Jr., C.A., Travers, P., Walport, M., Shlomchik, M.J., 2001. The major histocompatibility complex and its functions. In: Immunobiology: The Immune System in Health and Disease. fifth ed. Garland Science, New York.

Jeney, G., Yin, G., Ardo, L., Jeney, Z., 2009. The use of immunostimulating herbs in fish. An overview of research. Fish Physiology and Biochemistry 35, 669–676.

Jha, A.K., Pal, A., Sahu, N., Kumar, S., Mukherjee, S., 2007. Haemato-immunological responses to dietary yeast RNA, ω-3 fatty acid and β-carotene in *Catla catla* juveniles. Fish and Shellfish Immunology 23 (5), 917–927.

Jorgensen, J.B., 2014. The innate immune response of fish. In: Gudding, R., Lillehaug, A., Evensen, Ø. (Eds.), Fish Vaccination. John Wiley and Sons, Ltd, pp. 85–103. Chapter 8.

Kaattari, S.L., Irwin, M.J., 1985. Salmonid spleen and anterior kidney harbor populations of lymphocytes with different B cell repertoires. Developmental and Comparative Immunology 9 (3), 433–444.

Katzenback, B.A., Belosevic, M., 2009. Isolation and functional characterization of neutrophil-like cells, from goldfish (*Carassius auratus* L.) kidney. Developmental and Comparative Immunology 33, 601–611.

Katzenback, B.A., 2015. Antimicrobial peptides as mediators of innate immunity in teleosts. Biology 4, 607–639.

Kenari, A.A., Mahmoudi, N., Soltani, M., Abediankenari, S., 2013. Dietary nucleotide supplements influence the growth, haemato-immunological parameters and stress responses in endangered Caspian brown trout (*Salmo trutta caspius* Kessler, 1877). Aquaculture Nutrition 19 (1), 54–63.

Kesarcodi-Watson, A., Kaspar, H., Lategan, M.J., Gibson, L., 2008. Probiotics in aquaculture: the need, principles and mechanisms of action and screening processes. Aquaculture 274, 1–14.

Kim, D., Austin, B., 2006. Innate immune responses in rainbow trout (*Oncorhynchus mykiss*, Walbaum) induced by probiotics. Fish and Shellfish Immunology 21, 513–524.

Klein, J., Klein, D., Figueroa, F., Sato, A., O'hUigin, C., 1997. Major histocompatibility complex genes in the study of fish phylogeny. In: Kocher, T.D., Stepien, C.A. (Eds.), Molecular Systematics of Fishes. Academic Press, San Diego, CA, pp. 271–283.

Kocher, A., 2004. The potential for immunosaccharides to maximise growth performance a review of six published meta-analyses on Bio-Mos. In: Tucker, L.A., Taylor-Pickard, J.A. (Eds.), Interfacing Immunity, Gut Health and Performance. Nottingham University Press, pp. 107–116.

Koppang, E.O., Hordvik, I., Bjerkås, I., Torvund, J., Aune, L., Thevarajan, J., Endresen, C., 2003. Production of rabbit antisera against recombinant MHC class II β chain and identification of immunoreactive cells in Atlantic salmon (*Salmo salar*). Fish and Shellfish Immunology 14, 115–132.

Koppang, E.O., Fischer, U., Moore, L., Tranulis, M.A., Dijkstra, J.M., Köllner, B., Aune, L., Jirillo, E., Hordvik, I., 2010. Salmonid T cells assemble in the thymus, spleen and in novel interbranchial lymphoid tissue. Journal of Anatomy 217, 728–739.

Korkea-aho, T.L., Heikkinen, J., Thompson, K.D., von Wright, A., Austin, B., 2011. *Pseudomonas* Sp. M174 inhibits the fish pathogen *Flavobacterium psychrophilum*. Journal of Applied Microbiology 111, 266–277.

Korkea-aho, T., Papadopoulou, A., Heikkinen, J., von Wright, A., Austin, A., Adams, B., Thompson, K.D., 2012. *Pseudomonas* M162 confers protection against rainbow trout fry syndrome by stimulating immunity. Journal of Applied Microbiology 113, 24–35.

Kumar, H., Kawai, T., Akira, S., 2011. Pathogen recognition by the innate immune system. International Reviews of Immunology 30, 16–34.

Leonardi, M., Sandino, A., Klempau, A., 2003. Effect of a nucleotide-enriched diet on the immune system, plasma cortisol levels and resistance to infectious pancreatic necrosis (IPN) in juvenile rainbow trout (*Oncorhynchus mykiss*). Bulletin of the European Association of Fish Pathologists 23 (2), 52–59.

Li, P., Gatlin III, D.M., 2006. Nucleotide nutrition in fish: current knowledge and future applications. Aquaculture 251 (2–4), 141–152.

Li, J., Barreda, D.R., Zhang, Y.A., Boshra, H., Gelman, A.E., Lapatra, S., Tort, L., Sunyer, J.O., 2006. B lymphocytes from early vertebrates have potent phagocytic and microbicidal abilities. Nature Immunolology 7, 1116–1124.

Li, P., Gatlin, D.M., Neill, W.H., 2007. Dietary supplementation of a purified nucleotide mixture transiently enhanced growth and feed utilization of juvenile red drum, *Sciaenops ocellatus*. Journal of the World Aquaculture Society 38 (2), 281–286.

Lin, Y., Wang, H., Shiau, S., 2009. Dietary nucleotide supplementation enhances growth and immune responses of grouper, *Epinephelus malabaricus*. Aquaculture Nutrition 15 (2), 117–122.

Low, C., Wadsworth, S., Burrells, C., Secombes, C., 2003. Expression of immune genes in turbot (*Scophthalmus maximus*) fed a nucleotide-supplemented diet. Aquaculture 221 (1), 23–40.

Ma, C., Ye, J., Kaattari, S.L., February 2013. Differential compartmentalization of memory B cells versus plasma cells in salmonid fish. European Journal of Immunology 43 (2), 360–370.

Makrinos, D.L., Bowden, T.J., June 2016. Natural environmental impacts on teleost immune function. Fish and Shellfish Immunology 53, 50–57.

Malina, A., Tassakka, A., Sakai, M., 2005. Current research on the immunostimulatory effects of CpG oligodeoxynucleotides in fish. Aquaculture 246 (1–4), 25–36.

Mancuso, M., 2013. Probiotics in aquaculture. Journal of Fisheries and Livestock Production 2 (1), e107. http://dx.doi.org/10.4172/2332-2608.1000e107.

Manning, M.J., 1994. Fishes. In: Turner, R.J. (Ed.), Immunology: A Comparative Approach. John Wiley & Sons Ltd, Chichester, UK. ISBN: 0471944009, pp. 69–100.

Martin, S.A.M., Douglas, A., Houlihan, D.F., Secombes, C.J., 2010. Starvation alters the liver transcriptome of the innate immune response in Atlantic salmon (*Salmo salar*). BMC Genomics 11, 418. http://dx.doi.org/10.1186/1471-2164-11-418.

Masso-Silva, J.A., Diamond, G., 2014. Antimicrobial peptides from fish. Pharmaceuticals 7, 265–310. http://dx.doi.org/10.3390/ph7030265.

Meena, D., Das, P., Kumar, S., Mandal, S., Prusty, A., Singh, S., Akhtar, M., Behera, B., Kumar, K., Pal, A., 2013. Beta-glucan: an ideal immunostimulant in aquaculture (a review). Fish Physiology and Biochemistry 39 (3), 431–457.

Midtlyng, P.J., Reitan, L.J., Lillehaug, A., Ramstad, A., 1996. Protection, immune responses and side effects in Atlantic salmon (*Salmo salar* L.) vaccinated against furunculosis by different procedures. Fish and Shellfish Immunology 6, 559–613.

Miethke, M., Marahiel, M.A., 2007. Siderophore-based iron acquisition and pathogen control. Microbiology and Molecular Biology Reviews 71, 413–451.

Millán, A., Gómez-Tato, A., Pardo, B.G., Fernández, C., Bouza, C., Vera, M., Alvarez-Dios, J.A., Cabaleiro, S., Lamas, J., Lemos, M.l., Martínez, P., 2011. Gene expression profiles of the spleen, liver, and head kidney in turbot (*Scophthalmus maximus*) along the infection process with *Aeromonas salmonicida* using an immune-enriched oligomicroarray. Marine Biotechnology (New York) 13, 1099–1114.

Miller, N.W., Deuter, A., Clem, L.W., 1986. Phylogeny of lymphocyte heterogeneity: the cellular requirements for the mixed leukocyte reaction in channel catfish. Immunology 59, 123–128.

Morrison, R.N., Nowak, B.F., 2002. The antibody response of teleost fish. Seminars in Avian and Exotic Pet Medicine 11, 46–54.

Mulero, I., García-Ayala, A., Meseguer, J., Mulero, V., 2007. Maternal transfer of immunity and ontogeny of autologous immunocompetence of fish: a minireview. Aquaculture 268, 244–250.

Mutoloki, S., Alexandersen, S., Evensen, O., 2004. Sequential study of antigen persistence and concomitant inflammatory reactions relative to side-effects and growth of Atlantic salmon (*Salmo salar* L.) following intraperitoneal injection with oil-adjuvanted vaccines. Fish and Shellfish Immunology 16, 633–644.

Mutoloki, S., Jørgensen, J.B., Evensen, Ø., 2014. The adaptive immune response in fish. In: Gudding, R., Lillehaug, A., Evensen, Ø. (Eds.), Fish Vaccination. John Wiley and Sons, Ltd, pp. 104–115. Chapter 9.

Nakanishi, T., Fischer, U., Dijkstra, J.M., Hasegawa, S., Somamoto, T., Okamoto, N., 2002. Cytotoxic T cell function in fish. Developmental and Comparative Immunology 26, 131–139.

Nakao, M., Kajiya, T., Sato, Y., Somamoto, T., Kato-Unoki, Y., Matsushita, M., Nakata, M., Fujita, T., Yano, T., 2006. Lectin pathway of bony fish complement: identification of two homologs of the mannose-binding lectin associated with MASP2 in the common carp (*Cyprinus carpio*). Journal of Immunology 177, 5471–5479.

Nakao, M., Tsujikura, M., Ichiki, S., Vo, T.K., Somamoto, T., 2011. The complement system in teleost fish: progress of post-homolog-hunting researches. Developmental and Comparative Immunology 35 (12), 1296–1308.

Nayak, S.K., 2010. Probiotics and immunity: a fish perspective. Fish and Shellfish Immunology 29, 2–14.

Nemeth, E., Baird, A.W., O'Farrelly, C., 2009. Microanatomy of the liver immune system. Seminars in Immunopathology 31, 333–343.

Newaj-Fyzul, A., Austin, B., 2015. Probiotics, immunostimulants, plant products and oral vaccines, and their role as feed supplements in the control of bacterial fish diseases. Journal of Fish Diseases 38, 937–955.
Newaj-Fyzul, A., Adesiyun, A.A., Mutani, A., Ramsubhag, A., Brunt, J., Austin, B., 2007. *Bacillus subtilis* AB1 controls *Aeromonas* infection in rainbow trout (*Oncorhynchus mykiss*, Walbaum). Journal of Applied Microbiology 103, 1699–1706.
Nikoskelainen, S., Ouwehand, A.C., Bylund, G., Salminen, S., Lilius, E., 2003. Immune enhancement in rainbow trout (*Oncorhynchus mykiss*) by potential probiotic bacteria (*Lactobacillus rhamnosus*). Fish and Shellfish Immunology 15, 443–452.
Palić, D., Andreasen, C.B., Ostojić, J., Tell, R.M., Roth, J.A., 2007. Zebrafish (*Danio rerio*) whole kidney assays to measure neutrophil extracellular trap release and degranulation of primary granules. Journal of Immunological Methods 319, 87–97.
Palti, Y., 2011. Toll-like receptors in bony fish: from genomics to function. Developmental and Comparative Immunology 35 (12), 1263–1272.
Pandiyan, P., Balaraman, D., Thirunavukkarasu, R., Jothi-George, E.G., Subaramaniyan, K., Manikkam, S., Sadayappan, B., 2013. Probiotics in aquaculture. Drug Invention Today 1, 55–59.
Panigrahi, A., Kiron, V., Puangkaew, J., Kobayashi, T., Satoh, S., Sugita, H., 2005. The viability of probiotic bacteria as a factor influencing the immune response in rainbow trout *Oncorhynchus mykiss*. Aquaculture 243, 241–254.
Panigrahi, A., Kiron, V., Satoh, S., Hirono, I., Kobayashi, T., Sugita, H., Puangkaew, J., Aoki, T., 2007. Immune modulation and expression of cytokine genes in rainbow trout *Oncorhynchus mykiss* upon probiotic feeding. Developmental and Comparative Immunology 31, 372–382.
Parra, D., Takizawa, F., Sunyer, J.O., 2013. Evolution of B cell immunity. Annual Review of Animal Biosciences 1, 65–97.
Passer, B.J., Chen, C.H., Miller, N.W., Cooper, M.D., 1996. Identification of a T lineage antigen in the catfish. Developmental and Comparative Immunology 20 (6), 441–450.
Peng, M., Xu, W., Ai, Q., Mai, K., Liufu, Z., Zhang, K., 2013. Effects of nucleotide supplementation on growth, immune responses and intestinal morphology in juvenile turbot fed diets with graded levels of soybean meal (*Scophthalmus maximus* L.). Aquaculture 392–395, 51–58.
Peterson, R.A., 2012. Regulatory T-cells: diverse phenotypes integral to immune homeostasis and suppression. Toxicological Pathology 40 (2), 186–204.
Pickering, A.D., Pottinger, T.G., 1989. Stress responses and disease resistance in salmonid fish: effects of chronic elevation of plasma cortisol. Fish Physiology and Biochemistry 7, 253–258.
Pijanowski, L., Golbach, L., Kolaczkowska, E., Scheer, M., Verburg-van Kemenade, B.M.L., Chadzinska, M., 2013. Carp neutrophilic granulocytes form extracellular traps via ROS-dependent and independent pathways. Fish and Shellfish Immunology 34, 1244–1252.
Plouffe, D.A., Hanington, P.C., Walsh, J.G., Wilson, E.C., Belosevic, M., 2005. Comparison of select innate immune mechanisms of fish and mammals. Xenotransplantation 12, 266–277.
Praveen, K., Evans, D.L., Jaso-Friedmann, L., 2004. Evidence for the existence of granzyme-like serine proteases in teleost cytotoxic cells. Journal of Molecular Evolution 58, 449–459.
Press, C.McL., Evensen, Ø., 1999. The morphology of the immune system in teleost fishes. Fish and Shellfish Immunology 9, 309–318.
Priyadarshini, P., Deivasigamani, B., Rajasekar, T., Jothi, G.E.G., Kumaran, S., Sakthivel, M., Balamurugan, S., 2013. Probiotics in aquaculture. Drug Invention Today 5 (1), 55–59.
Pulsford, A.L., Lemairegony, S., Tomlinson, M., Collingwood, N., Glynn, P.J., 1994. Effects of acute stress on the immune system of the dab *Limanda limanda*. Comparative Biochemistry and Physiology C: Pharmacology, Toxicology and Endocrinology 109 (2), 111–217.
Ramirez-Gomez, F., Greene, W., Rego, K., Hansen, J.D., Costa, G., Kataria, P., Bromage, E.S., 2012. Discovery and characterization of secretory IgD in rainbow trout: secretory IgD is produced through a novel splicing mechanism. Journal of Immunology 188, 1341–1349.
Ranjan, R., Prasad, K.P., Vani, T., Kumar, R., 2012. Effect of dietary chitosan on haematology, innate immunity and disease resistance of Asian seabass *Lates calcarifer* (Bloch). Aquaculture Research 45 (6), 983–993.
Reyes-Cerpa, S., Maisey, K., Reyes-López, F., Toro-Ascuy, D., Sandino, A.M., Imarai, M., 2013. Fish Cytokines and Immune Response. InTech. Chapter 1 http://dx.doi.org/10.5772/53504.
Rieger, A.M., Konowalchuk, J.D., Grayfer, L., Katzenback, B.A., Havixbeck, J.J., Kiemele, M.D., Belosevic, M., Barreda, D.R., 2012. Fish and mammalian phagocytes differentially regulate pro-inflammatory and homeostatic responses in vivo. PloS One 2012 (7), e47070.
Ringø, E., Song, S.K., 2016. Application of dietary supplements (synbiotics and probiotics in combination with plant products and β-glucans) in aquaculture. Aquaculture Nutrition 22, 4–24.
Ringø, E., Olsen, R.E., Gifstad, T.O., Dalmo, R.A., Amlund, H., Hemre, G.I., Bakke, A.M., 2010. Prebiotics in aquaculture: a review. Aquaculture Nutrition 16, 117–136.
Ringø, E., Olsen, R.E., Vecino, J.L.G., Wadsworth, S., Song, S.K., 2012. Use of immunostimulants and nucleotides in aquaculture: a review. Journal of Marine Science: Research and Development 1, 104. http://dx.doi.org/10.4172/2155-9910.1000104.
Ringø, E., Zhou, Z., Vecino, J.L.G., Wadsworth, S., Romero, J., Krogdahl, Å., Olsen, R.E., Dimitroglou, A., Foey, A., Davies, S., Owen, M., Lauzon, H.L., Martinsen, L.L., De Schryver, P., Bossier, P., Sperstad, S., Merrifield, D.L., 2016. Effect of dietary components on the gut microbiota of aquatic animals. A never-ending story? Aquaculture Nutrition 22, 219–282.
Roberfroid, M., 1993. Dietary fibre, inulin and oligofructose: a review comparing their physiological effects. Critical Reviews in Food Science and Nutrition 33, 103–148.
Roberfroid, M.B., 2007. Prebiotics: the concept revisited. Journal of Nutrition 137 (3), 830S–837S.
Robertsen, B., 2006. The interferon system of teleost fish. Fish and Shellfish Immunology 20, 172–191.
Rodriguez-Estrada, U., Satoh, S., Haga, Y., Fushimi, H., Sweetman, J., 2009. Effects of single and combined supplementation of *Enterococcus faecalis*, mannan oligosaccharides and polyhydroxybutyrate acid on growth performance and immune response of rainbow trout, *Oncorhynchus mykiss*. Aquaculture Science 57, 609–617.

Saeij, J.P., Stet, R.J., de Vries, B.J., van Muiswinkel, W.B., Wiegertjes, G.F., 2003. Molecular and functional characterization of carp TNF: a link between TNF polymorphism and trypanotolerance. Developmental and Comparative Immunology 27 (1), 29–41.

Sakai, M., Taniguchi, K., Mamoto, K., Ogawa, H., Tabata, M., 2001. Immunostimulant effects of nucleotide isolated from yeast RNA on carp, *Cyprinus carpio* L. Journal of Fish Diseases 24 (8), 433–438.

Sakai, M., 1999. Current research status of fish Immunostimulants. Aquaculture 172, 63–92.

Salinas, I., Abelli, L., Bertoni, F., Picchietti, S., Roque, A., Furones, D., Cuesta, A., Meseguer, J., Esteban, M.Á., 2008. Monospecies and multispecies probiotic formulations produce different systemic and local immunostimulatory effects in the gilthead seabream (*Sparus aurata* L.). Fish and Shellfish Immunology 25, 114–123.

Salinas, I., 2015. The mucosal immune system of teleost fish. Biology 4 (3), 525–539.

Samuel, C.E., 2001. Antiviral actions of interferons. Clinical Microbiolical Reviews 14 (4), 778–809.

Schubiger, C.B., Orfe, L.H., Sudheesh, P.S., Cain, K.D., Shah, D.H., Call, D.R., 2015. Entericidin is required for a probiotic treatment (*Enterobacter* sp. strain C6-6) to protect trout from cold-water disease challenge. Applied Environmental Microbiology 81, 658–665.

Secombes, C.J., Fletcher, T.C., 1992. The role of phagocytes in the protective mechanisms of fish. Annual Review of Fish Diseases 2, 53–71.

Secombes, C., Wang, T., 2012. The innate and adaptive immune system of fish. In: Austin, B. (Ed.), Infectious Disease in Aquaculture: Prevention and Control. Woodhead Publishing Limited, pp. 3–68. Chapter 1.

Secombes, C.J., Manning, M.J., Ellis, A.E., 1982. The effect of primary and secondary immunization on the lymphoid tissue of the carp, *Cyprinus carpio* L. Journal of Experimental Zoology 220, 277–287.

Secombes, C., Wang, T., Hong, S., Peddie, S., Crampe, M., Laing, K., Cunningham, C., Zou, J., 2001. Cytokines and innate immunity of fish. Developmental and Comparative Immunology 25, 713–723.

Secombes, C.J., Wang, T., Bird, S., 2011. The interleukins of fish. Developmental and Comparative Immunology 35 (12), 1336–1345.

Sharifuzzaman, S.M., Austin, B., 2010. Development of protection in rainbow trout (*Oncorhynchus mykiss*, Walbaum) to *Vibrio anguillarum* following use of the probiotic Kocuria SM1. Fish and Shellfish Immunolology 29, 212–216.

Shen, L., Stuge, T.B., Zhou, H., Khayat, M., Barker, K.S., Quiniou, S.M., Wilson, M., Bengtén, E., Chinchar, V.G., Clem, L.W., Miller, N.W., 2002. Channel catfish cytotoxic cells: a mini-review. Developmental and Comparative Immunology 26 (2), 141–149.

Shen, L., Stuge, T.B., Bengtén, E., Wilson, M., Chinchar, V.G., Naftel, J.P., Bernanke, J.M., Clem, L.W., Miller, N.W., 2004. Identification and characterization of clonal NK-like cells from channel catfish (*Ictalurus punctatus*). Developmental and Comparative Immunology 28, 139–152.

Sirimanapong, W., Adams, A., Ooi, E.L., Green, D.G., Nguyen, D.K., Browdy, C.L., Collet, B., Thompson, K.D., 2015. The effects of feeding immunostimulant β-glucan on the immune response of *Pangasianodon hypophthalmus*. Fish and Shellfish Immunology 45 (2), 357–366.

Solem, S.T., Stenvik, J., 2006. Antibody repertoire development in teleosts – a review with emphasis on salmonids and *Gadus morhua* L. Developmental and Comparative Immunology 30 (1–2), 57–76.

Somamoto, T., Nakanishi, T., Okamoto, N., 2000. Specific cell-mediated cytotoxicity against a virus-infected syngeneic cell line in isogeneic ginbuna crucian carp. Developmental and Comparative Immunology 24, 633–640.

Song, K.S., Beck, B.R., Kim, D., Park, J., Kim, J., Kim, H.D., Ringø, E., 2014. Prebiotics as immunostimulants in aquaculture: a review. Fish and Shellfish Immunology 40 (1), 40–48.

Spanggaard, B., Huber, I., Nielsen, J., Sick, E.B., Pipper, C.B., Martinussen, T., Slierendrecht, W.J., Gram, L., 2001. The probiotic potential against vibriosis of the indigenous microflora of rainbow trout. Environmental Microbiology 3, 755–765.

Ström-Bestor, M., Wiklund, T., 2011. Inhibitory activity of *Pseudomonas* sp. on *Flavobacterium psychrophilum*, in vitro. Journal of Fish Diseases 34, 255–264.

Stuge, T.B., Wilson, M.R., Zhou, H., Barker, K.S., Bengtén, E., Chinchar, G., Miller, N.W., Clem, L.W., 2000. Development and analysis of various clonal alloantigen-dependent cytotoxic cell lines from channel catfish. Journal of Virology 164, 2971–2977.

Sun, W., Li, Y., Chen, L., Chen, H., You, F., Zhou, X., Zhou, Y., Zhai, Z., Chen, D., Jiang, Z., 2009. ERIS, an endoplasmic reticulum IFN stimulator, activates innate immune signalling through dimerization. Proceedings of the National Academy of Sciences of the United States of America 106, 8653–8658.

Sun, B., Skjaeveland, I., Svingerud, T., Zou, J., Jorgensen, J., Robertsen, B., 2011. Antiviral activity of salmonid gamma interferon against infectious pancreatic necrosis virus and salmonid alphavirus and its dependency on type I interferon. Journal of Virology 85 (17), 9188–9198.

Svingerud, T., Solstad, T., Sun, B., Nyrud, M.L., Kileng, Ø., Greiner-Tollersrud, L., Robertsen, B., 2012. Atlantic salmon type I IFN subtypes show differences in antiviral activity and cell-dependent expression: evidence for high IFNb/IFNc-producing cells in fish lymphoid tissues. Journal of Immunology 189 (12), 5912–5923.

Tacchi, L., Musharrafieh, R., Larragoite, E.T., Crossey, K., Erhardt, E.B., Martin, S.A.M., LaPatra, S.E., Salinas, I., 2014. Nasal immunity is an ancient arm of the mucosal immune system of vertebrates. Nature Communications 5, 5205. http://dx.doi.org/10.1038/ncomms6205.

Tafalla, C., Bogwald, J., Dalmo, R.A., 2013. Adjuvants and immunostimulants in fish vaccines: current knowledge and future perspectives. Fish and Shellfish Immunology 35 (6), 1740–1750.

Tahmasebi-Kohyani, A., Keyvanshokooh, S., Nematollahi, A., Mahmoudi, N., Pasha-Zanoosi, H., 2011. Dietary administration of nucleotides to enhance growth, humoral immune responses, and disease resistance of the rainbow trout (*Oncorhynchus mykiss*) fingerlings. Fish and Shellfish Immunology 30 (1), 189–193.

Tahmasebi-Kohyani, A., Keyvanshokooh, S., Nematollahi, A., Mahmoudi, N., Pasha-Zanoosi, H., 2012. Effects of dietary nucleotides supplementation on rainbow trout (*Oncorhynchus mykiss*) performance and acute stress response. Fish Physiology and Biochemistry 38 (2), 431–440.

Tatner, M.F., Findlay, C., 1991. Lymphocyte migration and localization patterns in rainbow trout, *Onchorhynchus mykiss*, studies using the tracer sample method. Fish and Shellfish Immunology 1, 107–117.

Trichet, V.V., 2010. Nutrition and immunity: an update. Aquaculture Research 41 (3), 356–372.

Tsujii, T., Seno, S., 1990. Melano-macrophage centers in the aglomerular kidney of the sea horse (teleosts): morphologic studies on its formation and possible function. Anatomical Record 226 (4), 460–470.

Utke, K., Bergmann, S., Lorenzen, N., Kollner, B., Ototake, M., Fischer, U., 2007. Cell mediated cytotoxicity in rainbow trout, *Oncorhynchus mykiss*, infected with viral haemorrhagic septicaemia virus. Fish and Shellfish Immunology 22, 182–196.

Van Muiswinkel, W.B., Lamers, C.H., Rombout, J.H., 1991. Structural and functional aspects of the spleen in bony fish. Research Immunology 142 (4), 362–366.

Wang, G., Li, X., Wang, Z., 2009. APD2: the updated antimicrobial peptide database and its application in peptide design. Nucleic Acids Research 37, 933–937.

Wang, T., Gorgoglione, B., Maehr, T., Holland, J.W., Vecino, J.L., Wadsworth, S., Secombes, C.J., 2011. Fish Suppressors of Cytokine Signaling (SOCS): gene discovery, modulation of expression and function. Journal of Signal Transduction. 2011:905813. http://dx.doi.org/10.1155/2011/905813.

Wen, Y., Fang, W., Xiang, L.X., Pan, R.L., Shao, J.Z., 2011. Identification of Treg-like cells in tetraodon: insight into the origin of regulatory T subsets during early vertebrate evolution. Cellular and Molecular Life Sciences 68, 2615–2626.

Wendelaar Bonga, S.E., 1997. The stress response in fish. Physiological Reviews 77, 591–625.

Whyte, S.K., 2007. The innate immune response of finfish – a review of current knowledge. Fish and Shellfish Immunology 23, 1127–1151.

Xu, Z., Parra, D., Gómez, D., Salinas, I., Zhang, Y.A., von Gersdorff Jørgensen, L., Heinecke, R.D., Buchmann, K., LaPatra, S., Sunyer, J.O., 2013. Teleost skin, an ancient mucosal surface that elicits gut-like immune responses. Proceedings of the National Academy of Sciences of the United States of America 110, 13097–13102.

Yadav, M., Schorey, J.S., 2002. The β-glucan receptor dectin-1 functions together with TLR2 to mediate macrophage activation by mycobacteria. Blood 108, 3168–3175.

Zapata, A.G., Chibá, A., Varas, A., 1996. Cells and tissues of the immune system of fish. In: Iwama, G., Nakanishi, T. (Eds.), The Fish Immune System Organism: Pathogen and Environment. Academic Press, pp. 1–62.

Zhang, Y.A., Salinas, I., Li, J., Parra, D., Bjork, S., Xu, Z., LaPatra, S.E., Bartholomew, J., Sunyer, J.O., 2010. IgT, a primitive immunoglobulin class specialized in mucosal immunity. Nature Immunology 11, 827–835.

Zhu, L.Y., Nie, L., Zhu, G., Xiang, L.X., Shao, J.Z., 2013. Advances in research of fish immune-relevant genes: a comparative overview of innate and adaptive immunity in teleosts. Developmental and Comparative Immunology 39 (1–2), 39–62.

Zou, J., Secombes, S.J., 2011. Teleost fish interferons and their role in immunity. Developmental and Comparative Immunology 35 (12), 1376–1387.

Zou, J., Wang, T., Hirono, I., Aoki, T., Inagawa, H., Honda, T., Soma, G.-I., Ototake, M., Nakanishi, T., Ellis, A.E., Secombes, C.J., 2002. Differential expression of two tumor necrosis factor genes in rainbow trout, *Oncorhynchus mykiss*. Developmental and Comparative Immunology 26 (2), 161–172.

Chapter 2

Improvement of Disease Resistance by Genetic Methods

Nicholas Andrew Robinson[1,2], Trygve Gjedrem[1], Edwige Quillet[3]
[1]*Nofima, Ås, Norway;* [2]*The University of Melbourne, Parkville, VIC, Australia;* [3]*INRA, AgroParisTech, Université Paris-Saclay, France*

Robustness can be defined as the ability of an individual, or a population, to face a range of biotic and abiotic stressors throughout its productive lifespan (Knap, 2009). Abiotic stress factors can be physical (e.g., temperature, turbidity, and ambient lighting) or chemical (e.g., salinity, pH, and pollutants) characteristics of water and environment, as well as stressors associated with farming operations (confinement or other manipulations). Biotic stress factors can be nutritious, social, or infectious. An animal's robustness may be affected by its genetic background, history of exposure to the above factors, and consequent physiology (e.g., immune or cardiovascular fitness). Although it would be desirable to improve an animal's overall robustness, the complexity of this trait and its causes are likely to make it difficult to achieve significant progress, and therefore, research efforts generally focus more on narrow traits such as survival when exposed to specific stressors.

Some studies have tried to evaluate the possibility of selecting fish for increased survival over the rearing cycle. For instance, after a multigenerational survey of 10-year classes of rainbow trout reared in Finland for one growing season in three test stations (fresh or seawater), Vehviläinen et al. (2008) concluded that overall survival has limited potential to predict general resistance across generations and environments because there are different mortality factors among years and locations that do not share a common genetic control. Several other studies led to the same conclusion (discussed in later sections). However, the Vehviläinen et al. study also demonstrated that within a homogeneous environment, there is large genetic variability and great potential for selection for separate components of survival.

Infectious diseases, caused by viruses, bacteria, fungi, or parasites, are recognized as a major cause of mortality in farmed finfish worldwide, and constitute a major global threat for the development and sustainability of intensive farming. Disease control strategies combine preventive (prophylaxis, vaccination) and curative measures (use of antimicrobial drugs and antibiotics). However, commercial vaccines are available for a limited number of diseases, and the use of drugs raises environmental and consumer health issues. Accumulating evidence indicates that in fish and shellfish species resistance to pathogens is partly under genetic control, as in terrestrial domestic animals, and that selection of brood stock with genetically improved resistance is a promising strategy. In this chapter, we will mainly focus on the genetic variability of resistance to specific infectious diseases and how this natural variability can be used in selective breeding programs and long-term health management strategies in aquaculture. Possible approaches for increased overall robustness will be reported in the last section of this chapter.

2.1 HOW TO ASSESS DISEASE RESISTANCE IN FISH?

Reliable and quantifiable phenotypes are the key for the estimation and exploitation of genetic variation among individuals. However, getting relevant phenotypic data may be tricky for complex traits such as resistance or susceptibility to diseases.

Conceptually, individual host response to infectious pathogens involves two distinct and complementary mechanisms, *resistance* and *tolerance*. Resistance refers to the host's ability to reduce pathogen invasion (limitation of pathogen entry into the target tissue and replication). Tolerance refers to the host's ability, once infected, to limit the detrimental effect of the pathogen (limiting the physiopathogenic effects and promoting repair) which translates into the capacity to maintain production performance, or to minimize its reduction during the course of the infectious episode. Typically, resistance/susceptibility can be described by the measure of pathogen burden, while tolerance can be measured as the net impact on the performance of a given level of infection (Doeschl-Wilson and Kyriazakis, 2012; Bishop and Woolliams, 2014). At the population level, characteristics involved in disease transmission constitute additional key components of host response and

Fish Diseases. http://dx.doi.org/10.1016/B978-0-12-804564-0.00002-8

disease risk control. These include traits such as shedding, asymptotic carriage and pathogen persistence (an animal has recovered but not cleared the disease).

In practice, however, it is not easy to disentangle resistance and tolerance. In this chapter, we will use "disease resistance" as a generic term (see Bishop and Woolliams, 2014). At the individual level, the term "resistance" encompasses the way an individual is less likely to be infected (blocking pathogen entry into host target tissue), limits pathogen proliferation once infected and prevents the pathogenic effects of the disease and, consequently, overlaps some attributes of tolerance.

2.1.1 Which Traits Can Be Used to Score Disease Resistance at the Individual Level?

Once infected, the host organism will express a range of symptoms that may often not be specific to the pathogen, such as loss of appetite and reduction of activity, abnormal swimming, skin darkening, skin ulcerations, gill or muscular lesions, hemorrhages, exophthalmia, etc. Internal organs will also be damaged depending on the nature and severity of the disease. Ultimately, the fish may die or recover, as a result of the combined efficacy of host defense components acting at different stages of infection.

There are few diseases in fish where the pathogen burden (level of infection) can be easily measured. For sea louse infection of Atlantic salmon, for instance, it is possible to count the number of copepods on the body surface (Kolstad et al., 2005). The pathogen burden in specific tissues (serum, spleen) is also measured for viral or bacterial infections. However, sampling is often invasive and assays may be expensive and time-consuming. Consequently, pathogen burden is normally not measured for large-scale experiments, but restricted to thorough analyses aiming to study the host response to the disease.

The number or extent of lesions on target organs is also indicative of infection severity. For amoebic gill disease (AGD) caused by the parasite *Neoparamoeba* sp., a visual scoring of gill lesions proved to be an efficient proxy of more sophisticated histopathology or gill image analyses (Taylor et al., 2009). When available, such external evaluations of infection level are very convenient as they allow the researcher to carry out large-scale and/or longitudinal studies.

In practice, the dead versus alive status at the end of an infectious challenge is the trait most widely used to assess individual resistance or susceptibility to a disease, especially in the case of viral or bacterial diseases. Fish that die during a challenge are classified as susceptible, and fish that survive as resistant. For this measurement, resistance is therefore scored as a binary qualitative trait. It is often complemented by the measurement of time-until-death (sometimes referred to as endurance) that aims at better discriminating the potential variation in ability to restrict the disease kinetics and development of pathogenic effects among the dead fish (i.e., a combination of resistance and tolerance characteristics). It should be emphasized that in many situations, the actual states of surviving fish (infected and recovered, uninfected, or having cleared the pathogen) remain unknown. We will discuss how this missing information can reduce the accuracy of genetic parameters in breeding programs further in a later section.

Phenotypic characteristics correlated to resistance that would predict the resistance of an individual without the need of disease exposure would be convenient alternatives to infectious challenges. In rainbow trout, Quillet et al. (2001, 2007a) demonstrated that the in vitro proliferation of viral hemorrhagic septicemia (VHS) virus on fin tissue explants is positively correlated to fish survival after immersion infection with the virus. Similarly, Green et al. (2014) showed that an assay measuring the antiviral activity of factors present in the hemolymph of Pacific oysters was heritable, and proposed that such assays can be used as useful markers for selection to improve disease resistance. Such indirect indicators would also solve the problem of pathogen carriage in surviving challenged animals and the risk that breeders transmit the disease to progeny. Resistance phenotypes specified at the individual level could then be included in genetic evaluation to improve selection accuracy. However, although a number of studies have attempted to find such indicators, many of them still do not meet the requirements for effective use (i.e., consistently correlated to resistance, easy to measure on large number of candidates, and cheap). We will see in Section 2.4 that high-throughput genomics has the potential to exploit individual genetic variation.

2.1.2 Testing for Resistance: Natural Outbreaks Versus Controlled Challenge Testing

The core issue for genetic evaluation is to be able to compare the selected candidates (individuals or families) for a target trait. Making use of survival data from a natural outbreak, where thousands of individuals can have phenotypes, is an attractive option as it provides the large datasets required for meaningful genetic analyses. However, controlling experimental noise is difficult to achieve with field tests. Previous life history of the fish and environmental conditions (diet, water characteristics, social interactions, stress, previous infections, and immune status) are likely to interfere with the host response to infection and should be standardized. It may also be difficult to make a precise diagnostic of the disease responsible for the epidemic under such conditions (a single pathogen or coinfection with different pathogens) and equally difficult to determine when the infection of an individual occurred. Two other important factors should also be considered; incomplete exposure to infection and incomplete diagnosis of the individual status. Incomplete exposure results in some animals not being exposed to the disease. In most cases, when incomplete exposure occurs, it is not possible to distinguish unexposed individuals from "true" resistant or recovered ones. Such individuals will be classified as resistant, regardless of their actual genetic value for resistance. By masking the true extent of genetic variation within the population, incomplete exposure and imprecise diagnoses of the actual states of animals regarding infection (healthy, diseased, and recovered) result in lower heritability estimates and lower accuracy of selective breeding when survival data after a natural outbreak is used (Bishop and Woolliams, 2010, 2014; Bangera et al., 2014). In Atlantic salmon, estimates of the heritability of resistance to infectious pancreatic necrosis (IPN) virus were shown to increase from 0.32 to 0.97 as the level of exposure increased (mortality rates from 10% to 30%, Bishop and Woolliams, 2010). Other examples are shown in Table 2.1. In some cases, treatments are likely to be administrated to rapidly curb the epidemic, which may result in the same bias as incomplete exposure.

For a number of pathogens (mainly viruses or bacteria), experimental infectious models have been developed, which allow disease challenge testing in controlled conditions. Experimental infection can be induced by introducing a small number of diseased fish into tanks with healthy fish (cohabitation challenge), by introducing a large dose of infectious particles (virus or bacteria) into the rearing water (waterborne or immersion challenge) or by directly inoculating fish through intramuscular or intraperitoneal injection (injection challenge). In practice, immersion and cohabitation challenges are frequently used to test young fish with parasites, viral, and some bacterial diseases. Injection is efficient for most of the viral and bacterial diseases, and is usually preferred for testing large-sized individuals.

While cohabitation or immersion challenges are expected to be representative of natural infection, this is clearly not true for injection challenge because injections bypass the natural external barriers such as mucus, skin, or gills that are likely to play a protective role during the very first steps of infection (Gomez et al., 2013). However, injection is a tractable route of infection. It requires low quantities of inoculum and it is usually more efficient for inducing infection in large-sized fish. Moreover, the process of pathogen entry into the host after immersion is subjected to stochastic variation that contributes to higher heterogeneity in subsequent pathogen loads than the process of infection by injection (Wargo et al., 2012). In any case, residual experimental noise (due to individual life history and/or stochastic events during infection) is unavoidable, as illustrated by the spread of viral load or time-until-death observed within fish clonal lines, despite the fact that all fish within a line share the same genetic background and history (Quillet et al., 2007b; Wargo et al., 2012).

Controlled challenge testing presents many advantages. Most of the experimental noise can be controlled (water quality or temperature, health status of fish prior infection, etc.) and a standardized challenge can be optimized and reproduced in the same manner over time. In a number of host–pathogen interactions, there is a dose–effect response, i.e., an increased infectious dose leads to increased host morbidity (Kinnula et al., 2015). Therefore, performing a controlled challenge test makes it feasible to modulate the infectious dose and control the severity of the challenge. This is a critical feature for the accuracy of genetic studies. When resistance was measured as a binary trait, Vandeputte et al. (2009) demonstrated that the highest precision of heritability estimates for resistance were obtained with intermediate survival rates (30–70%). In a controlled infectious challenge, it is also expected that all fish are infected at the same time, making individual time-until-death relevant for comparing individual relative resistance.

TABLE 2.1 Effect of Exposure Levels on Heritability Estimates for Resistance to Different Diseases: Examples in Atlantic Salmon

	Experimental Infection			Natural Field Outbreak		
Pathogen	Infection Route	Heritability	Prevalence[a]	Heritability	Prevalence[a]	References
IPN	Immersion	0.31 ± 0.06	~36%	0.10	~15%	Wetten et al. (2007)
Aeromonas salmonicida	Cohabitation	0.34 ± 0.13	~70%	0.23 ± 0.05	~35%	Gjøen et al. (1997)
Sea louse, *Lepeophtheirus salmonis*	Immersion	0.26 ± 0.07	75 parasites/fish	0.06 ± 0.04	<5 parasites/fish	Kolstad et al. (2005)

[a]*Level of exposure is measured as the overall mortality rate (%) observed in the population or by the infection level (number of parasites per individual).*

There are two main concerns with the use of experimental challenge tests. The first is whether the resistance observed during the experimental challenge can be correlated with the on-field performance (refer to Section 2.3.4 for examples). The second point concerns the practical implementation of infectious challenge tests. Tests have to be performed in specific facilities that respect strict ethical and environment regulations (e.g., decontamination of effluents). Implementing large scale tests with hundreds or thousands individuals for meaningful genetic studies is expensive.

2.2 BASIC GENETIC PRINCIPLES

2.2.1 DNA, Chromosomes, and Inheritance

The smallest unit in an organism is the cell and each cell (apart from a few types, such as the red blood cells of some organisms) has a nucleus containing DNA molecules. Each individual's DNA sequence is the basic unit affecting the genetic or heritable component of all phenotypic traits. Watson and Crick (1953) showed that the structure of the DNA molecule resembles a ladder twisted into a right-handed double helix. DNA consists of a string of nucleotides, adenine (A), guanine (G), cytosine (C), and thymine (T). The order of this sequence of nucleotides provides the code, or blueprint, for the development and function of all living organisms. Overlaid on top of this basic unit of inheritance are some epigenetic factors, which will be described in more detail below, and which act to modulate how the code of nucleotide bases is read. The DNA nucleotide sequence codes for genes (which are transcribed into RNA that is then translated into proteins). The sections of DNA sequence that code for proteins make up a small proportion of the overall DNA sequence. DNA sequence close to the protein-coding sections consists of untranslated but transcribed sequence, which is believed to be important for regulating the efficiency of translation, stability of the mRNA, and other functions. A large portion of the noncoding DNA consists of intervening sequence that is never transcribed, but which could be involved in "packaging" of the DNA, or have other functions which are not known as yet.

2.2.2 Genetic Variation

Most of the DNA in living organisms is packed into bodies called chromosomes (Fig. 2.1). The DNA in each chromosome consists of two strands which split apart and are replicated during mitosis and meiosis. Most fish and shellfish are diploid, meaning that the chromosomes exist in matching pairs, one of each pair is inherited from the father and one of each pair from the mother. With the division of cells during mitosis, each new cell receives an entire copy of all of the chromosome pairs from the parent cell. With meiotic divisions to form gametes, each gamete receives just one of each chromosome pair (half the DNA from the parent cell). During the process of meiosis, the pairs of chromosomes align, and a shuffling of sections from each chromosome pair can occur in a process known as recombination. Therefore, there are effectively three main processes that affect what variants of each gene are inherited by an individual from its parents:

1. the random splitting of chromosome pairs into gametes;
2. recombination during meiosis shuffling corresponding sections of DNA sequence between chromosome pairs; and
3. gamete fertilization where new combinations of chromosome pairs form to create a new diploid individual.

In addition, there are occasional mutations that occur to change the sequence of particular genes, and there may be gross rearrangements of the order, or splitting/fusion of chromosomes, that all could induce genetic variability within a population and the phenotypic variability of particular individuals within the population.

In diploid organisms the chromosomes, and therefore the genes the chromosomes contain, appear in pairs, one copy of each gene is inherited from the female parent and one copy from the male parent. The effect of the two alleles for each gene is sometimes additive, in that the total effect of the alleles at each genetic locus approximates the average of the effects caused by each allele. In other cases, one allele may show complete dominance over another (known as *dominant* or *recessive* alleles, respectively) to influence the effect on the trait.

In most instances, traits such as the ability of an individual to resist and/or tolerate disease are likely to be affected by many genes, most of which have a small overall effect on resistance and a few of which have a moderate to large overall effect on resistance (Hayes and Goddard, 2001). Because the additive effect of numerous genetic loci determines the extent of resistance, and because the extent of resistance is continuous, like a sliding scale, the disease-resistance trait is known as a quantitative trait, even though it can result in a binary phenotype (e.g., dead or alive).

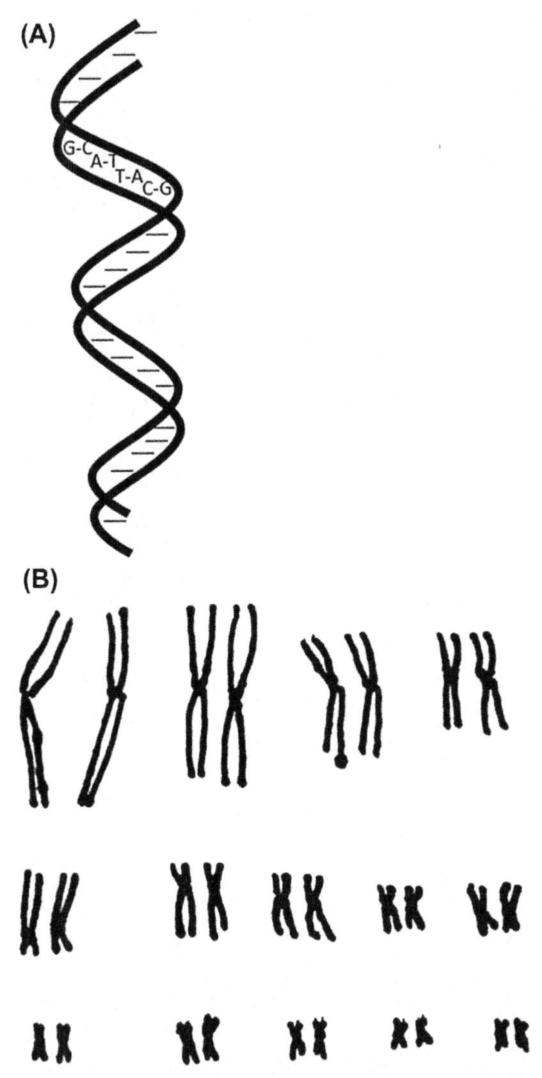

FIGURE 2.1 Schematic showing a DNA molecule (A) and a set of chromosomes (B). The DNA molecule consists of two polynucleotide strands that are held and aligned by hydrogen bonds that form between adjacent bases (adenine with guanine and cytosine with thymine which are represented as bases A, G, C, and T, respectively). In diploid organisms the chromosomes exist in pairs. In the schematic the organism has $2n=14$ pairs of chromosomes. During normal cell division (mitosis) each chromosome ($n=28$ in schematic) can be seen to consist of two chromatids held together by their centromeres.

The development of an animal depends not only on its genes or genotype but also on the environmental conditions it experiences during its lifespan. So the phenotype (*P*, which is the appearance of the trait) may be simply described as follows:

$$\text{Phenotype}\,(P) = \text{Genotype}\,(G) + \text{Environment}\,(E)$$

where *G* is the genotypic effect and *E* is the environmental effect on the trait. To characterize a trait (*x*) in a population of animals the mean or average $(\dot{X})$ is used to describe its value. The average alone does not give sufficient information about the trait. We need to know about the difference or variation between animals which is symbolized by sigma (σ^2) and estimated as follows:

$$\sigma^2 = \text{sum of } (\dot{X} - x)^2 / N - 1$$

where *N* is the number of individuals measured.

The total phenotypic variance $(\sigma^2{}_{\text{P}})$ can be simply described as the sum of the genetic $(\sigma^2{}_{\text{G}})$ and environmental variance $(\sigma^2{}_{\text{E}})$ components as follows:

$$\sigma^2{}_{\text{P}} = \sigma^2{}_{\text{G}} + \sigma^2{}_{\text{E}}$$

The total genetic variance can be further divided into several parts:

$$\sigma^2_G = \sigma^2_A + \sigma^2_D + \sigma^2_I$$

where σ^2_A is the additive genetic variance, σ^2_D is the dominance variance, and σ^2_I is the epistatic variance (allelic interactions between different loci). The sum of dominance and epistatic variance is called nonadditive genetic variance, the majority of which is due to dominance (Falconer and Mackay, 1996).

2.2.3 Heritability

Traits included in the breeding goal are usually quantitative traits controlled by a number of genes each with small effects. However, some major genes have been recently found for some diseases. Heritability (h^2) which ranges from 0 to 1, describes the ratio of genetic variance to total phenotypic variance:

$$h^2 = \sigma^2_G / \sigma^2_P$$

The heritability in the broad sense will not describe breeding values correctly since a part of σ^2_G can include nonadditive genetic variance, which is not transmitted to offspring. Of greater relevance to animal breeding is heritability in the narrow sense, which describes the ratio of the additive genetic variance to the phenotypic variance:

$$h^2 = \sigma^2_A / \sigma^2_P$$

Heritability is an important parameter in breeding theory because it expresses the amount of additive (heritable) genetic variation for a trait. High heritability indicates that it is possible to obtain high genetic gain for a trait when that trait is included in the goal of a breeding program. Some estimates are shown in Table 2.2.

2.2.4 Inbreeding

Inbreeding is defined as the mating of animals that share a recent common ancestor, e.g., mating parents with offspring, mating full- or half-siblings. Mating relatives leads to an increased homozygosity in the population (an animal is homozygous at a particular locus when its genotype has identical alleles, e.g., *AA* or *aa*). Inbreeding has two detrimental effects. It makes the animals more similar and thus reduces the genetic variation in the population such that the population has a reduced potential for genetic gain. Inbreeding also results in reduction of overall performance (including robustness and survival) and may uncover undesirable recessive genes, which may be lethal when homozygous (Fjalestad, 2005). It is therefore important to keep records of relationships between animals and avoid mating relatives in a breeding population.

2.2.5 Correlation Between Traits

In some cases, the variation in one trait is associated (positively or negatively) with the variation of another trait, i.e., traits are correlated. This correlation between phenotypes (phenotypic correlation) is caused by the underlying correlation between genotypes (genetic correlation) and between environments (environmental correlation). Some estimates are given in Table 2.3. A high genetic correlation means that selection for one trait will modify the genetic value for the second trait, in a favorable or adverse direction according to the sign of the correlation.

2.3 SELECTIVE BREEDING TO IMPROVE RESISTANCE

2.3.1 General Processes in Relation to Diseases

For a breeding program, we first need to determine if the traits of interest are heritable and can be selected, and we then need to estimate the underlying additive genetic effect on disease resistance for the animals that have been "earmarked" for breeding (known as the breeding candidates). It is not possible to determine the relative resistance or susceptibility of animals if they have not been exposed to the disease. It is therefore particularly difficult to assess the level of resistance of potential breeders because such animals need to be maintained in a healthy (preferably, disease-free) condition for reproduction and to avoid the transmission of the disease to the broader selected population. The resistance measured for the near relatives (full- and half-siblings) of potential breeders after controlled exposure to the disease (a *challenge* test) provides data that can be used to predict the genetic value of the potential breeders for resistance.

Challenge testing of both fish and shellfish yields useful data for making breeding value predictions because such species typically have high fertility and can produce both full- and half-sibling families.

TABLE 2.2 Heritability for Survival and Resistance to Diseases

Species and Traits	Heritability	References
General Survival		
Atlantic cod:	0.00	Gjerde et al. (2004)
Atlantic salmon:	0.08	Standal and Gjerde (1987)
	0.08 (0.02)	Rye et al. (1990)
Common carp:	0.34 (0.09)	Nielsen et al. (2010)
Nile tilapia:	0.08	Eknath et al. (1998)
	0.03–0.14	Charo-Karisa et al. (2006)
	0.12	Rezk et al. (2009)
Rainbow trout:	0.16 (0.03)	Rye et al. (1990)
Rohu carp:	0.16	Gjerde et al. (2002)
Black tiger shrimp, *Penaeus monodon*:	0.21 (0.06)	Krishna et al. (2011)
Pacific white shrimp, *Penaeus vannamei*:	0.04	Gitterle et al. (2005a)
	0.00–0.06	Caballero-Zamora et al. (2015)
Red abalone, *Haliotis rufescens*:	0.11	Jonasson et al. (1999)
Resistance to Specific Diseases		
Atlantic Cod:		
Vibriosis, *Vibrio anguillarum*	0.08–0.17	Kettunen et al. (2007a)
VNN	0.43–0.75	Ødegård et al. (2010b)
Atlantic Salmon:		
ISAV	0.21–0.32	Ødegård et al. (2007a)
Vibriosis:	0.12 (0.05)	Gjedrem and Aulstad (1974)
Vibrio salmonicida	0.13 (0.08)	Gjedrem and Gjøen (1995)
Furunculosis, *Aeromonas salmonicida*	0.48 (0.17)	Gjedrem et al. (1991)
Furunculosis	0.16 (0.12)	Gjedrem and Gjøen (1995)
Furunculosis	0.59–0.63	Ødegård et al. (2006)
Furunculosis	0.43 (0.02)	Ødegård et al. (2007b)
Furunculosis	0.59 (0.06)	Olesen et al. (2007)
Furunculosis	0.62	Kjøglum et al. (2008)
Furunculosis	0.47	Gjerde et al. (2009)
BKD	0.23 (0.10)	Gjedrem and Gjøen (1995)
Sea louse, *Lepeophtheirus salmonis*	0.25 (0.07)	Kolstad et al. (2005)
Sea louse, *Caligus elongatus*	0.22	Mustafa and MacKinnon (1999)
AGD, first infection	0.14 (0.02)	Kube et al. (2012)
AGD, second infection	0.23–0.40	Kube et al. (2012)
Brook Char:		
Furunculosis, *A. salmonicida*	0.51 (0.03)	Perry et al. (2004)

Continued

TABLE 2.2 Heritability for Survival and Resistance to Diseases—cont'd

Species and Traits	Heritability	References
Chinook Salmon:		
Furunculosis, *A. salmonicida*	0.14 (0.11)	Beacham and Evelyn (1992)
Vibriosis, *V. anguillarum*	0.08 (0.05)	Beacham and Evelyn (1992)
Vibriosis, *V. ordalii*	0.10 (0.06)	Beacham and Evelyn (1992)
Chum Salmon:		
Furunculosis, *A. salmonicida*	0.07 (0.05)	Beacham and Evelyn (1992)
Vibriosis, *V. anguillarum*	0.06 (0.16)	Beacham and Evelyn (1992)
Vibriosis, *V. ordalii*	0.14 (0.11)	Beacham and Evelyn (1992)
Coho Salmon:		
Furunculosis, *A. salmonicida*	0.00 (0.12)	Beacham and Evelyn (1992)
Vibriosis, *V. anguillarum*	0.00 (0.07)	Beacham and Evelyn (1992)
Vibriosis, *V. ordalii*	0.11 (0.10)	Beacham and Evelyn (1992)
Rainbow Trout:		
VHSV	0.69 (0.25)	Chevassus and Dorson (1990)
NHIV	0.05–0.51	Yamamoto et al. (1991)
Red mouth disease, *Yersinia ruckeri*	0.42	Henryon et al. (2005)
BCWD, *Flavobacterium psychrophilum*	0.43	Henryon et al. (2005)
BCWD	0.39	Silverstein et al. (2009)
BCWD	0.22 (0.03)	Leeds et al. (2010)
BCWD	0.45	Vallejo et al. (2010)
Columnare disease, *F. columnare*	0.17 (0.09)	Evenhuis et al. (2015)
Helminth, *Diplostomum* spp.	0.35 (0.09)	Kuukka-Anttila et al. (2010)
Common Carp:		
Aeromonas hydrophila	0.04 (0.03)	Ødegård et al. (2010a)
Rohu Carp:		
A. hydrophila	0.03–0.39	Das Mahapatra et al. (2008)
Gilt-head Sea Bream:		
Pasteurellosis, *Photobacterium damselae*	012–0.45	Antonello et al. (2009)
***Pacific White Shrimp,* P. vannamei:**		
Taura syndrome virus	0.28 (0.14)	Argue et al. (2002)
Taura syndrome virus:		
• Susceptibility	0.41 (0.07)	Ødegård et al. (2011)
• Endurance	0.07 (0.03)	Ødegård et al. (2011)
White spot syndrome virus	0.03–0.07	Gitterle et al. (2005b)
Pacific Oyster:		
Survival[a]	0.52 (0.03)	Degremont et al. (2015b)
OsHV-1 virus	0.49–0.61	Degremont et al. (2015a)

[a]*Survival associated to OsHV-1 viral infection.*

TABLE 2.3 Genetic Correlations Among Disease Resistances and Between Production and Disease Resistance Traits in Some Species

Traits Diseases	ISA	IHN	Red Mouth Disease	Furunculosis	Vibriosis, *Vibrio salmonicida*	BCWD	Growth	Feed Efficiency
IPN	−0.10[a]			−0.11[a]				
VHS		NS[l]	−0.06 to −0.11[b]			0.12–0.15[b]	−0.14 to −0.33[c]	−0.01 to −0.22[c]
ISA				0.07[a]; 0.15[i] −0.05 to −0.11[d]	−0.24[d]		−0.03[g]	
WSSV							−0.55 to −0.64[k]	
Furunculosis					0.10 to 0.18[d]		0.15[e]	
Vibriosis, *Vibrio anguillarum*	−0.05[d]			0.10–0.36[d]	0.91[d]			
BCWD			−0.07 to −0.23[b]				0.09 to −0.10, NS[f]; −0.15 to −0.18[l]	
Columnaris						0.35[m]	−0.16 to −0.19, NS[m]	
Pasteurellosis, *Photobacterium damselae*							0.61[h]	
Eye fluke, *Diplostomum* sp.							0.08 to 0.72[j]	

[a]*Atlantic salmon (Kjøglum et al., 2008).*
[b]*Rainbow trout (Henryon et al., 2005).*
[c]*Rainbow trout (Henryon et al., 2002).*
[d]*Atlantic salmon (Gjøen et al., 1997).*
[e]*Brook trout (Perry et al., 2004).*
[f]*Rainbow trout (Silverstein et al., 2009).*
[g]*Atlantic salmon, Sissel & Kløglum, cited in (Moen et al., 2007).*
[h]*Gilt-head bream (Antonello et al., 2009).*
[i]*Atlantic salmon (Ødegård et al., 2007b).*
[j]*Rainbow trout (Kuukka-Anttila et al., 2010).*
[k]*White shrimp (Gitterle et al., 2007).*
[l]*Rainbow trout (Verrier et al., 2013b).*
[m]*Rainbow trout (Evenhuis et al., 2015).*

2.3.2 Heritability of Disease Resistance and Survival Traits

Table 2.2 provides some estimates of heritability for resistance to different diseases for a number of species. Overall heritability of general survival in the absence of controlled challenge testing is low (0.12, 14 estimates), which is not surprising because, as discussed previously, many environmental factors affect the health of animals during their lifetime.

Of the 45 estimates of heritability for resistance to specific diseases listed in Table 2.2 most are for bacterial or viral diseases, but some parasitic diseases have also been studied (AGD, sea lice, etc.). Altogether, the diseases exhibit moderate to high heritability (overall average 0.27, ranging from 0.00 to 0.69) in most cases giving good prospects for efficient selection to improve resistance.

2.3.3 Response to Selection for Survival and Disease Resistance

Most estimations of response to selection have been obtained for experiments using family-based selection and controlled infectious challenge tests. However, the first successful selection for resistance to a disease may be the mass selection carried out in Japan, after a sudden outbreak of IPN virus among rainbow trout in 1965 (Okamoto et al., 1993). More than 90% mortality was recorded among the affected brood stock. Survivors and their progeny were used for reproduction until 1973. At that time, no mortality was observed, despite the virus being still detected in the farm environment. A highly resistant line was created using a single pair of parents in 1973, and testing of five generations derived from this isolate (with comparison to control lines) has shown that the resistance to IPNV is stable and has a strong genetic basis. The narrow genetic base that has come about as a result of this work could have negative consequences (loss of genetic variation and inbreeding depression for fitness) and this may be why further dissemination of this line was hindered.

Table 2.4 gives estimates of genetic gain for some disease-resistance traits. The genetic gain achieved with selection for survival averages 7.6% per generation. The average genetic gain achieved for resistance to the six diseases listed is relatively high at 14.3% per generation, excluding resistance to the white spot syndrome virus (WSSV) in shrimp (1.7% per generation, which can be explained by the strong negative genetic correlation between growth rate and WSSV, see Section 2.3.4). This high average genetic gain in disease resistance is consistent with heritability estimates and is very promising for the future development of aquaculture production.

2.3.4 Combining Resistance to Diseases and Production Traits in Breeding Objectives

It is important to consider the genetic correlations among traits of interest when designing a selection strategy. In the ideal case, if two target traits share some common genetic determinism, then the correlation will be positive and selection for one trait will provide genetic gain for the other. Genetic progress is still possible when the genetic correlation is negative, but this requires special care and, in any case, will lower the genetic gain for both traits.

2.3.4.1 Correlations With Production Traits

In most selective breeding programs production traits (such as growth rate, fillet yield, and meat quality) are the primary traits considered for selection. Fortunately, most reports of correlations between growth rate and resistance to different diseases are positive (such that as growth rate increases, disease resistance also increases, Table 2.3).

2.3.4.2 Correlations Among Resistance to Different Diseases

The nature of the genetic correlation between resistances to different diseases is another issue. It is possible that different pathogens will trigger different host immune mechanisms with possible antagonism depending on the pathogens involved. In such a situation, selecting for increased resistance to a given pathogen might increase the genetic susceptibility to another pathogen. However, as shown in Table 2.3, there is no current evidence to support the occurrence of such negative correlations in fish or shellfish. In most of the studied cases, correlations are weak, indicating that resistances for different diseases (whether viral, bacterial, or parasitic) can be selected independently. Two cases of positive correlation have been recorded, between diseases caused by close bacterial species [*Flavobacterium* sp. responsible for bacterial cold water disease (BCWD) and columnaris, and *Vibrio* sp., responsible for vibriosis, Table 2.3].

Another important consideration is that genetic variability is present not only in the host populations but also in the pathogen populations. It is well known that viral or bacterial isolates exhibit variable virulence (affecting the severity of the disease in a given host). Differences in virulence reflect differing virulence mechanisms (McBeath et al., 2015, for example)

TABLE 2.4 Examples of Genetic Gain Obtained

Trait, Species	Genetic Gain (% per Generation)	Number of Generations	References
General Survival:			
Pacific white shrimp, *Litopenaeus vannamei*	5.7	4	Gitterle et al. (2007)
Nile tilapia	8.8	2	Rezk et al. (2009)
Blue tilapia	8.4	4	Thodesen et al. (2012)
Red tilapia	5.0	4	Thodesen et al. (2013)
Giant freshwater prawn, *Macrobrachium rosenbergii*	1.1	6	Luan et al. (2014)
Small abalone, *Haliotis diversicolor*	4.1	4	Liu et al. (2015a)
Pacific oyster, *C. gigas*[a]	20.2	4	Degremont et al. (2015b)
Resistance to Specific Diseases:			
Atlantic Salmon:			
IPN	18.7	1	Storset et al. (2007)
Rainbow Trout:			
Vibriosis, *Vibrio salmonicida*	19.0	2	Leeds et al. (2010)
L. vannamei:			
White spot syndrome	1.7	4	Gitterle et al. (2007)
White spot syndrome	6.3	4	Huang et al. (2012)
Taura syndrome	12.4	1	Fjalestad et al. (1997)
Taura syndrome	18.4	1	Argue et al. (2002)
Sydney Rock Oyster, Saccostrea glomerata:			
QX parasite, *Marteilia sydneyi*	11.0	2	Nell and Hand (2003)

[a]*Survival associated to OsHV-1 viral infection.*

which are likely to trigger specific host immune responses. Genetically improved stocks selected for resistance may therefore not express all the expected improvement if they are distributed in geographic areas where the genetic characteristics of the pathogen differ to the isolate used for experimental challenge tests. So far, only a few studies have addressed this issue.

2.3.4.3 Correlation Between Survival in the Field and Survival After Experimental Disease Challenge

Breeding values for resistance estimated using challenge test data will be of little value if they are not highly correlated with the performance of animals exposed to the disease under farm conditions. Although few results are available so far, high correlations (~0.8 or higher) have been recorded between field mortality and viral (Wetten et al., 2007; Storset et al., 2007), bacterial (Gjøen et al., 1997), and parasitic (Kolstad et al., 2005) diseases. In the same line, a farm trial performed with rainbow trout lines selectively bred for varying resistance to BCWD in laboratory challenge provided evidence of genetic improvement under production conditions (Wiens et al., 2013a). Lower correlations between response to experimental and field challenges may occur because of differences in the overall environment, or due to differences in the conditions under which the animals became infected (prevalence of the pathogen, uniformity of exposure, differences between pathogen strains or isolates, and the possibility of the existence of specific virulence mechanisms) as exemplified by Wiens et al. (2013a).

TABLE 2.5 Disease Resistance in Breeding Objectives for Selection of Atlantic Salmon: The Example of AquaGen AS Breeding Program

Year	Disease Traits	Selection Criteria
1990	Furunculosis	Sibs survival (experimental challenge)
1995	Furunculosis, ISA	Sibs survival (experimental challenge)
2001	Furunculosis, ISA, IPN	Sibs survival (experimental challenge)
2013	Furunculosis, ISA, IPN, sea lice	Sibs survival and lice count (experimental challenge)
2015	ISA, IPN, sea lice, Piscirickettsiosis, winter sore	Sibs survival (experimental challenge), lice count (experimental challenge) + individual molecular information for IPN

In 2002, there had been seven generations of selection. Altogether, disease resistance represented 30% of selection index. The first molecular information (QTL markers) was introduced in 2007 for IPN and the first generation of QTL-IPN fish from AquaGen was introduced in 2009.
According to Midtlyng, P.J., Storset, A., Michel, C., Slierendrecht, W.J., Okamoto, N., 2002. Breeding for disease resistance in fish. Bulletin of the European Association of Fish Pathologists 22, 166–172; Moen, T., Baranski, M., Kent, M., Kjøglum, S., Lien, S., 2009b. Quantitative trait loci (QTL) for use in the AquaGen breeding programme for Atlantic salmon. In: Tenth International Symposium on Genetics in Aquaculture. Bangkok, Thailand; and http://aquagen.no/en/.

2.3.5 Resistance to Diseases in Selection Programs Worldwide

The first large-scale breeding programs for aquaculture species were implemented in Norway for Atlantic salmon and rainbow trout (Gjedrem, 2010). Rye et al. (2010) and Neira (2010) surveyed fish breeding programs worldwide and cataloged multiple breeding programs running for Atlantic salmon (Norway, Canada, Ireland, Scotland, Chili, and Iceland) and rainbow trout (Norway, France, USA, Chili, and Finland). Breeding programs have also been implemented for coho and chinook salmon, Arctic char, and for several marine species (sea bass, sea bream, cod, turbot, etc.). For warm water species, tilapia (25–30 programs) and common carp (8 programs) were the most selected species, but selection is also being implemented for other species such as catfish and rohu. 1 mussel, 11 shrimp, and 5 oyster breeding programs were identified by these surveys. Gjedrem et al. (2012) estimated based on the previous reviews that about 8.2% of aquaculture production was based on family-based breeding programs.

The first generations of the Atlantic salmon and rainbow trout selective breeding programs focused on growth rate and quality traits. Disease-resistance traits were first included in the 1990s, with the beginning of family selection for furunculosis in Atlantic salmon and more diseases (infectious salmon anemia or ISA, IPN, Piscirickettsiosis, and winter sores) have been gradually included since then (Table 2.5).

Selection for disease resistance is most often a form of family selection, where the breeding value of candidate breeders is estimated using data from the survival of siblings during an experimental infectious challenge. The first trait, for which individual molecular information was introduced, is resistance to IPN in Atlantic salmon, for which candidates are genotyped for markers associated to resistance (Table 2.5, discussed in Section 2.4.6).

2.4 APPLICATION OF NEW BIOTECHNOLOGIES

Major advances over the last decade have made it more feasible to search the entire genome for genetic associations with disease resistance and other traits. A number of companies offer services to sequence whole genomes, or transcriptomes, at a price and speed that was unimaginable 10–20 years ago. The new sequencing technologies can also be used to resequence or genotype large families of individuals for multiple single nucleotide polymorphisms (SNPs) which are single base variants in the genome, some of which could be associated with, or directly affecting, the trait of interest. Modern computer systems have capacity and power of analyzing many terabytes of data using complex algorithms and many software packages for the analysis of genomic data are becoming available.

These developments in high-throughput genomic biotechnologies have great potential for improving the accuracy and accelerating the rate of genetic improvement. They could make it possible to directly assess the genetic value of breeding candidates. For traits that can be directly measured on candidate breeding individuals at low cost, and before animals reach sexual maturity (such as growth rate in most instances), the benefit–cost ratio associated with the development and application of the new technologies is probably low (Jonas and De Koning, 2015; Hayes et al., 2007). However, in the case of disease resistance, individuals that were challenge-tested with the stress, even if they survive the challenge, are generally compromised, and cannot normally be used for breeding. Therefore, the economic benefits from the improved genetic gain in disease resistance achieved by applying these technologies could outweigh the costs in instances where disease resistance is concerned. Such technologies could also lead to a greater understanding of the genetic architecture of disease resistance and help to identify genes involved in immune defense mechanisms and consequently result in new developments for the treatment of sick fish, prevention of infection, or reduced transmission of disease.

In this chapter, we will use examples to outline some of the major approaches that are under development or being taken to apply new biotechnologies to the improvement of disease resistance in fish. First, it is necessary to perform experiments to find genes or markers that are associated with disease resistance. The most effective way to do this is by using linkage mapping analysis or genome-wide association testing. Gene expression analysis could be informative as the expression of whole pathways of genes can be regulated in response to infection, or could differ between susceptible and resistant individuals. Indirect ways through which the molecular data can facilitate the improvement of disease resistance include utilizing information about genomic relationships to apply tighter limits on inbreeding (Sonesson et al., 2012) and provision of population genetic data which can be utilized to maximize the genetic diversity captured when initiating a breeding program (Gibson and Bishop, 2005).

2.4.1 Measuring Variation in the Nucleotide Sequence of DNA

Variation in the DNA sequence can directly affect the phenotype or disease resistance of an individual. It can also be used as a "marker" to map where on the chromosomes the causative mutations affecting the phenotype occur and to select animals containing favorable variants at the loci affecting disease resistance. A number of methods have been used to assay DNA variation in the genome. Probably the most popular way to find DNA variation is now to use one of the numerous forms of high-throughput genome sequencing to compare the nucleotide sequence at the same positions of the genome between a number of different individuals in the population. Complete genome sequencing might be the most popular option in the future for both finding genetic variants, and for genotyping large numbers of individuals for these variants, depending on how low the cost of sequencing can go. Options that are often employed for detecting variation (at the time of writing this chapter) are to sequence all the transcribed sequences (e.g., mRNA-seq used to detect polymorphisms in rohu, Robinson et al., 2012), to limit the sequencing to the ends of a set of fragments created using restriction enzyme digestion of the DNA (e.g., RAD-sequencing), or combinations of such approaches (Houston et al., 2014).

The sequencing can be used to identify variation in the occurrence of bases (A, C, G, or T) at single base positions in the DNA molecule, otherwise known as single nucleotide polymorphisms (SNPs). In some cases (such as RAD-sequencing) it might be worth sequencing all the individuals in the experiment in a single pass and using the sequence itself as genotypes, in other cases (such as mRNA-seq) it is usually more efficient to use the first pass sequence from a few individuals to design a number of assays, and to test the variants inherited by numerous individuals in the experiment using an alternative methodology (e.g., SNP chip methods).

The output from the DNA sequencer consists of many short contiguous stretches of DNA sequence. If paired ends are sequenced, the DNA (or RNA) fragment is sequenced from both ends. The stretches of sequence overlap in many cases, and the areas of common overlap allow the sequence to be assembled into contiguous longer stretches of sequence known as contigs (Fig. 2.2A). In cases where paired ends of the fragments of DNA have been sequenced, although there may not be complete overlap of sequenced ends, some of the contigs can be further assembled into common scaffolds. The assembly process can make use of an existing "reference" sequence, which might be a whole genome or transcriptome sequence for the same or a related species, to provide a template for guiding the assembly, or the assembly can be performed "de novo," in which case there is no reference sequence, and each stretch of sequence must be compared with every other stretch of sequence to look for areas of alignment. There needs to be some allowance made for redundancy built into the assembly criteria so that sequence containing polymorphisms (e.g., at single base positions) can be aligned, and so that occasional sequencing errors do not disrupt the alignment. After the assembly, and some quality control, information will be available about the sequence called at each position in each unique contiguous stretch of sequence.

Variation in the call of the sequence at particular positions in a contig could be due to true genetic variation between chromosome pairs or individuals, or could be due to sequencing or assembly error. A number of criteria are used to select positions that are most likely to contain true genetic variants (or single nucleotide polymorphisms, SNPs). These include limits on the frequency of the less frequent allele, number of occurrences of the less frequent allele and quality score at the sequence position. Animals can then be genotyped using a set of putative SNPs to determine which are true genetic variants and these validated SNPs can be used for further testing and analysis. A putative SNP is highlighted by a yellow box in Fig. 2.2B. The minor allele frequency

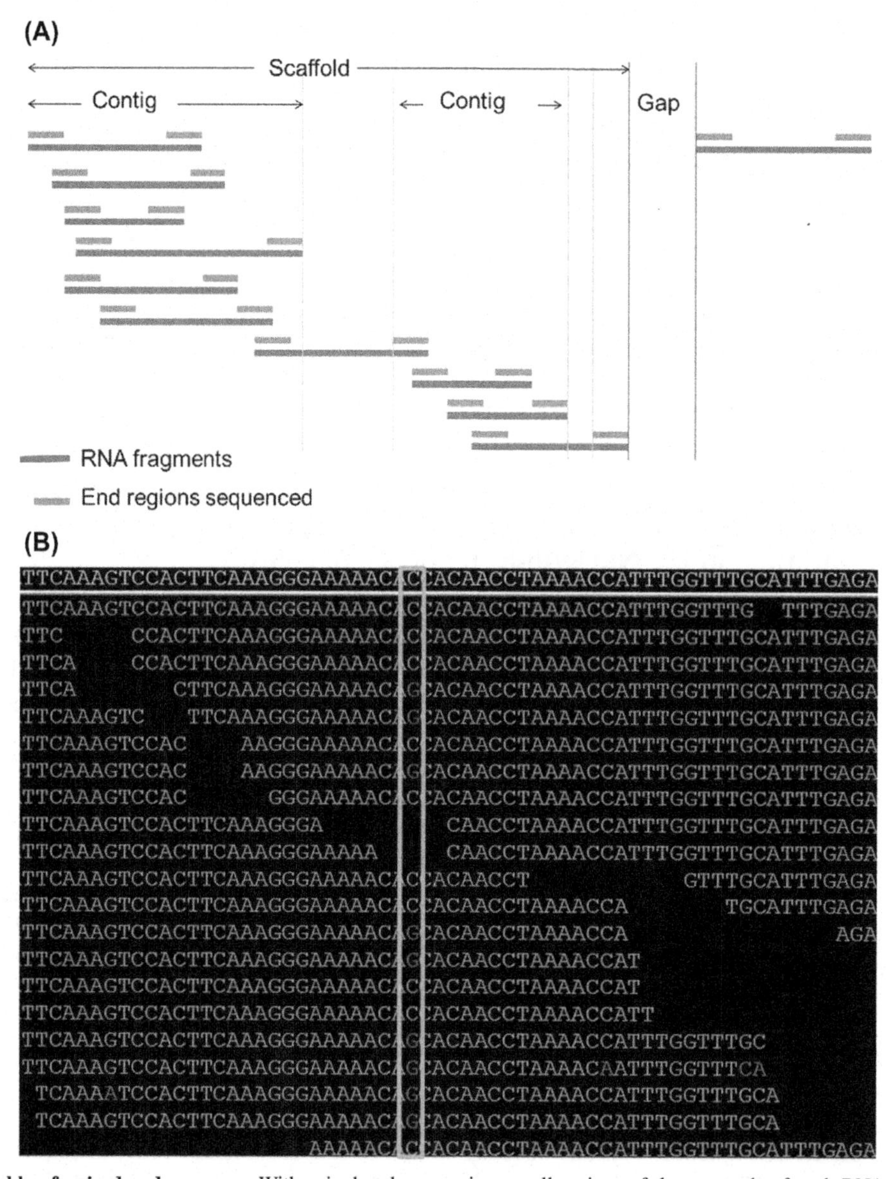

FIGURE 2.2 Assembly of paired end sequence. With paired end sequencing, small regions of the two ends of each RNA or DNA fragment are sequenced [shown in green (gray in print versions)]. Assembly software searches through the sequenced end regions of each DNA fragment for areas of alignment with other sequenced fragments. The connection between each paired end to the same RNA or DNA fragment is taken into account when assembling the sequence. The final assembly consists of contiguous stretches of sequence (contigs) which are sometimes linked to other contigs by a matching pair end into what is called a scaffold (A). The result for each contig (B) is a consensus sequence (top) and its member paired end sequence fragments (below) which can be used to detect likely single nucleotide polymorphisms [highlighted in a yellow box (light gray in print versions) with alternative bases in red (dark gray in print versions) and green (gray in print versions)].

(of the G allele in this example) is 0.47, close to 50%. Also, the minor allele was sequenced nine times, giving some confidence that this is a true SNP. Fourteen bases to the right, the consensus base is a C, and a G is called for one sequence fragment at this position (minor allele frequency 0.05). This difference is therefore less likely to be a true SNP and may be caused by a sequencing error or error in the call of the base by the program.

2.4.2 Linkage Mapping and Genome-Wide Association

An undergraduate student of Thomas H. Morgan, Alfred H. Sturtevant, discovered that genes are positioned along chromosomes and made the first genetic linkage map (Sturtevant, 1913). This was one of the key discoveries that now allows us to find genetic variation associated with traits of interest such as disease resistance. The closer two SNPs are positioned on the chromosome, the less the chance that recombination will occur between the two, and the tighter the linkage between the SNPs. This relationship between the distance along the chromosome and the frequency of recombination, coupled with the ability to find and genotype different individuals for large numbers of genetic variants spaced throughout the genome, is what allows us to map the position of genes, SNPs, or quantitative trait loci (QTL) affecting disease resistance along the chromosome (the unit of measurement known as a centimorgan, cM, is the distance at which the frequency of recombination is 1%). Two general forms of analysis are commonly used to map variation affecting traits such as disease resistance, linkage analysis (LA), and genome-wide association (GWAS) (Londin et al., 2013).

Fig. 2.3 gives a simplified case to illustrate how LA or GWAS works. In the case of linkage analysis, large full- or half-sibling families are used to look for associations between the measure of disease resistance (e.g., days survival) and the inheritance of alleles at loci spread throughout the entire genome (e.g., evenly spaced SNPs mapping to every chromosome). The example shows a chromosome containing an SNP and its inheritance in relation to a gene affecting the trait disease resistance. The sire possesses SNP alleles C and G. One allele (e.g., C) occurs on the paternally derived chromosome. The alternative allele (G) occurs on the maternally derived chromosome. Mapping near the position of the SNP on this chromosome is a gene affecting disease resistance. The progeny of this sire inherit either the favorable (+ve) allele which boosts disease resistance or the unfavorable (−ve) allele which makes animals more susceptible to the disease. If the SNP on this chromosome maps very closely to the gene affecting resistance, such that there is no recombination breaking the link between the SNP and the gene, then all of the progeny inheriting the C allele at the SNP locus will also inherit the +ve allele at the gene locus. Conversely, all progeny inheriting the G allele at the SNP locus will inherit the −ve allele at the gene locus. If the size of the effect of this particular gene on disease resistance is large, it should be possible to detect a strong association between the level of disease resistance (number of days of survival postinfection) and the inheritance of alleles at the SNP locus. If the SNP and the gene are very close, there may be no or little recombination between the SNP and the gene, and this lack of recombination between the SNP and the gene could go back several generations, so the same relationship between the SNP and the gene (same linkage phase) exists even among distant relatives in the general population.

However, if the SNP is located further away from the gene on this chromosome there may be a low level of recombination breaking the linkage between the SNP and the gene. For instance, if recombination occurs in 10% of cases of meiosis, such that the SNP and the gene map 10 cM apart, then about 10% of the progeny inheriting the C allele will receive the −ve form of the gene and about 10% of progeny inheriting the G allele will inherit the +ve form of the gene. So, the further the SNP maps from the gene, the greater the chance that recombination will break up the linkage between the SNP and the gene, and the weaker the association will be with disease resistance. Linkage analysis uses information about the distance between the markers and the gene, and the strength of association between the SNPs and the gene to estimate the position of the gene on a chromosome map. Fig. 2.4 shows an example of QTL associated with hours of survival to white spot syndrome virus resistance infection in *Penaeus vannamei* (Robinson et al., 2014b). One QTL was found to map to a position 55 cM along linkage group 17 and was closely linked to an SNP coding for mitogen activated

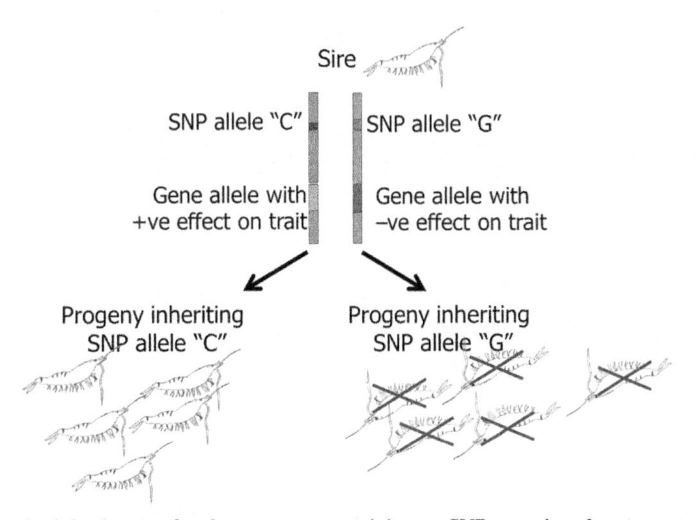

FIGURE 2.3 Schematic showing the inheritance of a chromosome containing an SNP mapping close to a gene affecting disease resistance. In this example, progeny inheriting the "C" allele at the SNP are likely to also inherit a closely linked gene allele with a positive effect on disease resistance, whereas, progeny inheriting the "G" allele at the SNP inherit a gene allele making them more susceptible to the disease.

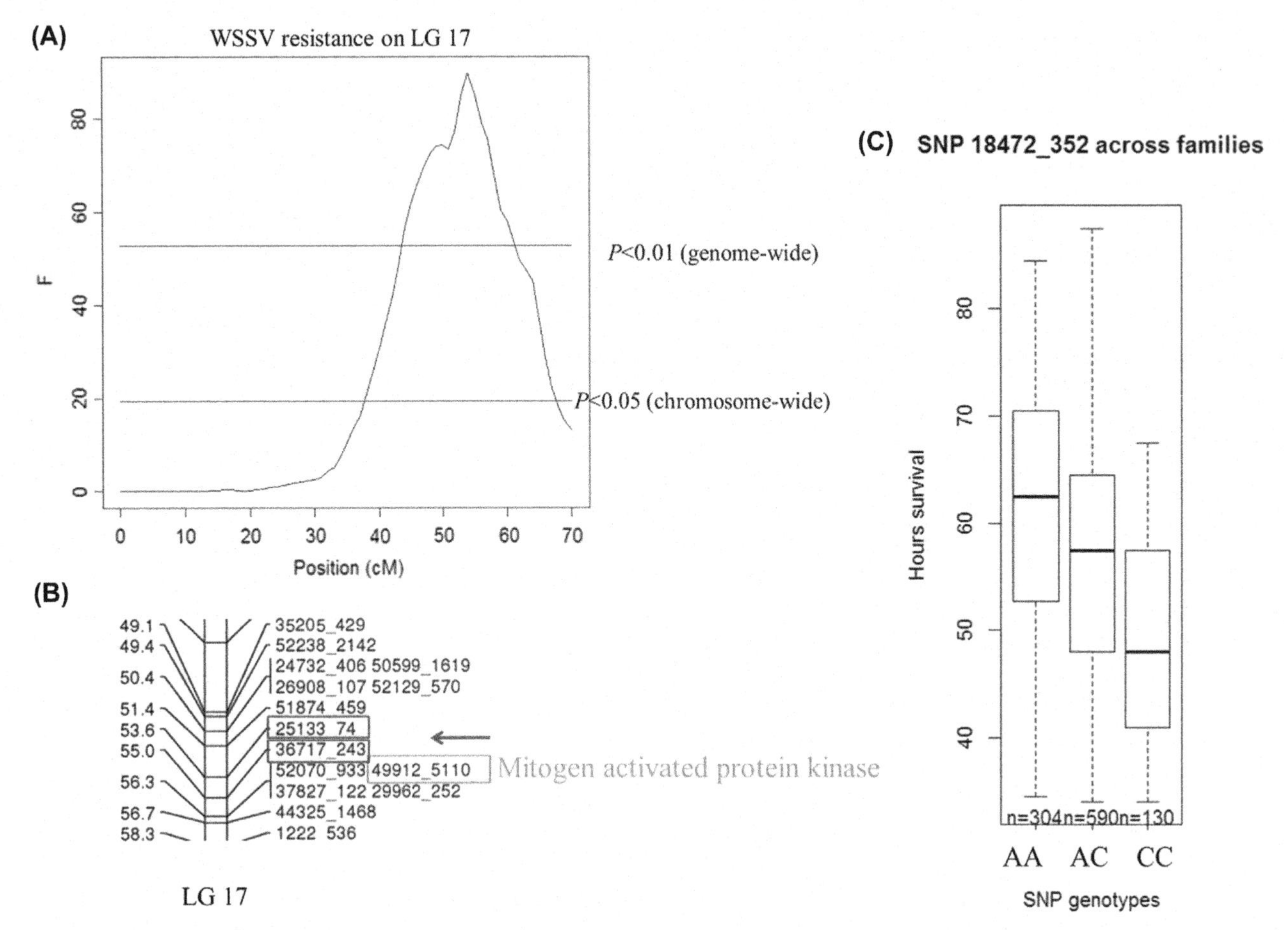

FIGURE 2.4 Detection of quantitative trait loci affecting white spot syndrome virus resistance in *Penaeus monodon*. Result of linkage analysis on linkage group 17 showing statistical significance marked on the *y*-axis and position in centimorgans on the *x*-axis (A), position [red boxes (dark gray in print versions)] of SNP loci significantly associated with resistance on linkage group 17 relative to an SNP in the mitogen activated protein kinase gene [green box (light gray in print versions)] (B) and boxplot of hours survival for animals with alternative SNP genotypes (AA, AC, or CC) for an SNP mapping to linkage group 22 (C). *Data from Robinson, N.A., Gopikrishna, G., Baranski, M., Katneni, V.K., Shekhar, M.S., Shanmugakarthik, J., Jothivel, S., Gopal, C., Ravichandran, P., Gitterle, T., Ponniah, A.G., 2014b. QTL for white spot syndrome virus resistance and the sex-determining locus in the Indian black tiger shrimp (Penaeus monodon). BMC Genomics 15, 731.*

protein kinase (Fig. 2.4A and B). Resistance to this particular disease is likely to be controlled by many genes and a noticeable trend in hours survival for different genotype combinations across families was observed for some SNPs (e.g., SNP, *18472_352*, Fig. 2.4C).

Because there are likely to be many genes affecting resistance, and because most of these genes will have weak effects on the trait, we need to genotype many animals for many SNPs to detect associations with disease resistance. This increases the chance of having an SNP that maps near the gene affecting disease resistance and increases the power of the experiment for detecting association between disease resistance and the inheritance of alleles for one or more SNPs.

If a very dense set of markers is available, then it will be likely that the genes with a strong effect on the trait will have a nearby mapping SNP in complete linkage disequilibrium with the gene. This means that if we trace back through history to the common ancestors of the individuals in this population, there will have been no recombination between the gene and the SNP. The history of the population determines the extent of the linkage disequilibrium in the mapping population. If the extent of linkage disequilibrium is relatively high, and if large numbers of individuals and SNPs are included in the experiment, it will be possible to use a population for mapping where there are quite distant relationships between individuals to find what are known as genome-wide associations (reviewed by Slatkin, 2008).

Research to find QTL associated with disease resistance in fish and shellfish over the last 15 years is summarized in Table 2.6. Most published research to date has been concerned with viral diseases in Atlantic salmon, but the same approaches are applicable to all species. Many QTL controlling disease resistance have been identified so far. In a number of cases, genes that are believed to affect immune function map closely to QTL, although in the vast majority, the causative genes/mutations are not precisely identified yet.

TABLE 2.6 Summary of Research Over the Last 15 Years (2000–15) to Find QTL Affecting Disease Resistance in Fish and Shellfish

Disease Agent	Approach	Associations Detected	References
Atlantic Salmon (*Salmo salar*)			
Gyrodactylus salaris	B1 backcrosses.	10 Regions.	Gilbey et al. (2006)
	39 Microsatellite markers.	Explained 27.3% of total phenotypic variation.	
Infectious pancreatic necrosis virus (IPNV)	Linkage disequilibrium–based test for deducing QTL allele.	Putative causal SNP identified in gene cdh1.	Moen et al. (2015)
	Complete genome sequencing of individuals with deduced QTL genotypes.		
	Genome-wide scans: **1.** Sire-based QTL analysis used to detect linkage groups with significant effects, 2–3 microsatellites/linkage group. **2.** Female-based analysis using additional markers for significant linkage groups.	A single QTL on linkage group 21 was found to explain almost all the genetic variation in IPN mortality.	Houston et al. (2008a,b), Gheyas et al. (2010), and Houston et al. (2010)
	Genome-wide scan using microsatellite markers distributed across the genome.	1 Major QTL.	
		Explained 29% and 83% of the phenotypic and genetic variances, respectively.	
Infectious salmon anemia virus (ISAV)	340 amplified fragment length polymorphisms. Three stage screen: **1.** Transmission disequilibrium test (TDT) using dead offspring from a challenge test. **2.** Mendelian segregation test using genotyped survivors. **3.** Significant TDT markers used for within-family survival analyses using all offspring.	2 Markers were significant at $p < .05$.	Moen et al. (2004)
	Follow-up validation of Moen et al. (2004) using more family material and interval mapping.	Previous QTL confirmed in the new data set.	Moen et al. (2007)
	Follow-up of Moen et al. (2004, 2007) using bacterial artificial chromosome (BAC) clones and three contigs from the Atlantic salmon physical map.	Identified candidate genes based on differential expression profiles from ISA challenges or on the putative biological functions of proteins. Gene HIV-EP2/MBP-2 was implicated.	Li et al. (2011)
Infectious salmon anemia virus (ISAV) and furunculosis (*Aeromonas salmonicida* bacteria)	Looked for association between MHC genotypes and survival of semi-wild Atlantic salmon.	Significant association between survival to both pathogens and MH class I and class II polymorphisms.	Grimholt et al. (2003)

Continued

TABLE 2.6 Summary of Research Over the Last 15 Years (2000–2015) to Find QTL Affecting Disease Resistance in Fish and Shellfish—cont'd

Disease Agent	Approach	Associations Detected	References
Pancreas disease (salmonid alphavirus)	2 Populations genotyped for 69 SNPs.	6 QTL identified. One mapped to the same location on chromosome 3 in both populations.	Gonen et al. (2015)
		Each explained 4–9% of the within family phenotypic variance.	
Proliferative kidney disease-induced mortality	Natural outbreak.	5 Markers.	Cauwelier et al. (2010)
	34 Microsatellites and five RFLPs from different genetic linkage groups in the species' genome.		
Black Tiger Shrimp (*Penaeus monodon*)			
White spot syndrome virus (WSSV, *Whispovirus* sp.)	30 Random RAPD primers in disease resistant and disease susceptible populations.	1 Marker.	Dutta et al. (2014)
	7 Full-sibling tiger shrimp families.	9 Regions.	Robinson et al. (2014b)
	3959 Mapped transcribed SNPs.	Genes with putative immune functions of interest.	
Channel Catfish and Blue Catfish			
Columnaris (*F. columnare*)	F2 backcross.	1 Region.	Geng et al. (2015)
	250,000 SNP array.	Genes mapping to region with known immune function.	
Gilthead Sea Bream (*Sparus aurata*)			
Fish pasteurellosis [*Photobacterium damselae* subsp. *piscicida* (Phdp)]	8 Sires and 6 dams in a single mass-spawning event.	2 Regions.	Massault et al. (2011)
	151 Microsatellite loci.		
Japanese Flounder (*Paralichthys olivaceus*)			
Lymphocystis disease	F2 backcross.	1 Region.	Fuji et al., (2006, 2007)
	50 Microsatellite markers.		
Vibriosis, *Vibrio anguillarum*	12,712 Mapped SNPs.	9 Regions.	Shao et al. (2015)
		Explained 5.1–8.38% of the total phenotypic variation.	
	F1 family, pooled DNA.	2 Markers.	Wang et al. (2014)
	170 Simple sequence repeat (SSR) markers.	Explained >60% phenotypic variance.	

Rainbow Trout (*Oncorhynchus mykiss*)			
Bacterial Cold Water Disease (*Flavobacterium psychrophilum*)	322 Pedigreed families (n=25,369 fish).	1 Region.	Wiens et al., 2013b
	336 Microsatellites.		
	Follow-up of Wiens et al. (2013b). QTL further evaluated in additional F2 back-cross families. 270 Microsatellites.	9 Region. Largest effect explained 40% of the phenotypic variance.	Vallejo et al. (2014a)
	Follow-up of Vallejo et al. (2014a) and Wiens et al. (2013b).	Confirmed QTL.	Vallejo et al. (2014b)
	7 Microsatellite markers for one region.		
	800 Offspring from 4 full-sib matings.		
	Follow-up of Vallejo et al. (2014a,b) and Wiens et al. (2013b). 2 Families. Selective genotyping of SNPs to validate QTL.	Validated previous QTL detected. More QTL detected.	Palti et al. (2015)
	Follow-up of Palti et al. (2015), Vallejo et al. (2014a,b) and Wiens et al. (2013b). 298 Offspring from the 2 half-sibling families. RAD-seq resulted in 7849 informative SNPs.	18 SNPs. 3 Regions.	Liu et al. (2015b)
Crowding Stress	F-2 generation of two families.	3 Regions.	Rexroad et al. (2013)
	321 Microsatellites.		
Fishing stress	1 Double-haploid family.	2 Regions.	Drew et al. (2007)
	432 AFLP markers.	Explained 43% of phenotypic variation of cortisol level.	
Confinement stress	5 F2 full-sibling families (F2 cross between high and low cortisol responsive lines).	10 Significant or highly significant and 10 sugges-tive QTL. Individual QTL explained no more than 10% of phenotypic variance.	Quillet et al. (2014)
	~1000 Offspring genotyped with 268 micro-satellite and SNP markers.		
Infectious pancreatic necrosis virus (IPNV)	Backcross.	2 Regions.	Ozaki et al. (2001)
	51 Mapped microsatellites.		
Viral Haemorrhagic Septicaemia Virus or Rhabdovirus (VHSV)	2 Induced mitogynogenetic doubled haploid F2 families.	1 Region.	Verrier et al. (2013a)
	Selective genotyping.	Explained up to 65% of phenotypic variance.	
Whirling disease (*Myxobolus cerebralis*)	F(2) family (n=480) and results confirmed in three outbred F(2) families (n=96 per family).	1 Region.	Baerwald et al. (2011)
		Explained 50–86% of the phenotypic variance.	

Continued

TABLE 2.6 Summary of Research Over the Last 15 Years (2000–2015) to Find QTL Affecting Disease Resistance in Fish and Shellfish—cont'd

Disease Agent	Approach	Associations Detected	References
Rohu Carp (*Labeo rohita*)			
Aeromonas hydrophila	21 Full-sibling families.	21 SNPs. Several homologous to genes of known immune function or in close linkage to such genes.	Robinson et al. (2014a)
	6000 SNPs.		
Turbot (*Scophthalmus maximus*)			
Furunculosis (*Aeromonas salmonicida*)	4 Turbot families.	7 Regions.	Rodriguez-Ramilo et al. (2011)
		Some explained up to 17% of the phenotypic variance. Genes of putative immune function located in regions.	
Philasterides dicentrarchi	4 Turbot families.	Several markers.	Rodriguez-Ramilo et al. (2013)
		One explained up to 22% of the phenotypic variance.	
Viral Haemorrhagic Septicaemia (VHSV)	3 Full-sibling families (90 individuals each).	Several regions.	Rodriguez-Ramilo et al. (2014)
	93 Microsatellites.	Explained up to 14% of the phenotypic variance.	
Yellowtail Kingfish (*Seriola quinqueradiata*)			
Monogenean Parasite (*Benedenia seriolae*)	F-1 (n=90 per family). 860 Microsatellite and 142 SNP markers.	2 Regions.	Ozaki et al. (2013)
		Explained 32.9%–35.5% of the phenotypic variance.	

2.4.3 Gene Expression

Transcription of the gene (conversion of DNA into RNA, or gene expression) is controlled in a temporal and spatial fashion so that the quantity of RNA produced by each gene varies depending on tissue type, cell type within the same tissue, stage of development, or occurrence of environmental stressors. In this way, the phenotype "disease resistance" is not only affected by the types and configuration of the proteins coded by genes, but also by where, when, and how much of each gene is transcribed and translated into protein. It is the interaction of all of these different quantities and types of proteins in different tissues and cell types at various stages of development, or in response to different stimuli, that determines the final phenotype. Because of this complexity at multiple levels, we are still a long way from understanding the underlying mechanisms affecting disease resistance, or any other quantitative trait for that matter. However, there are some very effective tools available for studying the simultaneous expression of genes in the genome, and when such approaches are combined with positional information for QTL, they can take us closer to identifying the causative mutations affecting disease resistance.

2.4.4 Epigenetics

Most genetic studies have focused on how the order of nucleotide bases in the genome influences, or is associated with, phenotype. However, there is evidence to suggest that other genomic modifications can affect the transcriptional regulation of genes in somatic and/or germ cells, thereby affecting phenotype. These processes are generally referred to as "epigenetic." Epigenetic changes are triggered by environmental stimuli which modify the structure of histone or chromatin, or result in changes to levels of DNA methylation. Epigenetic changes have been shown to persist throughout the life or across multiple generations under some circumstances (e.g., control of expression of agouti locus in mice, Morgan et al., 1999).

There is a growing interest in the study of epigenetics and its effect on major traits of importance, such as disease resistance. Epigenetic modification could affect the penetrance of particular causative variants, for instance with the "masking" or "unmasking" of the Hsp90 gene mutation in *Drospohila melanogaster* (Rutherford and Lindquist, 1998). Masking due to the epigenetic state of the animals in a gene mapping experiment might affect the association between the disease resistance phenotype and the inheritance at marker loci, and consequently our ability to detect and utilize marker-assisted or genomic selection. Genomic selection accounting for epigenetic effects was proposed by (Moghadam et al., 2015; Goddard and Whitelaw, 2014). New technologies are found to be accurate for measuring genome-wide epigenetic changes, such as relative levels of DNA methylation (Harris et al., 2010; Bock et al., 2010), and our understanding of the influence of epigenetics is therefore likely to significantly advance in the near future.

2.4.5 Marker-Assisted and Genomic Selection

The value of a marker depends on heritability, proportion of the total genetic variance explained by the marker (or set of markers), and how closely linked each marker is to the causative mutation affecting the trait. In most cases disease resistance is likely to be affected by many genes of small effect and few genes of large to moderate effect, and thus markers for more than one gene (preferably those with the largest affects) will be needed for marker-assisted selection to be significantly more beneficial than phenotypic selection. There are three general ways in which genotype information can be applied to speed the rate of genetic improvement. If the causative mutations affecting disease resistance are known (for example, with the discovery of the causative mutations for epithelial cadherin affecting resistance to infectious pancreatic necrosis as described in more detail below, Moen et al., 2015), then a form of gene-assisted selection can be applied. This is a best-case scenario, as there is no need to calibrate marker tests to determine linkage phase and only a few tests (one per causative mutation) are needed. If the causative mutations are unknown, but markers that are linked to the causative mutations have been identified, then marker-assisted selection can be used. With marker-assisted selection it is necessary to determine the phase of linkage between the marker alleles and the favorable allele for the causative gene within each family. This involves some additional testing to calibrate the markers. Normally 2–3 markers known to flank each locus affecting the trait are used. In this section we will focus on the third and latest approach which is known as genomic selection (GS). GS is starting to be routinely used for the genetic improvement of some livestock (such as dairy cattle, Dekkers, 2012), but is only at its infancy in aquaculture.

GS uses genome-wide marker associations to estimate the breeding value and genomic relationships of candidate individuals (Meuwissen et al., 2001). In the first step, validation using a population of disease challenge–tested and genotyped individuals (a "training" population) is necessary to estimate the marker effects at each genomic interval. In a second step, those estimates are used to select breeding candidates according to their genotype at markers. The accuracy of GS is dependent on the availability of precise genomic breeding values for training population individuals (Taylor, 2014; Meuwissen et al., 2001). Greatest accuracy can be achieved by updating or reforming training populations in each generation and ensuring that the training population is closely related to the breeding candidates.

For most aquatic species, the breeding populations are large and both sexes are considered as selection candidates. This makes it costly to perform GS with a dense marker coverage across large numbers of candidate and training (reference) individuals. Therefore, innovative approaches are required to reduce the costs of GS for aquatic species. Selective genotyping of candidates and training animals using sparse marker panels to estimate identity-by-descent (IBD, where identity is determined using knowledge about common ancestry or family structure), is a low cost way of performing GS for aquaculture applications. IBD–GS has been predicted to improve the rate of genetic gain by 13–32% compared to traditional selection for traits like disease resistance (Ødegård and Meuwissen, 2014). IBD–GS is less sensitive to marker density than is identity-by-state GS (IBS–GS, comparison without knowledge of family structure) (Luan et al., 2012).

Computer simulations have also predicted that the use of GS could provide 15% reduction of inbreeding over conventional methods due to its high accuracy and ability to distinguish the best individuals within families (Lillehammer et al., 2013). Inbreeding depression may be exacerbating the susceptibility and spread of disease through some aquaculture sectors (e.g., shrimp in Asia, Doyle, 2016), so that the ability with the use of GS to place more stringent limitations on inbreeding could be especially important in some cases.

Example: Discovery and Application of a Suspected Causative Gene Mutation With a Major Effect on Resistance to Infectious Pancreatic Necrosis

The most successful application of gene mapping and marker-assisted selection to date is for resistance to infectious pancreatic necrosis (IPN) which is a viral disease causing significant mortality and loss of production in salmonid aquaculture. Early work by researchers in the United Kingdom and Norway using microsatellite markers simultaneously detected a locus on linkage group 21 of Atlantic salmon whose effect on resistance explained approximately 83% of the genetic variance for resistance to this disease, indicating that there was one major gene of importance affecting the trait (Houston et al., 2008b; Moen et al., 2009a). Using a four-marker haplotype, the genotype of the underlying gene could be deduced in 72% of animals in the Norwegian breeding nucleus (Moen et al., 2009a). The mean mortality rate was reduced from 49% to 13% when offspring inherited two copies of the favorable underlying allele (Q) compared to those inheriting two copies of the alternative allele (q). The frequency of Q allele in the Norwegian population at this time was approximately 0.3.

Research then focused on finer mapping in an effort to find closer markers in complete linkage disequilibrium with the gene. The advantage of having closer markers, or knowledge about the causative mutation affecting IPN resistance, is that marker-assisted selection could be performed across many generations without having to assess the linkage phase between the marker and the gene. A linkage disequilibrium–based test for the gene was developed, applied to a selective breeding program for Atlantic salmon in Norway, and shown to reduce the incidence of IPN mortality in freshwater from 47% (seven outbreaks) to 0%, to reduce mortality during the crucial first 90 days at sea from 6.4% to 1.1% for transfers in autumn 2010, and from 9% to 4.6% for transfers in spring 2011 (Moen et al., 2015).

In an effort to find the causative gene mutation affecting IPN resistance in Atlantic salmon, a 7.8 Mb long scaffold containing the QTL was sequenced (Moen et al., 2015). Two-time coverage was obtained for 22 individuals deduced to be homozygous for the resistant allele at the QTL, 23 individuals were deduced to be homozygous for the susceptible allele of the QTL, and the sequence was analyzed to identify polymorphisms showing large allele differences between the resistant and susceptible fish. The polymorphisms showing the strongest associations with the QTL mapped to a region containing two epithelial cadherin (*cdh1*) genes. One of these polymorphisms (in the second extracellular cadherin domain of *cdh1-1*) causes a serine to proline amino acid shift which was suspected to be the causative mutation for the effect on resistance. Cellular studies found that IPNV binds to *cdh1-1* and clathrin light chains on the cell membrane of hepatocytes of susceptible qq_{QTL} individuals in pits (developing endosomes), implying that IPNV enters the cell bound to *cdh1-1* and clathrin by clathrin-mediated endocytosis (Moen et al., 2015). As this occurrence of the bound IPNV in pits was not observed in QQ_{QTL} resistant individuals, this finding provided strong support for a role of the QQ_{QTL} genotype of *cdh1-1* in providing IPN resistance.

2.5 ROLE OF SELECTIVE BREEDING IN AN OVERALL HEALTH IMPROVEMENT STRATEGY

Genetic improvement of disease resistance has obvious economic benefits for fish and shellfish aquaculture. If the incidence and severity of disease outbreaks can be reduced by making genetic improvement in disease resistance, there will be a greater willingness to reduce the use of chemical treatments. This will benefit consumer health and aquatic environments, preserve water quality, and limit the threat of disseminating antibiotic resistance in microorganism populations. By improving the health and welfare of the fish in this way we not only generate a more positive perception of aquaculture, but also produce a more consistent and higher quality product of higher value in the market place. Selective breeding for improved disease resistance is undoubtedly an important piece of the puzzle for health management in aquaculture, though not the only one. On the one hand, genetic improvement is cumulative over generations and substantial gains are expected. On the other hand, the rate of genetic progress depends on the generation interval of species of interest, which may last 2 years or more for several of them (e.g., salmonids, sea bass, and turbot). It is therefore important that the role of genetic improvement in health improvement strategies is given attention and that aquaculture sectors anticipate what should be the objectives of selection in order to implement sustainable improvement in the resistance of fish that are farmed for food production.

2.5.1 Combining Different Health Management Strategies

A selection index is a combination of several traits for which genetic improvement is desired. The total selection pressure (the proportion of selected individuals/families in the population of candidates) is then spread according to the weight given to each trait in the overall breeding objective. Consequently, the greater the number of traits in a breeding objective, the lower the selection pressure for each of them. Giving priority to the most beneficial disease and robustness-related traits considering the tool box available for health management (prophylactics, vaccinations, use of probitotics, etc.) is therefore needed.

2.5.1.1 Genetic Resistance and Vaccination

Interaction between genetic resistance and vaccination deserves special attention. Breeding for resistance could focus initially on diseases for which the development of vaccines is difficult, such as BCWD in salmonids (Gomez et al., 2014). Combined strategies, such as the selection of fish that develop a high antibody response to vaccination so that the protective effect of vaccines is enhanced, could therefore be considered in cases where the protective effects of vaccination are found to be incomplete. Little has been done in this field, although past data showed low heritabilities for antibodies levels after vaccination against *Aeromonas* and *Vibrio* in Atlantic salmon (Fjalestad et al., 1996; Strømsheim et al., 1994a,b). More recently, intermediate heritabilities were recorded for negative vaccine side effects (visceral and peritoneal adhesions and melanin deposits) but there was no correlation between these responses and survival after infection (Gjerde et al., 2009; Drangsholt et al., 2011b). Moreover, in Atlantic salmon, Drangsholt et al. (2011a) showed a low correlation between resistance to furunculosis of vaccinated and unvaccinated fish. In this case a strategy that selects for resistance on the basis of data from challenge-tested unvaccinated fish would produce limited improvements in farm performance while vaccination of the farmed fish population continues. However, after some generations of selection, the genetic improvement in resistance might become greater than the resistance afforded by vaccination, and the genetic improvement program could therefore render vaccinations unnecessary. Such results underline the need for a careful analysis before implementing a global strategy of health management.

2.5.1.2 Stress Response, Cardiovascular Fitness and Disease Susceptibility

It is well established that exposure to stressors (social interactions, environment changes, and rearing operations) interfere with immune response and occurrence of disease outbreaks (Tort, 2011). In particular, elevation of plasma cortisol level due to stress activation of the hypothalamic–pituitary–interrenal (HPI) axis is suspected to have suppressive effects on immune response components. A number of studies have investigated the potential relationship between cortisol response to stress and susceptibility to bacterial or viral diseases in rainbow trout (Refstie, 1982; Fevolden et al., 1992, 1993, 1994; Weber et al., 2008), Atlantic salmon (Kittilsen et al., 2009), catfish (Praveen et al., 2001) and Atlantic cod (Kettunen et al., 2007b). With one exception in Atlantic salmon, where a strong and negative genetic correlation (approximately −0.7) between stress cortisol and behavior responsiveness and resistance to IPNV (Kittilsen et al., 2009) was found, no clear evidence of consistent genetic correlations between stress and disease responses has yet been established in fish. Altogether, it seems that disease resistance and stress response are different traits, and that it should be possible to genetically improve resistance to specific diseases without impairing susceptibility to stress.

It has also been shown that animals gain a greater ability to resist disease as their general level of cardiovascular fitness is improved by exercise and training (Castro et al., 2011) and that resistance to disease is positively associated with swimming performance (Castro et al., 2013). More research is needed to determine whether there is a genetic component to this relationship and whether selection for cardiovascular fitness traits could improve resistance.

2.5.2 Selection for Sustainable Resistances: Integrating New Facets of Host Response

2.5.2.1 Targeting General or Specific Immune Responses?

In farmed conditions, fish are generally exposed to a range of pathogens. Improving resistance to one or two specific diseases, even the most prevalent ones, may not be the most sustainable solution. The pathogen may evolve to adapt to the increasing selective pressure inflicted by the host resistance or new pathogens may arise. An option, therefore, could be to select for "general" components of the immune response that could improve resistance to a wide range of pathogens, rather than for resistance to specific diseases. Research in terrestrial vertebrates has established that distinct components of innate or adaptive immunity are heritable (Biozzi et al., 1984 in mouse, Minozzi et al., 2008 in chicken, Flori et al., 2011 in pig). However, the actual benefit for health improvement and disease resistance in the farm environment has not yet been established, and this is probably why little attention has been paid so far to the genetic control of such immune effectors in fish.

2.5.2.2 Toward New Desirable Traits to Improve Resistance

In the first section of the chapter, we described the different components of the host response to pathogens, including traits important for individual fitness, such as resistance and tolerance, and traits like infectivity.

As described previously, tolerance is the host ability, for a given level of infection, to reduce the detrimental effects of pathogen on performance. There is an increasing interest in the use of tolerance as a relevant trait for selection to limit the effects of diseases on farmed animals (Doeschl-Wilson and Kyriazakis, 2012; Kause and Ødegård, 2012; Kause et al., 2012). Tolerance captures host immune mechanisms other than those related to the reduction of the pathogen burden (resistance). By uncoupling resistance and tolerance we would be better able to discriminate between the effects of different components of host defense against pathogens. This could lead to a more accurate estimation of breeding values and more efficient breeding strategies. The measurement of tolerance might allow us to quantifying the effects of genotype-by-pathogen burden interactions. These interactions are likely to contribute to phenotypic variation for production traits depending on environmental and infectious farm conditions. Finally, by breeding for greater tolerance the host is likely to put less pressure on the pathogen to evolve into a more virulent form than if breeding was focused on improving resistance mechanisms (Gibson and Bishop, 2005).

Tolerance and resistance have been found to be poorly, or even negatively, correlated in some studies (Råberg et al., 2009 in mouse, Kause et al., 2012 in chicken). The genetic correlation between susceptibility and endurance to the Taura syndrome virus in shrimp was found to be 0.22 ± 0.25 (Ødegård et al., 2011), while in the Atlantic salmon, the correlation was strong and negative (-0.61 ± 0.15) between resistance (survival after infection with *Piscirikettsia salmonis*) and tolerance (growth rate difference between infected and uninfected host, Yàñez et al., 2010).

However, disentangling resistance and tolerance remains a challenge, especially at the individual level because it is not possible to measure the performance of an individual for two alternative disease statuses (infected/noninfected) at the same time point. In practice, studies could be performed at the group (family) level. A simulation study indicates that the large-sized families that can be produced in fish would make this type of study possible and reasonably powerful (Kause, 2011). Studies are underway to develop mathematical and statistical models to distinguish resistance and tolerance from phenotypes that can be collected in practice, the aim being to devise experiments that will provide informative parameters amenable for genetic analyses (Doeschl-Wilson et al., 2012).

Other desirable traits relate to disease control at the population level. Asymptotic carriage (an animal has recovered but not cleared the disease) may constitute a threat as carriers may be a starting point for future disease outbreaks. However, the relevance of carriage should be assessed in fish and shellfish according to the biology of the pathogen and conditions of production. In the open environment, the main source of pathogens may well be the environment itself (such as other fish species, sediments, or biofilms). Infectivity is the propensity of an infected individual to transmit the infection to other individuals in the population. Simulation studies have shown that selection incorporating resistance and infectivity traits would provide higher reduction of disease risk than selection for resistance traits only (Lipschutz-Powell et al., 2012). Again, measuring infectivity and the underlying genetic variability is challenging, but theoretical studies are being developed (Lipschutz-Powell et al., 2014) that should help incorporating this attribute in future breeding objectives. In the future, it will be necessary to obtain more knowledge on the variability and genetic control of these desirable traits in order to take advantage of their potential for making genetic improvement and to be able to devise more sophisticated breeding objectives.

2.6 CONCLUSION

Disease is a major issue affecting the profitability of aquaculture worldwide. The use of genetic methods to improve resistance to disease should be a major focus for consideration in any animal health program for aquaculture species. Moderate heritability exists for disease resistance in most instances, and significant genetic improvement has been achieved using experimental challenge-testing to estimate breeding values for selective breeding. Benefits from new technologies, such as high-throughput sequencing, are promising, should accelerate the rate of genetic improvement achievable in the future, and should greatly improve our understanding of fish and shellfish immune function. There is a need for research that will lead to a better understanding of more detailed and complex traits such as tolerance, infectivity, and robustness. Facets of the biology and culture of fish and shellfish, such as the ability to produce and challenge-test large full-sibling families under controlled environmental conditions, make these species more amenable to the study and application of technologies and methodologies for improving disease resistance than livestock. Genetic progress should therefore be large, and this should reduce the need for chemicals and antibiotics, make aquaculture more profitable and sustainable and lead to greater health benefits, public acceptance, and consumption in the future.

REFERENCES

Antonello, J., Massault, C., Franch, R., Haley, C., Pellizzari, C., Bovo, G., Patarnello, T., De Koning, D.-J., Bargelloni, L., 2009. Estimates of heritability and genetic correlation for body length and resistance to fish pasteurellosis in the gilthead sea bream (*Sparus aurata* L.). Aquaculture 298, 29–35.

Argue, B.J., Arce, S.M., Lotz, J.M., Moss, S.M., 2002. Selective breeding of Pacific white shrimp (*Litopenaeus vannamei*) for growth and resistance to Taura Syndrome Virus. Aquaculture 204, 447–460.

Baerwald, M.R., Petersen, J.L., Hedrick, R.P., Schisler, G.J., May, B., 2011. A major effect quantitative trait locus for whirling disease resistance identified in rainbow trout (*Oncorhynchus mykiss*). Heredity 106, 920–926.

Bangera, R., Ødegård, J., Mikkelsen, H., Nielsen, H.M., Seppola, M., Puvanendran, V., Gjoen, H.M., Hansen, O.J., Mortensen, A., 2014. Genetic analysis of francisellosis field outbreak in Atlantic cod (*Gadus morhua* L.) using an ordinal threshold model. Aquaculture 420, S50–S56.

Beacham, T.D., Evelyn, T.P., 1992. Genetic variation in disease resistance and growth of chinook, coho, and chum salmon with respect to vibriosis, furunculosis, and bacterial kidney disease. Transactions of the American Fisheries Society 121, 456–485.

Biozzi, G., Mouton, D., Stiffel, C., Bouthillier, Y., 1984. A major role of macrophage in quantitative genetic regulation of immune responsiveness and anti-infectious immunity. Advances in Immunology 36, 189–234.

Bishop, S.C., Woolliams, J.A., 2010. On the genetic interpretation of disease data. PLoS One 5, e8940.

Bishop, S.C., Woolliams, J.A., 2014. Genomics and disease resistance studies in livestock. Livestock Science 166, 190–198.

Bock, C., Tomazou, E.M., Brinkman, A.B., Muller, F., Simmer, F., Gu, H., Jager, N., Gnirke, A., Stunnenberg, H.G., Meissner, A., 2010. Quantitative comparison of genome-wide DNA methylation mapping technologies. Nature Biotechnology 28, 1106–1114.

Caballero-Zamora, A., Cienfuegos-Rivas, E.G., Montaldo, H.H., Campos-Montes, G.R., Martínez-Ortega, A., Castillo-Juárez, H., 2015. Genetic parameters for spawning and growth traits in the Pacific white shrimp (*Penaeus (Litopenaeus) vannamei*). Aquaculture Research 46, 833–839.

Castro, V., Grisdale-Helland, B., Helland, S.J., Kristensen, T., Jorgensen, S.M., Helgerud, J., Claireaux, G., Farrell, A.P., Krasnov, A., Takle, H., 2011. Aerobic training stimulates growth and promotes disease resistance in Atlantic salmon (*Salmo salar*). Comparative Biochemistry and Physiology A: Molecular and Integrative Physiology 160, 278–290.

Castro, V., Grisdale-Helland, B., Jørgensen, S.M., Helgerud, J., Claireaux, G., Farrell, A.P., Krasnov, A., Helland, S.J., Takle, H., 2013. Disease resistance is related to inherent swimming performance in Atlantic salmon. BMC Physiology 13, 1–12.

Cauwelier, E., Gilbey, J., Jones, C.S., Noble, L.R., Verspoor, E., 2010. Genotypic and phenotypic correlates with proliferative kidney disease-induced mortality in Atlantic salmon. Diseases of Aquatic Organisms 89, 125–135.

Charo-Karisa, H., Komen, H., Rezk, M.A., Ponzoni, R.W., Van Arendonk, J.A.M., Bovenhuis, H., 2006. Heritability estimates and response to selection for growth of Nile tilapia (*Oreochromis niloticus*) in low-input earthen ponds. Aquaculture 261, 479–486.

Chevassus, B., Dorson, M., 1990. Genetics of resistance to disease to disease in fishes. Aquaculture 85, 83–107.

Das Mahapatra, K., Gjerde, B., Sahoo, P.K., Saha, J.N., Barat, A., Sahoo, M., Mohanty, B.R., Ødegård, J., Rye, M., Salte, R., 2008. Genetic variations in survival of rohu carp (*Labeo rohita*, Hamilton) after *Aeromonas hydrophila* infection in challenge tests. Aquaculture 279, 29–34.

Degremont, L., Lamy, J.B., Pepin, J.F., Travers, M.A., Renault, T., 2015a. New insight for the genetic evaluation of resistance to Ostreid herpesvirus infection, a worldwide disease, in *Crassostrea gigas*. PLoS One 10, e0127917.

Degremont, L., Nourry, M., Maurouard, E., 2015b. Mass selection for survival and resistance to OsHV-1 infection in *Crassostrea gigas* spat in field conditions: response to selection after four generations. Aquaculture 446, 111–121.

Dekkers, J.C.M., 2012. Application of genomics tools to animal breeding. Current Genomics 13, 207–212.

Doeschl-Wilson, A.B., Kyriazakis, I., 2012. Should we aim for genetic improvement in host resistance or tolerance to infectious pathogens? Frontiers in Genetics 3, 272.

Doeschl-Wilson, A.B., Bishop, S., Kyriazakis, I., Villanueva, B., 2012. Novel methods for quantifying host response to infectious pathogens for genetic analyses. Frontiers in Genetics 3, 266.

Doyle, R.W., 2016. Inbreeding and disease in tropical shrimp aquaculture: a reappraisal and caution. Aquaculture Research 47, 21–35.

Drangsholt, T.M.K., Gjerde, B., Odegard, J., Finne-Fridell, F., Evensen, O., Bentsen, H.B., 2011a. Quantitative genetics of disease resistance in vaccinated and unvaccinated Atlantic salmon (*Salmo salar* L.). Heredity 107, 471–477.

Drangsholt, T.M.K., Gjerde, B., Odegard, J., Fridell, F., Bentsen, H.B., 2011b. Quantitative genetics of vaccine-induced side effects in farmed Atlantic salmon (*Salmo salar*). Aquaculture 318, 316–324.

Drew, R.E., Schwabl, H., Wheeler, P.A., Thorgaard, G.H., 2007. Detection of QTL influencing cortisol levels in rainbow trout (*Oncorhynchus mykiss*). Aquaculture 272, S183–S194.

Dutta, S., Biswas, S., Mukherjee, K., Chakrabarty, U., Mallik, A., Mandal, N., 2014. Identification of RAPD-SCAR marker linked to white spot syndrome virus resistance in populations of giant black tiger shrimp, *Penaeus monodon* Fabricius. Journal of Fish Diseases 37, 471–480.

Eknath, A.E., Dey, M.M., Rye, M., Gjerde, B., Abella, T.A., Sevilleja, R., Tayamen, M.M., Reyes, R.A., Bentsen, H.B., 1998. Selective breeding of Nile tilapia for Asia. In: Sixth World Congress Genetics Applied Livestock Production, pp. 89–96 Leipzig, Germany.

Evenhuis, J.P., Leeds, T.D., Marancik, D.P., Lapatra, S.E., Wiens, G.D., 2015. Rainbow trout (*Oncorhynchus mykiss*) resistance to columnaris disease is heritable and favorably correlated with bacterial cold water disease resistance. Journal of Animal Science 93, 1546–1554.

Falconer, D.S., Mackay, T.F.C., 1996. An Introduction to Quantitative Genetics. Addison Wesley Longman Limited, Edinburgh Gate, United Kingdom.

Fevolden, S.E., Røed, K.H., Gjerde, B., 1994. Genetic components of post stress cortisol and lysozyme activity in Atlantic salmon: correlations to disease resistance. Fish and Shellfish Immunology 4, 507–514.

Fevolden, S.E., Refstie, T., Roed, K.H., 1992. Disease resistance in rainbow-trout (*Oncorhynchus mykiss*) selected for stress response. Aquaculture 104, 19–29.

Fevolden, S.E., Refstie, T., Gjerde, B., 1993. Genetic and phenotypic parameters for cortisol and glucose stress-response in Atlantic salmon and rainbow-trout. Aquaculture 118, 205–216.

Fjalestad, K.T., Larsen, H.J.S., Red, K.H., 1996. Antibody response in Atlantic salmon *(Salmo salar)* against *Vibrio anguillarum* and *Vibrio salmonicida* O-antigens: heritabilities, genetic correlations and correlations with survival. Aquaculture 145, 77–89.

Fjalestad, K.T., Gjedrem, T., Carr, W.H., Sweeney, J.N., 1997. The Shrimp Breeding Program. Selective Breeding of *Penaeus vannamei*. AKVAFORSK, Ås, Norway.

Fjalestad, K., 2005. Breeding strategies. In: Gjedrem, T. (Ed.), Selection and Breeding Programs in Aquaculture. Springer, Dordrecht, The Netherlands.

Flori, L., Gao, Y., Lemonnier, G., Leplat, J.J., Teillaud, A., Cossalter, A.M., Laffitte, J., Piton, P., De Vaureix, C., Bouffaud, M., Mercat, M.J., Lefèvre, F., Oswald, I.P., Bidanel, J.P., Rogel-Gaillard, C., 2011. Immunity traits in pigs: substantial genetic variation and limited covariation. PLoS One 6, e22717.

Fuji, K., Kobayashi, K., Hasegawa, O., Coimbra, M.R.M., Sakamoto, T., Okamoto, N., 2006. Identification of a single major genetic locus controlling the resistance to lymphocystis disease in Japanese flounder (*Paralichthys olivaceus*). Aquaculture 254, 203–210.

Fuji, K., Hasegawa, O., Honda, K., Kumasaka, K., Sakamoto, T., Okamoto, N., 2007. Marker-assisted breeding of a lymphocystis disease-resistant Japanese flounder (*Paralichthys olivaceus*). Aquaculture 272, 291–295.

Geng, X., Sha, J., Liu, S., Bao, L., Zhang, J., Wang, R., Yao, J., Li, C., Feng, J., Sun, F., Sun, L., Jiang, C., Zhang, Y., Chen, A., Dunham, R., Zhi, D., Liu, Z., 2015. A genome-wide association study in catfish reveals the presence of functional hubs of related genes within QTLs for columnaris disease resistance. BMC Genomics 16.

Gheyas, A.A., Houston, R.D., Mota-Velasco, J.C., Guy, D.R., Tinch, A.E., Haley, C.S., Woolliams, J.A., 2010. Segregation of infectious pancreatic necrosis resistance QTL in the early life cycle of Atlantic salmon (*Salmo salar*). Animal Genetics 41, 531–536.

Gibson, J.P., Bishop, S.C., 2005. Use of molecular markers to enhance resistance of livestock to disease: a global approach. Revue Scientifique et Technique 24, 343–353 Office international des Epizooties (International Office of Epizootics).

Gilbey, J., Verspoor, E., Mo, T.A., Sterud, E., Olstad, K., Hytterd, S., Jones, C., Noble, L., 2006. Identification of genetic markers associated with *Gyrodactylus salaris* resistance in Atlantic salmon *Salmo salar*. Diseases of Aquatic Organisms 71, 119–129.

Gitterle, T., Rye, M., Salte, R., Cock, J., Johansen, H., Lozano, C., Suarez, J.A., Gjerde, B., 2005a. Genetic (co)variation in harvest body weight and survival in *Penaeus (Litopenaeus) vannamei* under standard commercial conditions. Aquaculture 243, 83–92.

Gitterle, T., Salte, R., Gjerde, B., Cock, J., Johansen, H., Salazar, M., Lozano, C., Rye, M., 2005b. Genetic (co)variation in resistance to White Spot Syndrome Virus (WSSV) and harvest weight in *Penaeus (Litopenaeus) vannamei*. Aquaculture 246, 139–149.

Gitterle, T., Johansen, H., Erazo, C., Lozano, C., Cock, J., Salazar, M., Rye, M., 2007. Response to multi trait selection for harvest weight, overall survival, and resistance to white spot syndrome virus (WSSV) in *Penaeus (Litopenaeus) vannamei*. Aquaculture 272, S262.

Gjedrem, T., Aulstad, D., 1974. Selection experiments with salmon: I. Differences in resistance to vibrio disease of salmon parr (*Salmo salar*). Aquaculture 3, 51–59.

Gjedrem, T., Gjøen, H.M., 1995. Genetic variation in susceptibility of Atlantic salmon, *Salmo salar* L., furunculosis, BKD and cold water vibriosis. Aquaculture Research 26, 129–134.

Gjedrem, T., Salte, R., Gjoen, H.M., 1991. Genetic variation in susceptibility of Atlantic salmon to furunculosis. Aquaculture 97, 1–6.

Gjedrem, T., Robinson, N., Rye, M., 2012. The importance of selective breeding in aquaculture to meet future demands for animal protein: a review. Aquaculture 350–353, 117–129.

Gjedrem, T., 2010. The first family-based breeding program in aquaculture. Reviews in Aquaculture 2, 2–15.

Gjerde, B., Reddy, P., Mahapatra, K.D., Saha, J.N., Jana, R.K., Meher, P.K., Sahoo, M., Lenka, S., Govindassamy, P., Rye, M., 2002. Growth and survival in two complete diallele crosses with five stocks of Rohu carp (*Labeo rohita*). Aquaculture 209, 103–115.

Gjerde, B., Terjesen, B.F., Barr, Y., Lein, I., Thorland, I., 2004. Genetic variation for juvenile growth and survival in Atlantic cod (*Gadus morhua*). Aquaculture 236, 1–4.

Gjerde, B., Evensen, O., Bentsen, H.B., Storset, A., 2009. Genetic (co)variation of vaccine injuries and innate resistance to furunculosis (*Aeromonas salmonicida*) and infectious salmon anaemia (ISA) in Atlantic salmon (*Salmo salar*). Aquaculture 287, 52–58.

Gjøen, H.M., Refstie, T., Ulla, O., Gjerde, B., 1997. Genetic correlations between survival of Atlantic salmon in challenge and field tests. Aquaculture 158, 277–288.

Goddard, M.E., Whitelaw, E., 2014. The use of epigenetic phenomena for the improvement of sheep and cattle. Frontiers in Genetics 5, 247.

Gomez, D., Sunyer, O.J., Salinas, I., 2013. The mucosal immune system of fish: the evolution of tolerating commensals while fighting pathogens. Fish and Shellfish Immunology 35, 1729–1739.

Gomez, E., Mendez, J., Cascales, D., Guijarro, J.A., 2014. *Flavobacterium psychrophilum* vaccine development: a difficult task. Microbial Biotechnology 7, 414–423.

Gonen, S., Baranski, M., Thorland, I., Norris, A., Grove, H., Arnesen, P., Bakke, H., Lien, S., Bishop, S.C., Houston, R.D., 2015. Mapping and validation of a major QTL affecting resistance to pancreas disease (salmonid alphavirus) in Atlantic salmon (*Salmo salar*). Heredity 115, 405–414.

Green, T.J., Robinson, N., Chataway, T., Benkendorff, K., O'connor, W., Speck, P., 2014. Evidence that the major hemolymph protein of the Pacific oyster, *Crassostrea gigas*, has antiviral activity against herpesviruses. Antiviral Research 110, 168–174.

Grimholt, U., Larsen, S., Nordmo, R., Midtlyng, P., Kjoeglum, S., Storset, A., Saeb, S., Stet, R.J.M., 2003. MHC polymorphism and disease resistance in Atlantic salmon (*Salmo salar*); facing pathogens with single expressed major histocompatibility class I and class II loci. Immunogenetics 55, 210–219.

Harris, R.A., Wang, T., Coarfa, C., Nagarajan, R.P., Hong, C., Downey, S.L., Johnson, B.E., Fouse, S.D., Delaney, A., Zhao, Y., Olshen, A., Ballinger, T., Zhou, X., Forsberg, K.J., Gu, J., Echipare, L., O'geen, H., Lister, R., Pelizzola, M., Xi, Y., Epstein, C.B., Bernstein, B.E., Hawkins, R.D., Ren, B., Chung, W.Y., Gu, H., Bock, C., Gnirke, A., Zhang, M.Q., Haussler, D., Ecker, J.R., Li, W., Farnham, P.J., Waterland, R.A., Meissner, A., Marra, M.A., Hirst, M., Milosavljevic, A., Costello, J.F., 2010. Comparison of sequencing-based methods to profile DNA methylation and identification of monoallelic epigenetic modifications. Nature Biotechnology 28, 1097–1105.

Hayes, B., Goddard, M.E., 2001. The distribution of the effects of genes affecting quantitative traits in livestock. Genetics Selection Evolution 33, 209–229.

Hayes, B., Baranski, M., Goddard, M.E., Robinson, N., 2007. Optimisation of marker assisted selection for abalone breeding programs. Aquaculture 265, 61–69.

Henryon, M., Jokumsen, A., Berg, P., Lund, I., Pedersen, P.B., Olesen, N.J., Slierendrecht, W.J., 2002. Genetic variation for growth rate, feed conversion efficiency, and disease resistance exists within a farmed population of rainbow trout. Aquaculture 209, 59–76.

Henryon, M., Berg, P., Olesen, N.J., Kjaer, T.E., Slierendrecht, W.J., Jokumsen, A., Lund, I., 2005. Selective breeding provides an approach to increase resistance of rainbow trout (*Onchorhynchus mykiss*) to the diseases, enteric redmouth disease, rainbow trout fry syndrome, and viral haemorrhagic septicaemia. Aquaculture 250, 621–636.

Houston, R.D., Gheyas, A., Hamilton, A., Guy, D.R., Tinch, A.E., Taggart, J.B., Mcandrew, B.J., Haley, C.S., Bishop, S.C., 2008a. Detection and confirmation of a major QTL affecting resistance to infectious pancreatic necrosis (IPN) in Atlantic salmon (*Salmo salar*). Developments in Biologicals 132, 199–204.

Houston, R.D., Haley, C.S., Hamilton, A., Guyt, D.R., Tinch, A.E., Taggart, J.B., Mcandrew, B.J., Bishop, S.C., 2008b. Major quantitative trait loci affect resistance to infectious pancreatic necrosis in Atlantic salmon (*Salmo salar*). Genetics 178, 1109–1115.

Houston, R.D., Haley, C.S., Hamilton, A., Guy, D.R., Mota-Velasco, J.C., Gheyas, A.A., Tinch, A.E., Taggart, J.B., Bron, J.E., Starkey, W.G., Mcandrew, B.J., Verner-Jeffreys, D.W., Paley, R.K., Rimmer, G.S.E., Tew, I.J., Bishop, S.C., 2010. The susceptibility of Atlantic salmon fry to freshwater infectious pancreatic necrosis is largely explained by a major QTL. Heredity 105, 318–327.

Houston, R.D., Taggart, J.B., Cézard, T., Bekaert, M., Lowe, N.R., Downing, A., Talbot, R., Bishop, S.C., Archibald, A.L., Bron, J.E., Penman, D.J., Davassi, A., Brew, F., Tinch, A.E., Gharbi, K., Hamilton, A., 2014. Development and validation of a high density SNP genotyping array for Atlantic salmon (*Salmo salar*). BMC Genomics 15, 1–13.

Huang, Y.C., Yin, Z.X., Weng, S.P., He, J.G., Li, S.D., 2012. Selective breeding and preliminary commercial performance of *Penaeus vannamei* for resistance to white spot syndrome virus (WSSV). Aquaculture 364, 111–117.

Jonas, E., De Koning, D.-J., 2015. Genomic selection needs to be carefully assessed to meet specific requirements in livestock breeding programs. Frontiers in Genetics 6, 49.

Jonasson, J., Stefansson, S.E., Gudnason, A., Steinarsson, A., 1999. Genetic variation for survival and shell length of cultured red abalone (*Haliotis rufescens*) in Iceland. Journal of Shellfish Research 18, 621–625.

Kause, A., Ødegård, J., 2012. The genetic analysis of tolerance to infections: a review. Frontiers in Genetics 3, 262.

Kause, A., Vandalen, S., Bovenhuis, H., 2012. Genetics of ascites resistance and tolerance in chicken: a random regression approach. G3 2, 527–535.

Kause, A., 2011. Genetic analysis of tolerance to infections using random regressions: a simulation study. Genetic Research 93, 291–302.

Kettunen, A., Serenius, T., Fjalestad, K.T., 2007a. Three statistical approaches for genetic analysis of disease resistance to vibriosis in Atlantic cod (*Gadus morhua* L.). Journal of Animal Science 85, 305–313.

Kettunen, A., Westgard, J.I., Peruzzi, S., Fevolden, S.E., 2007b. Genetic parameters for post-stress cortisol activity and vibriosis resistance in Atlantic cod (*Gadus morhua* L.). Aquaculture 272, S275–S276.

Kinnula, H., Mappes, J., Valkonen, J.K., Sundberg, L.-R., 2015. The influence of infective dose on the virulence of a generalist pathogen in rainbow trout (*Oncorhynchus mykiss*) and zebra fish (*Danio rerio*). PLoS One 10, e0139378.

Kittilsen, S., Ellis, T., Schjolden, J., Braastad, B.O., Øverli, Ø., 2009. Determining stress-responsiveness in family groups of Atlantic salmon (*Salmo salar*) using non-invasive measures. Aquaculture 298, 146–152.

Kjøglum, S., Henryon, M., Aasmundstad, T., Korsgaard, I., 2008. Selective breeding can increase resistance of Atlantic salmon to furunculosis, infectious salmon anaemia and infectious pancreatic necrosis. Aquaculture Research 39, 498–505.

Knap, P.W., 2009. Robustness. In: RAUW, W.M. (Ed.), Resource Allocation Theory Applied to Farm Animal Production. CABI Publishing, Wallingford, UK.

Kolstad, K., Heuch, P.A., Gjerde, B., Gjedrem, T., Salte, R., 2005. Genetic variation in resistance of Atlantic salmon (*Salmo salar*) to the salmon louse *Lepeophtheirus salmonis*. Aquaculture 247, 145–151.

Krishna, G., Gopikrishna, G., Gopal, C., Jahageerdar, S., Ravichandran, P., Kannappan, S., Pillai, S.M., Paulpandi, S., Kiran, R.P., Saraswati, R., 2011. Genetic parameters for growth and survival in *Penaeus monodon* cultured in India. Aquaculture 318, 74–78.

Kube, P.D., Taylor, R.S., Elliott, N.G., 2012. Genetic variation in parasite resistance of Atlantic salmon to amoebic gill disease over multiple infections. Aquaculture 364–365, 165–172 Apr.

Kuukka-Anttila, H., Peuhkuri, N., Kolari, I., Paananen, T., Kause, A., 2010. Quantitative genetic architecture of parasite-induced cataract in rainbow trout, *Oncorhynchus mykiss*. Heredity 104, 20–27.

Leeds, T.D., Silverstein, J.T., Weber, G.M., Vallejo, R.L., Palti, Y., Rexroad, C.E., Evenhuis, J., Hadidi, S., Welch, T.J., Wiens, G.D., 2010. Response to selection for bacterial cold water disease resistance in rainbow trout. Journal of Animal Science 88, 1936–1946.

Li, J., Boroevich, K.A., Koop, B.F., Davidson, W.S., 2011. Comparative genomics identifies candidate genes for infectious salmon anemia (ISA) resistance in Atlantic salmon (*Salmo salar*). Marine Biotechnology 13, 232–241.

Lillehammer, M., Meuwissen, T.H., Sonesson, A., 2013. A low-marker density implementation of genomic selection in aquaculture using within-family genomic breeding values. Genetics Selection Evolution 45, 39.

Lipschutz-Powell, D., Woolliams, J.A., Bijma, P., Doeschl-Wilson, A.B., 2012. Indirect genetic effects and the spread of infectious disease: are we capturing the full heritable variation underlying disease prevalence? PLoS One 7, e39551.

Lipschutz-Powell, D., Woolliams, J.A., Doeschl-Wilson, A.B., 2014. A unifying theory for genetic epidemiological analysis of binary disease data. Genetics Selection Evolution 46, 15.

Liu, J., Lai, Z., Fu, X., Wu, Y., Bao, X., Hu, Z., Lai, M.M., 2015a. Genetic parameters and selection responses for growth and survival of the small abalone *Haliotis diversicolor* after four generations of successive selection. Aquaculture 436, 58–64.

Liu, S., Vallejo, R.L., Palti, Y., Gao, G., Marancik, D.P., Hernandez, A.G., Wiens, G.D., 2015b. Identification of single nucleotide polymorphism markers associated with bacterial cold water disease resistance and spleen size in rainbow trout. Frontiers in Genetics 6, 298.

Londin, E., Yadav, P., Surrey, S., Kricka, L.J., Fortina, P., 2013. Use of linkage analysis, genome-wide association studies, and next-generation sequencing in the identification of disease-causing mutations. Methods in Molecular Biology 1015, 127–146.

Luan, T., Woolliams, J., Ødegård, J., Dolezal, M., Roman-Ponce, S., Bagnato, A., Meuwissen, T., 2012. The importance of identity-by-state information for the accuracy of genomic selection. Genetics Selection Evolution 44, 28.

Luan, S., Yang, G., Wang, J., Luo, K., Chen, X., Gao, Q., Hu, H., Kong, J., 2014. Selection responses in survival of *Macrobrachium rosenbergii* after performing five generations of multitrait selection for growth and survival. Aquaculture International 22, 993–1007.

Massault, C., Franch, R., Haley, C., De Koning, D.J., Bovenhuis, H., Pellizzari, C., Patarnello, T., Bargelloni, L., 2011. Quantitative trait loci for resistance to fish pasteurellosis in gilthead sea bream (*Sparus aurata*). Animal Genetics 42, 191–203.

McBeath, A., Aamelfot, M., Christiansen, D.H., Matejusova, I., Markussen, T., Kaldhusdal, M., Dale, O., Weli, S.C., Falk, K., 2015. Immersion challenge with low and highly virulent infectious salmon anaemia virus reveals different pathogenesis in Atlantic salmon, *Salmo salar* L. Journal of Fish Diseases 38, 3–15.

Meuwissen, T., Hayes, B., Goddard, M., 2001. Prediction of total genetic value using genome-wide dense marker maps. Genetics 157, 1819–1829.

Midtlyng, P.J., Storset, A., Michel, C., Slierendrecht, W.J., Okamoto, N., 2002. Breeding for disease resistance in fish. Bulletin of the European Association of Fish Pathologists 22, 166–172.

Minozzi, G., Bidanel, J.P., Minvielle, F., Bed'hom, B., Gourichon, D., Baumard, Y., Pinard-Van Der Laan, M.H., 2008. Crossbreeding parameters of general immune response traits in White Leghorn chickens. Livestock Science 119, 221–228.

Moen, T., Fjalestad, K.T., Munck, H., Gomez-Raya, L., 2004. A multistage testing strategy for detection of quantitative trait loci affecting disease resistance in Atlantic salmon. Genetics 167, 851–858.

Moen, T., Sonesson, A.K., Hayes, B., Lien, S., Munck, H., Meuwissen, T.H.E., 2007. Mapping of a quantitative trait locus for resistance against infectious salmon anaemia in Atlantic salmon (*Salmo salar*): comparing survival analysis with analysis on affected/resistant data. BMC Genetics 8.

Moen, T., Baranski, M., Sonesson, A., Kjoglum, S., 2009a. Confirmation and fine-mapping of a major QTL for resistance to infectious pancreatic necrosis in Atlantic salmon (*Salmo salar*): population-level associations between markers and trait. BMC Genomics 10, 368.

Moen, T., Baranski, M., Kent, M., Kjøglum, S., Lien, S., 2009b. Quantitative trait loci (QTL) for use in the AquaGen breeding programme for Atlantic salmon. In: Tenth International Symposium on Genetics in Aquaculture. Bangkok Thailand.

Moen, T., Torgersen, J., Santi, N., Davidson, W.S., Baranski, M., Ødegård, J., Kjøglum, S., Velle, B., Kent, M., Lubieniecki, K.P., Isdal, E., Lien, S., 2015. Epithelial cadherin determines resistance to infectious pancreatic necrosis virus in Atlantic salmon. Genetics 200, 1313–1326.

Moghadam, H., Mørkøre, T., Robinson, N., 2015. Epigenetics—potential for programming fish for aquaculture? Journal of Marine Science and Engineering 3, 175.

Morgan, H.D., Sutherland, H.G., Martin, D.I., Whitelaw, E., 1999. Epigenetic inheritance at the agouti locus in the mouse. Nature Genetics 23, 314–318.

Mustafa, A., Mackinnon, B.M., 1999. Genetic variation in susceptibility of Atlantic salmon to the sea louse *Caligus elongatus* Nordmann, 1832. Canadian Journal of Zoology-Revue Canadienne de Zoologie 77, 1332–1335.

Neira, R., 2010. Breeding in aquaculture species: genetic improvement programs in developing countries. In: Ninth World Congress Genetics Applied Livestock Production Leipzig. abstract 0062.

Nell, J.A., Hand, R.E., 2003. Evaluation of the progeny of second-generation Sydney rock oyster *Saccostrea glomerata* (Gould, 1850) breeding lines for resistance to QX disease *Marteilia sydneyi*. Aquaculture 228, 27–35.

Nielsen, H.M., Ødegård, J., Olesen, I., Gjerde, B., Ardo, L., Jeney, G., Jeney, Z., 2010. Genetic analysis of common carp (*Cyprinus carpio*) strains. I: genetic parameters and heterosis for growth traits and survival. Aquaculture 304, 14–21.

Ødegård, J., Meuwissen, T.H.E., 2014. Identity-by-descent genomic selection using selective and sparse genotyping. Genetics Selection Evolution 46, 3.

Ødegård, J., Olesen, I., Gjerde, B., Klemetsdal, G., 2006. Evaluation of statistical models for genetic analysis of challenge test data on furunculosis resistance in Atlantic salmon (*Salmo salar*): prediction of field survival. Aquaculture 259, 116–123.

Ødegård, J., Olesen, I., Gjerde, B., Klemetsdal, G., 2007a. Evaluation of statistical models for genetic analysis of challenge-test data on ISA resistance in Atlantic salmon (*Salmo salar*): prediction of progeny survival. Aquaculture 266, 70–76.

Ødegård, J., Olesen, I., Gjerde, B., Klemetsdal, G., 2007b. Positive genetic correlation between resistance to bacterial (furunculosis) and viral (infectious salmon anaemia) diseases in farmed Atlantic salmon (*Salmo salar*). Aquaculture 271, 173–177.

Ødegård, J., Olesen, I., Dixon, P., Jeney, Z., Nielsen, H.M., Way, K., Joiner, C., Jeney, G., Ardo, L., Ronyai, A., Gjerde, B., 2010a. Genetic analysis of common carp (*Cyprinus carpio*) strains. II: resistance to koi herpesvirus and *Aeromonas hydrophila* and their relationship with pond survival. Aquaculture 304, 7–13.

Ødegård, J., Sommer, A.I., Kettunen, P., 2010b. Heritability of resistance to viral nervous necrosis in Atlantic cod (*Gadus morhua* L.). Aquaculture 300, 59–64.

Ødegård, J., Gitterle, T., Madsen, P., Meuwissen, T.H.E., Yazdi, H.Y., Gjerde, B., 2011. Quantitative genetics of taura syndrome resistance in pacific white shrimp (*Penaeus vannamei*): a cure model approach. Genetics Selection and Evolution 43, 14.

Okamoto, N., Tayama, T., Kawanobe, M., Fujiki, N., Yasuda, Y., Sano, T., 1993. Resistance of a rainbow trout strain to infectious pancreatic necrosis. Aquaculture 117, 71–76.

Olesen, I., Hung, D., Odegard, J., 2007. Genetic analysis of survival in challenge tests of furunculosis and ISA in Atlantic salmon. Genetic parameter estimates and model comparisons. Aquaculture 272, S297–S298.

Ozaki, A., Sakamoto, T., Khoo, S., Nakamura, K., Coimbra, M.R.M., Akutsu, T., Okamoto, N., 2001. Quantitative trait loci (QTLs) associated with resistance/susceptibility to infectious pancreatic necrosis virus (IPNV) in rainbow trout (*Oncorhynchus mykiss*). Molecular Genetics and Genomics 265, 23–31.

Ozaki, A., Yoshida, K., Fuji, K., Kubota, S., Kai, W., Aoki, J.-Y., Kawabata, Y., Suzuki, J., Akita, K., Koyama, T., Nakagawa, M., Hotta, T., Tsuzaki, T., Okamoto, N., Araki, K., Sakamoto, T., 2013. Quantitative trait loci (QTL) associated with resistance to a Monogenean parasite (*Benedenia seriolae*) in yellowtail (*Seriola quinqueradiata*) through genome wide analysis. PLoS One 8, e64987.

Palti, Y., Vallejo, R.L., Gao, G., Liu, S., Hernandez, A.G., Rexroad III, C.E., Wiens, G.D., 2015. Detection and validation of QTL affecting bacterial cold water disease resistance in rainbow trout using restriction-site associated DNA sequencing. PLoS One 10, e0138435.

Perry, G.M.L., Tarte, P., Croisetiere, S., Belhumeur, P., Bernatchez, L., 2004. Genetic variance and covariance for 0+ brook charr (*Salvelinus fontinalis*) weight and survival time of furunculosis (*Areomonas salmonicida*) exposure. Aquaculture 235, 263–271.

Praveen, K., Goodwin, E., Wolters, W.R., Davis, K.B., 2001. Selective breeding for disease resistance in channel catfish: phenotypic correlations between immune function assays and stress response. In: Tenth E.A.F.P International Conference. Book of Abstracts, Dublin.

Quillet, E., Dorson, M., Aubard, G., Torhy, C., 2001. In vitro viral haemorrhagic septicaemia virus replication in excised fins of rainbow trout: correlation with resistance to waterborne challenge and genetic variation. Diseases of Aquatic Organisms 45, 171–182.

Quillet, E., Dorson, M., Aubard, G., Torhy, C., 2007a. An in vitro assay to select rainbow trout with variable resistance/susceptibility to viral haemorrhagic septicaemia virus. Diseases of Aquatic Organisms 76, 7–16.

Quillet, E., Dorson, M., Leguillou, S., Benmansour, A., Boudinot, P., 2007b. Wide range of susceptibility to rhabdoviruses in homozygous clones of rainbow trout. Fish and Shellfish Immunology 22, 510–519.

Quillet, E., Krieg, F., Dechamp, N., Hervet, C., Berard, A., Le Roy, P., Guyomard, R., Prunet, P., Pottinger, T.G., 2014. Quantitative trait loci for magnitude of the plasma cortisol response to confinement in rainbow trout. Animal Genetics 45, 223–234.

Råberg, L., Graham, A.L., Read, A.F., 2009. Decomposing health: tolerance and resistance to parasites in animals. Philosophical Transactions of the Royal Society Series B, Biological Sciences 364, 37–49.

Refstie, T., 1982. Preliminary results: differences between rainbow trout families in resistance against vibriosis and stress. Developmental and Comparative Immunology (Suppl. 2), 205–209.

Rexroad, C.E., Vallejo, R.L., Liu, S., Palti, Y., Weber, G.M., 2013. Quantitative trait loci affecting response to crowding stress in an F-2 generation of rainbow trout produced through phenotypic selection. Marine Biotechnology 15, 613–627.

Rezk, M.A., Ponzoni, R.W., Khaw, H.L., Kamel, E., Dawood, T., John, G., 2009. Selective breeding for increased body weight in a synthetic breed of Egyptian Nile tilapia, *Oreochromis niloticus*: response to selection and genetic parameters. Aquaculture 293, 187–194.

Robinson, N., Sahoo, P.K., Baranski, M., Das Mahapatra, K., Saha, J.N., Das, S., Mishra, Y., Das, P., Barman, H.K., Eknath, A.E., 2012. Expressed sequences and polymorphisms in rohu carp (*Labeo rohita*, Hamilton) revealed by mRNA-seq. Marine Biotechnology 14, 620–633.

Robinson, N.A., Baranski, M., Das Mahapatra, K., Saha, J.N., Das, S., Mishra, Y., Das, P., Kent, M., Arnyasi, M., Sahoo, P.K., 2014a. A linkage map of transcribed SNPs in rohu (*Labeo rohita*) and QTL associated with resistance to *Aeromonas hydrophila*. BMC Genomics 15, 541.

Robinson, N.A., Gopikrishna, G., Baranski, M., Katneni, V.K., Shekhar, M.S., Shanmugakarthik, J., Jothivel, S., Gopal, C., Ravichandran, P., Gitterle, T., Ponniah, A.G., 2014b. QTL for white spot syndrome virus resistance and the sex-determining locus in the Indian black tiger shrimp (*Penaeus monodon*). BMC Genomics 15, 731.

Rodriguez-Ramilo, S.T., Toro, M.A., Bouza, C., Hermida, M., Pardo, B.G., Cabaleiro, S., Martinez, P., Fernandez, J., 2011. QTL detection for *Aeromonas salmonicida* resistance related traits in turbot (*Scophthalmus maximus*). BMC Genomics 12.

Rodriguez-Ramilo, S.T., Fernandez, J., Toro, M.A., Bouza, C., Hermida, M., Fernandez, C., Pardo, B.G., Cabaleiro, S., Martinez, P., 2013. Uncovering QTL for resistance and survival time to *Philasterides dicentrarchi* in turbot (*Scophthalmus maximus*). Animal Genetics 44, 149–157.

Rodriguez-Ramilo, S.T., De La Herran, R., Ruiz-Rejon, C., Hermida, M., Fernandez, C., Pereiro, P., Figueras, A., Bouza, C., Toro, M.A., Martinez, P., Fernandez, J., 2014. Identification of quantitative trait loci associated with resistance to viral haemorrhagic septicaemia (VHS) in turbot (*Scophthalmus maximus*): a comparison between bacterium, parasite and virus diseases. Marine Biotechnology 16, 265–276.

Rutherford, S.L., Lindquist, S., 1998. Hsp90 as a capacitor for morphological evolution. Nature 396, 336–342.

Rye, M., Lillevik, K.M., Gjerde, B., 1990. Survival in early life of Atlantic salmon and rainbow trout: estimates of heritabilities and genetic correlations. Aquaculture 89, 209–216.

Rye, M., Gjerde, B., Gjedrem, T., 2010. Genetic improvement programs for aquaculture species in developed countries. In: Ninth World Congress Genetics Applied Livestock Production Leipzig. abstract 0963.

Shao, C., Niu, Y., Rastas, P., Liu, Y., Xie, Z., Li, H., Wang, L., Jiang, Y., Tai, S., Tian, Y., Sakamoto, T., Chen, S., 2015. Genome-wide SNP identification for the construction of a high-resolution genetic map of Japanese flounder (*Paralichthys olivaceus*): applications to QTL mapping of *Vibrio anguillarum* disease resistance and comparative genomic analysis. DNA Research 22, 161–170.

Silverstein, J.T., Vallejo, R.L., Palti, Y., Leeds, T.D., Rexroad III, C.E., Welch, T.J., Wiens, G.D., Ducrocq, V., 2009. Rainbow trout resistance to bacterial cold-water disease is moderately heritable and is not adversely correlated with growth. Journal Animal Science 87, 860–867.

Slatkin, M., 2008. Linkage disequilibrium—understanding the evolutionary past and mapping the medical future. Nature Reviews Genetics 9, 477–485.

Sonesson, A.K., Woolliams, J.A., Meuwissen, T.H.E., 2012. Genomic selection requires genomic control of inbreeding. Genetics Selection Evolution 44, 27.

Standal, M., Gjerde, B., 1987. Genetic variation in survival of Atlantic salmon during the sea-rearing period. Aquaculture 66, 197–207.

Storset, A., Strand, C., Wetten, M., Sissel, K., Rarnstad, A., 2007. Response to selection for resistance against infectious pancreatic necrosis in Atlantic salmon (*Salmo salar* L.). Aquaculture 272, S62–S68.

Strømsheim, A., Eide, D.M., Fjalestad, K.T., Larsen, H.J.S., Rø, K.H., 1994a. Genetic variation in the humoral immune response in Atlantic salmon (*Salmo salar*) against *Aeromonas salmonicida* A-layer. Veterinary Immunology and Immunopathology 41, 341–352.

Strømsheim, A., Eide, D.M., Hofgaard, P.O., Larsen, H.J.S., Refstie, T., Roed, K.H., 1994b. Genetic variation in the humoral immune-response against *Vibrio salmonicida* and in antibody titer against *Vibrio anguillarum* and total IgM in Atlantic salmon (*Salmo salar*). Veterinary Immunology and Immunopathology 44, 85–95.

Sturtevant, A.H., 1913. The linear arrangement of six sex-linked factors in Drosophila, as shown by their mode of association. Journal of Experimental Zoology 14, 43–59.

Taylor, R.S., Kube, P.D., Muller, W.J., Elliott, N.G., 2009. Genetic variation of gross gill pathology and survival of Atlantic salmon (*Salmo salar* L.) during natural amoebic gill disease challenge. Aquaculture 294, 172–179.

Taylor, J.F., 2014. Implementation and accuracy of genomic selection. Aquaculture 420, S8–S14.

Thodesen, D.-Y.M.J., Rye, M., Wang, Y.X., Li, S.J., Bentsen, H.B., Gjedrem, T., 2012. Genetic improvement of tilapias in China: genetic parameters and selection responses in growth, pond survival and cold-water tolerance of Blue tilapia (*Oreochromis aureus*) after four generations of multi-trait selection. Aquaculture 396–399, 32–42.

Thodesen, D.-Y.M.J., Rye, M., Wang, Y.X., Li, S.J., Bentsen, H.B., Gjedrem, T., 2013. Genetic improvement of tilapias in China: genetic parameters and selection responses in growth, survival and external colour of red tilapia (*Oreochromis* spp.) after four generations of selections of multi-trait selection. Aquaculture 416–417, 354–366.

Tort, L., 2011. Stress and immune modulation in fish. Developmental and Comparative Immunology 35, 1366–1375.

Vallejo, R.L., Wiens, G.D., Rexroad III, C.E., Welch, T.J., Evenhuis, J.P., Leeds, T.D., Janss, L.L.G., Palti, Y., 2010. Evidence of major genes affecting resistance to bacterial cold water disease in rainbow trout using Bayesian methods of segregation analysis. Journal of Animal Science 88, 3814–3832.

Vallejo, R.L., Palti, Y., Liu, S., Evenhuis, J.P., Gao, G., Rexroad III, C.E., Wiens, G.D., 2014a. Detection of QTL in rainbow trout affecting survival when challenged with *Flavobacterium psychrophilum*. Marine Biotechnology 16, 349–360.

Vallejo, R.L., Palti, Y., Liu, S., Marancik, D.P., Wiens, G.D., 2014b. Validation of linked QTL for bacterial cold water disease resistance and spleen size on rainbow trout chromosome Omy19. Aquaculture 432, 139–143.

Vandeputte, M., Chapuis, H., Dupont-Nivet, M., 2009. Optimisation of factorial mating designs to estimate genetic parameters for threshold traits in DNA-pedigreed mixed families of fish. In: Tenth International Symposium on Genetics in Aquaculture (Bangkok).

Vehviläinen, H., Kause, A., Quinton, C., Koskinen, H., Paananen, T., 2008. Survival of the currently fittest: genetics of rainbow trout survival across time and space. Genetics 180, 507–516.

Verrier, E.R., Dorson, M., Mauger, S., Torhy, C., Ciobotaru, C., Hervet, C., Dechamp, N., Genet, C., Boudinot, P., Quillet, E., 2013a. Resistance to a rhabdovirus (VHSV) in rainbow trout: identification of a major QTL related to innate mechanisms. PLoS One 8, e55302.

Verrier, E.R., Ehanno, A., Biacchesi, S., Le Guillou, S., Dechamp, N., Boudinot, P., Brémont, M., Quillet, E., 2013b. Lack of correlation between the genetic resistances to two rhabdovirus infections in rainbow trout. Fish and Shellfish Immunology 35, 9–17.

Wang, L., Fan, C., Liu, Y., Zhang, Y., Liu, S., Sun, D., Deng, H., Xu, Y., Tian, Y., Liao, X., Xie, M., Li, W., Chen, S., 2014. A genome scan for quantitative trait loci associated with *Vibrio anguillarum* infection resistance in Japanese flounder (*Paralichthys olivaceus*) by bulked segregant analysis. Marine Biotechnology 16, 513–521.

Wargo, A.R., Kell, M.A., Scotta, R.J., Thorgaard, G.H., Kurath, G., 2012. Analysis of host genetic diversity and viral entry as sources of between-host variation in viral load. Virus Research 165, 71–80.

Watson, J., Crick, F.H.C., 1953. Molecular structure of nucleic acids. Nature 121, 737–738.

Weber, G.M., Vallejo, R.L., Lankford, S.E., Silverstein, J.T., Welch, T.J., 2008. Cortisol response to a crowding stress: heritability and association with disease resistance to *Yersinia ruckeri* in rainbow trout. North American Journal of Aquaculture 70, 425–433.

Wetten, M., Aasmundstad, T., Kjøglum, S., Storset, A., 2007. Genetic analysis of resistance to infectious pancreatic necrosis in Atlantic salmon (*Salmo salar* L.). Aquaculture 272, 111–117.

Wiens, G.D., Lapatra, S.E., Welch, T.J., Evenhuis, J.P., Rexroad, C.E.I., Leeds, T.D., 2013a. On-farm performance of rainbow trout (*Oncorhynchus mykiss*) selectively bred for resistance to bacterial cold water disease: effect of rearing environment on survival phenotype. Aquaculture 388-391, 128–136.

Wiens, G.D., Vallejo, R.L., Leeds, T.D., Palti, Y., Hadidi, S., Liu, S., Evenhuis, J.P., Welch, T.J., Rexroad III, C.E., 2013b. Assessment of genetic correlation between bacterial cold water disease resistance and spleen index in a domesticated population of rainbow trout: identification of QTL on chromosome Omy19. PLoS One 8, e75749.

Yamamoto, S., Sanjyo, I., Sato, R., Kohara, M., Tahara, H., 1991. Estimation of the heritability for resistance to infectious hematopoietic necrosis in rainbow trout. Nippon Suisan Gakkaishi (Bullettin of the Japanese Society for the Science of Fish) 57, 1519–1522.

Yàñez, J.M., Smith, P.A., Manneschi, G., Guarjado, A., Rojas, M.E., Larenas, J., Diaz, S., Valdes, M., Martinez, V.A., 2010. Unravelling genetic co-variation for resistance and tolerance against *Piscirickettsia salmonis* infection in Atlantic salmon (*Salmo salar*). In: Proceedings of the Ninth World Congress Genetics Applied Livestock Production, p. 125 Leipzig. abstract 0965.

Part II

Pathogens

Chapter 3

Types of Pathogens in Fish, Waterborne Diseases

Ian Bricknell
The University of Maine, Orono, ME, United States

3.1 INTRODUCTION

Diseases in fish are often considered as the poor relations of veterinary clinical microbiologists. Although fish are the oldest extant group of true vertebrates on the planet (Long, 2010) and have experienced around 420 million years of evolution along with their pathogens, this has allowed many unique and intimate host–pathogen interactions to evolve. Even with this exceptional evolutionary advantage many fish diseases remain poorly studied. Although a few fish species have made it to the mainstream [Zebra fish *Danio rerio* (Hamilton, 1822), medaka, *Oryzias latipes* (Temminck and Schlegel, 1846), and puffer fish, *Tetraodon nigroviridis*, Marion de Procé, 1822], research on these fish equivalents of white mice can often be criticized for the poor understanding of the health status of them, especially before disease or immunity experiments take place (D'Angelo et al., 2016; Lin et al., 2016a; Pederzoli and Mola, 2016; Secombes, 2016).

This lack of understanding comes about for many reasons, for example, fish are the most numerous vertebrates on earth with around 33,300 described species (Fishbase, 2016), and each one having its own unique clade of pathogens and parasites. Some fish diseases are indiscriminate (Austin and Austin, 2016), however, each fish species will have a unique group of pathogens that are specific to that species. Species-specific pathogens may be bacterial, viral, or fungal, but are far more likely to be specialized obligate parasites unique to that host. These tend to be the pathogens we know least about, yet are the most important from the biodiversity point of view.

Due to the vast number of fish species, there are many pathogens that will remain undescribed to science simply due to lack of resources. It is estimated that with every animal that goes extinct, 10 unique metazoan parasite species go into extinction with it (Moir et al., 2014; Campiao et al., 2015; Farrell et al., 2015; Jorgensen, 2015; Rozsa and Vas, 2015; Cuthill et al., 2016). If this is the case, there are around 330,000 parasites that have just one fish host, as well as many parasites, such as *Caligus elongatus*, that have a very catholic taste in hosts—parasitizing not just fish but mammals too, such as Minke whales (Olafsdottir and Shinn, 2013). Many of the species of parasites that parasitize only one host have never been descried. A good example of this diversity is the Atlantic cod (*Gadus morhua* L.) with 107 described parasites (Mackenzie and Hemmingsen, 2015), which is probably a realistic number for most fish species. However, the recently described blue walking snakehead (*Channa andrao*; Britz, 2013) currently has no known parasites or microbial pathogens associated with it. Although this is just playing with numbers, it is an example of how diverse fish are as a group and how little we know about the disease fauna and flora of nonmammalian vertebrates.

3.2 HOST SPECIFICITY OF PATHOGENS

Pathogens are classically dived into three categories: obligate pathogens that require a host as a fundamental part of their life style; facultative pathogens that can live free in the environment or become a pathogen if a suitable host is available; and opportunistic pathogens. This latter group is rarely pathogenic, only becoming so in unusual circumstances, such as severe stress events or an immunocompromised host. This group tend to be the most difficult to predict and diagnose, as the aquatic environment has such a rich fauna and flora of pathogens there are many organisms that can take on the opportunistic pathogen role.

Many pathogens including viruses, bacteria, protists, and metazoans have a degree of host specificity. Infectious salmon anemia has the ability to infect salmonids (Aamelfot et al., 2014; Pettersen et al., 2015) and little else while the Caligid parasite, *C. elongatus*, is reported in over 111 different fish species (Piasecki and Mackinnon, 1995; Heuch et al., 2007;

Fish Diseases. http://dx.doi.org/10.1016/B978-0-12-804564-0.00003-X

Jensen et al., 2016) and one mammal (Olafsdottir and Shinn, 2013). Why is there such a difference? One of the reasons why pathogens are limited to one host is often due to a close evolutionary relationship developing between the pathogen and its host. The retroviruses are a case in point. These viruses have evolved such a close relationship with their hosts that they can insert their genetic material into the hosts' genome. Sitting there for years, or generations, waiting for a time when physiological conditions within the host change sufficiently to allow the virus to replicate and spread to new hosts. However, because an infected host has the retroviruses genes encoded into its somatic DNA the virus is also transmitted vertically during reproduction.

What causes these triggers is a complex phenomenon and will be touched upon later when we consider specific types of diseases. However, the host specificity is often due to the close evolution of the disease with the host. So much so that many pathogens have evolved molecules on their surface that mimic the active sites of receptors on specific cells within their hosts or suppress specific components of the immune system.

3.3 VIRAL PATHOGENS

Viruses as we are so often told are merely entities, neither alive nor dead but hovering in a strange existence in between the worlds of biochemistry and biology. They are too small, both in size and the actual number of genes in their genetic information codes, to perform the physiological requirement of a living organism. Yet, they can meet many of the requirements of a living organism.

Living things are characterized by six rules (Table 3.1).

TABLE 3.1 Definition of Living Organisms

Rules of Life	Viral Characteristics
Living things are made of cells.	Viruses are not cells.
Living things grow and develop.	Viruses only grow within other cells by using the infected cells organelles.
Living things adapt to their environment.	Viruses do not adapt to the environment, either physiologically or environmentally. Unless they are in a host cell.
Living things obtain and use energy.	They only use the host cell's energy while replicating in that cell.
Living things reproduce.	By taking over a cell, viruses hijack the cells mechanism to grow and develop.
Living things respond to their environment.	Viruses do not respond to the environments, either physiologically or environmentally.

One would think that sitting in this strange zombie world between living and dead things, viruses would only be transmitted passively between susceptible hosts. Although they may not directly transmit infectious virions between hosts, unlike parasites that can actively track down their hosts, viruses have evolved many complex strategies to ensure a successful transmission strategy.

3.3.1 Virally Induced Host Behavioral Changes

There are few people who have not heard of the hydrophobia induced by the rhabdovirisus that causes rabies. As the virus replicates in the central nervous system, it damages the brain triggering a strange behavior in which the last stage of the disease can cause the classic symptom of frothing at the mouth, where saliva rich in the infectious virus is ejected into the environment. This highly infectious material can enter a new host via mucus membranes, or following a bite as the infected host also becomes very aggressive, often biting conspecifics.

In the aquatic world this type of transmission is not well studied, that does not mean to does not happen, no one has yet demonstrated any such change in a fish host.

3.3.2 Viral Shedding

By far the commonest method of transmission of viruses in the aquatic environment is by shedding into the aquatic environment through the biological fluids of the fish, such as semen or ovarian fluid (Smail and Munro, 2008; Gurcay et al., 2013; Marshall et al., 2014), mucus (Griffiths and Melville, 2000; Giray et al., 2005; Graham et al., 2011, 2012; Nita-Lazar et al.,

2016), feces (Gonzalez et al., 2011; Molloy et al., 2013), urine (Totland et al., 1996), or during the last stages of disease, such as VHSV or ISAV when the infected fish may undergo severe loss of body fluids as their tissues begin to degenerate.

An excellent example of virus shedding in fish is infectious salmon anemia (ISA) caused by infectious salmon anemia virus (ISAV). A member of orthomyxoviruses, a group that includes six genera of single-stranded RNA; influenza virus A, B, and C, isavirus, thogotovirus and quaranjavirus, ISAV belongs to the Isavirus genus.

ISAV preferentially infects epithelial, hemoptoic, and red blood cells of salmonids. During the initial stages of the disease, the fish appears well and continue eating. However, the first clinical sign is a gradual drop in hematocrit as the hemoptoic tissues fail when the fish becomes increasingly infected. The vast majority of hemoptoic tissues of salmonids are located in the pronephros or head kidney and the heamoptoic tissues of the pronephros (Fig. 3.1) epithelial cells lining blood vessels, and the kidney tubules are infected at the same time. This often leads to release of virions into the bloodstream and urine. Often the first shedding event during an ISAV infection into the surrounding water is via urine. The number of virus-like particles per/mL of urine was estimated to be 6×10^6 (Totland et al., 1996). Shedding via urine is probably responsible for initial spreading of the virus from fish to fish in the early stages of an ISA outbreak.

ISAV also readily infects the epithelia cell lining of blood vessels (Fig. 3.1). As ISA progresses the epithelial cells lining the vessels of the circulation system become infected. This leads to two things. First, as the infected cells lyse, the viral titer of the blood increases, and second, the vessels themselves become permeable to the fluids they contain. These leaky circulatory vessels permit considerable ascetic fluid to accumulate which is released into the environment via Ellis's pore, near the cloaca. The gills also become permeable, and due to collective osmosis and hydrostatic pressures in the blood vessels of the gills, the virion-rich tissue fluid and plasma leak into the environment. Both of these pathways increase the shedding of active virus into the environment.

This process is compounded, as there is evidence that the red blood cells of an infected fish are hijacked and lysed by the virus further reducing the hematocrit and increasing the plasma viral load. In addition to the leakiness of the circulatory system, the fish also become edemic and there is an increase of secretion of fluid into the interstitium. This causes the fish to appear swollen and soft to the touch. At this stage the fish often begin shedding the virus across the epithelia surface, into mucus, and into the surrounding environment.

Ultimately, the infected fish dies when the hematocrit falls below 5% and during period when the fish is required to exert itself physiologically. Due to the severe lack of red blood cells in the blood and acute ischemia occurs and the fish dies due to anoxia (Hovland et al., 1994; Prost, 1997) simply because it cannot transport sufficient oxygen around its body to maintain homeostasis.

The shedding of the virus stops after death of the fish. This is due to several factors. The autolysis of the body releases a suite of enzymes that degrade the capsule, proteins, and RNA of the virus, and render them nonviable. Of course, the death of the host also stops viral replication as the host cells cannot support it any more. It can be hypothesized that the salmon are probably not infectious a few days after death as the processes of decay within their bodies inactivates any remaining

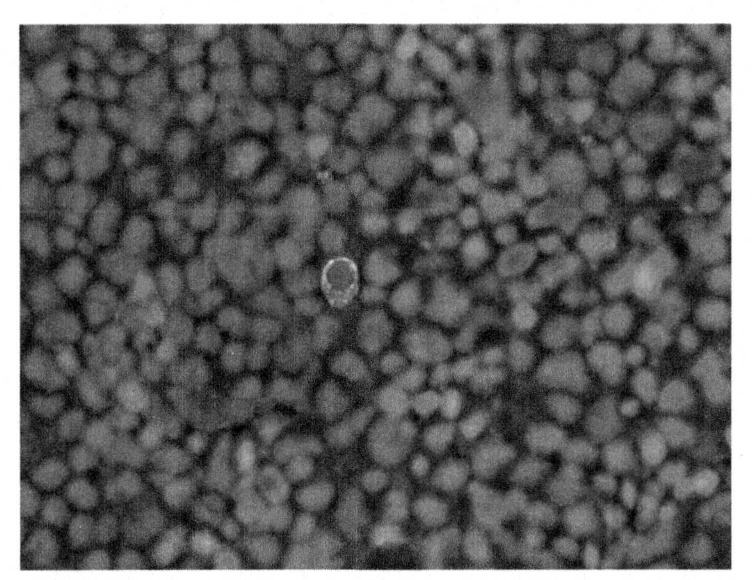

FIGURE 3.1 IFAT of an ISAV-infected kidney of an Atlantic salmon, the green cell is an antibody labeled infected cell. *Photo Bricknell, I., Crown Copyright.*

virion. However, this is not the case for all viruses where injection of infectious tissues dead or alive can transmit the infection as we shall see.

3.3.3 The Carrier Status

Viruses do not only rely on shedding as a route of infection. Many viruses rely on a carrier status to maintain an infectious pressure within a population. The best example of this, in fish, is infectious pancreatic necrosis virus, an Aquabirnavirus, a genus of viruses in the family Birnaviridae (Dobos and Roberts, 1983; Bernard and Bremont, 1995; Roberts and Pearson, 2005). It is a small, 70 nm double-stranded RNA viruse and is a common pathogen of salmonid fish in North America and Europe. It is often asymptomatic only causing mortality in very young yolk sack fish (Biering et al., 1994; Biering and Bergh, 1996a,b; Bergh et al., 2001; Roberts and Pearson, 2005) or following sea transfer, particularly in Europe, of Atlantic salmon smolts [*Salmo salar* L. (Roberts and Pearson, 2005)]. The acute disease is very characteristic, usually associated with a necrosis of the acinar tissue of the pancreas (Fig. 3.2) and often causes catarrhal enteritis which typically consists of the sloughed epithelial cells of the digestive tract. Typically, the disease is self-limiting with the exceptions mentioned above and the infected fish often recover, the acinar tissue regenerates and the fish return to a normal health status and appear to be IPNV-negative (Samuelsen et al., 2006).

Recurrence of disease was reported in IPNV-infected populations (Roberts and Pearson, 2005) even when it was not in contact with other populations of fish that may have caused reexposure to IPNV. However, these populations may have tested negative after repeated virus culture which is typically carried out using head kidney samples [Pronephros (Roberts and Pearson, 2005)]. This observation poses a diagnostic quandary. If the fish are head kidney tissue culture–negative, where is the IPNV infection coming from? The answer is viruses can develop a carrier status where the virus "hides" in specific tissues that may not be associated with the initial target tissues.

Possibly, the most well-studied carrier in the virus world is human herpes simplex virus, a virus that causes cold and genital sores in humans. The classic HSV causes an unsightly and painful lesion in the mucus membranes and skin of an infected individual. Again, these lesions are typically self-limiting in a healthy host and resolves with time. After resolution of the lesion, the patient appears to be virus-free, as it cannot be re-isolated from the site of the lesion or an excreted tissue fluid. This "virus-free" postinfection period is often called the latent period (Rimstad et al., 1991). If the patient is then exposed to a stressful event, either physiologically or emotionally, or becomes immunosuppressed the initial lesions reoccur.

During the latent HSV phase, the virus remains viable within the host, but resides in the dorsal root ganglions of the central nervous system. However, HSV is classified as a neurotropic and neuro-invasive virus. In the subsequent outbreaks, the virus in the neurons replicates and can be transported via the neuron's axon to the infectious site in the skin or mucus membranes, where virus replication and shedding occur and cause clinical symptoms. In a rare case, the reactivation of HSV from the latent state permits the virus to travel along neurons and it can cause HSV viral meningitis which is frequently fatal.

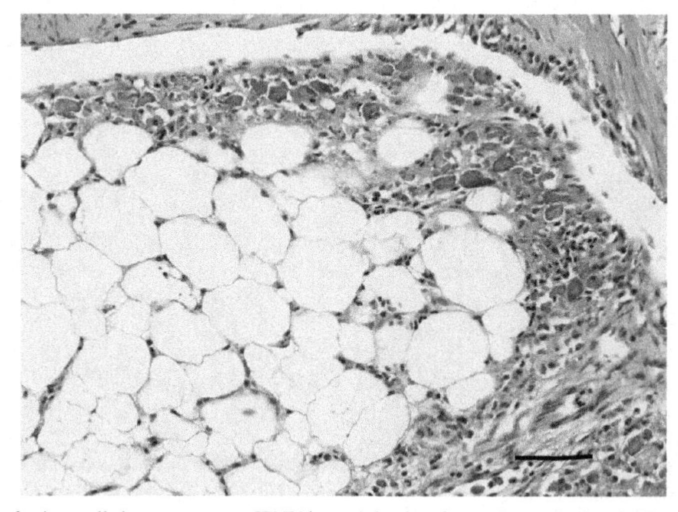

FIGURE 3.2 Diffused necrosis of acinar cells in an acute case IPNV in an Atlantic salmon. *Image Bailey, J., University of Stirling. Scale Bar 50 μm.*

For many years, IPNV infections seemed to lead to a carrier state but the tissue where the virus underwent its latent phase was unknown. The first research that casts some light on this identified that blood could transmit the virus from recovered and apparently healthy fish to naïve fish (Johansen and Sommer, 1995; Rodriguez et al., 2001). Accompanying this research was the observation that ovarian fluid (McAllister et al., 1993; Munro and Ellis, 2008; Smail and Munro, 2008), especially the cells contained within the ovarian fluid, was an excellent tissue to screen for positive female brood stock animals. Many researchers identified the peripheral blood leukocytes as an excellent tissue for detecting IPNV (Gahlawat et al., 2004; Munro et al., 2004; Cutrin et al., 2005; Garcia et al., 2006; Ronneseth et al., 2006, 2012, 2013; Munro and Ellis, 2008; Lopez-Jimena et al., 2010) followed this line of enquiry and found that the macrophages of infected fish carry a viable viral titer, even when the other tissues tested negative for IPNV. There was one interesting observation associated with this finding that the macrophages were only positive for about 9 months of the year. In the winter, around December–March, the macrophages had undetectable levels of virus in them. Indeed, if the fish were sexually mature then the macrophages from spawning fish were negative. Yet the pronephros and ovarian fluid became positive as the macrophage titer decreased. This led them to propose that IPNV carrier status is closely linked to the reproductive cycle. The mechanism that causes the decrease in virus titer in circulating macrophages and the subsequent reappearance in the head kidney or ovarian fluid is unknown. It could be environmental (Bowden et al., 2002, 2003), associated with cooler temperature and/or shortening day lengths. The trigger may be the changes that occur during spawning, such as elevation of sex hormones, or a diversion of energy resources away from the immune system to the reproductive system causing the fish to become immunocompromised during reproduction.

Whatever the cause of the outbreak from the circulating macrophages, the strategy is very successful for the virus. During reproduction there is an opportunity to infect the eggs either in ovo or during fertilization. In a contaminated spawning habitat with high levels of virus and IPNV persistence in the environment, it is quite feasible for the virus to persist for many months on the spawning ground before the eggs hatch, providing a new population of fish to infect. Of course the spawning animals may begin to shed virus during spawning too, depending on the intensity of the IPN flare-up. Given the close proximity of fish during spawning, this gives the virus an excelled chance of being transmitted horizontally between spawning adult fish.

3.3.4 Vectors

Many viruses use arthropods or other invertebrates as disease vectors, one of the best studied is yellow fever. Yellow fever (YF) is caused by the yellow fever virus (YFV), an enveloped RNA virus in the range of 40–50 nm which belongs to the viral family Flaviviridae. YFV has two hosts, mammals (often humans or other primates) and mosquito *Aedes aegypti* (Linnaeus in Hasselquist, 1762) but other *Aedes* mosquito species including *A. albopictus*, Skuse, 1894, can be vectors (Gratz, 2004; Eritja et al., 2005; Hill, 2012; Medlock et al., 2012; Olson and Blair, 2012; Alto and Lounibos, 2013).

When an *Aedes* mosquito bites an infected host, the virus invades and replicates in the salivary glands, brain, and occasionally in the cells of the suboesophageal ganglion of the mosquito. In *Aedes* mosquitoes the infection is often subclinical and the host mosquitoes appear fit and well, feeds and reproduces normally. Indeed, there is some evidence that YFV is vertically transmitted in *A. aegypti* (Fontenille et al., 1997; Diallo et al., 2000) When an infected mosquito infects a susceptible host, the virus is injected along with the saliva into the host tissues where it initially targets and replicates. After lysis of the dendritic macrophages, the virus then invades the liver, kidneys, spleen, and eventually the heart of the primate host. During the acute phase of the disease, viral titers are high in the blood and if another *Aedes* mosquito bites an infected host, it gets infected, and continues the life cycle further. The mortality rate of YF is around 15% in primates.

Arthropods have long been suspected of being vectors for viral diseases in fish, unfortunately the evidence is not conclusive. It has been demonstrated the molecular signature of ISAV in the Salmon louse (*Lepeophtheirus salmonis*, Kroyer, 1837) and other Siphonostomatoida (Fontenille et al., 1997; Diallo et al., 2000) and argulid and branchiuran crustaceans have been indicated in transmitted spring viremia of carp and carp pox to cyprinid fish (Ahne et al., 2002; Overstreet et al., 2009) freshwater turtles have been indicated in the transmission of Viral Hemorrhagic Septicemia virus (Goodwin and Merry, 2011) as have freshwater isopods (Al-Hussinee et al., 2011). Although, these studies always looked for the presence of the virus in the parasitic hosts, none of them successfully demonstrated the transmission by the presumptive vector. Thus, it is impossible to demonstrate that whether the potential vector is acting as a disease-transmitting organism or the virus is simply being detected because it has recently fed on or been in contact with an infected host.

Given the long evolutionary relationship between fish and their parasites, arthropod parasites were detected in fossil fish for at least 160 million years. Well-preserved, lower cretaceous fossil dichelesthioidea copepods were found in the gill chambers of *Cladocyclus gardneri* and early ichthyodectid fish (Forey and Cavin, 2007; Sousa et al., 2016). It is highly likely that arthropods, and other vectors, of viruses have evolved in the aquatic ecosystem. However, the area is so poorly studied these mysteries still remain to be uncovered.

3.3.5 Predator–Prey Relationships

Many fish viruses rely on a predator–prey relationship for transmission between hosts. This can be considered as a variation of the carrier status discussed above.

One of the classic examples of this in fish is viral hemorrhagic septicemia virus (VHSV) the causative agent of viral hemorrhagic septicemia (VHS; Fig. 3.3A and B). VHSV is a single-stranded RNA virus of genus Novirhabdovirus and family Rhabdoviridae. VHSV was originally found in freshwater salmonids in Europe (Schlotfeldt and Kleingeld, 1993; Dixon, 1999) causing major outbreaks of clinical disease in species such as rainbow trout [*Oncorhynchus mykiss* (Walbaum, 1792)]. More recently, VHSV has been found in marine fish in the North Atlantic Ocean, North Sea, Pacific Ocean, and Baltic Sea.

In 2005, sudden die-offs occurred in many freshwater fish species found in the Great Lakes watershed and, subsequently, the Mississippi river in America (Faisal and Winters, 2011; Frattini et al., 2011). VHSV remains a serious disease of farmed fish and is reportable under the OIE regulations (OIE, 2016).

During the acute phase of the disease, the virus VHSV undergoes rapid replication in the endothelial cells of the circulator system causing the classic swelling of infected fish and hemorrhage of blood and bodily fluids into the tissues and environment (Fig. 3.3). At this stage mortality can be high. However, there is an evolutionary pressure for VHSV to become less virulent within the host (Snow et al., 2005). Surviving animals and fish infected with a less virulent strain, or genotype, typically go on to develop a long-term carrier state in their central nervous system. The lesions caused by the virus often change the behavior of the carrier state animals, often making the animal less effective at predator avoidance, due to impaired vision or reduced motor control. These fish are regularly eaten by predators and it is assumed that the virus can also be transmitted by the oral route. Although the field evidence for this is limited, there is experimental evidence of oral transmission of VHSV by gastric lavage (Snow et al., 2005) and food (Snow et al., 2005). Molecular evidence also suggests that the genotype of VHS associated with Baltic herring is transmitted to gadoids by ingestion, as both species were found to be culture positive for VHSV genotypes-II and-III although they do not shoal together.

Another evidence of ingestion being an important mechanism for the transmission of VHSV was the outbreak of VHSV genotype-III in turbot [*Scophthalmus maximus* (L.); Stone et al., 1997; Snow and Smail, 1999] in Scotland and Ireland in the mid-1990s. In both of these cases, the turbot showed the classical signs of acute VHS (Fig. 3.3) and were culled due to

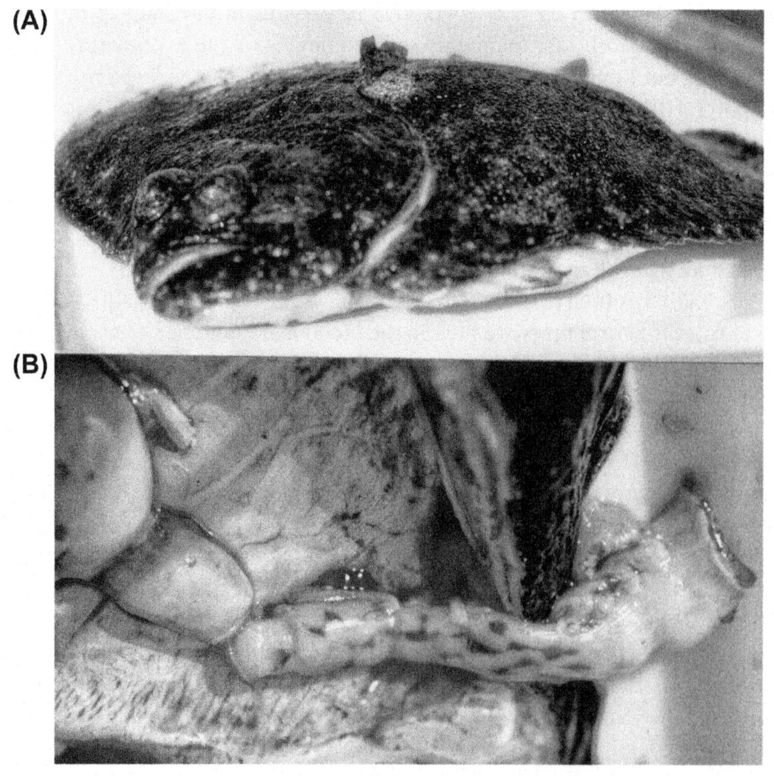

FIGURE 3.3 Acute VHSV in a Turbot. Note the swollen body due to edema from the leaky blood vessels (A) and the internal hemorrhages (B). *Photo credit Bricknell, I., Crown Copyright.*

the severity of the disease outbreak. Epidemiological investigation recovered the virus from the frozen herring diet the fish were being fed at both locations, and this was proposed as the route of entry of the virus to the farmed turbot.

3.3.6 Modes of Transmission of Bacterial Pathogens Between Fish

If viruses are the simplest known biological entities (Harper, 2011) bacteria are the oldest living group of organisms. The earliest evidence of bacteria are found in rocks nearly 3.5 billion years old, and appear to be related to cyanobacteria (Schopf et al., 2002). Unlike viruses, bacteria contain both DNA and RNA suspended in a biochemically and physiologically active cytoplasm actively carrying out all of the processes required for life (Table 3.1). Bacteria dominated the history of life on earth since then and form the ecological basis that permits the amazing diversity of organisms that exist today.

3.3.7 Bacteria

Although the evidence for bacterial infections are rare in the fossil record, as bacteria do not fossilize very well and many bacteria, even pathogens, ultimately contribute to the decomposition of a dead organism, thereby breaking down any evidence of soft tissue pathology an infection may have caused. Evidence for paleo-infections usually involves the organism suffering from an infection within it's hard tissues, bone, cosmine, dentine, wood, etc. which leaves evidence of an infection. Possibly one of the best known example of this is a series of fatal abscesses evident in a skull from the Broken Hill formation in Zimbabwe. The fossil skull (Fig. 3.4) shows evidence of pyorrhea or alveolar abscesses that eroded into the skull and breached the cranium. The individual probably died of sepsis or an intercranial bleed as the abscess appears to breach the middle meningeal artery (Cooper, 1988).

Unlike higher vertebrates, there are no reported bacterial diseases of fish in the fossil record. This does not mean that bacterial diseases did not occur in ancient fish. It is almost certain that they did, however, they left little evidence in fossil fish, and if any evidence was left, it probably stays overlooked in museum cabinets.

3.3.8 Bacterial Pathogens

Bacterial pathogens are probably the most common cause of diseases currently found in fish. Bacteria are often classified as obligate pathogens (those that cause infections), facultative pathogens (those that can be pathogens), opportunistic pathogens (those that can take advantage of certain conditions within the host to cause disease), and nonpathogenic [bacteria that have never been associated with disease (Hurst, 2016)]. See Table 3.2 for examples in fish.

Although many bacteria fall into the category of opportunistic pathogens (Austin and Austin, 2016) fish have excellent examples of all three types of pathogens. However, one gray area is the question when does a nonpathogenic organism become an opportunistic pathogen? It is challenging to define this. Table 3.2 shows the *Nitrosomonas* spp. as a group of bacteria that converts ammonia to nitrite, none of which have been reported as pathogenic in any organism so far. However, other examples are not so clear. Under certain circumstances many bacteria that are considered nonpathogenic may become pathogenic if

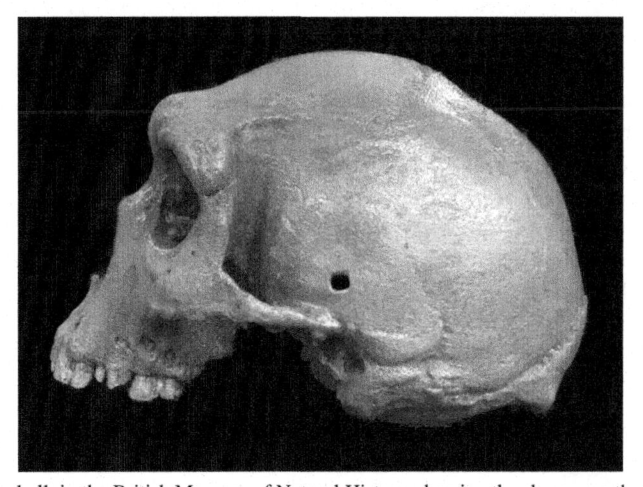

FIGURE 3.4 *Homo heidelbergensis* skull, in the British Museum of Natural History, showing the abscess on the left side of the cranium that ruptured the meningeal artery. *Photo Bricknell, I.*

TABLE 3.2 Examples of Pathogen Types in Fish

Classification	Pathogen	Disease
Obligate	*Renibacterium salmonariun*	Bacterial kidney disease
Facultative	*Aeromonas salmonicida*	Furunculosis
Opportunistic	*Vibrio* spp.	Vibriosis
Nonpathogenic	*Nitrosomonas* spp.	–

certain criteria are met (Austin and Austin, 2016). These circumstances may be as simple as an enrichment of nutrients in the water that allows the microbial burden to increase beyond the normal bacterial load (Wedekind et al., 2010), or if the fish become immunosuppressed. Immunosuppression can occur naturally due to many reasons, such as the stress of spawning, extreme environmental conditions such as high or low temperature (Bowden et al., 2007; Lamkova et al., 2007; Breckels and Neff, 2013), or the active immunosuppression due to immunomodulation by a parasite (Holm et al., 2015). Under such circumstances the bacteria may utilize this opportunity to ultimately infect and cause sepsis in a host.

3.3.9 Methods of Bacterial Transmission

3.3.9.1 Behavioral Changes

There are many behavioral change associated with bacterial infections in fish. Infected fish often darken, which allows them to absorb more solar radiation and warm up their bodies, allowing their immune systems to function at their optimal rate. They may also seek out warmer habitats for the same reason. Some fish species may also become less active to conserve energy, hypothetically allowing more energy to be available to be utilized by the immune system. Other individuals may seek out more oxygenated water or gasp at the surface to compensate for gill pathology or to provide a better oxygen saturation of the tissues. However, none of these changes in behavior are directly caused by the bacteria to permit transmission, but are parts of the innate immune system being used by the fish to increase its chances of surviving the infection.

So far there are no confirmed mechanisms of bacterial fish pathogens changing the behavior of the host to increase transmission. However, there are examples where bacterial pathogens are triggered to replicate or increase the shedding rate at certain times, such as an increase of sex hormone associated with the reproductive cycle, but these are related to fish's physiology and are not induced by the pathogens.

3.3.9.2 Shedding

By far the commonest way of pathogenic fish bacteria spreading between hosts is by shedding into the environment followed by subsequent adherence, penetration, and entrance into a susceptible host. There are many examples for this, but one of the best studied is *Aeromonas salmonicida* subsp. *salmonicida*, the causative agent of furunculous in Atlantic salmon (*Salmo salar* L.) and other salmonids (Menanteau-Ledouble et al., 2016).

Once a susceptible host is infected with this pathogen, the bacteria rapidly multiply at the site of the infection causing a localized focal infection. The infection can then take two courses. First, it can break out from the focal infection and cause a widespread acute sepsis which typically leads to the death of the animal within a few days. Death usually occurs due to circulatory collapse and respiratory failure (Fig. 3.5A). Second, the infection does not break out from the initial focal infection and goes on to form an abscess. As the abscess progresses, the center of the mass liquefies due to the action of powerful proteases secreted by *A. salmonicida* subsp. *salmonicida*, forming the classical furuncle from which the infection gets its common name. An excellent review of the pathology of *A. salmonicida* can be found in "Furunculosis: Multidisciplinary Fish Disease Research" (Bernoth et al., 1997). Typically, the furuncle form of the disease causes a long-term chronic infection allowing the infected fish to shed bacteria into the environment for a long period (Fig. 3.5).

During the acute phase of the disease shedding can reach very high numbers of bacteria per unit time (Rose et al., 1989). The actual mechanism of the shedding is poorly known for *A. salmonicida* subsp. *salmonicida*. The bacteria were detected in the bodily secretions of the fish such as feces, mucus, urine, and blood (from petechial hemorrhages in the skin). Rose et al. (1989) in a series of elegant experiments demonstrated that the rate of shedding from an acutely infected Atlantic salmon was in the order of 2×10^3 CFUs/g/hr. Once dispersed in the environment, the bacterium has two opportunities. First, it can adhere to another fish by hydrophobic interactions (Bricknell, 1995; Bricknell et al., 1996) or by ingestion of an infectious particle. Second, it can enter a biofilm and exist as an environmental organism for long periods of time (Bricknell,

FIGURE 3.5 Chronic furunculous in a brown trout, *Salmo truttae* L.: (A) note the large swelling full of fluid (B) the acute version of the disease with a superficial skin lesion with numerous petechial hemorrhages and lacking the large furuncule. *Photo Bricknell, I., Crown Copyright.*

1995; Mainous et al., 2011; San et al., 2012). Earlier work identified a viable, nonculturable state of *A. salmonicida* subsp. *salmonicida* as well (Effendi and Austin, 1994, 1995). Here the bacteria enters into a, what can best be described as, resting state. The mechanisms that trigger the bacteria's entrance into or emergence from the viable nonculturable state of *A. salmonicida* subsp. *salmonicida,* are not well understood, and research in this aspect of *A. salmonicida* subsp. *salmonicida* ecology needs further work.

It is hypothesized that the resuspension of bacteria into the water column by biophysical action on the biofilm suspends the viable infectious particles into water with potential to adhere to a new susceptible host.

A third infectious pathway was postulated in which an infectious particle is ingested by the host and colonizes the digestive tract (Hiney et al., 1994; Hiney and Smith, 1998). Once in the digestive tract, it adheres and starts to replicate, forming a nonpathological carrier state. However, carriers were shown to shed considerable amounts of bacteria in their feces (Gustafson et al., 1992; Obrien et al., 1994).

Unlike viruses, once the host dies the bacteria can utilize the dead animal tissues as a nutrient source and continue to grow and replicate. This state represents a switch of the bacteria from a pathogenic physiology to a necrophytic stage. This may require the induction of a new set of enzymes to digest the dead host, but this strategy permits many more potentially infectious bacteria to be shed often until the dead host is fully decomposed, or the conditions within the corpse become unsuitable for replication.

3.3.9.3 Adhesion

In addition to shedding, adhesion is another very important mechanism to transmit infection among fish. The adhesion factors can be very varied (see Table 3.3). However, the variations can be simple hydrophobic interactions depending on the negative or positive charge the bacteria carry on their surface that permits the pathogen to be attracted by the host. Various types of molecular interactions take place once the bacteria reach the host, such as the bacteria may mimic molecules that bind to the receptors on the host (or vice versa). The bacteria can also grow fimbriae that physically anchor the bacteria to the host, or proteins called lectins that bind to the carbohydrate moiety of the host cell surface glycoproteins. Specific mechanisms of adhesion will be discussed later in this book in chapter dealing with specific diseases.

3.3.9.4 Host-Seeking

Host-seeking behavior is not often considered an important mechanism for the transmission of bacteria between hosts, although bacteria are known to possess structures and mechanisms that permit locomotion. *Vibrio anguillarum*, for example, has well-developed flagella (Toranzo and Barja, 1990) and *Flavobacterium psychrophilum* has the ability to glide over surfaces (Vatsos et al., 2001; Vatsos et al., 2006). There are many observations of motile bacteria being attracted to the semiochemical cues from a host organism. These have taken the form of seeing bacteria move down a concentration gradient of semiochemicals secreted from the host, where they can attach and then invade the tissues.

TABLE 3.3 Adhesion Factors of Bacteria

Adherence Factor	Description
Adhesin	A surface structure or macromolecule that binds a bacterium to a specific surface
Receptor	A complementary macromolecular binding site on a (eucaryotic) surface that binds specific adhesins or ligands
Lectin	Any protein that binds to a carbohydrate
Ligand	A surface molecule that exhibits specific binding to a receptor molecule on another surface
Fimbriae	Filamentous proteins on the surface of bacterial cells that may behave as adhesins for specific adherence
Type 1 fimbriae	Fimbriae in *Enterobacteriaceae* which bind specifically to mannose-terminated glycoproteins on eucaryotic cell surfaces
S-layer	Proteins that form the outermost cell envelope component of a broad spectrum of bacteria, enabling them to adhere to host cell membranes and environmental surfaces in order to colonize them
Glycocalyx	A layer of exopolysaccharide fibers on the surface of bacterial cells which may be involved in adherence to a surface—sometimes a general term for a capsule
Capsule	A detectable layer of polysaccharide on the surface of a bacterial cell which may mediate specific or nonspecific attachment
Lipopolysaccharide (LPS)	A distinct cell wall component of the outer membrane of Gram-negative bacteria with the potential structural diversity to mediate specific adherence. Probably functions as an adhesin
Teichoic acids and lipoteichoic acids (LTA)	Cell wall components of Gram-positive bacteria that may be involved in nonspecific or specific adherence

Modified from Todar, K., 2011. Textbook of Microbiology. Mechanisms of Bacterial Pathogenicity (2016).

One of the best studied mechanisms of this is the ability of *F. psychrophilum* to glide over the surface of fish eggs and then enter the micropile to infect the embryo inside. Adams et al. first reported this in rainbow trout eggs and produced the first real-time observations of host-seeking behavior in a pathogenic bacteria of fish. Fig. 3.6 shows the sequence of infection of *F. psychrophilum* in a rainbow trout egg.

3.3.9.5 Carriers

The first carrier state ever documented was the famous case of Typhoid Mary. An individual who after recovering from clinical typhoid caused by the virulent pathogens *Salmonella typhi* continued to shed the pathogen in her feces for the rest of her life. Her story is quite a sad one. Mary Mallon (later to be nicknamed Typhoid Mary) was born in 1869 and in 1900 moved to New York as a cook. It is recorded that in 1900, within 2 weeks of arriving in New York, she contracted typhoid fever and subsequently recovered. During her convalescence from this infection, the bacterium colonized her gall bladder, where it would remain until her death in 1938. The first outbreak of typhoid attributed to her occurred in Manhattan in 1906 in a household where she worked as a cook. In 1906, she moved to Oyster Bay, Long Island, with a new family as cook. Two weeks later, 10 of the 11 family members had contacted typhoid. After leaving this post she changed jobs a further 3 times and each time an outbreak of typhoid affected her new employers (Dex and McCaff, 2000).

When New York City Health Department investigated these cases of typhoid, Mallon was identified as the major epidemiological link between all the cases, which led to her being quarantined for the first time in 1907 and the publicity earned the name of "typhoid Mary." She was released from quarantine in 1910 under the understanding she would not work as a cook anymore. She was then employed as a laundress, but finding the lower rate of pay unacceptable she changed her name to Mary Brown and resumed her career as a cook. She worked as a cook for the next 5 years and numerous typhoid outbreaks occurred in the localities where she worked, but she changed jobs frequently and her true identify went undetected. In 1915, Mallon started another typhoid outbreak at the Sloane Hospital for Women in New York. At least 25 people were infected with two confirmed deaths (Dex and McCaff, 2000). This led to her true identity being discovered and a second period of quarantine being enforced. This incarceration turned out to be a life sentence for Mary Mallon as she died in

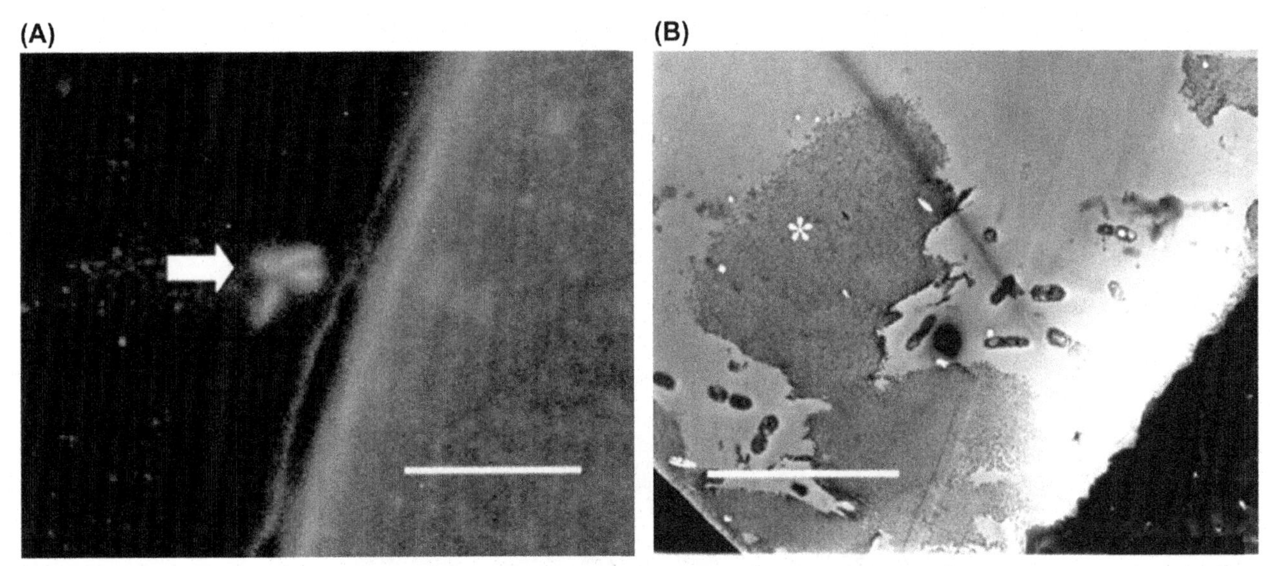

FIGURE 3.6 (A) Identification of *Flavobacterium psychrophilum* (*white arrow*) on the surface of an infected fertilized rainbow trout egg using immunofluorescent antibody technique 30 days postfertilization (bar = 10 μm). (B) Transmission electron microscopy. Mixed microflora on the surface of an infected fertilized rainbow trout egg; destruction of mucous layer by proteolytic bacteria, 30 days postinfection (bar = 10 μm). *Photo Credit Vatsos, I.N., Thompson, K.D., Adams, A. Used with permission.*

1938 of pneumonia still isolated from the general public. Cultures from her gall bladder taken at autopsy were positive for *S. typhi*, even 38 years after she had the clinical disease.

As mentioned above, carrier status is also important for the continued maintenance of *A. salmonicida* subsp. *salmonicida* in the environment. For many years the gold standard method of detection was the stress test (Hiney et al., 1994; Cipriano et al., 1996, 1997; Bullock et al., 1997). The test population of fish were injected with prednisolone and temperature stressed at 18°C. Although moderately successful in catching asymptomatic populations, the method was considered to be challenging to carry out and the test fish often succumbed to opportunistic pathogens, causing considerable welfare issues. It also failed to identify where the *A. salmonicida* was residing in the host. Mainly because when the symptoms appeared the *A. salmonicida* infection had already become systemic.

One of the most interesting outcomes of molecular ecology in fish pathobiology was the use of polymerase chain reaction (PCR) test to identify the location of the tissues that *A. salmonicida* colonizes to form the carrier sate, and the likely method of how carries infect naïve fish. Hiney et al. (1994) (Obrien et al., 1994; Hiney and Smith, 1998) used PCR to test the tissues of fish from populations that had been identified as *A. salmonicida* carriers using stress tests. During this series of experiments, they identified the intestine as the organ that was colonized by *A. salmonicida* in asymptomatic carries. Not only did they confirm this by PCR, they were also able to recover *A. salmonicida* by culture for the PCR-positive individuals. It appears that during periods of environmental or physiological stress, the number of viable *A. salmonicida* increases in the intestine and is shed into the environment via feces. These infectious particles then go on to infect naïve fish either by injection via the classical fecal–oral route (as we saw with Typhoid Mary) or by adhesion onto the mucosal surface of a susceptible fish.

3.3.9.6 Vertical Transmission

Vertical transmission of bacterial pathogens is a well-known phenomenon, e.g., Hutchison's teeth in syphilic neonates or Brucellosis in Caprinae. Fish have several examples of vertically transmitted bacterial disease. Two of the best studies are *Fransicella* (Pradeep et al., 2016) in tilapia and *Renibacterium salmoninarum* in salmonids (Fryer and Lannan, 1993; Pradeep et al., 2016). Although the outcome of vertical transmission is the same there are two main mechanisms to achieve this. First, the gametes become infected and the egg becomes diseased at fertilization with pathogen either coating the egg or getting attached to the spermatozoids. The pathogen enters the egg through micropile usually at the time of fertilization. Second, during spermatogenesis or oogenesis, when the gamete-forming cells are infected with the pathogen, transferring the pathogen to the gametes, and infecting the eggs prior to fertilization (Fryer and Lannan, 1993). *Francisella noatunensis* subsp. *orientalis* has been shown to favor the former mechanism while *R. salmoninarum* the latter.

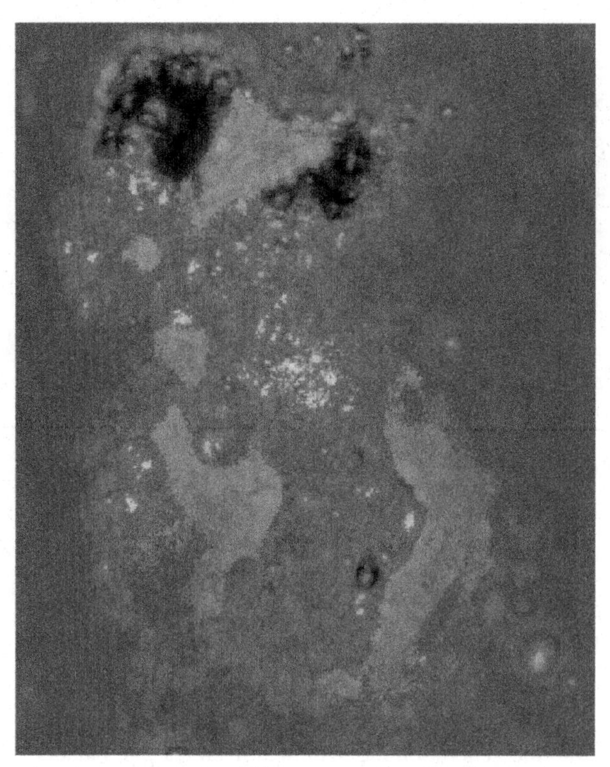

FIGURE 3.7 Confocal image of an *R. salmoninarum* infected head kidney macrophage. Red (*light gray in print version*) is antibody labeled bacterial cells and blue (*gray in print version*) is the nuclear material of the macrophage. *Photo Credit Bron, J./Bricknell, I.*

Although vertical transmission of *F. noatunensis* subsp. *orientalis* has long been suspected in populations of Tilapia (*Oreochromis* spp.), it was only conclusively demonstrated by Pradeep et al. (2016). This study was the first to demonstrate vertical transmission of *F. noatunensis* subsp. *orientalis* and found the pathogen in the milt, unfertilized eggs, fertilized eggs, and offspring up to 30days old. Prevalence in the examined populations was between 40% and 60%, and probably contributes to the very high incidents of *F. noatunensis* outbreaks in farmed tilapia (Brevik et al., 2011; Leal et al., 2014; Lin et al., 2016b). However, although long suspected, a carrier state had not been described in tilapia, as in Atlantic cod (Ottem et al., 2008).

R. salmoninarum has been long considered to be an obligate pathogen and an obligate intracellular organism (Fryer and Lannan, 1993). Classically, it is found in the cytoplasm of the cells of infected fish, epically the kidney tissues (Fig. 3.7), while chronically infected fish shed *R. salmoninarum* cells in their urine (McKibben and Pascho, 1999; Faisal et al., 2012), but it is also found in the gametocytes that go on to form the sperm and eggs (Yoshimizu, 2016). This somatic infection often means that the gametes themselves are infected as they are produced and hence an infected sperm fertilizing an uninfected egg, or vice versa, will lead to an infected embryo. The pathobiology of the infected embryo then has two ways to go forward. First, the bacteria overwhelm the embryo which dies and sheds *R. salmoninarum* into the environment (Yoshimizu, 2016). Second, *R. salmoninarum* infects the tissues and the embryo goes on to develop normally but with *R. salmoninarum* in the vast majority of its cells (Sudheesh et al., 2007). These infected animals can be considered to be chronic shedders shedding *R. salmoninarum* cells constantly throughout their lives.

3.3.9.7 Vectors

Arthropods are a major vector of bacterial diseases. Ticks spread typhus and Lyme's disease, but the best known example of this is Bubonic plague where there is a close association between the pathogens (*Yersinia pestis*) and its insect host the fleas, particularly *Xenopsylla* spp. (Lotfy, 2015; Pechous et al., 2016). The disease is transmitted by a flea vector. When the flea bites an infected host it becomes infected with *Y. pestis*. The organism colonizes the proventriculus organ of the flea, a structure that fleas use to rupture red blood cells of its meal. The bacteria form an extensive biofilm on the proventriculus and occlude the entry of food into the flea's digestive tract (Fig. 3.8). This in turn causes the flea to starve, thereby increasing an infected flea's attempts at feeding on a host. When the infected flea attempts to bite a host in a vain to gain a meal, it

FIGURE 3.8 *Xenopsylla cheopis* infected with the *Y. pestis* bacterium: The proventriculus [red (*light gray in print version*)] of this flea is occluded and during feeding the pathogen is vomited into the wound. *Image https://en.wikipedia.org/wiki/Yersinia_pestis#/media/File:Flea_infected_with_yersinia_pestis.jpg.*

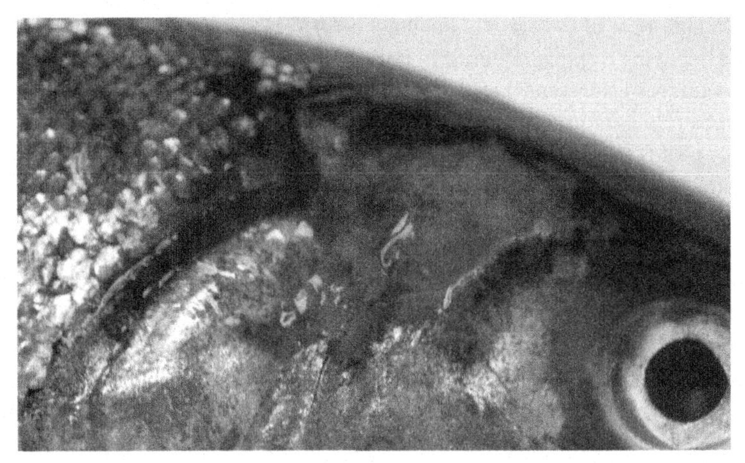

FIGURE 3.9 Sea lice lesion breaching the epithelia of an infected Atlantic salmon. *Photo credit Bricknell, I.*

regurgitates some of its upper digestive tract contents into the bite, along with an infectious dose of *Y. pestis*, thus continuing the infectious cycle.

No such close association of a bacterium with its vector has been described in fish. It has been suggested that sea lice (*L. salmonis*) may transmit *A. salmonicida* between infected and noninfected hosts via their digestive tracts or on their surfaces (Barker et al., 2009) transmission has not been confirmed. Many ectoparasites of fish (such as the argulid crustaceans and siphonostomatoida) cause significant lesions in fish (Fig. 3.9) and these lesions may become infected with the opportunistic infections including *Vibrio* sp., and it is not inconceivable that the source of these opportunistic bacteria are the parasites' epidermis or fecal material.

3.3.9.8 Predator–Prey Relationships

Similar to viruses (Section 3.3.5), predator–prey relationship play an important role in the transmission of bacteria between fish and between trophic levels. Although not as well studied as VHSV, oral transmission of furunculosis and mycobacteriosis are also shown to occur. Usually the debilitated fish shows marked behavioral changes from its conspecifics making it a more obvious target for predation. Surprisingly, this was not as well studied in fish bacterial pathogens as in viral diseases.

Researchers have (Ringo et al., 1992; Hjelm et al., 2004) demonstrated that the larval turbot were invaded through the intestine by *Vibrio anguillarium* that was attached to their cuticle. More recently, Fouz et al. (2010) showed the fecal–oral ingestion of *Vibrio vulnificus* biotype 2 in the diet. This is not all that surprising as *Vibrio* sp. has long been suggested as having a major uptake route into the host via the intestine.

3.4 TRANSMISSION OF FUNGAL DISEASE: THE WATER MOLDS

The characteristic fungal pathogen of fish *Saprolegnia parasitica* was taxonomically classified under fungi during my undergraduate days, while now it falls under algae. However, this section will attempt to outline the mechanisms this eclectic group of eukaryotic organisms uses to remain the number one cause of all nonanthropogenic fish deaths globally.

One thing that simplifies this part of the chapter is the mode of transmission of this complex group of pathogens as they share similar ways of transmission to their fish hosts. However, the study of transmission in fish remains poorly understood and many fish models for fungal challenge often require severe near-surgical methods to infect the host. There are no descriptions of host-seeking in the literature, and the behavioral changes described were induced by their pathology rather than the induction of specific lesions in the neurological tissue as described for nodaviruses (Muroga, 1995; Nakajima et al., 1998; Munday et al., 2002; Bricknell et al., 2006). These include scarification or some other method of breaching the integument before the pathogen is introduced (Pottinger and Day, 1999; Stueland et al., 2005; Das et al., 2013; Firouzbakhsh et al., 2014).

3.4.1 External "Fungal" Pathogens

The classical white, fluffy fungus growing on eggs or sick fish is the bane of many freshwater fish farmers and is usually the last stage of a complex series of events where a saprophytic organism has become an opportunistic pathogen. Normally the water molds are a common component of the aquatic habitats' biota. These organisms spend most of their time harmlessly breaking down the dead and decaying organic material in the environment, such as plant matter or the old dead bacteria in biofilms. The breakdown products of the water molds diet often go on to form mulm, or, as also termed, detritus—a key component of many food webs (Holgerson et al., 2016).

What triggers these water molds to change from a saprophyte into a pathogen and ultimately a necrophyte is complex. This series of events along this path typically starts with a drastic change in water quality, i.e., water often becomes polluted, usually with high amounts of dissolved organic compounds and organic particles; sediment becomes suspended; or diatoms bloom in the water column. The pathogenic fish "fungi" either utilizes these nutrients to reach unnaturally high population densities where they can colonize the host, infect by ingestion, or the sediment or silica skeletons of diatoms damage the epithelium to cause the initial infection (Yang and Albright, 1992; Kent et al., 1995).

Here we have two mechanisms going on; one is the classics uptake of an infectious particle from the aquatic environment and the other is by adhesion onto the hosts epithelia, inhalation, or ingestion. The classical adhesion method will be considered when we discuss the shedding of spores of *S. parasitica* a little later. However, when water conditions cause a diatom bloom or there is an environmental event, such as a storm, that suspends considerable amounts of sediment, a new mechanism occurs that can be considered a major factor in permitting the water molds to change from saprophytes to opportunistic pathogens (Hurst, 2016).

As mentioned above, one of the great difficulties of working with the water molds is the issue with infecting the host fish. Often healthy fish are very refractory toward infection with this group of organisms unless the integument is breached. The skin forms a part of the innate immune system (Biller-Takahashi and Urbinati, 2014; Sirisinha, 2014; Dezfuli et al., 2016) and is very effective at keeping many pathogens out by forming a physical barrier that has to be breached by the invading organism. The water molds seem to be very poor at doing this under normal circumstances (Hardham and Hyde, 1997; Ali, 2005). However, a widely accepted hypothesis states that after diatom blooms or a storm event that suspends sediment, the siliceous skeletons of the diatoms or sharp sediment particles, like fine silt or sand particles, abrade the skin and breach the integument, effectively scarify the surface of the fish, bypassing the innate immunity provided by the mucus and the skin. As diatom blooms and storm events are associated with a high organic load, disturbed detritus and other potentially infective fomites, respectively these provide a mechanism for the water molds to adhere and invade the host.

3.4.2 Egg, Fry, and Larvae Infections

Although there are no known vertically transmitted water molds, eggs, fry, and larvae remain the most vulnerable stages of a fish's life cycle. This is partly due to the relatively poorly developed immune system.

Again, water quality often plays a part in egg infections. Typically, fish eggs are laid in very specific environments depending on the species, e.g., the open ocean for Bluefin tuna, in gravel redds for Atlantic salmon, and even incubated in the mouth of one of the parents as is the case in *Cyphotilapia frontosa* (Balon, 1991). The eggs and, more specifically, the maternal investment in the eggs plays an important role in preventing the opportunistic water molds from infecting them. Infection usually occurs when things go wrong with water quality. For example, salmon eggs become vulnerable to infection if the water supply to them is interrupted, depriving them of the oxygen, fast-flowing, oxygen-rich waters they need to develop normally; or if a large number of the eggs in a spawn die for another reason such as vertically transmitted *R. salmoninarum* infection (Ali et al., 2013; Jiang et al., 2013; Liu et al., 2014; Songe et al., 2016). In the latter scenario the large numbers of dead eggs provide an excellent substrate for water molds to germinate on as a natural part of the decay processes. The poor water quality allows spores of the water mold in particular *Saprolegnia* spp. to settle out of the water column onto the surface of the egg where they germinate. The vegetative hyphae initially form a complex biofilm on the egg along with other bacteria. The hyphae then go on to penetrate the chorion and invade the developing embryo (Fig. 3.10). Once the chorion is breached, the embryo has only a few hours to live until the proteolytic enzymes secreted by the pathogen dissolve it (Songe et al., 2016).

Fry and larvae have rather different issues to overcome. Often a fishes' immune system is not fully functional until metamorphosis when the thymus appears and immune cell differentiation begins (Bricknell and Dalmo, 2005). Additionally, the integument of larval fish is often very thin, only one- or two-cell thick layer. If an infectious particle of one of these water molds lodges in the gills or attaches to the skin of the larval fish the hyphae meet very little resistance from the integument and the poorly developed immune system of the larvae does not prevent the pathogen penetrating and ultimately killing the larval fish.

3.4.3 Shedding

Shedding is an important method for transmission for this group of pathogens. However, few studies have been carried out on the mechanisms of shedding and the subsequent uptake by naïve hosts. What is clear is that there are two mechanisms of shedding that the water molds do effectively.

Vegetative shedding is common in these pathogens. Usually vegetative growth leads to the development of a mass of hyphae that grows externally from the tissues of the fish (or egg) into the environment (Fig. 3.10). This mass of tissues is subjected to the physical changes of the environment, such as hydrodynamics, as the fish swims through the water. If the pathogen infects the gills, such as *Branchiomyces sanguinis* the causative agent of gill rot, the movement of the gill tissues during respiration will increase the likelihood of hyphae breaking off or necrotic infected tissue breaking away from the gill and entering the environment. *B. sanguinis* is an interesting water mold from the pathogenesis point of view because it specifically infects the gill tissues causing the fish disease gill rot. The vegetative shedding of viable hyphal material from the gills is an important method of transmitting *B. sanguinis* between susceptible hosts. Here inhalation is the key mechanism

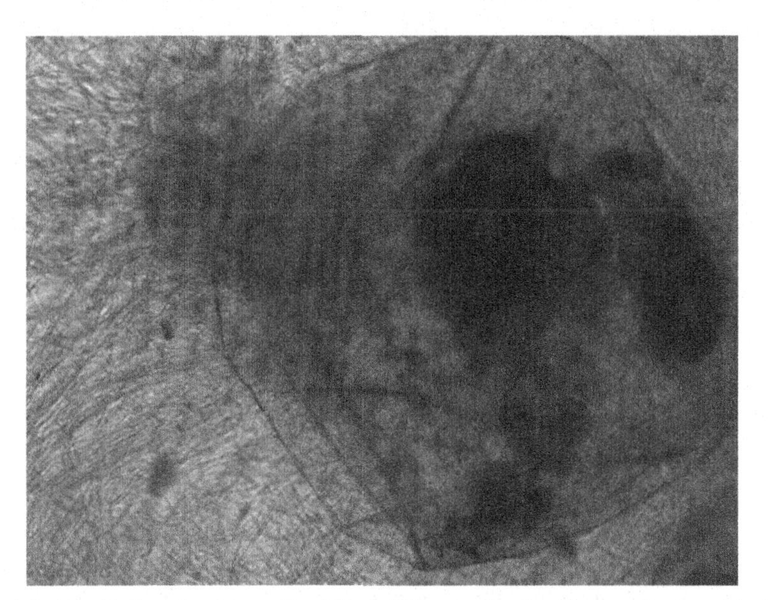

FIGURE 3.10 Zebra fish egg infected with *Saprolegnia parasitica* 24 hpi at 10× magnification. The mycelia were stained for 15 min with 250 μm fluorescent brightener 28 (CW). Scale bar is 200 μm. *Photo credit Liberman, K., University of Maine. Used with permission.*

for uptake of the vegetative tissue from the water column and lodging it in the gill where it continues to grow infecting the new host (Paperna and Smirnova, 1997; Khoo et al., 1998; Bocklisch and Otto, 2000).

Another method that the water molds use to shed infectious particles is sporulation. Sporulation of the water molds is unpredictable as the mechanisms that trigger sexual reproduction are not well understood or described. Indeed, the problems with inducing sporulation have hampered the taxonomy of the water molds, at least until the advent of molecular biology and the induction of gametes, a key step in sporulation, is required for the morphological identification of many water molds (Hardham and Hyde, 1997).

If *B. sanguinis* becomes sexual during an infection, it produces its sexual structures at the terminal end of the hyphae. These structures are called the oogonium and the antheridium—the female and male gamete producing structures. The antheridium wraps itself around the oogonium and fertilizes the female gametes, which divide to produce the oospores. The oogonium ruptures when it becomes mature and releases the mature spores in the water column. The spores then germinate when they encounter a suitable environment. In the case of *B. sanguinis*, this could be a gill, but it could also be decaying organic matter equally. However, inhalation remains the primary mechanism through which *B. sanguinis* infects a new host (Paperna and Smirnova, 1997; Khoo et al., 1998; Bocklisch and Otto, 2000).

If the spore, or vegetative tissue, does not encounter a potential fish host it can germinate on decomposing organic matter or biofilms, and return to a saprophytic life style.

3.4.4 Biofilms and Secondary Infection of Open Wounds

The water molds rarely cause primary infections, one of the exceptions is discussed above. However, they are common component of the complex microbial environment seen in open lesions following a wound or another infection. Again, the integument is often breached by another pathogen, or predator, and becomes colonized by a complex microbial community which is the classic beloved environment of *S. parasitica.* Vegetative tissue or a spore lands on the lesion, germinates, and invades the superficial tissues of the host. Although not the primary cause of the infection, colonization with *S. parasitica* is frequently the terminal stage as the hyphae invade the superficial tissues of the host and prevent the lesion resolving. As the *S. parasitica* grows the epidermis is eroded, and often the lesion becomes so large the fish dies of osmotic failure rather than sepsis (Ali et al., 2013).

The role biofilms play in the pathogenicity of the water molds, and other diseases, and has only recently been identified as a major factor in pathogenesis; it is expected that these complex microbial interactions will become significant in controlling the major impact these organisms have in aquaculture in the near future (Ali et al., 2013).

3.4.5 Internal Infections and Behavioral Changes

Some fish "fungal" infections only appear to have an internal etiology in fish. One such example is *Exophiala salmonis*, (Richards et al., 1966). This is an enigmatic disease first described in 1966 (Richards et al., 1978) and seem to be a common disease of Atlantic salmon reared in seawater, and is uncommon in wild fish. When compared to most of the marine water molds, it does not form external vegetative structures on an infected host, possibly due to the high osmotic pressure exerted by seawater. Classically, salmon suffering from *E. salmonis* have a distended abdomen and a dry peritonitis (Otis et al., 1985; Madan et al., 2006; Yoon et al., 2012). The peritoneum is distended with a mass of *E. salmonis* vegetative tissue. Typically, the posterior kidney is chronically infected and becomes enlarged, a large amount of hard nodules and granulomas form in the liver, spleen, and the majority of other tissues. This changes the behavior of the fish simply by debilitating it due to chronic renal failure and morphological changes.

Fish in the last stage of *E. salmonis* infection are readily predated by piscivorous animals. However, transmission by this route has not been demonstrated to occur naturally. Experimental challenges have indicated that the uptake of *E. salmonis* into a naïve host is by ingestion (Richards et al., 1978; Otis et al., 1985; Uijthof et al., 1997; Madan et al., 2006; Yoon et al., 2012). However, why it is such a common disease of farmed salmon, especially in Europe, remains a mystery as these fish do not have the opportunity to predate the infected fish, at least not in the numbers required to see the high prevalence reported in some outbreaks (Richards et al., 1978; Rameshkumar et al., 2013), suggesting that *E. salmonis* has another, as yet undescribed, mechanism of transmission, possible with an intermediate host, vector, or via a spore.

3.5 PARASITE TRANSMISSION

Parasites are the undisputed experts at novel methods of transmitting themselves between hosts. Parasites can transmit themselves using variations of all the methods described above. They also do some absolutely wonderful things to ensure their survival and reproduction. Some will change the behavior of the intermediate host making them effectively suicidal so they are eaten by the definitive host; some are sexually transmitted, not only infecting the gametes and embryos by vertical transmission, but also infecting adults during the act of reproduction; others secrete hormone analogues to sex to reverse their hosts, usually castrating males and making them nonfunctional females. It is fair to say that parasite represent the pinnacle of evolutionary biology when it comes to diverse reproductive strategies.

Parasite infections are not considered to be carriers, although in the strictest sense a host with a parasite is a carrier of that parasite. However, it may not be infectious to others individuals of the same species during that time and cannot be considered as an infectious carrier in the same way an asymptomatic carrier of a virus or a pathogen is.

Parasites can be vectors for other pathogens as discussed above, but they are rarely vectors for other parasites although they can change the behavior of intermediates (or definitive hosts) so they act as a vector for the juvenile stages till they find the definitive host. They can also change the behavior of the definitive host to increase the risk of shed eggs or larvae finding a new intermediate host.

3.5.1 Parasite-Induced Host Behavior Changes

The acanthocephalan worms are a very common group of fish parasites with evolutionary affinities with the rotifers (Near, 2002; Mark Welch, 2005). They also have some of the most complex reproductive cycles known. Indeed some are so diverse, the adults have a different morphology in different hosts. The acanthocephalan *Neoechinorhynchus rutili* (Walkey, 1962) has a completely different morphology in cyprinid fish and salmonid fish that for decades they were considered to be different species. It was not until molecular biological techniques were applied to *N. rutile* that this phenomenon was understood.

Acanthocephalans are often considered to be effective brain jackers for changing the behavior of the intermediate host. The intermediate host is an organism that a parasite infects leading to an eventual uptake by the definitive host (the definitive host is where reproduction takes place). *N. rutile* typically infects two types of intermediate hosts, ostracods and gammarids. These two groups of crustacean are infected when they eat eggs from the infected fish that enter the water column from infected feces. The eggs hatch in the digestive system of the crustacean and the larval *N. rutile* migrates into the tissues. Here it forms a cyst and waits for the intermediate host to be eaten. To speed up this process the cyst interferes with the behavior of the crustacean intermediate host. Infected ostracods and gammarids change their behavior instead of being cryptic animals that avoid open water and bright light, their normal behavior, an infected intermediate host happily enters open water and are not deterred by bright light. The mechanism for this is not well understood in *N. rutile* but it is believed that the parasite suppresses the neuroendocrinological pathways that trigger flight responses. It may also suppress the production of the photosensitive pigment in the eyes of the copepods (Dezfuli, 1996). Through whichever of these mechanisms it exploits the biology of its host, it significantly changes the host. Parasitized copepods and gammarids are estimated to be predated at a rate of three times higher than an uninfected intermediate host (Walkey, 1967).

This is just one example of the parasite changing the behavior of the host to permit transmission between intermediate and definitive hosts. As we shall see later, some parasites that use fish as an intermediate host can also affect the fish's behavior making them more susceptible to predation.

3.5.2 Shedding

Parasites with direct and indirect life cycles often shed large amount of infectious stages into the environment. This strategy is not just limited to single cell parasites such as myzosporidian, *Kudoa thyrsites*, (Gilchrist, 1924) that causes kudoa, or the classic ciliate parasite of fish *Ichthyophthirius multifiliis*, Fouquet, 1876, the cause of white spot, one of the most common freshwater parasite infections. It is also adopted by metazoans too like the salmon louse (*L. salmonis*, Kroyer, 1837) or the three-spined stickleback (*Gasterosteus aculeatus* L.) tapeworm *Schistocephalus solidus*.

The *r*/*k* selection hypothesis examines the trade-off between producing a high number of offspring of low quality, e.g., with very low reserve of nutrients, such as yolk, or at an early developmental stage. While *k* strategists produce a few offspring with a considerable amount of parental investment. These may take the form of large well-developed babies, extensive parental care requirements, for example. All of these organisms have adopted an *r* life strategy where little investment is put into the eggs, but to compensate for low parental investment in the offspring large numbers of eggs are produced. This is the opposite of *k* strategy organisms, where there is considerable amount of paternal investment, such as long gestation periods, or long periods when the juvenile is dependent on the parents for shelter and food, such as high-school students.

Shedding is particularly common in parasites that have a direct life cycle. That's when the parasite leaves its host, its eggs or offspring develop in the environment, and they reattach to host species.

One of the best examples for this is marine velvet disease caused by dinoflaggelate organism *Amyloodinium ocellatum*, Brown (1931).

A. ocellatum has a classical direct life cycle (Fig. 3.11), there is no intermediate host. The trophont lives on the skin of its fish host and develops a network of rhizomes that pierce the epidermis and take up nutrition directly from the host. Once

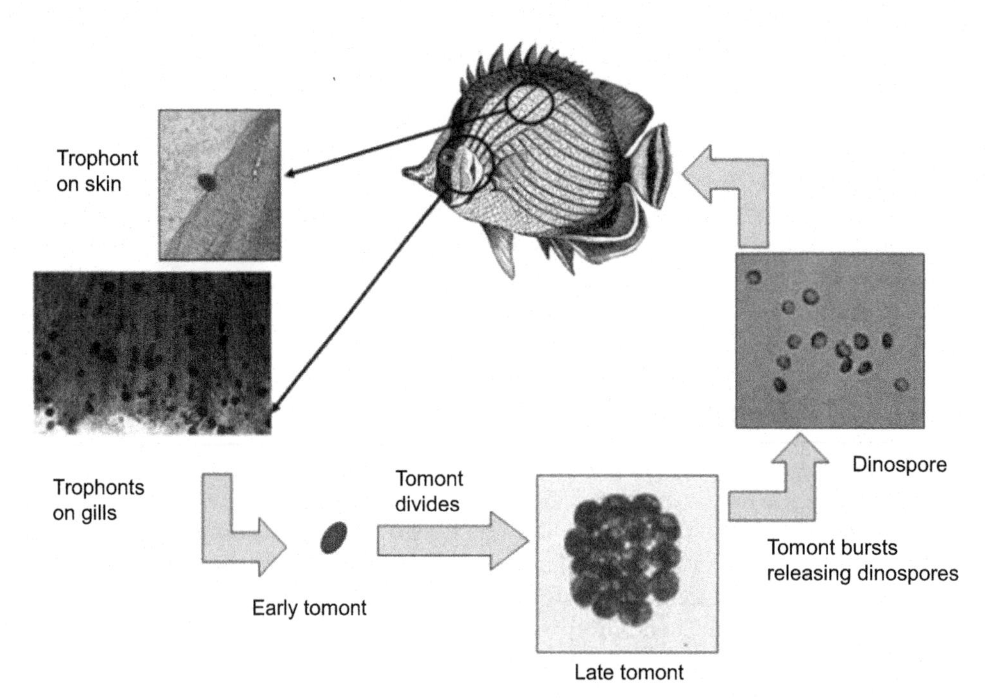

FIGURE 3.11 Direct life cycle of *Amyloodinium ocellatum*.

mature the trophont detaches from the network of rhizomes and falls through the water column onto the substrate, where it divides asexually. The late tomont stage can contain up to 256 dinospores representing 8 asexual divisions. When it is fully matured, the tomont ruptures releasing the motile dinospores into the water column (Brown, 1946; Lawler, 1980; Paperna, 1980). The dinospores are motile having flagella and are attracted to new fish hosts by a tail of semiochemicals that it swims toward (Aaen et al., 2014). *A. ocellatum* has a large host range (Cruz-Lacierda et al., 2004) so the dinospores have a good chance (relatively speaking) of finding a new, susceptible fish host.

3.5.3 Predatory–Prey Relationships

Many parasites use the predation by definite hosts to ensure transmission between the intermediate and definitive hosts. Indeed, some use a phenomenon known as polycyclic transmission to ensure this mechanism is efficient. Polycyclic transmission is the transmission of an immature state of a parasite between intermediate hosts, and we will see an excellent example of this as given below.

The nematode parasite complex *Anisakis simplex* (Rudolphi, 1809), commonly known as the herring worm or whale worm, is a group of closely related species of nematodes (Mattiucci and Nascetti, 2006) with a wide range of fish hosts (Smith, 1983). Typically, the eggs are shed in the feces of infected piscivorous marine mammal such as a whale or dolphin. Once they enter the water the eggs become embryonated, the L1 stage develops in the eggs, and they hatch to release the L2 larvae. The L2 larvae actively swim the water column and make themselves very attractive to crustacean predators such as mysids, copepods, and malacostraca. Once eaten by a crustacean predator, the L2 larvae encyst in the crustaceans' tissues and mature into an L3 larvae. Here it waits for the infected crustacean to be ingested by a fish. It has been suggested that the L3 larvae debilitate the crustacean host in some way to assist in the predation process, however this has been poorly studied (Kuhn et al., 2011; Gregori et al., 2015). Once inside the fish's digestive system, the L3 larvae break out of the cyst and migrate through the stomach or intestine wall into the tissues of the viscera or muscle of the fish host (Smith, 1983). Here it encysts again. The encysted L3 larvae are usually considered to be fairly benign and many marine fish will have numerous L3 larvae in their tissues (Fig. 3.12) with little ill effect.

When the fish with the L3 larvae is eaten by a predator, either another fish or a marine mammal, the nematode migrates through the now dead intermediate host's tissues, provided it has been consumed by a marine mammal, to the digestive tract where it matures into an adult worm and begins its sexual stage (Kuhn et al., 2011). However, if it has been consumed by a predatory fish, the L3 larvae undergo a polycyclic transmission (Dallares et al., 2014). It again migrates through the tissues of the fish hosts' digestive system and encysts in either the muscles or organs of the host where it begins the waiting game again, awaiting the time when that fish too is consumed by a marine mammal or another fish to continue its life cycle.

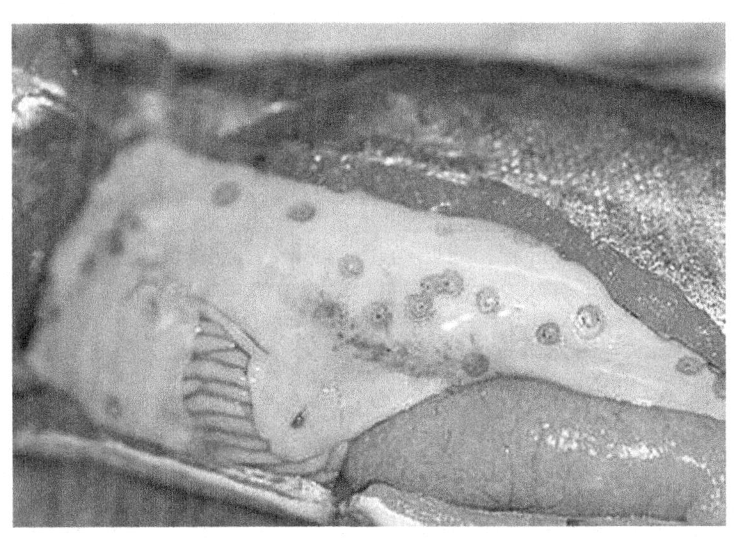

FIGURE 3.12 Encysted L3 larval *Anisakis simplex* in a wild brown trout's viscera. This animal had an excess of 200 L3 larvae, yet appeared healthy when caught. *Photo Pert, C., Bricknell, I., Crown Copyright.*

Other parasites improve their chances of transmission through predator–prey relationship by debilitating their intermediate hosts. An excellent example of this is the eye fluke *Diplostomum spathaceum* (Rudolphi, 1819). This group of digenean trematodes has a complex life cycle involving two intermediate hosts aquatic snails such as *Lymnaea stagnalis*, L. and fish (Sukhdeo and Sukhdeo, 2004; Zbikowska, 2011). The snail is infected by the miracidium stage which emerges from the egg excreted in an infected bird's feces. It undergoes asexual division and maturation into the cercaria larvae within the snail. After maturation the cercaria larvae erupt through the skin of the snail and into the water column. The cercaria are attracted to fish, probably due to a combination of environmental cues that are not very specific. Indeed many species of digenean trematodes will invade many unsuitable hosts where they die. If a human is invaded by cercaria, the larvae die in the skin where they trigger an immune response. This contributes to the medical condition "swimmer's itch" seen in people who go swimming in warm shallow lakes in the summer when large numbers of cercaria are released from snails. These cercaria larvae bore into the unfortunate swimmers' skin causing an unpleasant rash.

If the cercaria larva is lucky and encounters a fish, it penetrates through the skin and migrates to the eye where it encysts in the lens of the eye. The cyst of fluke *D. spathaceum* is opaque and combined with the scar tissue it causes a cataract to develop (Barber, 2007; Lagrue and Poulin, 2010; Mikheev, 2011). Often a fish will be infected with several fluke *D. spathaceum* cercaria larvae every year and depending on the degree of the infection the function of the eye will decline effectively blinding the fish. Once the fish has lost the ability to see it is much more easily predated by the definitive host, a fish eating bird, where it develops into the adult fluke to complete its life cycle. Although this parasite does not change the host's behavior by biochemical means, by destroying the ability of the fish to see it certainly increases its risk of predation (Barber, 2007; Lagrue and Poulin, 2010; Mikheev, 2011).

3.5.4 Host-Seeking Behavior

Many parasites actively seek out their hosts (Thomas et al., 2010). This can involve making an intermediate host less predator aware, so the intermediate host is more easily predated as we saw above. However, parasites can actively seek out their hosts on their own accord. One of the most obvious example of this in the aquatic environment are the leaches. Leaches are classical ectoparasites attaching to the epidermis of fish taking a blood meal and subsequently digesting it (Karlsbakk and Nylund, 2006). However, they do not need to detach from the host, as the classic life history of leaches as portrayed in movies such as the "African Queen." Fish leeches can demonstrate an attach, feed, and leave strategy [e.g., *Piscicola geometra* (L.) (Rost-Roszkowska et al., 2012)], or be permanently attached to its host *Branchellion torpedinis* (Savigny, 1822) (Williams et al., 1994; Marancik et al., 2012).

Both *P. geometra* and *B. torpedinis* will actively move toward their hosts' in the environment before attaching to them by a mouth that is full of chitenous teeth and a basal sucker. The triggers that cause an aquatic leech to migrate toward the host are various. The first two factors that usually trigger response are photoreception and physical disturbance. Leeches do not have eyes in the true sense of the word but have photosensitive cells throughout their skin. In some species such as *P. geometra,* a shadow passing over the leech will cause it to rear upwards and attempt to attach to the object that has triggered the response (Dziekonska-Rynko et al., 2009; Rost-Roszkowska et al., 2012, 2015). If the mouth makes contact with a fish, the leech grabs the passing fish, relaxes its sucker from the substrate, and begins to take a blood meal. A similar process with vibration or pressure waves from a passing fish will also cause *P. geometra* to start host attachment behavior due to the activation of mechanoreceptors within the epidermis of the leech. Cues from the pressure waves in the water also determine which direction the leech needs to direct its attachment behavior toward.

Semiochemicals also play an important role in host location in leeches. These may be as crude as detecting metabolites, such as ammonia or urea in the water column or migration down carbon dioxide gradients (Hildebrandt, 1992; Rose and Deitmer, 1995), or far more specific compounds such as the presence of haem, glucose, or amino acids in the buccal cavity after an attempt at attachment on a host trigger the leech to take full meal. If absent, for example, if the leech inadvertently attaches to a crustacean which would lack haem, it does not continue to feed, but drops off from the deadend host back into the environment where it continues to seek out a new host.

3.5.5 Vectors

Parasites can be vectors for many viral and bacterial diseases as well as causing lesions that open up the host to infections with the water molds or opportunistic bacteria. It is, however, relatively rare for a parasite to be a vector of another parasite in an aquatic ecosystem. However, there are examples of parasites that use other parasites as vectors.

Parasites that use other parasites as vectors are usually hyperparasites (a parasite of a parasite) of the vectors. Typically, they may cause minimal impact on the vector, although they may subtly change its behavior by making it more aggressive, attempting more feeding attacks per unit time than an uninfected vector (Barber and Huntingford, 1995; Scholz, 1999; Baldacchino et al., 2013).

One well-researched parasite that infects fish as a hyperparasite of leeches is *Trypanosoma danilewskyi* (Laveran and Mesnil, 1904) a common parasite of fish, especially common carp (*Cyprinus carpio* L.) (Qadri, 1952, 1962; Woo and Black, 1984; Jones and Woo, 1992). *T. danilewskyi* will often use *P. geometra* as a vector to ensue transmission between infected and uninfected hosts. This parasite causes a widespread parasitemia in common carp (*C. carpio* L.) and other cyprinid fishes often leading to a chronic anemia and long-term infection. Although it is not very host-specific, infecting perciforms and ictyluriforms too. *T. danilewskyi* transform into the procyclic trypasome stage in the midgut of the leech (Zintl et al., 2000) and then migrates to the salivary glands where it undergoes binary fission to increase in numbers. When an infected leech bites a host it injects saliva into the wound, the saliva contains anticoagulants, analgesics, and proteases as well as an infectious dose of *T. danilewskyi.* It is not known if the aggressins injected at the time of the bite help *T. danilewskyi* establish an infection in the host, or if *T. danilewskyi* contributes an advantage to the infected leech by producing additional component that aid in feeding. However, *T. danilewskyi* is a very successful parasite that hyperparasitizes leeches to provide an infectious vector.

3.5.6 Sexually Transmitted Parasites

A unique method parasites use to achieve successful transmission between fish is as a sexually transmitted disease (STD). Typically STD's are a highly diverse group of conditions affecting many species of vertebrates consisting of viruses, bacteria, protozoans, fungi, and arthropods. Fish have an unusual group of parasites that have an endoparasitic lifestyle and are only transmitted sexually.

The Cnidarian parasite *P. hydriforme* (Ussov, 1885) infects the eggs of sturgeon (Acipenseridae) and paddlefish (Polyodontidae). Its classification is uncertain, it is the only member of its order, and its taxonomic position in the Cnidarians' is based on it possession of nematocysts. Evans et al.'s (2008) molecular analysis of *P. hydriforme* seems to confirm this phylogenetic relationship.

The life cycle of *P. hydriforme* is unusual, most of its life is spent in the oocytes of its hosts. *P. hydriforme* initially develops as a large double nucleated cell within the oocyte. From here it develops into a planuliform larva not dissimilar to other cniderians, although Evans et al. (2008) describe it as having an inside out morphology with the epidermal tissue located internally and the ectoderm tissue located where the epidermis would be expected. The embryo, larva, and stolon stages are encased by a large polyploid cell. It is believed that this polypoid cell takes up nutrients for the developing larvae from the oocyte.

As the host matures sexually and just prior to spawning the stolon evert to normal cnidarian morphology and actually possess tentacles like a typical cnidarian. During the inversion event, yolk from the maturing egg is taken up by *P.*

hydriforme for use by the later medusa stage. When the egg is laid, the exposure to freshwater triggers emergence of the medusoid stage from infected eggs into the environment, which go onto mature and develop mature reproductive organs. After fertilization the gametophores seek out and infect naïve fish to continue its amazing life cycle (Evans et al. (2008)).

It could be argued that this is a form of vertical transmission. However, *P. hydriforme* only infects female fish, infection is dependent on the reproductive cycle to complete transmission and there is no known horizontal transmission mechanism. In fact, no vertically transmitted parasites have been recorded in fish to date.

3.5.7 Hormonal Manipulation

Some fish parasites manipulate the endocrine levels with the host to improve transmission between susceptible animals. One of the most amazing examples of this belongs to the rhizocephalan barnacles and in particular, *Sacculina carcini*, Thompson, 1836. *S. carcini* develops as a typical planktonic barnacle larva after it hatches from the eggs, progressing through the free-living nauplius larvae molts and the cypris stage. As the female cypris stage matures, it locates a crab (usually *Carcinus meanus* L.), lands on its surface, and crawls across the surface to look for a suitable location to infect the host on the first antennae. Once it attaches to the first antennae, the cypris then sheds the thorax and abdomen, and injects a small mass of stem cells into the host. The parasite then grows throughout the coelom of the crab, forming a diffuse network of parasitic tissues in the host (Powell and Rowley, 2008). Not only does the parasite manipulate the hosts immune system to prevent rejection by the crab, it also highjacks the reproductive cycle of its host. It castrates and feminizes male crabs, hormonally, and suppresses the natural reproductive cycle of female crabs (Mathieson et al., 1998; Kristensen et al., 2012). Once sexually mature, the female *S. carcini* extrude a small bleb of tissue through the arthrodial membrane of the crab and develops a genital opening (Rubiliani et al., 1980; Hoeg, 1984, 1987a,b). Using pheromones, the female *S. carcini* attracts the planktonic male to the infected hosts and mates with him (Pasternak et al., 2005). Once mated *S. carcini* then takes over the reproductive structure of both male and female crabs, and extrudes an egg mass which the crab cares for as if it were their own to begin the cycle over again.

Although nothing quite as spectacular as *S. carcini* is described in fish, there are some very astonishing examples of hormonal manipulation of a fish's endocrine system to ensure successful transmission of a parasite. The tapeworm *Ligula intestinalis* L. also highjacks the reproductive cycle of its host, usually a cyprinid fish-like bream, *Blicca bjoerkna* L. Once in the host, it has two choices to make. In a female fish the parasite manipulates the hormone levels so it reabsorbs any reproductive tissue allowing the plerocercoid larvae to use this as a resource to grow (Vanacker et al., 2012). In a male fish *L. intestinalis* first hormonally feminizes the host and then mobilizes the hosts' fat reserves to provide sufficient nutritional resources to permit growth (Vanacker et al., 2012). Both these strategies allow the pluracid larvae to reach a very large size quickly, often accounting for 30% or more of the host's body weight (Vanacker et al., 2012). Once it has reached its mature size the pluracoid larvae then triggers a second hormonal change suppressing adrenalin and corticosteroids making the host less fearful of predators and spending more time in the open where it is more likely to be predated by its definitive hosts—a fish eating bird.

This is just one of many examples where fish parasites can manipulate the hormone levels of their hosts to increase the change of transmission. There is some evidence that another tapeworm, *S. solidus*, Müller, 1776, feminizes male sticklebacks and makes them more cryptic in nature. In turn this increases their life expectancy, bringing it in line with female sticklebacks, allowing the plerocercoid larvae to grow to a size where it is optimum for transmission into the definitive host (Nordeide and Matos, 2016). *S. solidus* can also change the temperature preference of an infected host. Infected sticklebacks prefer water temperature of 16°C or higher which is the optimum temperature for the parasite's growth (Macnab and Barber, 2012). Although the mechanism is not clearly understood, somehow the pluracoid larvae must by controlling the function of the hypothalamus and its neurosecretions.

3.6 CONCLUSION

As we have seen pathogens use a wide variety of mechanism to ensure successful transmission between susceptible hosts. These can vary from simply being shed into the water in the hope of encountering another host to a sophisticated mechanism where the pathogen actively seeks a host and undergoes very sophisticated behavior. It is worth remembering that the pathogens do not have this all their own way. They are locked in an evolutionary war with their hosts. Each evolutionary change in a pathogens' virulence triggers a selective pressure on the host to evolve an immunological counter measure. A useful infection can simply be seen as Darwinian process to remove unsuccessful genes from a hosts' population while honing the pathogens genes and vice versa.

REFERENCES

Aaen, S.M., Aunsmo, A., Horsberg, T.E., 2014. Impact of hydrogen peroxide on hatching ability of egg strings from salmon lice (*Lepeophtheirus salmonis*) in a field treatment and in a laboratory study with ascending concentrations. Aquaculture 422, 167–171.

Aamelfot, M., Dale, O.B., Falk, K., 2014. Infectious salmon anaemia – pathogenesis and tropism. Journal of Fish Diseases 37 (4), 291–307.

Ahne, W., Bjorklund, H.V., Essbauer, S., Fijan, N., Kurath, G., Winton, J.R., 2002. Spring viremia of carp (SVC). Diseases of Aquatic Organisms 52 (3), 261–272.

Al-Hussinee, L., Lord, S., Stevenson, R.M.W., Casey, R.N., Groocock, G.H., Britt, K.L., Kohler, K.H., Wooster, G.A., Getchell, R.G., Bowser, P.R., Lumsden, J.S., 2011. Immunohistochemistry and pathology of multiple Great Lakes fish from mortality events associated with viral hemorrhagic septicemia virus type IVb. Diseases of Aquatic Organisms 93 (2), 117–127.

Ali, E.H., 2005. Morphological and biochemical alterations of oomycete fish pathogen *Saprolegnia parasitica* as affected by salinity, ascorbic acid and their synergistic action. Mycopathologia 159 (2), 231–243.

Ali, S.E., Thoen, E., Vralstad, T., Kristensen, R., Evensen, O., Skaar, I., 2013. Development and reproduction of *Saprolegnia* species in biofilms. Veterinary Microbiology 163 (1–2), 133–141.

Alto, B.W., Lounibos, L.P., 2013. Vector competence for arboviruses in relation to the larval environment of mosquitoes. In: Takken, W., Koenraadt, C.J.M. (Eds.). Takken, W., Koenraadt, C.J.M. (Eds.), Ecology of Parasite-Vector Interactions, vol. 3. Wageningen Academic Publishers, Wageningen, pp. 81–101.

Austin, B., Austin, D.A., 2016. Bacterial Fish Pathogens: Disease of Farmed and Wild Fish. Springer, London.

Baldacchino, F., Muenworn, V., Desquesnes, M., Desoli, F., Charoenviriyaphap, T., Duvallet, G., 2013. Transmission of pathogens by *Stomoxys* flies (Diptera, Muscidae): a review. Parasite 20.

Balon, E.K., 1991. Probable evolution of the coelacanths reproductive style – lecithotrophy and orally feeding embryos in cichlid fishes and in *Latimeria chalumnae*. Environmental Biology of Fishes 32 (1–4), 249–265.

Barber, I., 2007. Parasites, behaviour and welfare in fish. Applied Animal Behaviour Science 104 (3–4), 251–264.

Barber, I., Huntingford, F.A., 1995. The effect of *Schistocephalus solidus* (Cestoda: Pseudophyllidea) on the foraging and shoaling behaviour of three-spined sticklebacks, *Gasterosteus aculeatus*. Behaviour 132, 1223–1240.

Barker, D.E., Braden, L.M., Coombs, M.P., Boyce, B., 2009. Preliminary studies on the isolation of bacteria from sea lice, *Lepeophtheirus salmonis*, infecting farmed salmon in British Columbia, Canada. Parasitology Research 105 (4), 1173–1177.

Bergh, O., Nilsen, F., Samuelsen, O.B., 2001. Diseases, prophylaxis and treatment of the Atlantic halibut *Hippoglossus hippoglossus*: a review. Diseases of Aquatic Organisms 48 (1), 57–74.

Bernard, J., Bremont, M., 1995. Molecular-biology of fish viruses – a review. Veterinary Research 26 (5–6), 341–351.

Bernoth, E.M., Ellis, A.E., Midtlyng, P.J., Olivier, G., Smith, P. (Eds.), 1997. Furunculosis: Multidisciplinary Fish Disease Research, Kindle ed. Academic Press.

Biering, E., Bergh, O., 1996a. Experimental infection of Atlantic halibut, *Hippoglossus hippoglossus* L., yolk-sac larvae with infectious pancreatic necrosis virus: detection of virus by immunohistochemistry and in situ hybridization. Journal of Fish Diseases 19 (4), 261–269.

Biering, E., Bergh, O., 1996b. Experimental infection of Atlantic halibut, *Hippoglossus hippoglossus* L., yolk-sac larvae with infectious pancreatic necrosis virus: detection of virus by immunohistochemistry and in situ hybridization. Journal of Fish Diseases 19 (6), 405–413 p. 261.

Biering, E., Nilsen, F., Rodseth, O.M., Glette, J., 1994. Susceptibility of Atlantic halibut *Hippoglossus hippoglossus* to infectious pancreatic necrosis virus. Diseases of Aquatic Organisms 20 (3), 183–190.

Biller-Takahashi, J.D., Urbinati, E.C., 2014. Fish immunology. The modification and manipulation of the innate immune system: Brazilian studies. Anais Da Academia Brasileira De Ciencias 86 (3), 1483–1495.

Bocklisch, H., Otto, B., 2000. Mycotic diseases in fish. Mycoses 43, 76–78.

Bowden, T.J., Smail, D.A., Ellis, A.E., 2002. Development of a reproducible infectious pancreatic necrosis virus challenge model for Atlantic salmon, *Salmo salar* L. Journal of Fish Diseases 25 (9), 555–563.

Bowden, T.J., Lockhart, K., Smail, D.A., Ellis, A.E., 2003. Experimental challenge of post-smolts with IPNV: mortalities do not depend on population density. Journal of Fish Diseases 26 (5), 309–312.

Bowden, T.J., Thompson, K.D., Morgan, A.L., Gratacap, R.M.L., Nikoskelainen, S., 2007. Seasonal variation and the immune response: a fish perspective. Fish and Shellfish Immunology 22 (6), 695–706.

Breckels, R.D., Neff, B.D., 2013. The effects of elevated temperature on the sexual traits, immunology and survivorship of a tropical ectotherm. Journal of Experimental Biology 216 (14), 2658–2664.

Brevik, O.J., Ottem, K.F., Nylund, A., 2011. Multiple-locus, variable number of tandem repeat analysis (MLVA) of the fish-pathogen *Francisella noatunensis*. BMC Veterinary Research 7.

Bricknell, I.R., 1995. A reliable method, for the induction of experimental furunculosis. Journal of Fish Diseases 18 (2), 127–133.

Bricknell, I., Dalmo, R.A., 2005. The use of immunostimulants in fish larval aquaculture. Fish and Shellfish Immunology 19 (5), 457–472.

Bricknell, I.R., Bruno, D.W., Stone, J., 1996. *Aeromonas salmonicida* infectivity studies in goldsinny wrasse, *Ctenolabrus rupestris* (L.). Journal of Fish Diseases 19 (6), 469–474.

Bricknell, I.R., Bron, J.E., Bowden, T.J., 2006. Diseases of gadoid fish in cultivation: a review. ICES Journal of Marine Science 63 (2), 253–266.

Britz, R., 2013. *Channa andrao*, a new species of dwarf snakehead from West Bengal, India (Teleostei: Channidae). Zootaxa 3731 (2), 287–294.

Brown, E.M., 1946. *Amyloodinium ocellatum* (brown), a peridinian parasitic on marine fishes – a complementary study. Proceedings of the Zoological Society of London 116 (1), 33–46.

Bullock, G.L., Cipriano, R.C., Schill, W.B., 1997. Culture and serodiagnostic detection of *Aeromonas salmonicida* from covertly-infected rainbow trout given the stress-induced furunculosis test. Biomedical Letters 55 (219–20), 169–177.

Campiao, K.M., Ribas, A.C.D., Cornell, S.J., Begon, M., Tavares, L.E.R., 2015. Estimates of coextinction risk: how anuran parasites respond to the extinction of their hosts. International Journal for Parasitology 45 (14), 885–889.

Cipriano, R.C., Bullock, G.L., Schill, W.B., Kretschmann, R., 1996. Enhanced culture of *Aeromonas salmonicida* from covertly-infected rainbow trout following administration of the stress induced furunculosis test. Biomedical Letters 54 (214), 105–112.

Cipriano, R.C., Ford, L.A., Smith, D.R., Schachte, J.H., Petrie, C.J., 1997. Differences in detection of *Aeromonas salmonicida* in covertly infected salmonid fishes by the stress-inducible furunculosis test and culture-based assays. Journal of Aquatic Animal Health 9 (2), 108–113.

Cooper, W., 1988. Rhodesian man. En Tech Journal 3, 137–151.

Cruz-Lacierda, E.R., Maeno, Y., Pineda, A.J.T., Matey, V.E., 2004. Mass mortality of hatchery-reared milkfish (*Chanos chanos*) and mangrove red snapper (*Lutjanus argentimaculatus*) caused by *Amyloodinium ocellatum* (Dinoflagellida). Aquaculture 236 (1–4), 85–94.

Cuthill, J.F.H., Sewell, K.B., Cannon, L.R.G., Charleston, M.A., Lawler, S., Littlewood, D.T.J., Olson, P.D., Blair, D., 2016. Australian spiny mountain crayfish and their temnocephalan ectosymbionts: an ancient association on the edge of coextinction? Proceedings of the Royal Society B-Biological Sciences 283 (1831), 10.

Cutrin, J.M., Lopez-Vaquez, C., Oliveira, J.G., Castro, S., Dopazo, C.P., Bandin, I., 2005. Isolation in cell culture and detection by PCR-based technology of IPNV-like virus from leucocytes of carrier turbot, *Scophthalmus maximus* (L.). Journal of Fish Diseases 28 (12), 713–722.

D'Angelo, L., Lossi, L., Merighi, A., de Girolamo, P., 2016. Anatomical features for the adequate choice of experimental animal models in biomedicine: I. Fishes. Annals of Anatomy-Anatomischer Anzeiger 205, 75–84.

Dallares, S., Constenla, M., Padros, F., Cartes, J.E., Sole, M., Carrasson, M., 2014. Parasites of the deep-sea fish *Mora moro* (Risso, 1810) from the NW Mediterranean Sea and relationship with fish diet and enzymatic biomarkers. Deep-Sea Research Part I-Oceanographic Research Papers 92, 115–126.

Das, B.K., Pattnaik, P., Debnath, C., Swain, D.K., Pradhan, J., 2013. Effect of beta-glucan on the immune response of early stage of *Anabas testudineus* (Bloch) challenged with fungus *Saprolegnia parasitica*. Springerplus 2.

Dex and McCaff, August 14, 2000. Dex and McCaff, "Who Was Typhoid Mary?" the Straight Dope. Available from: http://www.straightdope.com/columns/read/1816/who-was-typhoid-mary.

Dezfuli, B.S., 1996. *Cypria reptans* (Crustacea: Ostracoda) as an intermediate host of *Neoechinorhynchus rutili* (Acanthocephala: Eoacanthocephala) in Italy. Journal of Parasitology 82 (3), 503–505.

Dezfuli, B.S., Bosi, G., DePasquale, J.A., Manera, M., Giari, L., 2016. Fish innate immunity against intestinal helminths. Fish and Shellfish Immunology 50, 274–287.

Diallo, M., Thonnon, J., Fontenille, D., 2000. Vertical transmission of the yellow fever virus by *Aedes aegypti* (Diptera, Culicidae): dynamics of infection in F-1 adult progeny of orally infected females. American Journal of Tropical Medicine and Hygiene 62 (1), 151–156.

Dixon, P.F., 1999. VHSV came from the marine environment: clues from the literature, or just red herrings? Bulletin of the European Association of Fish Pathologists 19 (2), 60–65.

Dobos, P., Roberts, T.E., 1983. The molecular-biology of infectious pancreatic necrosis virus – a review. Canadian Journal of Microbiology 29 (4), 377–384.

Dziekonska-Rynko, J., Bielecki, A., Palinska, K., 2009. Activity of selected hydrolytic enzymes from leeches (Clitellata: Hirudinida) with different feeding strategies. Biologia 64 (2), 370–376.

Effendi, I., Austin, B., 1994. Survival of the fish pathogen *Aeromonas salmonicida* in the marine-environment. Journal of Fish Diseases 17 (4), 375–385.

Effendi, I., Austin, B., 1995. Dormant unculturable cells of the fish pathogen *Aeromonas salmonicida*. Microbial Ecology 30 (2), 183–192.

Eritja, R., Escosa, R., Lucientes, J., Marques, E., Molina, R., Roiz, D., Ruiz, S., 2005. Worldwide invasion of vector mosquitoes: present European distribution and challenges for Spain. Biological Invasions 7 (1), 87–97.

Evans, N.M., Lindner, A., Raikova, E.V., Collins, A.G., Cartwright, P., 2008. Phylogenetic placement of the enigmatic parasite, *Polypodium hydriforme,* within the Phylum Cnidaria. BMC Evolutionary Biology 8, 139. http://dx.doi.org/10.1186/1471-2148-8-139. PMC: 2396633. PMID: 18471296.

Faisal, M., Winters, A.D., 2011. Detection of viral hemorrhagic septicemia virus (VHSV) from *Diporeia* spp. (Pontoporeiidae, Amphipoda) in the Laurentian great lakes, USA. Parasites Vectors 4, 4.

Faisal, M., Schulz, C., Eissa, A., Brenden, T., Winters, A., Whelan, G., Wolgamood, M., Eisch, E., VanAmberg, J., 2012. Epidemiological investigation of *Renibacterium salmoninarum* in three *Oncorhynchus* spp. in Michigan from 2001 to 2010. Preventive Veterinary Medicine 107 (3–4), 260–274.

Farrell, M.J., Stephens, P.R., Berrang-Ford, L., Gittleman, J.L., Davies, T.J., 2015. The path to host extinction can lead to loss of generalist parasites. Journal of Animal Ecology 84 (4), 978–984.

Firouzbakhsh, F., Mehrabi, Z., Heydari, M., Khalesi, M.K., Tajick, M.A., 2014. Protective effects of a synbiotic against experimental *Saprolegnia parasitica* infection in rainbow trout (*Oncorhynchus mykiss*). Aquaculture Research 45 (4), 609–618.

Fishbase, 2016. Available from: http://www.fishbase.org/ .

Fontenille, D., Diallo, M., Mondo, M., Ndiaye, M., Thonnon, J., 1997. First evidence of natural vertical transmission of yellow fever virus in *Aedes aegypti*, its epidemic vector. Transactions of the Royal Society of Tropical Medicine and Hygiene 91 (5), 533–535.

Forey, P.L., Cavin, L., 2007. A new species of *Cladocyclus* (Teleostei : Ichthyodectiformes) from the Cenomanian of Morocco. Palaeontologia Electronica 10 (3), 10.

Fouz, B., Llorens, A., Valiente, E., Amaro, C., 2010. A comparative epizootiologic study of the two fish-pathogenic serovars of *Vibrio vulnificus* biotype 2. Journal of Fish Diseases 33 (5), 383–390.

Frattini, S.A., Groocock, G.H., Getchell, R.G., Wooster, G.A., Casey, R.N., Casey, J.W., Bowser, P.R., 2011. A 2006 survey of viral hemorrhagic septicemia (VHSV) virus type IVb in New York state waters. Journal of Great Lakes Research 37 (1), 194–198.

Fryer, J.L., Lannan, C.N., 1993. The history and current status of *Renibacterium salmoninarum*, the causative agent of bacterial kidney-disease in Pacific salmon. Fisheries Research 17 (1–2), 15–33.

Gahlawat, S.K., Munro, E.S., Ellis, A.E., 2004. A non-destructive test for detection of IPNV-carriers in Atlantic halibut, *Hippoglossus hippoglossus* (L.). Journal of Fish Diseases 27 (4), 233–239.

Garcia, J., Urquhart, K., Ellis, A.E., 2006. Infectious pancreatic necrosis virus establishes an asymptomatic carrier state in kidney leucocytes of juvenile Atlantic cod, *Gadus morhua* L. Journal of Fish Diseases 29 (7), 409–413.

Giray, C., Opitz, H.M., MacLean, S., Bouchard, D., 2005. Comparison of lethal versus non-lethal sample sources for the detection of infectious salmon anemia virus (ISAV). Diseases of Aquatic Organisms 66 (3), 181–185.

Gonzalez, R.R., Ruiz, P., Llanos-Rivera, A., Cruzat, F., Silva, J., Astuya, A., Grandon, M., Jara, D., Aburto, C., 2011. ISA virus outside the cage: Ichthyofauna and other possible reservoirs to be considered for marine biosafety management in the far-southern ecosystems of Chile. Aquaculture 318 (1–2), 37–42.

Goodwin, A.E., Merry, G.E., 2011. Replication and persistence of VHSV IVb in freshwater turtles. Diseases of Aquatic Organisms 94 (3), 173–177.

Graham, D.A., Frost, P., McLaughlin, K., Rowley, H.M., Gabestad, I., Gordon, A., McLoughlin, M.F., 2011. A comparative study of marine salmonid alphavirus subtypes 1–6 using an experimental cohabitation challenge model. Journal of Fish Diseases 34 (4), 273–286.

Graham, D.A., Brown, A., Savage, P., Frost, P., 2012. Detection of salmon pancreas disease virus in the faeces and mucus of Atlantic salmon, *Salmo salar* L., by real-time RT-PCR and cell culture following experimental challenge. Journal of Fish Diseases 35 (12), 949–951.

Gratz, N.G., 2004. Critical review of the vector status of *Aedes albopictus*. Medical and Veterinary Entomology 18 (3), 215–227.

Gregori, M., Roura, A., Abollo, E., Gonzalez, A.F., Pascual, S., 2015. Anisakis simplex complex (Nematoda: Anisakidae) in zooplankton communities from temperate NE Atlantic waters. Journal of Natural History 49 (13–14), 755–773.

Griffiths, S., Melville, K., 2000. Non-lethal detection of ISAV in Atlantic salmon by RT-PCR using serum and mucus samples. Bulletin of the European Association of Fish Pathologists 20 (4), 157–162.

Gurcay, M., Turan, T., Parmaksiz, A., 2013. A study on the presence of infectious pancreatic necrosis virus infections in farmed rainbow trout (*Oncorhynchus mykiss* Walbaum, 1792) in Turkey. Kafkas Universitesi Veteriner Fakultesi Dergisi 19 (1), 141–146.

Gustafson, C.E., Thomas, C.J., Trust, T.J., 1992. Detection of *Aeromonas salmonicida* from fish by using polymerase chain-reaction amplification of the virulence surface array protein gene. Applied and Environmental Microbiology 58 (12), 3816–3825.

Hardham, A.R., Hyde, G.J., 1997. In: Andrews, J.H., Tommerup, I.C. (Eds.)Andrews, J.H., Tommerup, I.C. (Eds.), Asexual Sporulation in the Oomycetes. Advances in Botanical Research Incorporating Advances in Plant Pathology, vol. 24, pp. 353–398.

Harper, D.N., 2011. Viruses: Biology, Applications, and Control London. Garland Science.

Heuch, P.A., Oines, O., Knutsen, J.A., Schram, T.A., 2007. Infection of wild fishes by the parasitic copepod *Caligus elongatus* on the south east coast of Norway. Diseases of Aquatic Organisms 77 (2), 149–158.

Hildebrandt, J.P., 1992. External CO_2 levels influence energy yielding metabolic pathways under hypoxia in the leech, *Hirudo medicinalis*. Journal of Experimental Zoology 261 (4), 379–386.

Hill, D.R., 2012. Mapping the risk of yellow fever infection. Current Infectious Disease Reports 14 (3), 246–255.

Hiney, M.P., Smith, P.R., 1998. Validation of polymerase chain reaction-based techniques for proxy detection of bacterial fish pathogens: framework, problems and possible solutions for environmental applications. Aquaculture 162 (1–2), 41–68.

Hiney, M.P., Kilmartin, J.J., Smith, P.R., 1994. Detection of *Aeromonas salmonicida* in Atlantic salmon with asymptomatic furunculosis infections. Diseases of Aquatic Organisms 19 (3), 161–167.

Hjelm, M., Bergh, O., Riaza, A., Nielsen, J., Melchiorsen, J., Jensen, S., Duncan, H., Ahrens, P., Birkbeck, H., Gram, L., 2004. Selection and identification of autochthonous potential probiotic bacteria from turbot larvae (*Scophthalmus maximus*) rearing units. Systematic and Applied Microbiology 27 (3), 360–371.

Hoeg, J.T., 1984. Size and settling behavior in male and female cypris larvae of the parasitic barnacle *Sacculina carcini* Thompson (Crustacea, Cirripedia, Rhizocephala). Journal of Experimental Marine Biology and Ecology 76 (2), 145–156.

Hoeg, J.T., 1987a. Male cypris metamorphosis and a new male larval form, the trichogon, in the parasitic barnacle *Sacculina carcini* (Crustacea, Cirripedia, Rhizocephala). Philosophical Transactions of the Royal Society of London Series B-Biological Sciences 317 (1183), 47–63.

Hoeg, J.T., 1987b. The relation between cypris ultrastructure and metamorphosis in male and female *Sacculina carcini* (Crustacea, Cirripedia). Zoomorphology 107 (5), 299–311.

Holgerson, M.A., Post, D.M., Skelly, D.K., 2016. Reconciling the role of terrestrial leaves in pond food webs: a whole-ecosystem experiment. Ecology 97 (7), 1771–1782.

Holm, H., Santi, N., Kjoglum, S., Perisic, N., Skugor, S., Evensen, O., 2015. Difference in skin immune responses to infection with salmon louse (*Lepeophtheirus salmonis*) in Atlantic salmon (*Salmo salar* L.) of families selected for resistance and susceptibility. Fish and Shellfish Immunology 42 (2), 384–394.

Hovland, T., Nylund, A., Watanabe, K., Endresen, C., 1994. Observation of infectious salmon anemia virus in Atlantic salmon, *Salmo salar* L. Journal of Fish Diseases 17 (3), 291–296.

Jensen, A.J., Zydlewski, G.B., Barker, S., Pietrak, M., 2016. Sea lice infestations of a wild fish assemblage in the Northwest Atlantic Ocean. Transactions of the American Fisheries Society 145 (1), 7–16.

Jiang, R.H.Y., de Bruijn, I., Haas, B.J., Belmonte, R., Lobach, L., Christie, J., van den Ackerveken, G., Bottin, A., Bulone, V., Diaz-Moreno, S.M., Dumas, B., Fan, L., Gaulin, E., Govers, F., Grenville-Briggs, L.J., Horner, N.R., Levin, J.Z., Mammella, M., Meijer, H.J.G., Morris, P., Nusbaum, C., Oome, S., Phillips, A.J., van Rooyen, D., Rzeszutek, E., Saraiva, M., Secombes, C.J., Seidl, M.F., Snel, B., Stassen, J.H.M., Sykes, S., Tripathy, S., van den Berg, H., Vega-Arreguin, J.C., Wawra, S., Young, S.K., Zeng, Q.D., Dieguez-Uribeondo, J., Russ, C., Tyler, B.M., van West, P., 2013. Distinctive expansion of potential virulence genes in the genome of the oomycete fish pathogen *Saprolegnia parasitica*. PLoS Genetics 9 (6).

Johansen, L.H., Sommer, A.I., 1995. Multiplication of infectious pancreatic necrosis virus (IPNV) in head kidney and blood leukocytes isolated from Atlantic Salmon, *Salmo salar* L. Journal of Fish Diseases 18 (2), 147–156.

Jones, S.R.M., Woo, P.T.K., 1992. Vector specificity of *Trypanosoma catostomi* and its infectivity to fresh-water fishes. Journal of Parasitology 78 (1), 87–92.

Jorgensen, D., 2015. Conservation implications of parasite co-reintroduction. Conservation Biology 29 (2), 602–604.

Karlsbakk, E., Nylund, A., 2006. Trypanosomes infecting cod *Gadus morhua* L. in the North Atlantic: a resurrection of *Trypanosoma pleuronectidium* Robertson, 1906 and delimitation of *T. murmanense* Nikitin, 1927 (emend.), with a review of other trypanosomes from North Atlantic and Mediterranean teleosts. Systematic Parasitology 65 (3), 175–203.

Kent, M.L., Whyte, J.N.C., Latrace, C., 1995. Gill lesions and mortality in seawater pen-reared Atlantic salmon *Salmo salar* associated with a dense bloom of *Skeletonema costatum* and *Thalassiosira* species. Diseases of Aquatic Organisms 22 (1), 77–81.

Khoo, L., Leard, A.T., Waterstrat, P.R., Jack, S.W., Camp, K.L., 1998. Branchiomyces infection in farm-reared channel catfish, *Ictalurus punctatus* (Rafinesque). Journal of Fish Diseases 21 (6), 423–431.

Kristensen, T., Nielsen, A.I., Jorgensen, A.I., Mouritsen, K.N., Glenner, H., Christensen, J.T., Lutzen, J., Hoeg, J.T., 2012. The selective advantage of host feminization: a case study of the green crab *Carcinus maenas* and the parasitic barnacle *Sacculina carcini*. Marine Biology 159 (9), 2015–2023.

Kuhn, T., Garcia-Marquez, J., Klimpel, S., 2011. Adaptive radiation within marine anisakid nematodes: a zoogeographical modeling of cosmopolitan, zoonotic parasites. PLoS One 6 (12).

Lagrue, C., Poulin, R., 2010. Manipulative parasites in the world of veterinary science: implications for epidemiology and pathology. Veterinary Journal 184 (1), 9–13.

Lamkova, K., Simkova, A., Palikova, M., Jurajda, P., Lojek, A., 2007. Seasonal changes of immunocompetence and parasitism in chub (*Leuciscus cephalus*), a freshwater cyprinid fish. Parasitology Research 101 (3), 775–789.

Lawler, A.R., 1980. Studies on *Amyloodinium ocellatum* (Dinoflagellata) in Mississippi sound – natural and experimental hosts. Gulf Research Reports 6 (4), 403–413.

Leal, C.A.G., Tavares, G.C., Figueiredo, H.C.P., 2014. Outbreaks and genetic diversity of *Francisella noatunensis* subsp. *orientalis* isolated from farm-raised Nile tilapia (*Oreochromis niloticus*) in Brazil. Genetics and Molecular Research 13 (3), 5704–5712.

Lin, C.Y., Chiang, C.Y., Tsai, H.J., 2016a. Zebrafish and Medaka: new model organisms for modern biomedical research. Journal of Biomedical Science 23, 11.

Lin, Q., Li, N.Q., Fu, X.Z., Hu, Q.D., Chang, O.Q., Liu, L.H., Zhang, D.F., Wang, G.J., San, G.B., Wu, S.Q., 2016b. An outbreak of granulomatous inflammation associated with *Francisella noatunensis* subsp. *orientalis* in farmed tilapia (*Oreochromis niloticus* × *O. aureus*) in China. Chinese Journal of Oceanology and Limnology 34 (3), 460–466.

Liu, Y., de Bruijn, I., Jack, A.L.H., Drynan, K., van den Berg, A.H., Thoen, E., Sandoval-Sierra, V., Skaar, I., van West, P., Dieeguez-Uribeondo, J., van der Voort, M., Mendes, R., Mazzola, M., Raaijmakers, J.M., 2014. Deciphering microbial landscapes of fish eggs to mitigate emerging diseases. ISME Journal 8 (10), 2002–2014.

Long, J.A., 2010. The Rise of Fishes: 500 Million Years of Evolution Baltimore. Johns Hopkins University Press.

Lopez-Jimena, B., Garcia-Rosado, E., Infante, C., Cano, I., Manchado, M., Castro, D., Borrego, J.J., Alonso, M.C., 2010. Detection of infectious pancreatic necrosis virus (IPNV) from asymptomatic redbanded seabream, *Pagrus auriga* Valenciennes, and common seabream, *Pagrus pagrus* (L.), using a non-destructive procedure. Journal of Fish Diseases 33 (4), 311–319.

Lotfy, W.M., 2015. Plague in Egypt: disease biology, history and contemporary analysis: a minireview. Journal of Advanced Research 6 (4), 549–554.

Mackenzie, K., Hemmingsen, W., 2015. Parasites as biological tags in marine fisheries research: European Atlantic waters. Parasitology 142 (1), 54–67.

Macnab, V., Barber, I., 2012. Some (worms) like it hot: fish parasites grow faster in warmer water, and alter host thermal preferences. Global Change Biology 18 (5), 1540–1548.

Madan, V., Bisset, D., Harris, P., Howard, S., Beck, M.H., 2006. Phaeohyphomycosis caused by *Exophiala salmonis*. British Journal of Dermatology 155 (5), 1082–1084.

Mainous, M.E., Kuhn, D.D., Smith, S.A., 2011. Efficacy of common aquaculture compounds for disinfection of *Aeromonas hydrophila*, *A. Salmonicida* subsp. *salmonicida*, and *A. Salmonicida* subsp. *achromogenes* at various temperatures. North American Journal of Aquaculture 73 (4), 456–461.

Marancik, D.P., Dove, A.D., Camus, A.C., 2012. Experimental infection of yellow stingrays *Urobatis jamaicensis* with the marine leech *Branchellion torpedinis*. Diseases of Aquatic Organisms 101 (1), 51–60.

Mark Welch, D., 2005. Bayesian and maximum likelihood analyses of rotifer-acanthocephalan relationships. Hydrobiologia 546, 47–54.

Marshall, S.H., Ramirez, R., Labra, A., Carmona, M., Munoz, C., 2014. Bona fide evidence for natural vertical transmission of infectious salmon anemia virus in freshwater brood stocks of farmed Atlantic salmon (*Salmo salar*) in southern Chile. Journal of Virology 88 (11), 6012–6018.

Mathieson, S., Berry, A.J., Kennedy, S., 1998. The parasitic rhizocephalan barnacle *Sacculina carcini* in crabs of the Forth estuary, Scotland. Journal of the Marine Biological Association of the United Kingdom 78 (2), 665–667.

Mattiucci, S., Nascetti, G., 2006. Molecular systematics, phylogeny and ecology of anisakid nematodes of the genus *Anisakis* Dujardin, 1845: an update. Parasite-Journal De La Societe Francaise De Parasitologie 13 (2), 99–113.

McAllister, P.E., Schill, W.B., Owens, W.J., Hodge, D.L., 1993. Determining the prevalence of infectious pancreatic necrosis virus in asymptomatic brook trout *Salvelinus fontinalis* – a study of clinical-samples and processing methods. Diseases of Aquatic Organisms 15 (3), 157–162.

McKibben, C.L., Pascho, R.J., 1999. Shedding of *Renibacterium salmoninarum* by infected chinook salmon *Oncorhynchus tschawytscha*. Diseases of Aquatic Organisms 38 (1), 75–79.

Medlock, J.M., Hansford, K.M., Schaffner, F., Versteirt, V., Hendrickx, G., Zeller, H., Van Bortel, W., 2012. A review of the invasive mosquitoes in Europe: ecology, public health risks, and control options. Vector-Borne and Zoonotic Diseases 12 (6), 435–447.

Menanteau-Ledouble, S., Kumar, G., Saleh, M., El-Matbouli, M., 2016. *Aeromonas salmonicida*: updates on an old acquaintance. Diseases of Aquatic Organisms 120 (1), 49–68.

Mikheev, V.N., 2011. Monoxenous and heteroxenous parasites of fish manipulate behavior of their hosts in different ways. Zhurnal Obshchei Biologii 72 (3), 183–197.

Moir, M.L., Hughes, L., Vesk, P.A., Leng, M.C., 2014. Which host-dependent insects are most prone to coextinction under changed climates? Ecology and Evolution 4 (8), 1295–1312.

Molloy, S.D., Pietrak, M.R., Bricknell, I., Bouchard, D.A., 2013. Experimental transmission of infectious pancreatic necrosis virus from the blue mussel, *Mytilus edulis*, to cohabitating Atlantic salmon (*Salmo salar*) smolts. Applied and Environmental Microbiology 79 (19), 5882–5890.

Munday, B.L., Kwang, J., Moody, N., 2002. Betanodavirus infections of teleost fish: a review. Journal of Fish Diseases 25 (3), 127–142.

Munro, E.S., Ellis, A.E., 2008. A comparison between non-destructive and destructive testing of Atlantic salmon, *Salmo salar* L., broodfish for IPNV – destructive testing is still the best at time of maturation. Journal of Fish Diseases 31 (3), 187–195.

Munro, E.S., Gahlawat, S.K., Ellis, A.E., 2004. A sensitive non-destructive method for detecting IPNV carrier Atlantic salmon, *Salmo salar* L., by culture of virus from plastic adherent blood leucocytes. Journal of Fish Diseases 27 (3), 129–134.

Muroga, K., 1995. Viral and bacterial diseases in larval and juvenile marine fish and shellfish – a review. Fish Pathology 30 (1), 71–85.

Nakajima, K., Inouye, K., Sorimachi, M., 1998. Viral diseases in cultured marine fish in Japan. Fish Pathology 33 (4), 181–188.

Near, T.J., 2002. Acanthocephalan phylogeny and the evolution of parasitism. Integrative and Comparative Biology 42 (3), 668–677.

Nita-Lazar, M., Mancini, J., Feng, C.G., Gonzalez-Montalban, N., Ravindran, C., Jackson, S., de las Heras-Sanchez, A., Giomarelli, B., Ahmed, H., Haslam, S.M., Wu, G., Dell, A., Ammayappan, A., Vakharia, V.N., Vasta, G.R., 2016. The zebrafish galectins Drgal1-L2 and Drgal3-L1 bind in vitro to the infectious hematopoietic necrosis virus (IHNV) glycoprotein and reduce viral adhesion to fish epithelial cells. Developmental and Comparative Immunology 55, 241–252.

Nordeide, J.T., Matos, F., 2016. Solo *Schistocephalus solidus* tapeworms are nasty. Parasitology 143 (10), 1301–1309.

Obrien, D., Mooney, J., Ryan, D., Powell, E., Hiney, M., Smith, P.R., Powell, R., 1994. Detection of *Aeromonas salmonicida*, causal agent of furunculosis in salmonid fish, from the tank effluent of hatchery-reared Atlantic salmon smolts. Applied and Environmental Microbiology 60 (10), 3874–3877.

OIE, 2016. World Organisation for Animal Health. Available from: http://www.oie.int/.

Olafsdottir, D., Shinn, A.P., 2013. Epibiotic macrofauna on common minke whales, *Balaenoptera acutorostrata* Lacepede, 1804, in Icelandic waters. Parasites Vectors 6, 10.

Olson, K.E., Blair, C.D., 2012. Flavivirus-Vector Interactions. Caister Academic Press, Wymondham.

Otis, E.J., Wolke, R.E., Blazer, V.S., 1985. Infection of *Exophiala salmonis* in Atlantic salmon (*Salmo salar* L.). Journal of Wildlife Diseases 21 (1), 61–64.

Ottem, K.F., Nylund, A., Isaksen, T.E., Karlsbakk, E., Bergh, O., 2008. Occurrence of *Francisella piscicida* in farmed and wild Atlantic cod, *Gadus morhua* L., in Norway. Journal of Fish Diseases 31 (7), 525–534.

Overstreet, R.M., Jovonovich, J., Ma, H.W., 2009. Parasitic crustaceans as vectors of viruses, with an emphasis on three penaeid viruses. Integrative and Comparative Biology 49 (2), 127–141.

Paperna, I., 1980. *Amyloodinium ocellatum* (Brown, 1931) (Dinoflagellida) infestations in cultured marine fish at Eilat, Red-sea – Epizootiology and pathology. Journal of Fish Diseases 3 (5), 363–372.

Paperna, I., Smirnova, M., 1997. Branchiomyces-like infection in a cultured tilapia (*Oreochromis* hybrid, Cichlidae). Diseases of Aquatic Organisms 31 (3), 233–238.

Pasternak, Z., Garm, A., Hoeg, J.T., 2005. The morphology of the chemosensory aesthetasc-like setae used during settlement of cypris larvae in the parasitic barnacle *Sacculina carcini* (Cirripedia : Rhizocephala). Marine Biology 146 (5), 1005–1013.

Pechous, R.D., Sivaraman, V., Stasulli, N.M., Goldman, W.E., 2016. Pneumonic plague: the darker side of *Yersinia pestis*. Trends in Microbiology 24 (3), 190–197.

Pederzoli, A., Mola, L., 2016. The early stress responses in fish larvae. Acta Histochemica 118 (4), 443–449.

Pettersen, J.M., Osmundsen, T., Aunsmo, A., Mardones, F.O., Rich, K.M., 2015. Controlling emerging infectious diseases in salmon aquaculture. Revue Scientifique Et Technique-Office International Des Epizooties 34 (3), 923–938.

Piasecki, W., Mackinnon, B.M., 1995. Life-cycle of a sea louse, *Caligus elongatus* Vonnordmann, 1832 (Copepoda, Siphonostomatoida, Caligidae). Canadian Journal of Zoology-Revue Canadienne De Zoologie 73 (1), 74–82.

Pottinger, T.G., Day, J.G., 1999. A *Saprolegnia parasitica* challenge system, for rainbow trout: assessment of Pyceze as an anti-fungal agent for both fish and ova. Diseases of Aquatic Organisms 36 (2), 129–141.

Powell, A., Rowley, A.F., 2008. Tissue changes in the shore crab *Carcinus maenas* as it result of infection by the parasitic barnacle *Sacculina carcini*. Diseases of Aquatic Organisms 80 (1), 75–79.

Pradeep, P.J., Suebsing, R., Sirithammajak, S., Kampeera, J., Turner, W., Jeffs, A., Kiatpathomchai, W., Withyachumanarnkul, B., 2016. Vertical transmission and concurrent infection of multiple bacterial pathogens in naturally infected red tilapia (*Oreochromis* spp.). Aquaculture Research. http://dx.doi.org/10.1111/are.13102.

Prost, M., 1997. Infectious salmon anaemia – a review. Medycyna Weterynaryjna 53 (9), 487–489.

Qadri, S.S., 1952. Stages in the development of *Trypanosoma danilewskyi* found in a leech (*Hemiclepsis marginata*). Transactions of the Royal Society of Tropical Medicine and Hygiene 46 (1), 2.

Qadri, S.S., 1962. Experimental study of life cycle of *Trypanosoma danilewskyi* in leech, *Hemiclepsis marginata*. Journal of Protozoology 9 (3), 254.

Rameshkumar, G., Ravichandran, S., Sivasubramanian, K., 2013. Secondary microbial infection in carangid fishes due to cymothoid isopod parasites. National Academy Science Letters-India 36 (6), 591–597.

Richards, R.H., Holliman, A., Helgason, S., 1978. *Exophiala salmonis* infection in Atlantic salmon *Salmo salar* L. Journal of Fish Diseases 1 (4), 357–368.

Rimstad, E., Poppe, T., Evensen, O., Hyllseth, B., 1991. Inoculation of infectious pancreatic necrosis virus serotype Sp did not cause pancreas disease in Atlantic salmon (*Salmo salar* L.). Acta Veterinaria Scandinavica 32 (4), 503–510.

Ringo, E., Sinclair, P.D., Birkbeck, H., Barbour, A., 1992. Production of eicosapentaenoic acid (20:5 n-3) by *Vibrio pelagius* isolated from turbot (*Scophthalmus maximus* (L.)) larvae. Applied and Environmental Microbiology 58 (11), 3777–3778.

Roberts, R.J., Pearson, M.D., 2005. Infectious pancreatic necrosis in Atlantic salmon, *Salmo salar* L. Journal of Fish Diseases 28 (7), 383–390.

Rodriguez, S., Alonso, M., Perez-Prieto, S.I., 2001. Detection of infections pancreatic necrosis virus (IPNV) from leukocytes of carrier rainbow trout *Oncorhynchus mykiss*. Fish Pathology 36 (3), 139–146.

Ronneseth, A., Pettersen, E.F., Wergeland, H.I., 2006. Neutrophils and B-cells in blood and head kidney of Atlantic salmon (*Salmo salar* L.) challenged with infectious pancreatic necrosis virus (IPNV). Fish and Shellfish Immunology 20 (4), 610–620.

Ronneseth, A., Pettersen, E.F., Wergeland, H.I., 2012. Flow cytometry assay for intracellular detection of Infectious Pancreatic Necrosis virus (IPNV) in Atlantic salmon (*Salmo salar* L.) leucocytes. Fish and Shellfish Immunology 33 (6), 1292–1302.

Ronneseth, A., Haugland, G.T., Wergeland, H.I., 2013. Flow cytometry detection of infectious pancreatic necrosis virus (IPNV) within subpopulations of Atlantic salmon (*Salmo salar* L.) leucocytes after vaccination and during the time course of experimental infection. Fish and Shellfish Immunology 34 (5), 1294–1305.

Rose, C.R., Deitmer, J.W., 1995. Stimulus-evoked changes of extracellular and intracellular pH in the leech central-nervous-system.1. Bicarbonate dependence. Journal of Neurophysiology 73 (1), 125–131.

Rose, A.S., Ellis, A.E., Munro, A.L.S., 1989. The infectivity by different routes of exposure and shedding rates of *Aeromonas Salmonicida* subsp. *salmonicida* in Atlantic salmon, *Salmo salar* L., HELD in sea-water. Journal of Fish Diseases 12 (6), 573–578.

Rost-Roszkowska, M.M., Swiatek, P., Kszuk, M., Glowczyk, K., Bielecki, A., 2012. Morphology and ultrastructure of the midgut in *Piscicola geometra* (Annelida, Hirudinea). Protoplasma 249 (4), 1037–1047.

Rost-Roszkowska, M.M., Swiatek, P., Poprawa, I., Rupik, W., Swadzba, E., Kszuk-Jendrysik, M., 2015. Ultrastructural analysis of apoptosis and autophagy in the midgut epithelium of *Piscicola geometra* (Annelida, Hirudinida) after blood feeding. Protoplasma 252 (5), 1387–1396.

Rozsa, L., Vas, Z., 2015. Co-extinct and critically co-endangered species of parasitic lice, and conservation-induced extinction: should lice be reintroduced to their hosts? Oryx 49 (1), 107–110.

Rubiliani, C., Payen, G.G., Rubilianidurozoi, M., 1980. Action of root implants and homogenats of the Rhizocephala *Sacculina carcini* Thompson into the male crab *Carcinus maenas* (L.). Comptes Rendus Hebdomadaires Des Seances De L Academie Des Sciences Serie D 290 (4), 355–358.

Samuelsen, O.B., Nerland, A.H., Jorgensen, T., Schroder, M.B., Svasand, T., Bergh, O., 2006. Viral and bacterial diseases of Atlantic cod *Gadus morhua*, their prophylaxis and treatment: a review. Diseases of Aquatic Organisms 71 (3), 239–254.

San, N.O., Nazir, H., Donmez, G., 2012. Microbiologically influenced corrosion failure analysis of nickel-copper alloy coatings by *Aeromonas salmonicida* and *Delftia acidovorans* bacterium isolated from pipe system. Engineering Failure Analysis 25, 63–70.

Schlotfeldt, H.J., Kleingeld, D.W., 1993. A review of the results of 10 years state control service for fish epidemics in lower saxony and 17 years of the fish health-service in Hannover. Tierarztliche Umschau 48 (4), 230.

Scholz, T., 1999. Life cycles of species of *Proteocephalus*, parasites of fishes in the Palearctic region: a review. Journal of Helminthology 73 (1), 1–19.

Schopf, J.W., Kudryavtsev, A.B., Agresti, D.G., Wdowiak, H.J., Czaja, A.D., 2002. Laser–Raman imagery of Earth's earliest fossils. Nature 416, 73–76.

Secombes, C.J., 2016. What's new in fish cytokine research? Fish and Shellfish Immunology 53, 1–3.

Sirisinha, S., 2014. Evolutionary insights into the origin of innate and adaptive immune systems: different shades of grey. Asian Pacific Journal of Allergy and Immunology 32 (1), 3–15.

Smail, D.A., Munro, E.S., 2008. Isolation and quantification of infectious pancreatic necrosis virus from ovarian and seminal fluids of Atlantic salmon, *Salmo salar* L. Journal of Fish Diseases 31 (1), 49–58.

Smith, J.W., 1983. *Anisakis simplex* (Rudolphi, 1809, Det Krabbe, 1878) (Nematoda, Ascaridoidea) – morphology and morphometry of larvae from Euphausiids and fish, and a review of the life-history and ecology. Journal of Helminthology 57 (3), 205–224.

Snow, M., Smail, D.A., 1999. Experimental susceptibility of turbot *Scophthalmus maximus* to viral haemorrhagic septicaemia virus isolated from cultivated turbot. Diseases of Aquatic Organisms 38 (3), 163–168.

Snow, M., King, J.A., Garden, A., Raynard, R.S., 2005. Experimental susceptibility of Atlantic cod, *Gadus morhua* (L.), and Atlantic halibut, *Hippoglossus hippoglossus* (L.), to different genotypes of viral haemorrhagic septicaemia virus. Journal of Fish Diseases 28 (12), 737–742.

Songe, M.M., Willems, A., Sarowar, M.N., Rajan, K., Evensen, O., Drynan, K., Skaar, I., van West, P., 2016. A thicker chorion gives ova of Atlantic salmon (*Salmo salar* L.) the upper hand against *Saprolegnia* infections. Journal of Fish Diseases 39 (7), 879–888.

Sousa, F.E., da Silva, J.H., Saraiva, G.D., Abagaro, B.T.O., Barros, O.A., Saraiva, A.A.F., Viana, B.C., Freire, P.T.C., 2016. Spectroscopic studies of the fish fossils (*Cladocyclus gardneri* and *Vinctifer comptoni*) from the Ipubi formation of the cretaceous period. Spectrochimica Acta Part a-Molecular and Biomolecular Spectroscopy 157, 124–128.

Stone, D.M., Way, K., Dixon, P.F., 1997. Nucleotide sequence of the glycoprotein gene of viral haemorrhagic septicaemia (VHS) viruses from different geographical areas: a link between VHS in farmed fish species and viruses isolated from North Sea cod (*Gadus morhua* L.). Journal of General Virology 78, 1319–1326.

Stueland, S., Hatai, K., Skaar, I., 2005. Morphological and physiological characteristics of *Saprolegnia* spp. strains pathogenic to Atlantic salmon, *Salmo salar* L. Journal of Fish Diseases 28 (8), 445–453.

Sudheesh, P.S., Crane, S., Cain, K.D., Strom, M.S., 2007. Sortase inhibitor phenyl vinyl sulfone inhibits *Renibacterium salmoninarum* adherence and invasion of host cells. Diseases of Aquatic Organisms 78 (2), 115–127.

Sukhdeo, M.V.K., Sukhdeo, S.C., 2004. Trematode behaviours and the perceptual worlds of parasites. Canadian Journal of Zoology 82 (2), 292–315.

The Rasputin effect: when commensals and symbionts become parasitic. In: Hurst, C.J. (Ed.), 2016. Advances in Environmental Microbiology Springer, NY.

Thomas, F., Poulin, R., Brodeur, J., 2010. Host manipulation by parasites: a multidimensional phenomenon. Oikos 119 (8), 1217–1223.

Todar, K., 2011. Textbook of Microbiology. Mechanisms of Bacterial Pathogenicity. 2016.

Toranzo, A.E., Barja, J.L., 1990. A review of the taxonomy and seroepizootiology of *Vibrio anguillarum*, with special reference to aquaculture in the Northwest of Spain. Diseases of Aquatic Organisms 9 (1), 73–82.

Totland, G.K., Hjeltnes, B.K., Flood, P.R., 1996. Transmission of infectious salmon anaemia (ISA) through natural secretions and excretions from infected smelts of Atlantic salmon *Salmo salar* during their presymptomatic phase. Diseases of Aquatic Organisms 26 (1), 25–31.

Uijthof, J.M.J., Figge, M.J., de Hoog, G.S., 1997. Molecular and physiological investigations of *Exophiala* species described from fish. Systematic and Applied Microbiology 20 (4), 585–594.

Vanacker, M., Masson, G., Beisel, J.N., 2012. Host switch and infestation by *Ligula intestinalis* L. in a silver bream (*Blicca bjoerkna* L.) population. Parasitology 139 (3), 406–417.

Vatsos, I.N., Thompson, K.D., Adams, A., 2001. Adhesion of the fish pathogen *Flavobacterium psychrophilum* to unfertilized eggs of rainbow trout (*Oncorhynchus mykiss*) and n-hexadecane. Letters in Applied Microbiology 33 (3), 178–182.

Vatsos, I.N., Thompson, K.D., Adams, A., 2006. Colonization of rainbow trout, *Oncorhynchus mykiss* (Walbaum), eggs by *Flavobacterium psychrophilum*, the causative agent of rainbow trout fry syndrome. Journal of Fish Diseases 29 (7), 441–444.

Walkey, M., 1962. Observations on life history of *Neoechinorhynchus rutili* (Muller, 1776). Parasitology 52 (3–4), 18.

Walkey, M., 1967. Ecology of *Neoechinorhynchus rutili* (Muller). Journal of Parasitology 53 (4), 795.

Wedekind, C., Gessner, M.O., Vazquez, F., Maerki, M., Steiner, D., 2010. Elevated resource availability sufficient to turn opportunistic into virulent fish pathogens. Ecology 91 (5), 1251–1256.

Williams, E.H., Bunkleywilliams, L., Burreson, E.M., 1994. Some new records of marine and fresh-water leeches from Caribbean, Southeastern USA, Eastern Pacific, and Okinawan animals. Journal of the Helminthological Society of Washington 61 (1), 133–138.

Woo, P.T.K., Black, G.A., 1984. *Trypanosoma danilewskyi* – host specificity and hosts effect on morphometrics. Journal of Parasitology 70 (5), 788–793.

Yang, C.Z., Albright, L.J., 1992. Effects of the harmful diatom *Chaetoceros concavicornis* on respiration of rainbow-trout *Oncorhynchus mykiss*. Diseases of Aquatic Organisms 14 (2), 105–114.

Yoon, Y.A., Park, K.S., Lee, J.H., Sung, K.S., Ki, C.S., Lee, N.Y., 2012. Subcutaneous phaeohyphomycosis caused by *Exophiala salmonis*. Annals of Laboratory Medicine 32 (6), 438–441.

Yoshimizu, M., 2016. Bacterial kidney disease of salmonids. Fish Pathology 51 (2), 49–53.

Zbikowska, E., 2011. One snail – three Digenea species, different strategies in host-parasite interaction. Animal Biology 61 (1), 1–19.

Zintl, A., Voorheis, H.P., Holland, C.V., 2000. Experimental infections of farmed eels with different *Trypanosoma granulosum* life-cycle stages and investigation of pleomorphism. Journal of Parasitology 86 (1), 56–59.

Chapter 4

Prophylactic and Prevention Methods Against Diseases in Aquaculture

Parasuraman A. Subramani, R. Dinakaran Michael
Vels Institute for Science, Technology and Advanced Studies, Chennai, India

4.1 INTRODUCTION

Aquaculture is the fastest growing food industry that has increased its worldwide production annually at a rate of approximately 9.6% from 1980 to 2010 (The World Bank, 2013). Fish raised in aquaculture systems face various types of stressors that can be broadly classified into abiotic and biotic stressors. Effects of abiotic stressors in cultured fishes are very difficult to estimate (Fig. 4.1). Even if we estimate the effects, one cannot have control over it, e.g., temperature (Gudding and Van Muiswinkel, 2013) and rainfall. Interestingly, in a survey conducted at Alabama catfish farms, it is found that a very large scale farm (area of the farm >53 hectares) is more likely to get an *Aeromonas hydrophila* outbreak than a comparatively smaller farm (Bebak et al., 2015). On the other hand, some of the biotic factors can be readily controlled and a careful manipulation of certain biotic factors may successfully prevent or at least minimize disease loss in aquaculture. For successful control of diseases, one has to start from the quality of seeds, feed used, etc. to prophylactic measures such as vaccination or immunostimulation. Under unavoidable circumstances such as absence of any prophylactic measures, one may end up in using therapeutic agents, i.e., after eventual loss of some crop to disease. However, the latter was mostly the case with the farmers worldwide until recently.

Unfortunately, production of competent seeds through genetic manipulation of fish immune system is still in its initial stages as the candidate immune response genes are yet to be identified (Biller-Takahashi et al., 2014; Van Muiswinkel et al., 1999). Vaccines, on the other hand, confer protection only to a limited number of pathogens proportional to the number of vaccines developed and used. Literally no commercial vaccines are available to fish diseases in countries like India (Chandran et al., 2002). Another drawback of fish vaccine is the amount of finance and labor required for its procurement and administration respectively which may not be attractive to a fish farmer. Finally, fishes rely much on nonspecific immune responses, particularly during the early part of their lifetime. Thus, at present, immunostimulation seems to be the apt remedy for aquaculture disease problems till other better remedies come up in future (Bricknell and Dalmo, 2005).

The aim of this chapter is to review the relative merits of prophylactic measures in aquaculture, based on the previous studies.

4.2 GENERAL DISEASE PREVENTION AND CONTROL METHODS

4.2.1 Quality Seeds

Selection of cultivable organisms, both plants and animals, is one big discovery that has revolutionized modern, human life style. Historically, farmers selected crops on the basis of economically useful characters such as yield and disease resistance, which are followed even today. However, advancement in molecular biology and genomics provided us numerous technologies for marker-assisted breeding, genetic improvement, genetic modification, and so on. By using molecular biology techniques such as identification of quantitative trait loci (QTL), one can substantially reduce the number of animals killed during disease resistance experiments. High-throughput identification of multiple QTLs via microarray approach is also possible which can further disseminate disease resistance against multiple pathogens (Das and Sahoo, 2014; Laghari et al., 2014). As of 2010, QTLs of more than 20 fish species were reviewed by Yue (2014). The reviewer suggests that genome-wide screening and selection based on these QTLs are necessary in the future to associate these traits with disease resistance. Certain immunologically important genes that correlate with disease resistance are discussed in this section.

Fish Diseases. http://dx.doi.org/10.1016/B978-0-12-804564-0.00004-1

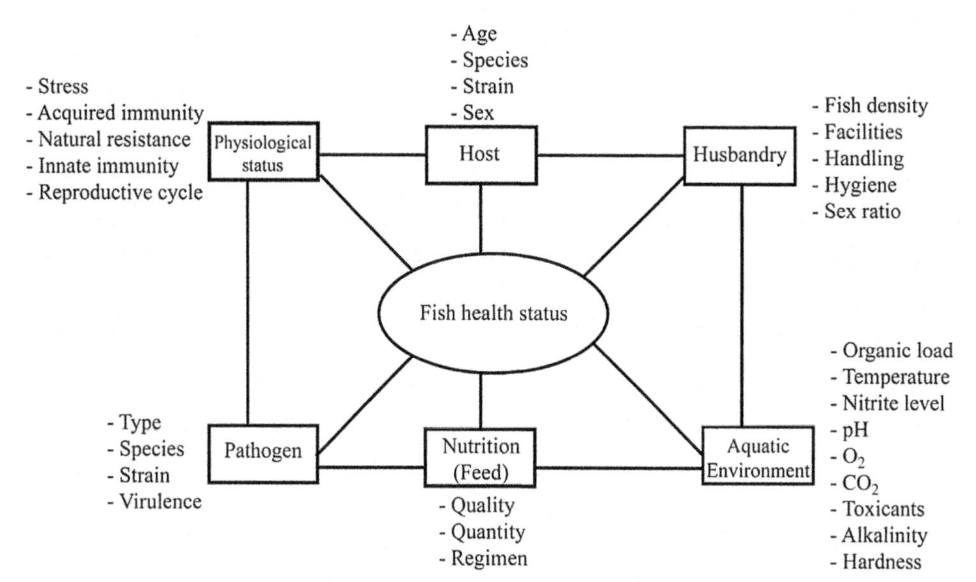

FIGURE 4.1 Factors affecting the health status of fish. *Adapted from Kum, C., Sekki, S., 2011. The immune system drugs in fish: immune function, immunoassay, drugs. In: Aral, F. (Ed.), Recent Advances in Fish Farms. InTech.*

Conventionally, traits for disease resistance were determined by injecting lethal doses of pathogens into different families of fishes and by accounting cumulative mortality of the individual families. In addition to mortality data, various nonspecific immune responses were studied. In one such study, interestingly, some of the important immunological characters of serum from rohu (*Labeo rohita*), such as bacterial agglutination titer, serum hemolysin titer, and hemagglutination titer were shown to be negatively correlated with disease resistance (Sahoo et al., 2008). Nonspecific immune parameters such as serum myeloperoxidase activity, superoxide production, and lysozyme activity were also shown not to correlate with disease resistance. However, only ceruloplasmin activity of *L. rohita* sera positively correlated with the disease resistance against *A. hydrophila* (Sahoo et al., 2008). In another study with *L. rohita* Mohanty et al. (2007) found that only NBT reduction (ROS production) by leukocytes was associated with disease resistance against *Edwardsiella tarda*.

Major histocompatibility complex (MHC)-II of fish is one of the well-explored QTL in fish. MHC-II molecules evolve due to positive pressure of the pathogens prevailing in particular situations (Edwards and Hedrick, 1998). Due to low recombination of MHC-II genes, it is essentially inherited in its entirety from parents in mammals. Many single nucleotide polymorphisms (SNPs) in MHC-II genes were found to be associated with disease resistance in various fishes (Li et al., 2012; Liu et al., 2014; Xu et al., 2010, 2008; Yang et al., 2016). Hence, selecting fishes with MHC-II molecules that can effectively bind and present pathogenic fragments may improve disease resistance in fishes.

Ødegård et al. (2011) and Yáñez et al. (2014) extensively reviewed the conventional as well as advanced methods and results of hybrid selection, QTL mapping, SNP array, marker- and genome-wide selection for successful production of disease resistant salmon seeds. This can be used as a reference guide for production of disease resistant seeds of other fish species.

4.2.2 Culture Hygiene

Diseases become an issue when hygienic practices are overlooked. Some of the important and easy approaches were taken for granted. For instance, lack of wash area for personnel in fish farms and fish marketing area may indirectly affect fish health status (Eltholth et al., 2015). Seining practices are another potential source of contamination in fish farms (Bebak et al., (2015). Eltholth et al. (2015) observed that the fish collecting crates were thrown in mud on the floor, and were seldom disinfected after harvest. Sharing of seining objects between farms is another issue in terms of hygiene that may lead to diseases and disease outbreaks (Bebak et al., 2015). Certain pathogens such as *Flavobacterium columnare* have the ability to sustain in contaminated waters by scavenging on dead fishes. Hence, accumulation of dead fishes and other organic matter must be periodically checked in order to prevent healthy fishes from acquiring diseases (Kunttu et al., 2009). This saprophytic adaptation of fish pathogens is very challenging even for countries that undertake aquaculture in large scale. For instance, periodic increase in the emergence of columnaris disease at a farm in northern Finland might have been due to an extremely adaptive *F. columnare* that acquired the ability to kill fish very quickly and enter saprophytic life cycle before any prophylactic or therapeutic measures could be taken (Pulkkinen et al., 2010). Hence, it is evident that proper hygienic conditions can prevent disease outbreaks in fish farms.

Governments of certain countries like Taiwan mandate their food industries including aquaculture sector for a quality assurance such as "Regulations on Taiwan Fish Products" in order to avert rejection by importing countries on the basis of hygiene (Jiang, 2010). Such mandates are necessary in order to make aquaculture production safe to the consumers.

4.2.3 Water Quality Parameters

Water is one of the limiting factors of aquaculture industry and its quality parameters are a difficult regimen that can ever be managed in fish farms. Water used for aquaculture is also dynamic and the parameters keep on changing within a matter of few hours. Fish farmers must thoroughly consult the seed providers or research laboratories for optimal concentrations of water quality parameters such as dissolved oxygen, nitrite, nitrate, and ammonia levels. One possible way to bypass this painstaking effort is to use biofloc technologies. Biofloc technology uses the heterotrophic bacteria to convert carbon and nitrogen wastes produced by fishes and enhance water quality (Crab et al., 2012). Biofloc technology also has other advantages such as zero water exchange, usage of inexpensive ingredients such as tapioca flour to produce invaluable microbial proteins. In addition to that, fishes grown in biofloc show better disease resistance than those grown in conventional water exchange systems (Ahmad et al., 2016). Alternatively, water reuse system can be used which recycles water by mechanical process in order to keep the ever-changing water quality parameters in check (Lang et al., 2012). Copper sulfate, the most commonly used disinfectant in aquaculture, can be harmful to fish upon prolonged exposure (Bebak et al., 2012). Similarly, increased use of sodium chloride in the US catfish farms led to *A. hydrophila* outbreak (Bebak et al., 2015). On the contrary, increased chloride concentration suppressed the disease outbreak (Bebak et al., 2015).

4.2.4 Nutrition

Aquaculture, just like agriculture, requires many micro- and macronutrients for the well-being of cultured species. Historically, fish farmers looked at fish feed only from the perspectives of growth, feed conversion ratio, and absence of deficiency syndromes. However, their perspective is revolutionized by the recent scientific information connecting the nutritional status with immunological status of fish. Today, fish farmers procure fish feed based on various aspects including normal health and disease resistance. Finding the effect of particular nutrient in fish immune system is challenging because (1) fishes live in water and whatever reaches as feed to a fish might have considerable amount of ingredients already leached by ambient water, and (2) poikilothermic nature of fishes may give different results for particular nutrient in question at different temperatures (Blazer, 1992). Different nutrient components were tested for their disease protecting efficacy in fishes. Some of the highly researched nutrients that have direct implications to fish farming are discussed in the next sections.

4.2.4.1 Vitamins

Ascorbic acid or vitamin C is one of the earliest and highly researched vitamins in terms of disease resistance conferred on different fishes. Transient depletion of ascorbic acid in interrenal tissues was shown to be a marker of stress in two salmonid fishes (Wedemeyer, 1969). Ascorbic acid is well tolerated by fishes (*Clarias batrachus*) even up to 2000 mg and conferred protection against *A. hydrophila* (Kumari and Sahoo, 2005). The authors also recommend that vitamin C be used as an immunostimulant for fishes. Vitamin C has been shown to enhance humoral immune response in *Oreochromis mossambicus* (Michael et al., 1998) and is also shown to boost the growth and immune system of other fishes including salmon (Waagbø et al., 1993), puffer fish (Eo and Lee, 2008), and cobia (Zhou et al., 2012). Pyridoxine (vitamin B_6) did not show appreciable disease resistance efficacy in fish species tested (Albrektsen et al., 1995). The fat soluble vitamin A did not increase resistance against bacterial challenge in Atlantic salmon (Thompson et al., 1994) and Nile tilapia (Guimarães et al., 2014, 2016). However, in one study, vitamin A has been shown to stimulate immune response in Mozambique tilapia (Rajan et al., 2013). Vitamin E also did not show any encouraging increase of resistance against bacterial challenge (Lim et al., 2009). However, in general, all the tested vitamins enhanced specific and innate immunity of fishes such as antibody production and phagocytosis. The other vitamins not listed here are yet to be tested for their efficacy of protecting fish from pathogenic infections.

4.2.4.2 Minerals

Unfortunately, studying the role of minerals in disease resistance of fish is very challenging. This is because of the passive absorption of minerals from water through gills and lateral lines. There are also evidences about toxicity of essential minerals, e.g., iron, copper, given in high quantities (Dalzell and Macfarlane, 1999; Ezeonyejiaku et al., 2011).

Iron is one of the important mineral species in fish similar to that of higher animals. Sheftel et al. (2012) explained the significance of iron in metabolism and pathological conditions of humans. Both host and pathogen compete for iron and their survival shifts toward the organism that procures and utilizes it better. For instance, families of Atlantic Salmon (*Salmo salar*) which depleted circulatory iron quickly by chelating showed disease resistance toward *Piscirickettsia salmonis* greater than families which did not deplete iron (Pulgar et al., 2015). In another classic study, eels (*Anguilla anguilla*) were exposed to bacteria along with sublethal doses of iron and copper. Results from the study show that co-exposure of metals and pathogens decreased the LD_{50} and also the time taken for fish death (Esteve et al., 2012). Thus, it can be recommended that metals such as iron and copper (which are naturally found in water) may be given in optimal level in feed. Selenium, another immunologically important mineral (Arthur et al., 2003), is not well studied in fishes (Sweetman et al., 2010). In one study, selenium supplementation at 4 mg/kg through feed to rainbow trout improved antiviral response by enhancing the expression of IFN-γ and downstream molecules involved in cell-mediated immune response (Pacitti et al., 2016). In a novel study involving azomite (volcanic ash that contains more than 70 minerals) enriched diet protected fishes *O. mossambicus* from *A. hydrophila* infection (Musthafa et al., 2016). In general, excess supplementation of certain minerals to fish feed has shown contradictory results in terms of fish growth and immune status (Sealey and Gatlin, 1999). Rigorous works in future must be done in order to define the role of minerals in boosting the immune system of fish and subsequently protecting them from possible infection with pathogens. Till then, care should be taken not to overdose minerals in fish feeds in order to prevent possible pathogen infections.

4.2.4.3 Proteins

Proteins are very important for proper functioning of immune system since most of the immune molecules such as antibodies, cytokines, serum enzymes, complements, and so on are proteins. However, very high protein diet also retards the immune system (Kiron et al., 1995b). Hence, a balance must be struck. Finfishes, like other vertebrates, need 10 dietary amino acids (FAO: Nutritional requirements, 2016). Farmers must optimize protein concentration in fish feed in order to get better results (Collins et al., 2012). According to the methods given by Kiron et al. (1995b), 20–35% protein in feed is considered to be optimal for protecting rainbow trout from viral infection.

Source of protein play an important role in protecting fishes from diseases. Currently plant-based protein sources are used widely in aquaculture, which include soy bean and cottonseed (Yue and Zhou, 2008). However, uses of conventional sources of proteins, e.g., fish meal, are becoming limited in aquaculture. On the other hand, effect of different sources of proteins on health status of fish is now being explored. For instance, fishes fed with protein hydrolyzates of animal origin (tilapia, krill, or shrimp) showed better resistance to *E. tarda* infection compared to that of fishes fed with protein hydrolyzates of plant (soybean) origin (Khosravi et al., 2015a, 2015b; Krogdahl et al., 2000). An interesting reason behind this fact is the difference in the composition of amino acids between plant and animal source. Arginine is one of the major components that differ in its concentration between the two sources. To add to the story, supplementation of arginine alone to fish diets augmented important immune parameters (Zhou et al., 2015). However, at very high concentration (3.3% w/w) arginine inhibited some of the immune functions. An optimum level of 2.7% arginine was recommended for catfish from this study. Another inexpensive alternative to fish hydrolyzate is poultry feather lysate which had better performances than that of soybean hydrolyzate (Zhang et al., 2014a). Methionine, another amino acid found in low levels in plant can impair growth and immune responses of fishes (Belghit et al., 2014).

4.2.4.4 Lipids and Sterols

Lipids are one of the macronutrients required for the formation of membranes, signaling molecules, and other structural roles in fishes. Deficiency of essential fatty acids may lead to decreased antibody production and bactericidal activity of macrophages (Kiron et al., 1995a). Inclusion of certain fatty acids to fish feed had in past led to improved disease resistance (Chakrabarti et al., 2012; Vargas et al., 2013). However, one should be careful in selecting the quality and source of lipids used for feed (Fracalossi and Lovell, 1994). For instance, Menhaden oil obtained from common marine foray fishes was used as one of the sole lipid source in one study. It was found that channel catfish feed containing only Menhaden oil (high concentration) showed increased mortality percentage against *Edwardsiella ictaluri* (Li et al., 1994). Replacement of fish oils with vegetable oil that are less expensive can be encouraged only up to a certain limit as higher ratio of vegetable oil in fish feed is detrimental to fish (Montero et al., 2003; Mourente et al., 2005). Hence, species-specific and predefined balanced lipid diet is the need of the hour to successfully protect finfish from infectious diseases.

Dietary linoleic acid is shown to increase disease resistance in dark barbel catfish (Li et al., 2015). Supplementation of n-3 highly unsaturated fatty acids improved nonspecific immune response and disease resistance of Japanese flounder against *E. tarda* challenge (Wang et al., 2006). Feed supplemented with coconut oil and cod-liver oil containing n-3 and n-6 unsaturated fatty acids protected fishes against *A. hydrophila* infection higher than control diet (Vargas et al., 2013). Lipids found in flaxseed oil enhance the performance of channel catfish at a suboptimal temperature (Thompson et al., 2015). Active principles of a plant (*Achyranthes aspera*)-derived immunostimulant is shown to be sterol ecdysterone and linoleic and oleic fatty acids (Chakrabarti et al., 2012). Interestingly, cholesterol supplementation in fish diet gave better protection for rainbow trout against *A. hydrophila* (Deng et al., 2013).

4.2.4.5 Carbohydrates

Like other nutrients, carbohydrates are essential for fish but their excessive use may harm the fish immune system. For instance, of the different doses of carbohydrates (0–28% w/w) mixed with feed and fed to golden pompano (*Trachinotus ovatus*), it was reported that 11.2–16.8% (w/w) carbohydrate in feed conferring increased protection against *Vibrio harveyi* (Zhou et al., 2014). Similar results were reported for black carp fed with different doses of carbohydrates and infected with *A. hydrophila* (Wu et al., 2016). Torrecillas et al. (2014) reviewed the use of mannan oligosaccharide supplemented feeds for aquaculture industry. Mannan oligosaccharides are very good stimulators of nonspecific immune system and are shown to protect fishes from experimental infections in a number of studies. However, optimum dose of mannan oligosaccharides is still debated (Torrecillas et al., 2014).

Nutrition is certainly an indispensable player of fish immune homeostasis. Though the exact figures are not known, optimum levels of nutrients are known for many species and for others it can be reached by a simple trial and error method. More research in future should focus on finding highly defined diets that can boost the immune status of fish and subsequently protect it from infection in addition to the enhanced growth parameters.

4.2.5 Chemicals Used to Control Diseases

4.2.5.1 Disinfectants

Disinfectants such as chlorine, ethyl alcohol, iodine, hydrogen peroxide, and formaldehyde are used regularly in finfish aquaculture.

Surface disinfection of marine fish eggs had shown to increase the performance of many hatcheries. Disinfecting marine fish eggs with 400 ppm glutaraldehyde for 10 min was found to be very effective (Salvesen and Vadstein, 1995). In several Norwegian hatcheries glutaraldehyde was used as a standard disinfectant (Skjermo and Vadstein, 1999). In one study, providon (iodophore) treatment at 50 mg/L for 15 min was shown to decrease bacterial counts associated with eggs of three marine fish but only with a very low hatching rate. On the other hand, formalin at concentration of 1000 mg/L did highly enhance the hatching rate though bacterial counts were unaltered (Stuart et al., 2010).

Hydrogen peroxide is considered to be the safest of the disinfectants. It readily decomposes into oxygen and water leaving no residue. It is also effective against most of the ectoparasites of fish (Pedersen and Pedersen, 2012). However, one has to optimize the concentration of hydrogen peroxide to be used depending on species of the fish and the parasite to be eliminated. Hydrogen peroxide is also susceptible to heat hence better compounds than hydrogen peroxide are currently being investigated (Pedersen and Pedersen, 2012).

Peracetic acid (PAA) is a powerful disinfectant with a wide spectrum of antimicrobial activity and a potential to replace harmful chemicals from aquaculture industry. PAA, like hydrogen peroxide, is rapidly degraded into water and oxygen causing no harm to environment and host (Pedersen et al., 2009). Peracetic acid exhibits very good antimicrobial and antiparasitic activities (Kitis, 2004; Pedersen et al., 2013). But peracetic acid is comparatively expensive and found to increase the organic content of water. This in turn leads to regrowth of microbes (Kitis, 2004).

In an in vitro study, ethyl alcohol (30, 50, or 70%), benzyl-4-chlorophenol/phenylphenol (1%), sodium hypochlorite (50, 100, 200, or 50,000 mg/L), *n*-alkyl dimethyl benzyl ammonium chloride (1:256 v/v), povidone iodine (50 or 100 mg/L), glutaraldehyde (2%), and potassium peroxymonosulfate/sodium chloride (1%) were found to be effective disinfectants, as each reduced or eliminated the number of detectable *E. ictaluri* and *E. tarda* within 1 min of contact time. However, neither Chloramine-T (15 mg/L) nor formalin (250 mg/L) was effective against these bacteria even after 60 min of contact time (Mainous et al., 2010). Accessory materials, e.g., seining nets, can be disinfected by soaking them in sodium hypochlorite for 30 min or in potassium peroxymonosulfate + sodium chloride for 5 or 30 min (Collymore et al., 2014; Garcia and Sanders, 2011). Ozone is another disinfection agent used in aquaculture. It was found that applying ozone at a dose of 20 g O_3/kg feed per day improved water quality parameters of recirculating seawater aquaculture systems (Park et al., 2013).

Some of the commonly used disinfectants mentioned in FAO report (GESAMP Reports and Studies No.65, 1997) are Chloramine-T, formalin, hypochlorite (sodium or calcium hypochlorite), iodophores, ozonization, and quaternary ammonium compounds such as *n*-alkyl dimethyl benzyl ammonium chloride. Potassium chloride is recommended by FAO for treating fish against various microbial diseases including scale erecting disease and saprolegniasis.

Since most of the disinfectants, especially chlorine, iodine, etc., are harmful to the hosts, they should be used minimally with utmost care. Ultimately, some disinfectants like formaldehyde are carcinogenic to the end users (human beings). Hence, it is not a surprise that some countries mandate a withdrawal period before harvesting (Burridge et al., 2010). Environmental concerns about the release of disinfectants treated water used for aquaculture water bodies into natural water bodies were also alarming. Particularly, toxic effect of chlorine was well documented (GESAMP Reports and Studies No.65, 1997).

4.2.5.2 Antiparasitic

Parasites such as worms and sea lice are conventionally controlled by the use of chemicals. Chemicals such as magnesium sulfate, copper sulfate, and potassium permanganate are routinely used for this purpose.

Copper can be used to control algal growth and most external parasites of fish, including "ich" (Weimin, 2004). Different forms of copper were used in aquaculture but the inexpensive form is its salt, copper sulfate. One concern about copper is that its toxicity toward fish increases when total alkalinity of water decreases. Thus, one has to be careful while using copper sulfate. One consensus is that copper treatment should not be done to fish reared in water with a total alkalinity less than 50 ppm (Watson and Yanong, 1989).

Potassium permanganate is a general biocide that can kill many fish parasites (Duncan, 1978; Schlenk et al., 2000). It is even used as a general water quality enhancer because of its capability to increase dissolved oxygen levels (Tucker, 1984).

Of the many antiparasitic drugs tested, praziquantel was found to be very effective in controlling the worm, *Lepidotrema bidyana*. Praziquantel exposure at 40 mg/L for 1 h killed 77% of the worms while the archaic formalin, sodium chloride, and sodium percarbonate were ineffective (Forwood et al., 2013). Avermectins were commonly used for treatment of sea lice in salmon aquaculture (Lees et al., 2008), but are associated with many side effects. Emamectin benzoate is the only licensed avermectin currently used in countries like Canada (Burridge et al., 2010). Organophosphorus compounds, such as malathion and trichlorfon, were widely used in aquaculture when they were routinely used for agriculture. As their adverse effects were evident, these drugs were withdrawn from aquaculture industry. Chitin synthesis inhibitors, e.g., teflubenzuron and diflubenzuron, are the two registered products used insecticides in aquaculture. They are most effective against sea lice particularly during their larval and copepodid stages (Branson et al., 2000).

4.2.5.3 Antibiotics

Fish farmers indiscriminately used antibiotics due to their anxiety to control the diseases and more so due to its easy availability over the counter. Fish farmers also have a notion that antibiotics can immediately cure fish and because of their specificity, they do not pose any threat to the hosts. Disregarding very rare reports of immunostimulating activity of the antibiotics (Bizzini and Fattal-German, 1989), most of the reports decipher the detrimental activity of antibiotics on immune system of animals. Despite an outcry to abolish antibiotics from aquaculture needs, investments in research and development of antibiotics were predicted to increase (Bondad-Reantaso et al., 2005). Antibiotics such as amoxicillin, florfenicol, oxolinic acid, oxytetracycline, and erythromycin are widely used in salmon aquaculture (Burridge et al., 2010) and other fish cultures (Zamri-Saad et al., 2014). Use of these antibiotics is regulated in developed countries by strict laws that demand prescription from registered veterinary practitioner to purchase them. Each of the antibiotics should also have a withdrawal period, only after which the fishes are harvested. Clearance from environment is another important factor that adds to the detrimental effects of antibiotics. For instance, natural β-lactam antibiotics are effectively degraded from aquatic environments while quinolones take very long time to degrade (Burridge et al., 2010). The time taken for degradation plays a vital role in inducing antibiotic resistance to bacteria and at the same time nonspecifically killing all the beneficial bacteria (Burridge et al., 2010). Antibiotics were often mixed with feed and administered to fishes. This may not give sufficient results if the infected fishes do not eat properly. There is also a risk that infection may recur if complete regimen of antibiotics is not properly followed. Upon recurrent infections, farmers tend to extend the regimen of antibiotics and also increase the dosage. Such indiscriminate use of antibiotics is detrimental to both fishes and humans (Zamri-Saad et al., 2014), and even to plants (Bártíková et al., 2016).

4.3 BIOLOGICS

"Prevention is better than cure" is an old and widely accepted maxim. We are aware that immunoprophylactic methods include vaccination and immunostimulation by different plant or microbial products or even microbes themselves (probiotics). Vaccines are preparations that mainly activate the host's specific immune responses with significant immunological memory against a particular pathogen. On the other hand, immunostimulants mainly activate the nonspecific immune responses of the host and protect the host from various pathogens. Use of probiotics is the voluntary administration of beneficial microorganisms to improve the health status of the host. The idea of all three prophylactic methods stemmed from their successful use in human beings against dreadful diseases. In principle, these methods are same in fishes and their mammalian counterparts, however, not without practical differences. In this section, we will describe in detail the three most important prophylactic measures namely vaccination, immunostimulants, and probiotics. Current prophylactic techniques against various pathogens are also described.

4.3.1 Vaccines

Vaccine by definition is a preparation that elicits acquired immunity against a disease caused by a particular pathogenic species. Though nonspecific mechanisms are also involved in the process, the specific antibody response mainly takes the credit for protection conferred by a vaccine. Fishes being lower vertebrates rely more upon nonspecific responses upon infection with a pathogen. However, pioneering studies by Snieszko, Duff, and others have made vaccines important in the field of aquaculture. The first experimental fish vaccination dates back to 1938 by a pioneering work done by Snieszko et al. that showed the formation of protective immunity against *A. hydrophila* upon injection of fish with the killed *A. hydrophila* (reviewed by Van Muiswinkel, 2008). However, the same authors later showed the antibacterial chemotherapy (antibiotics) to be better than vaccine (Snieszko and Friddle, 1949). The same school of thought was followed for another 20–25 years by leading researchers of aquaculture which was later termed as "the era of chemotherapy" by Evelyn (1997). However, due to the indiscriminate use and the negative side effects of antibiotics which we would discuss later, researchers recommended immediate curb on the use of antibiotics in the aquaculture industry. After this brief hiatus, scientists impressed by the success brought about by vaccines in humans, cattle, and poultry turned their attention toward the production of fish vaccines. This led to the issue of first licensed fish vaccine in 1976 against *Yersinia ruckeri* (Gudding and Van Muiswinkel, 2013). A successful point in the field of aquaculture vaccines was reached when vaccine use in Norway substantially reduced the usage of antibiotics (Brudeseth et al., 2013).

4.3.1.1 Types

Like their mammalian counterparts, both passive and active immunities are described in fishes. Passive immunity is obtained in a host from preformed antibody (horizontal transfer) against a pathogen in a different animal. An experiment conducted on rainbow trout showed that the preformed antibodies present in plasma of vaccinated fishes when transferred were able to confer protection to naive fishes (Evenhuis et al., 2014). Though it is fascinating to study the antisera-mediated passive immunity of the primitive vertebrate, it may not be suitable for large scale culture situations. Passive vaccination with antisera may be limited to otherwise incurable pathogen to which the host cannot mount a protective immunity upon vaccination. However, a promising alternate passive immunization exists and it happens naturally—the transfer of maternal antibodies to the offspring (vertical transfer). For instance, microinjection of anti-zebra fish IgM into zebra fish embryos decreased protection against the opportunistic pathogen, *A. hydrophila* compared to that of normal embryos (Wang et al., 2012) clearly indicating the role of maternal-derived antibodies in the survival of embryos. Zhang et al. (2013) in their review explained in detail the factors or molecules transferred from brood stock to larvae, mechanism of action of transferred factors, factors that affect vertical transfer, and the potential application of maternal immunity in aquaculture. They arrived at one important conclusion that fish larvae can be easily protected by immunizing the brood stock directly. An earlier review (Swain and Nayak, 2009) also explained the same point. Despite the advantages of maternal immunity, only limited work has been done in this area. Future research may lead to the transformation of this potential immunization into a widely followed method worldwide.

Active and passive immunities are opposite of each other. In the former, the host is exposed to the vaccine (either whole pathogen or pathogen-derived molecules) and the host mounts a protective immunity against the pathogen. The immunization against a pathogen may be either natural or artificial. Under natural conditions, infectious diseases follow after situations that cause stress such as overcrowding, temperature, and drought. Reports on natural active immunity mounted by fishes are very rare. However, in aquaculture farms where fishes essentially face these stressors may not get sufficient time to mount effective active immunity. In fishes, active immunity is developed in mucosal secretion of organs such as skin, gills, or intestine without an overall systemic immunity (Uribe et al., 2011). This is a hot research area in the field of fish immunology and a lot of work is being done mostly in comparison to the mammalian counterparts. As mentioned earlier, studies on active immunization date back to 1938 and it is currently the method of choice in aquaculture industries of the developed countries (Brudeseth et al., 2013). The earlier vaccines are simply either heat- or formalin-killed preparations (inactivated) of pathogenic bacteria. Modern vaccines such as recombinant DNA vaccines and nanoparticle-mediated vaccine delivery are also the products of research (Brudeseth et al., 2013).

There are many advantages and disadvantages of all vaccine types (Table 4.1). A live attenuated vaccine, for example, under certain conditions, may unleash its virulence upon exposure to a suitable host. Killed vaccines on the other hand lose their immunogenicity and may require repeated booster doses in order to give a longer protection.

4.3.1.2 Vaccination Methods

By and large, intraperitoneal injection of antigens into the fishes is considered to be the most effective route. At present, vaccination, in countries where practised, is done more by injection compared to other routes (Gudding and Van Muiswinkel, 2013). One of the first used fish vaccines, well-documented in Norway, was simple saline suspension (water-based) of bacterins injected intraperitoneally. However, these water-based vaccines are not very protective, and particularly after the sudden outbreak of *Aeromonas salmonicida* in Norwegian aquaculture they switched to oil-based vaccines (Freund's adjuvants). Since then, antigens suspended in oil are the method of choice injected to fish. Oil-based vaccines also conferred very long-term protection, from fingerling stage to harvest. Shift to oil-based vaccines was not without drawbacks. One of the important drawbacks was downgrading of salmons, i.e., poor growth rates. The other well-known effect being local injection-site reactions. Even now in some of the countries, fish vaccination is done manually which is economically daunting. Invention of automated vaccinating machines solved this problem to some extent though not without economic constraints (Brudeseth et al., 2013). Injection vaccination is also not feasible for larvae because it is difficult and labor intensive, but most of the mortalities in fishes occur at larval stages (Gudding and Van Muiswinkel, 2013).

Fishes can take up vaccine through their mucosal surfaces present in gills and skin. Fish kept in vaccine solution could take in vaccine and produce both systemic and mucosal antibodies leading to protection against the particular disease (Huising et al., 2003). Another method is called hyperosmotic infiltration (HI) method, commonly called immersion (if it is less than 2 min). In HI, fishes are first exposed to severe osmotic shock by keeping them in hypertonic solutions of salt and immediately transferring them to solution containing the vaccine. HI method is not economical for bigger fishes. For instance, an amount of vaccine that can vaccinate 1000 numbers of 10 g fish, can only effectively vaccinate 100 numbers of 100 g fish by HI method (Newman, 1993). Further, stress caused by HI treatment should be taken into consideration before vaccination. A popular variant of HI is the bath method in which fishes are exposed to diluted vaccine for a longer periods ranging from 20 min to several hours (Newman, 1993). Vaccination by bathing is an ideal method for immunizing large number of fishes without stressing them. However, bath immunization confers protection only to a limited period hence requiring subsequent repeated administration of antigens (Lillehaug, 1989).

Another modification of HI is the spray vaccination which is very economical and can be used to treat more number of fishes than by other immersion techniques. However, handling of fish and stress during administration may make spray vaccination formidable to fish farmers.

The fourth and last vaccination method is the oral (dry) vaccination. This is by far the most affordable method of vaccination, especially in large scale intensive cultures. In this method, bacterial suspensions are directly mixed with feed and given to fish. As no handling procedures and stressors are involved, this can be routinely used to immunize the entire fish populations even many times. However, ambiguous results about efficacy and duration of protection were reported for oral vaccination by different researchers (Vandenberg, 2004). Different fishes exhibit different mechanisms of antigen uptake and immune responses. For instance, cyprinids that lack functional stomach may uptake oral vaccines because of the lack of digestive proteolytic enzymes (Vandenberg, 2004). Thus, oral vaccination method can be used to administer boosters at regular intervals for vaccines that were administered by other means (Newman, 1993).

4.3.1.3 Current Status and Future Perspectives

A detailed list of vaccines available for finfish is tabulated for pathogen specificity and type of administration (Table 4.2). It should be noted that production of specific vaccine for every possible pathogen is very time-consuming and expensive. The future vaccination strategies may aim at increasing the efficacy of existing vaccines by supplementing with enhancing compounds called adjuvants. Only primitive adjuvants, if any, are used in aquaculture industry (Midtlyng et al., 1996). However,

TABLE 4.1 Advantages and Disadvantages of Different Types of Vaccines

Type of Vaccine	Advantages	Reference(s)	Disadvantages	Reference(s)
Live attenuated	**1.** Activates all branches of the immune system including nonspecific and mucosal immunity. **2.** Provides comparatively long lasting immunity. Boosters are seldom required. **3.** Protective immunity achieved rapidly.	Jones et al. (2005) Kaushal et al. (2015)	**1.** Pathogen can revert back to virulence. **2.** Can cause severe complications in immunocompromised patients. **3.** Can be difficult to transport due to requirement to maintain conditions. For instance, refrigeration.	Shimizu et al. (2004) and Kamboj and Sepkowitz (2007)
Killed or inactivated	**1.** Completely safe as the pathogen cannot revert back to virulence. **2.** Inexpensive. **3.** No special requirements necessary for carrying the vaccine. **4.** Almost all antigens present in pathogen are available to the host.	Baxter (2007)	**1.** Requirement of booster. **2.** Not effective against intracellular pathogens. **3.** Requirement of adjuvants leading to adjuvant-related complications.	Morefield et al. (2005) and Baxter (2007)
Protein subunit	Extremely safe and can be used even on immunocompromised, stressed and even brood stocks.		Problems associated with adjuvants, boosters and local-site reactions.	Baxter (2007)
DNA vaccine	**1.** Inexpensive, stable, easy to transport and handle. **2.** Long-term protection. **3.** If vaccine is a plasmid, a mixture of many plasmids can be used to form a broad spectrum vaccine. **4.** Considered to be safe although it activates different leukocyte population.	Sasaki et al. (2003) Alarcon et al. (1999) Khan (2013) Robinson and Pertmer (2000)	**1.** Exclusive for protein antigens. **2.** May induce immunological tolerance. **3.** Since the DNA is foreign it may induce production of antibodies against. **4.** Mostly by injection route.	Kindt et al. (2007) and Ferrera et al. (2007), Khan (2009)

TABLE 4.2 List of Commercial Vaccines Available Against Diseases of Farmed Finfish

	Causative Agent	Vaccine Type	Route of Administration	Manufacturer/References
Viral	Infectious hematopoietic necrosis virus (IHNV)	DNA vaccine	Intramuscular injection	Aqua Health Ltd., Novartis, Canada
	Irido virus	Inactivated	Intraperitoneal injection	Merck Animal Health, the USA
	Koi herpes virus (KHV)	Attenuated	Immersion or injection	KoVax Ltd., Jerusalem, Israel
	Red sea bream iridovirus	Inactivated	Intraperitoneal injection	Nakajima et al. (1999)
	Salmon alpha viruses	Inactivated	Intraperitoneal injection	Pharmaq AS, Norway Intervet-International BV, the Netherlands
	Spring viremia of carp virus	Subunit	Intraperitoneal injection	Pharos, S.A., Belgium
	Infectious salmon anemia virus	Subunit	Oral	Centrovet, Chile
		Inactivated	Intraperitoneal injection	**1.** Alpha Jects, Pharmaq AS, Norway **2.** Microtek International Inc., British Columbia, Canada **3.** FORTE VI, Aqua Health Ltd., Novartis, Canada
	Infectious pancreatic necrosis virus	Subunit	Intraperitoneal injection	**1.** Microtek International Inc., British Columbia, Canada **2.** Intervet-International BV, the Netherlands
			Oral	Merck Animal Health, New Jersey, the USA
		Inactivated	Intraperitoneal injection	**1.** Pharmaq AS, Norway **2.** Centrovet, Chile **3.** Aqua Health Ltd., Novartis, Canada
Bacterial	*Vibrio anguillarum* and *Vibrio ordalii*	Inactivated	Immersion or intraperitoneal injection	MSD Animal Health, the USA
	Yersinia ruckeri	Inactivated	Oral	Merck Animal Health, New Jersey, the USA
			Immersion	Merck Animal Health, New Jersey, the USA
	Aeromonas salmonicida	Subunit	Intraperitoneal injection	Merck Animal Health, New Jersey, the USA
	Vibrio salmonicida	Inactivated multivalent	Intraperitoneal injection	Merck Animal Health, New Jersey, the USA
	Photobacterium damselae	Inactivate	Immersion	Merck Animal Health, New Jersey, the USA
	Edwardsiella ictaluri	Live attenuated	Immersion	Merck Animal Health, New Jersey, the USA
	Moritella viscosa	Inactivated	Injection	Merck Animal Health, New Jersey, the USA
	Streptococcus iniae	Inactivated	Injection	Merck Animal Health, New Jersey, the USA
	Lactococcus garviae	Inactivated	Oral	Merck Animal Health, New Jersey, the USA
		Inactivated	Injection	Merck Animal Health, New Jersey, the USA
	Flavobacterium columnare	Live attenuated	Immersion	Merck Animal Health, New Jersey, the USA

TABLE 4.3 List of Selected DNA Vaccines Investigated

No	Pathogen	Antigen	References
1	*Streptococcus iniae*	Streptolysin (sag)	Sun et al. (2012a)
2	*Edwardsiella tarda*	Eta6 and FliC	Jiao et al. (2009)
3	Infectious pancreatic necrosis virus	preVP2, VP2 and VP3	Cuesta et al. (2010)
4	Infectious hematopoietic necrosis virus	pIRF1A-G	Ballesteros et al. (2015)
5	*Streptococcus iniae* and *Vibrio anguillarum*	Sia10 and OmpU	Sun et al. (2012b)
6	Many (multivalent)	EOmpAs-19	Li et al. (2016)

many side effects including local injection-site lesions to unwanted fillet colorations were reported (Midtlyng, 1996). These side effects should be avoided with improved adjuvants and perhaps with nanoparticles-mediated immunostimulating complexes (ISCOMs) that is capable of engaging both humoral and cellular branches of immune system in order to give better disease resistance and survival.

In addition to existing vaccines, many advanced ones were developed, or are being developed, by fish immunologists all over the world. One of the fascinating prototypes being developed is the DNA vaccines. Some of the DNA vaccines (Table 4.3) had shown outstanding results in laboratory settings especially against viral pathogens (Evensen and Leong, 2013). Nanoparticles-mediated delivery of vaccine is another promising prospect that needs more research in future (Zhao et al., 2014).

4.3.2 Immunostimulants

By definition, "an immunostimulant is a chemical, drug, stressor, or action that elevates the nonspecific defence mechanisms or the specific immune response if it (the treatment) is followed by vaccination or infection. Immunostimulants may be given by themselves to activate nonspecific defence mechanisms, or they may be administered with a vaccine to activate nonspecific defence mechanisms as well as heightening a specific immune response" (Anderson, 1992). Immunostimulants are the most sought after "wonder" products in aquaculture because of their ease of administration and the broad range of protection they provide compared to the vaccines. Immunostimulants from various sources were tested in laboratories and aquaculture farms; their mode of action (if known), and future prospects are described in this section. Although various candidates of different sources have been tested for their efficacy in stimulating nonspecific immune responses of fishes, only the ones that actually protected fishes from experimental infection are discussed in detail.

4.3.2.1 Sources

4.3.2.1.1 Plant-derived Immunostimulants

Plants are simple and direct sources of immunostimulants, and many of them were used for humans during the past few thousands of years (Tan and Vanitha, 2004). A list of traditional medicinal plants tested for their efficacy as immunostimulants in fish are tabulated (Table 4.4). Various medicinal and other plants were reported to have immunostimulatory and disease protective properties in fish in the laboratory and some of them are currently being investigated in detail. Crude extracts from plants were shown to enhance various components of innate immune parameters, e.g., lysozyme, complement, antiprotease, myeloperoxidase, reactive oxygen species, reactive nitrogen species, phagocytosis, respiratory burst activity, nitric oxide, total leukocytes, glutathione peroxidase, and subsequent protection against bacterial, fungal, viral, and parasitic diseases (Van Hai, 2015). The effect of various plant-derived immunostimulants in fish immune system is tabulated in Table 4.5 and their mechanisms of action are listed in Table 4.6. Unfortunately, not much is known about the bioactive compounds present in these plants. However, in aquaculture situations, purification of plant compounds is seldom required and even parts of a plant such as leaves are directly mixed with the feed and administered to fishes (Giri et al., 2015).

4.3.2.1.2 Microbial-derived Immunostimulants

Microbial-derived products are one of the first used immunostimulants. Mycobacterium-derived products present in commonly used Freund's complete adjuvant and bacterial LPS are good examples of them. Many such products are enlisted

TABLE 4.4 List of Plants That Protected Fish From Experimental Infection With Pathogen (Literature Published Mostly in the Last 10 Years)

No.	Name of the Plant	Fish Tested	Pathogen	References
1.	*Azadirachta indica*	*Lates calcarifer*	*Vibrio harveyi*	Talpur and Ikhwanuddin (2013)
2.	*Astragalus radix*	*Cyprinus carpio*	*Aeromonas hydrophila*	Yin et al. (2009)
3.	*Aloe vera*	*Oreochromis niloticus*	*Streptococcus iniae*	Gabriel et al. (2015)
4.	*Viscum album*	*Anguilla japonica*	*Aeromonas hydrophila*	Choi et al. (2008)
5.	*Avicennia marina*	*Amphiprion sebae*	*Vibrio alginolyticus*	Dhayanithi et al. (2015a)
6.	*Cynodon dactylon*	*Catla catla*	*Aeromonas hydrophila*	Kaleeswaran et al. (2011)
7.	*Sophora flavescens*	*Oreochromis niloticus*	*Streptococcus agalactiae*	Wu et al. (2013)
8.	*Sauropus androgynus*	*Epinephelus coioides*	*Vibrio alginolyticus*	Samad et al. (2014)
9.	*Toona sinensis*	*Oreochromis mossanbicus*	*Aeromonas hydrophila*	Wu et al. (2010)
10.	*Eriobotrya japonica*	*Epinephelus bruneus*	*Vibrio carchariae*	Kim et al. (2011)
11.	*Solanum trilobatum*	*Oreochromis mossambicus*	*Aeromonas hydrophila*	Divyagnaneswari et al. (2007)
12.	Mixed herbal extracts	*Carassius auratus*	*Aeromonas hydrophila*	Harikrishnan et al. (2010a)
13.	*Ficus benghalensis* and *Leucaena leucocephala*	*Clarias gariepinus*	*Aeromonas hydrophila*	Verma et al. (2013)
14.	*Astragalus, angelica, hawthorn, Licorice root* and *honeysuckle*	*Oreochromis niloticus*	*Aeromonas hydrophila*	Tang et al. (2014)
15.	*Azadirachta indica, Ocimum sanctum* and *Curcuma longa*	*Carassius auratus*	*Aeromonas hydrophila*	Harikrishnan et al. (2009b)
16.	*Kalopanax pictus*	*Epinephelus bruneus*	*Vibrio alginolyticus* and *Philasterides dicentrarchi*	Harikrishnan et al. (2011b)
17.	*Siegesbeckia glabrescens*	*Epinephelus bruneus*	*Vibrio parahaemolyticus*	Harikrishnan et al. (2012c)
18.	*Eclipta alba*	*Oreochromis mossambicus*	*Aeromonas hydrophila*	Christybapita et al. (2007)
19.	*Ixora coccinea*	*Carassius auratus*	*Aeromonas hydrophila*	Anusha et al., 2014
20.	*Viscum album*	*Oreochromis niloticus*	*Aeromonas hydrophila*	Park and Choi (2012)
21.	*Tinospora cordifolia*	*Oreochromis mossambicus*	*Aeromonas hydrophila*	Alexander et al. (2010)
22.	Astaxanthin	*Cyprinus carpio*	*Aeromonas hydrophila*	Jagruthi et al. (2014)
23.	Andrographolide	*Labeo rohita*	*Aeromonas hydrophila*	Basha et al. (2013)
24.	*Psidium guajava*	*L. rohita*	*Aeromonas hydrophila*	Giri et al. (2015)
25.	*Azadirachtin*	*Carassius auratus*	*Aeromonas hydrophila*	Kumar et al. (2013)
26.	Digestarom®	*Ictalurus punctatus*	*Edwardsiella ictaluri*	Peterson et al. (2015)
27.	*Astaxanthin and emodin*	*Pelteobagrus fulvidraco*	*Proteus mirabilis*	Liu et al. (2016)
28.	*Urtica dioica*	*Labeo victorianus*	*Aeromonas hydrophila*	Ngugi et al. (2015)

TABLE 4.4 List of Plants That Protected Fish From Experimental Infection With Pathogen (Literature Published Mostly in the Last 10 Years)—cont'd

No.	Name of the Plant	Fish Tested	Pathogen	References
29.	Soybean isoflavones	*Trachinotus ovatus*	*V. harveyi*	Zhou et al. (2015)
30.	*Ficus carica*	*Ctenopharyngodon idella*	*Flavobacterium columnare*	Yang et al. (2015)
31.	Emodin	*Megalobrama amblycephala*	*Aeromonas hydrophila*	Zhang et al. (2014a,b)
32.	*Allium sativum*	*Lates calcarifer*	*Neobenedenia* sp.	Militz et al. (2013)
33.	*Rehmannia glutinosa*	*Cyprinus carpio*	*Aeromonas hydrophila*	Wang et al. (2015)
34.	*Origanum heracleoticum*	*Ictalurus punctatus*	*Aeromonas hydrophila*	Zheng et al. (2009)
35.	*Citrus sinensis*	*Oreochromis mossambicus*	*Streptococcus iniae*	Acar et al. (2015)
36.	*Rhizophora apiculata*	*Amphiprion sebae*	*Vibrio alginolyticus*	Dhayanithi et al. (2015b)
37.	*Ginkgo biloba*	*Anguilla anguilla*	*Pseudodactylogyrus*	Wang et al. (2009)
38.	*Lactuca indica*	*Epinephelus bruneus*	*Streptococcus iniae*	Harikrishnan et al. (2011c)
39.	*Pueraria thunbergiana*	*Epinephelus bruneus*	*V. harveyi*	Harikrishnan et al. (2012d)
40.	*Dioscorea zingiberensis* and *Ginkgo biloba*	*Carassius auratus*	*Dactylogyrus* spp.	Jiang et al. (2014)
41.	*Rosmarinus officinalis*	*Oreochromis* sp.	*Streptococcus iniae*	Abutbul et al. (2004)
42.	*Ocimum sanctum*	*Oreochromis mossambicus*	*Aeromonas hydrophila*	Logambal et al. (2000)
43.	*Nyctanthes arbortristis*	*Oreochromis mossambicus*	*Aeromonas hydrophila*	Kirubakaran et al. (2016)
44.	*Echinacea purpurea* and *Allium sativum*	*Oreochromis niloticus*	*Aeromonas hydrophila*	Aly and Mohamed (2010)
45.	*Zingiber officinale*	*Oncorhynchus mykiss*	*Aeromonas hydrophila*	Nya and Austin (2009)
46.	*Punica granatum*	*Paralichthys olivaceus*	Lymphocystis disease virus	Harikrishnan et al. (2010b)
47.	Anthraquinone extract from *Rheum officinale*	*Megalobrama amblycephala*	*Aeromonas hydrophila*	Liu et al. (2012)
48.	*Astragalus membranaceus* + *Lonicera japonica* + boron	*Oreochromis niloticus*	*Aeromonas hydrophila*	Ardó et al. (2008)
49.	*Withania somnifera*	*L. rohita*	*Aeromonas hydrophila*	Sharma et al. (2010)
50.	Azadirachtin, camphor and curcumin	*Cirrhina mrigala*	*Aphanomyces invadans*	Harikrishnan et al. (2009a)
51.	*Punica granatum*, *Chrysanthemum cinerariaefolium* and *Zanthoxylum schinifolium*	*Paralichthys olivaceus*	*Uronema marinum*	Harikrishnan et al. (2010c)

TABLE 4.5 Immunostimulatory Activity/Effect of Different Plant Products

No.	Name of the Plant	Extraction Method Used	Fish Tested	Route of Administration	Immune Parameter Enhanced	References
1	*Aloe vera*	Whole plant exudate	*Brycon amazonicus*	Bathing	Leukocyte respiratory activity	Zanuzzo et al. (2015)
2	*Astragalus* sp.	Polysaccharide fraction	*Oreochromis niloticus*	Diet	**1.** Phagocytic activity **2.** Respiratory burst **3.** Plasma lysozyme **4.** Bactericidal activity **5.** Superoxide dismutase (SOD) **6.** Glutathione peroxidase (GPx) **7.** Amylase activity	Zahran et al. (2014)
3	Dihydroquercetin obtained from *Cedrus deodara*	Pure compound	*Sparus aurata*	Diet	**1.** Cellular phagocytosis and respiratory burst activities **2.** Serum complement activity, antiprotease, total protein, peroxidase, bactericidal activity and IgM level	Awad et al. (2015a)
4	*Achyranthes aspera*	Raw seed powder	*Cyprinus carpio*	Diet	**1.** Serum total protein, albumin, and globulin **2.** Myeloperoxidase activity	Chakrabarti et al. (2012)
5	*Ficus benghalensis*	Methanol extract	*Channa punctatus*	Diet	**1.** Serum lysozyme **2.** Tissue SOD **3.** Percentage phagocytosis, phagocytotic index, and nitric oxide (NO) **4.** Total serum protein and immunoglobulin	Verma et al. (2012)
6	*Astragalus* sp. and *Angelica sinensis*	Commercial mixture	*Cyprinus carpio* var. *Jian*	Diet (results significant only after 20 days)	**1.** Number of Nitroblue tetrazolium (NBT)-positive cells in the blood **2.** Lysozyme and complement activities in the serum	Jian and Wu (2004)
7	*Nigella sativa* oil and Quercetin from *Utrica dioica*	Commercial	*Oncorhynchus mykiss*	Diet	Serum lysozyme, antiprotease, total protein, myeloperoxidase, bactericidal activity, and IgM	Awad et al. (2013)
8	*Astragalus radix*	Commercial	*Oreochromis niloticus*	Diet	**1.** Serum lysozyme **2.** Phagocytosis **3.** Higher dose did not show significant effect **4.** Three weeks feeding was found to be optimal	Yin et al. (2006)
9	*Cynodon dactylon*	Ethanol extract	*Catla catla*	Diet	White blood cell counts, antibody response	Kaleeswaran et al. (2012)

10	*Viscum album*, *Urtica dioica* and *Zingiber officinale*	Lyophilized aqueous extract	*Oncorhynchus mykiss*	Diet	Phagocytosis and extracellular respiratory burst activity	Dügenci et al. (2003)
11	*Ananas comosus*, *Momordica charantia*, *Jatropha curcas*, *Sesbania aculeate*, *Cassia fistula*, etc.	Various	Various	Various	Various	Makkar et al. (2007)
12	*Urtica dioica*	Dried powder of leaves and stem	*Huso huso*	Diet	Lysozyme, respiratory burst activity	Binaii et al. (2014)
13	*Mentha piperita*	80% Ethanol extract	*Rutilus frisii*	Diet	**1.** Protein concentration, alkaline phosphatase, antimicrobial activity in skin and mucus **2.** Serum lysozyme and IgM **3.** Blood count and leukocyte respiratory burst activity	Adel et al. (2015)
14	*Allium cepa*	Dried powder	*Huso huso*	Diet	After 8 weeks of feeding serum lysozyme activity, SOD, respiratory burst activity and serum total immunoglobulin (Ig) showed a significant increase in treatment group with 1% (w/w diet) onion powder	Akrami et al. (2015)
15	*Uncaria tomentosa*	Ethanol extract	*Oreochromis niloticus*	Diet	**1.** White blood cells count in blood and exudate, respiratory burst activity **2.** Lysozyme activity **3.** Melanomacrophage centers count, villi length **4.** IgM in splenic tissue	Yunis-Aguinaga et al. (2015)
16	*Aloe barbadensis*	Commercial	*Oreochromis niloticus*	Diet	Hematocrit, total plasma protein, erythrocytes (RBC), leukocytes (WBC), differential leukocyte count, phagocytic activity, serum lysozyme activity, and serum antimicrobial activity, serum antimicrobial activity (evaluated against *Aeromonas hydrophila*, *Enterococcus durans* and *Escherichia coli*).	Dotta et al. (2014)
17	*Azadirachta indica*	Leaf powder	*Lates calcarifer*	Diet	**1.** Phagocytic activity, superoxide anion production **2.** Serum lysozyme, bactericidal activity, anti-protease	Talpur and Ikhwanuddin (2013)
18	*Astragalus radix*	Commercial powder	*Cyprinus carpio*	Diet	**1.** Respiratory burst activity and phagocytosis of phagocytic cells **2.** Lysozyme and circulatory antibody titers in plasma	Yin et al. (2009)
19	*Aloe vera*	Whole plant powder	*Oreochromis niloticus*	Diet	Increase in leukocyte levels	Gabriel et al. (2015)

Continued

TABLE 4.5 Immunostimulatory Activity/Effect of Different Plant Products—cont'd

No.	Name of the Plant	Extraction Method Used	Fish Tested	Route of Administration	Immune Parameter Enhanced	References
20	*Viscum album*	Aqueous extract of leaves	*Anguilla japonica*	Diet	Respiratory burst activity and phagocytosis of phagocytic cells	Choi et al. (2008)
21	*Avicennia marina*	Aqueous extract of leaves	*Amphiprion sebae*	Diet	**1.** Total white blood cell counts (WBC) **2.** Serum lysozyme activity, and alternative complement (ACH_{50}) **3.** Respiratory burst assay, assay, and phagocytic activity assay **4.** Gut bacteria	Dhayanithi et al. (2015a)
22	*Cynodon dactylon*	Ethanol extract	*Catla catla*	Diet	**1.** Serum lysozyme activity, antiprotease activity and hemolytic complement **2.** Production of reactive oxygen and nitrogen species, myeloperoxidase activity	Kaleeswaran et al. (2011)
23	*Sophora flavescens*	Commercial	*Oreochromis niloticus*	Diet	**1.** Serum lysozyme, antiprotease, and complement **2.** Cellular myeloperoxidase content	Wu et al. (2013)
24	*Sauropus androgynus*	70% ethanol extract	*Epinephelus coioides*	Diet	Respiratory burst activity, phagocytosis and reactive oxygen species	Samad et al. (2014)
25	*Toona sinensis*	Hot water extract	*Oreochromis mossanbicus*	Intraperitoneal injection	Respiratory burst, phagocytic activity and lysozyme activity	Wu et al. (2010)
26	*Eriobotrya japonica*	85% ethanol extract	*Epinephelus bruneus*	Diet	**1.** WBC count **2.** Serum lysozyme activity, bactericidal activity, and hemolytic complement **3.** Superoxide anion, and phogocytosis	Kim et al. (2011)
27	*Solanum trilobatum*	Water and hexane extract	*Oreochromis mossambicus*	Intraperitoneal injection	Serum lysozyme and cellular reactive oxygen species (ROS) production	Divyagnaneswari et al. (2007)
28	Mixed herbal extracts	Aqueous extract	*Carassius auratus*	Diet	Serum lysozyme and cellular ROS, phagocytic activity	Harikrishnan et al. (2010a)
29	*Ficus benghalensis* and *Leucaena leucocephala*	Prop root powder	*Clarias gariepinus*	Diet	**1.** Serum lysozyme and immunoglobulin levels **2.** Phagocytic index	Verma et al. (2013)
30	*Astragalus, angelica, hawthorn, Licorice root* and *honeysuckle*	Whole plant powder	*Oreochromis niloticus*	Diet	SOD and peroxidase activity in serum	Tang et al. (2014)

31	*Azadirachta indica, Ocimum sanctum* and *Curcuma longa*	Aqueous, ethanol and methanol	*Carassius auratus*	Intraperitoneal injection	Phagocytosis activity, respiratory burst activity, alternative complement activity and lysozyme activity	Harikrishnan et al. (2009b)
32	*Kalopanax pictus*	80% ethanol extract	*Epinephelus bruneus*	Diet	The respiratory activity, lysozyme activity, bactericidal activity, total protein level, and myeloperoxidase levels	Harikrishnan et al. (2011b)
33	*Siegesbeckia glabrescens*	85% ethanol extract	*Epinephelus bruneus*	Diet	**1.** ROS and reactive nitrogen intermediates (RNI) production **2.** Serum lysozyme, alternate complement, myeloperoxidase	Harikrishnan et al. (2012c)
34	*Eclipta alba*	Aqueous extract of leaves	*Oreochromis mossambicus*	Diet	**1.** Serum lysozyme **2.** Cellular ROS and myeloperoxidase content	Christybapita et al. (2007)
35	*Ixora coccinea*	Ethyl acetate extract	*Carassius auratus*	Diet	Phagocytic activity, serum bactericidal activity and lysozyme activity	Anusha et al. (2014)
36	*Viscum album*	Dry powder	*Oreochromis niloticus*	Diet	Respiratory burst activity, lysozyme activity, alternative complement hemolysis activity (ACH_{50}), and phagocytic activity	Park and Choi (2012)
37	*Tinospora cordifolia*	Water extract	*Oreochromis mossambicus*	Intraperitoneal injection	**1.** Cellular ROS, RNI and myeloperoxidase content **2.** Lysozyme, antiprotease and complement	Alexander et al. (2010)
38	Astaxanthin	Commercial	*Cyprinus carpio*	Diet	**1.** Phagocytic index and ROS production **2.** Serum lysozyme activity and bactericidal activity	Jagruthi et al. (2014)
39	Andrographolide	Commercial	*Labeo rohita*	Diet	**1.** Cellular NBT reduction, myeloperoxidase activity, and phagocytic activity **2.** Serum lysozyme, and antiprotease activity	Basha et al. (2013)
40	*Psidium guajava*	Leaves powder	*L. rohita*	Diet	**1.** Serum lysozyme, and alternative complement activity **2.** Plasma IgM levels **3.** Leukocyte phagocyte index	Giri et al. (2015)
41	Azadirachtin	Commercial	*Carassius auratus*	Diet	**1.** Total leukocyte count **2.** NBT reduction, and phagocytic activity **3.** Serum lysozyme, myeloperoxidase activity, and total immunoglobulin	Kumar et al. (2013)
42	*Astaxanthin* and *emodin*	Commercial	*Pelteobagrus fulvidraco*	Diet	Serum lysozyme activity	Liu et al. (2016)

Continued

TABLE 4.5 Immunostimulatory Activity/Effect of Different Plant Products—cont'd

No.	Name of the Plant	Extraction Method Used	Fish Tested	Route of Administration	Immune Parameter Enhanced	References
43	*Urtica dioica*	Leaves powder	*Labeo victorianus*	Diet	Serum immunoglobulins, lysozyme activity and respiratory burst	Ngugi et al. (2015)
44	Soybean isoflavones	Commercial powder	*Trachinotus ovatus*	Diet	Complement content, lysozyme activity as well as respiratory burst activity	Zhou et al. (2015)
45	*Ficus carica*	Polysaccharides fractio	*Ctenopharyngodon idella*	Diet	Serum complement C3, lysozyme, bactericidal activity	Yang et al. (2015)
46	Emodin	Commercial	*Megalobrama amblycephala*	Diet	Only four weeks feeding showed increase in White blood cell count (WBC), respiratory burst activity, SOD activity, and myeloperoxidase (MPO) activity	Zhang et al. (2014a,b)
47	*Rehmannia glutinosa*	Root powder and extract	*Cyprinus carpio*	Diet	Leukocyte phagocytic activity	Wang et al. (2015)
48	*Citrus sinensis*	Extracted essential oil	*Oreochromis mossambicus*	Diet	Lysozyme and myeloperoxidase activity	Acar et al. (2015)
49	*Rhizophora apiculata*	Leaves extract	*Amphiprion sebae*	Diet	Lysozyme, respiratory burst, complement, bactericidal, and phagocytic activities	Dhayanithi et al. (2015b)
50	*Lactuca indica*	95% ethanol extract	*Epinephelus bruneus*	Diet	**1.** NBT reduction level and phagocytic activity **2.** Serum lysozyme, and total immunoglobulin.	Harikrishnan et al. (2011c)
51	*Pueraria thunbergiana*	70% ethanol extract	*Epinephelus bruneus*	Diet	**1.** Serum lysozyme, bactericidal, anti-protease activity **2.** Cellular ROS production and phagocytic activity	Harikrishnan et al. (2012d)
52	*Dioscorea zingiberensis* and *Ginkgo biloba*	Ethanol extracts	*Carassius auratus*	Bath	Anthelminthic activity	Jiang et al. (2014)
53	*Ocimum sanctum*	Aqueous extract of leaves	*Oreochromis mossambicus*	Intraperitoneal injection and oral	Antibody production and NBT reduction	Logambal et al. (2000)
54	*Phyllanthus niruri*	Aqueous extract of leaves	*Oreochromis mossambicus*	Intraperitoneal injection	Antibody production and NBT reduction	Muthulakshmi et al. (2016)
55	*Nyctanthes arbortristis*	Methanol extract	*Oreochromis mossambicus*	Intraperitoneal injection	Lysozyme, myeloperoxidase, anti-protease, and antibody production	Kirubakaran et al. (2016)
56	*Echinacea purpurea* and *Allium sativum*	Commercial powders	*Oreochromis niloticus*	Diet	Neutrophil adherence and total leukocyte count	Aly and Mohamed (2010)
57	*Zingiber officinale*	Dried rhizome powder	*Oncorhynchus mykiss*	Diet	Total leukocyte count, phagocytic, respiratory burst, lysozyme, bactericidal, and	Nya and Austin (2009)

58	*Punica granatum*	Aqueous or methanol or ethanol extracts	*Paralichthys olivaceus*	Intraperitoneal injection	Phagocytic, respiratory burst, complement, and lysozyme activity	Harikrishnan et al. (2010b)
59	Anthraquinone extract from *Rheum officinale*	Commercial solution	*Megalobrama amblycephala*	Diet	Lysozyme activity	Liu et al. (2012)
60	*Astragalus membranaceus* + *Lonicera japonica* + boron	Commercial powder	*Oreochromis niloticus*	Diet	Respiratory burst and phagocytic activity	Ardó et al. (2008)
61	*Withania somnifera*	Root powder	*L. rohita*	Diet	**1.** NBT reduction, and phagocytic activity **2.** Lysozyme, and total immunoglobulin.	Sharma et al. (2010)
62	Azadirachtin, camphor and curcumin	Pure compounds	*Cirrhina mrigala*	Intramuscular injection	Serum lysozyme activity, ROS, and RNI	Harikrishnan et al. (2009a)
63	*Punica granatum*, *Chrysanthemum cinerariaefolium* and *Zanthoxylum schinifolium*	Aqueous or methanol or ethanol extracts	*Paralichthys olivaceus*	Intraperitoneal injection	**1.** Phagocytosis, and respiratory burst activity **2.** Serum alternative complement, and lysozyme	Harikrishnan et al. (2010c)
64	*Trigonella foenum*	Seed powder	*Sparus aurata*	Diet	**1.** Cellular respiratory burst activity and leukocyte peroxidase content **2.** Humoral complement activity, antiprotease, total protein, peroxidase, and IgM level **3.** Hematological parameters (WBC counts)	Awad et al. (2015b)
65	*Camellia sinensis*		*Oncorhynchus mykiss*	Diet	Serum bactericidal	Nootash et al. (2013)
66	*Caulerpa scalpelliformis*	N-oxide and quaternary alkaloids	*Channa striatus*	Intraperitoneal	**1.** Serum lysozyme, peroxidase, antiprotease and alternate complement activities **2.** Cellular ROS and RNI	Balasubramanian and Michael (2016)

TABLE 4.6 Mechanism of Action of Plant-derived Immunostimulants

No.	Name of the Plant	Extraction Method Used	Fish Tested	Route of Administration	Mechanism of Action in Terms of Gene Regulation	References
1	*Polygala tenuifolia*	Two purified compounds	*Ctenopharyngodon idella* kidney cells	Bath	↑Mx1	Yu et al. (2014)
					↑IL-1β	
					↑TNF-α	
					↑MyD88	
					↑IgM	
2	*Psidium guajava*	Flavonoid fraction	*Labeo rohita*	Diet	Inhibited LPS stimulated production of	Sen et al. (2015)
					↓Inducible nitric oxide synthase (iNOS)	
					↓Cyclooxygenase-2 (COX-2)	
3	*Heliotropium filifolium*	Filifolinone is a semisynthetic terpenoid	*Oncorhynchus mykiss*	Intramuscular injection	↑IFN-α	Valenzuela et al. (2016)
					↑IFN-γ	
					↑IL-4/13	
					↑IL-17	
4	Liposome-encapsulated cinnamaldehyde	—	*Danio rerio*	Immersion	↑Interleukins (IL-1β, IL-6, IL-15, IL-21)	Faikoh et al. (2014)
					↑TNF-α	
					↑Interferon (INF)-γ	
5	*Eriobotrya japonica*	85% ethanol extract	*Epinephelus bruneus*	Diet	↑Lymphokines production index	Kim et al. (2011)

6	*Astragalus, angelica, hawthorn, Licorice root* and *honeysuckle*	Whole plant powder	*Oreochromis niloticus*	Diet	↑TNF-α	Tang et al. (2014)
					↑IL-1β	
7	*Psidium guajava*	Leaves powder	*L. rohita*	Diet	↑IL-1β	Giri et al. (2015)
					↑TNF-α	
8	Digestarom®	Commercial	*Ictalurus punctatus*	Diet	↑Mannose binding lectin	Peterson et al. (2015)
9	*Ficus carica*	Polysaccharides fractio	*Ctenopharyngodon idella*	Diet	↑IL-1β	Yang et al. (2015)
					↑TNF-α	
10	Emodin	Commercial	*Megalobrama amblycephala*	Diet	↑TNF-α	Zhang et al. (2014a,b)
11	*Rehmannia glutinosa*	Root powder and extract	*Cyprinus carpio*	Diet	↑IL-1β	Wang et al. (2015)
					↑TNF-α	
					↑iNOS	
					↓IL-10	
					↓TGF-β	
12	*Trigonella foenum*	Seed powder	*Sparus aurata*	Diet	↑MHC1	Awad et al. (2015b)
					↑CSF-1R	
					↑IL-8	
					↑IgM	
13	*Camellia sinensis*		*Oncorhynchus mykiss*	Diet	↑IL-1β	Nootash et al. (2013)
					↑IL-6	
					↓IL-10	

TABLE 4.7 Microbial-derived Immunostimulants

No.	Species Tested	Compound	Host	References
1	—	β-glucan	*Scophthalmus maximus*	Miest et al. (2016)
2	*Bacillus thuringiensis*	Poly-β hydroxybutyrate–hydroxyvalerate	*Oreochromis mossambicus*	Suguna et al. (2014)
3	*Aeromonas hydrophila*	LPS	*Cyprinus carpio*	Selvaraj et al. (2009)
4	*Lactobacillus sakei* MN1 and *Leuconostoc mesenteroides* RTF10	Dextran	BF-2 and EPC fish cell-line monolayers	Nácher-Vázquez et al. (2015)
5	*Saccharomyces cerevisiae*	Zymosan	Carp macrophages	Pietretti et al. (2013)
6	Whole and lipid-extracted algae meals	—	*Sciaenops ocellatus*	Patterson and Gatlin III (2013)
7	*Chaetoceros mülleri*	Chrysolaminaran	*Gadus morhua*	Skjermo et al. (2006)
8	*Parietochloris incisa*	Hexane extract of algae containing arachidonic acid-rich triacylglycerols	*Poecilia reticulata*	Khozin-Goldberg et al. (2006)
9	*Euglena gracilis*	β-1,3-Glucan	*Oncorhynchus mykiss*	Skov et al. (2012)
10	*Phellinus linteus*	70% ethanol extract of whole mycelium	*Epinephelus bruneus*	Harikrishnan et al. (2011a)
11	*Inonotus obliquus*	85% ethanol extract of whole mycelium	*Epinephelus bruneus*	Harikrishnan et al. (2012a)
		70% ethanol extract of whole mycelium	*Paralichythys olivaceus*	Harikrishnan et al., 2012b
12	*Hericium erinaceum*	70% ethanol extract of whole mycelium	*Paralichthys olivaceus*	Harikrishnan et al. (2011d)
13	*Pleurotus ostreatus*	β-1,3/1,6-Glucan	*Cyprinus carpio*	Dobšíková et al. (2013)

in Table 4.7. Some of the microbial compounds such as yeast β-glucan were commercialized. For example, Macrogard®, a commonly used immunostimulant for pets, livestock, and poultry is now also being used in aquaculture. It is developed from yeast β-glucan (Engstad et al., 1992) and commercialized by Biotec Pharmacon, Norway.

4.3.2.1.3 Algal-Derived Immunostimulants

Algae are one of the important sources of carbohydrate polymers such as agarose, alginate, and carrageenan. They also produce pigments and oils which cannot be produced by other organisms. Despite being a very rich source of bioactive compounds, algae are poorly studied for their immunostimulating properties. Oral administration of dried cell suspension of *Spirulina platensis* to Nile tilapia enhanced the survival of fish upon challenge with *A. hydrophila* (Ragap et al., 2012). Alginate, an anionic polysaccharide present in brown algae, is sold in the name of AQUAVAC® Ergosan. Alginate was found to possess excellent immunostimulating properties in fish (Bagni et al., 2005; Huttenhuis et al., 2006; Mendoza et al., 2016). Recently, the immunostimulatory and disease resistance potential of N-oxide-quaternary alkaloid fraction of a marine chlorophycean macroalga in *Channa striata* against *A. hydrophila* has been reported (Balasubramanian and Michael, 2016). On the other hand, marine macroagla *Tetraselmis chuii* and *Phaeodactylum tricornutum* did not protect fish (*Sparusaurata* L.) from experimental infection with *Photobacterium damselae* subsp. *Piscicida* (Cerezuela et al., 2012). Futerpenol® is the licensed name for formulated immunostimulant that contains fucoidans (polysaccharides) and labdane diterpenes (Hernández et al., 2016). Upon administration of Futerpenol® in the diets of rainbow trout (Hernández et al., 2016) found that the fish were protected against *P. salmonis* challenge. A number of bioactive and immunostimulatory compounds from marine microalgae is reviewed by (de Morais et al., 2015). Future investigations must bring out algae that are potential warehouse of immunostimulants to limelight.

4.3.2.1.4 Synthetic Immunostimulants

Levamisole, actually used as antihelminthic drug for children (Lionel et al., 1969), was used as an immunostimulant in aquaculture industries (Anderson, 1992). However, its use is more controversial and there is ample literature talking about pros and cons of levamisole use (Li et al., 2004; Siwicki et al., 1990; Siwicki, 1989; Mulero et al., 1998a,b).

Polynucleotides like poly(I:C) that mimic dsRNA are very good immunostimulators reported from laboratory studies in fish. Many viral infections that were previously incurable were cured using poly(I:C) administration. Poly(I:C) is actually a double-stranded RNA that is made up of inosine on one strand and cytidine on the other. However, due to its low stability, poly(I:C) can only be injected. Previous studies with Atlantic salmon showed poly(I:C) injection protecting it from infectious salmon anemia virus (Jensen et al., 2002a). Poly(I:C) was also responsible for a general antiviral state of treated chinook salmon (Jensen et al., 2002b).

Muramyl dipeptides, repeating structure of both Gram-positive and Gram-negative bacteria is found to be an effective stimulator of nonspecific immunity. In humans, muramyl dipeptides directly activate the inflammasome complex through macrophage secretion of IL-1β independent of TLRs (Martinon et al., 2004). In one of the earliest studies, muramyl dipeptide was shown to act as an adjuvant to a formalin-killed *A. salmonicida* vaccine (Olivier et al., 1985). Synthetic immunostimulants, adjuvants, and vaccines (Irvine et al., 2013; Leclerc and Vogel, 1986) can be developed against important fish pathogens for which strong motivation of scientific community is warranted.

4.3.2.1.5 Animals and Other Sources

Rainbow trout injected with water extract of a mollusk called abalone, *Haliotis discus hannai,* showed increased resistance against *Vibrio anguillarum* (Sakai et al., 1991). Marine tunicate *Ecteinascidia turbinate* extracts protected American eel *Anguilla rostrata* from *A. hydrophila* (Davis and Hayasaka, 1984). ISK, a short-chain polypeptide derived from fish by-product was shown to protect rainbow trout from *A. salmonicida* infection upon injection (Anderson and Jeney, 1992). These investigators later showed that this immunostimulant could also be administered by bath treatment (Jeney and Anderson, 1993).

In addition to extracts from animals, purified animal proteins such as bovine lactoferrin were shown to activate nonspecific immune responses of fish protecting it from pathogens. For instance, Nile tilapia fed with bovine lactoferrin showed enhanced survival against challenge with *A. hydrophila* (Welker et al., 2007) and it protected channel catfish against challenge with *E. ictaluri* (Welker et al., 2010). Keyhole limpet hemocyanin (KLH) of the marine mollusk *Megathura crenulata* is a well-known immunogen and a stimulator of the lymphocytes (Lateef et al., 2007). Apart from this, a number of heavy metals were shown to stimulate the immune system of fish particularly at very low doses. Immunostimulation by heavy metals is discussed in Section 4.2.

4.3.2.2 Mechanism of Action of Immunostimulants

Of the many immunostimulants discussed above, the mechanism of action of β-glucans in fish immune system was studied in great detail. β-glucans are homopolysaccharides made up of repeating units of β (1,3) D-glucose. It is found in a wide variety of organisms including bacteria, algae, fungi, and plants. To host's immune cells, β-glucans naturally arrive as breakdown products of infecting microbes like bacteria or fungi. The process of β-glucan recognition is conserved from lower insects through finfish to human beings. One of the well-recognized effects of β-glucan is the augmentation of phagocytic activity of professional antigen presenting cells (Novak and Vetvicka, 2009). This enhanced phagocytosis is accomplished by a group of molecules called pattern recognition receptors (PRRs) present in them. These PRRs play a vital role in detecting pathogen-associated molecular patterns (PAMPS) that include that of β-glucan also. Activation of antigen presenting cells by β-glucan may lead to activation of downstream signaling cascade that is characterized by effector functions such as ROS, RNI production, and finally antibody production (Hodgkinson et al., 2015). β-glucan from various sources such as yeast and barley is shown to enhance phagocytosis and production of ROS and RNI by activating antigen presenting cells of fishes (Dalmo and Bøgwald, 2008). The receptors for β-glucan such as dectins and scavenger receptors are yet to be described in fish species.

Bacterial lipopolysaccharides (LPS) are another PAMP recognized by fish. Fishes recognize PAMPs in a different way than mammals. The lack of serum recognizing factors (that cause endotoxic shock in mammals) make monocyte-macrophage lineage of fishes less sensitive to LPS. The toll-like receptor TLR-4 is the key molecule involved in the transduction of LPS-mediated signaling in immune cells of a fish. TLR-4 was reported to be present in many species of fish. However, TLR-4 of fish is quite different from its mammalian counterpart and need to be explored in detail (Swain et al., 2008). LPS is capable of increasing total leukocyte counts, in particular neutrophil and mononuclear cells even at a dosage of 10 μg/

fish (Swain et al., 2008). LPS also stimulates the production of serum nonspecific immune enzymes such as lysozyme, cyclooxygenase, and overall cellular responses mediated by macrophages. In fishes, C3 complement (or alternate) pathway is directly activated by LPS (Boesen et al., 1999). Administration of LPS of pathogen as an immunostimulant showed better disease resistance compared to whole cell vaccine (Baba et al., 1988).

The main cells responsible for the detection of unmethylated polynucleotide PAMPs like CpG are dendritic cells, macrophages, and neutrophils. Again, recognition of CpG PAMPs is mediated by TLRs and related receptors. TLR9 of zebra fish is found to be the principle regulator of CpG-mediated signaling (Yeh et al., 2013). Synthetic oligo-DNAs such as CpGs are considered to be an activator of Th1 cell-type in mammals. Though Th1 responses are not well documented in fish, foreign DNAs such as oligo-CpG provoke the expression of pro-inflammatory cytokine, IL-1β, and the antiviral cytokine IFN-γ. These cytokines in turn lead to the proliferation of leukocytes and activation of natural-killer (NK) and T-killer lymphocytes (Carrington and Secombes, 2006) tabulated many CpG sequences and its effect on various fishes. CpG is long thought to act as adjuvant but is still not widely used.

Jantan et al. (2015) reviewed the mechanism of action of various plant-derived compounds in mammalian system. Most of the mechanisms deciphered were also found fitting for teleost fishes. At present, seldom pure compounds are studied for their immunostimulating properties in fishes but crude extracts were studied in detail. However, there is a concern about dose and duration of treatment particularly when the immunostimulant was given orally in the form of feed supplement. For instance, the synthetic immunostimulant, levamisole improved the chemiluminescent activity of phagocytes only at a concentration of 0.5 mg/kg feed and not by 5 mg/kg dose group (Kajita et al., 1990). Studies in this lab with tilapia show that only lower or moderate doses of plant extracts stimulate the immune system and subsequently protecting fish from pathogen challenge (Alexander et al., 2010; Divyagnaneswari et al., 2007; Kirubakaran et al., 2016). Further, the effect of immunostimulant is not long lasting (Matsuo and Miyazono, 1993; Yoshida et al., 1995) and so it has to be administered periodically.

4.3.2.3 Future Prospects

Immunostimulants can be a promising alternative to vaccines and antibiotics. Many immunostimulation studies conducted so far in finfish are with microbial products such as β-glucan and algal polysaccharides. However, these are also not pure homogenous compounds. Until purified to homogeneity, immunostimulation brought about by these agents can even be attributed to endotoxin and other contaminants. Hence, purification of plant or microbial products is warranted. This purification may be of indirect or even direct benefit to the ecosystem by abolishing overexploitation of natural resources. Once a purified compound is available with all its structural details, large scale microbial or chemical synthesis can minimize or even remove the need for plant material from nature.

Intraperitoneal route was found to be the best route for administering immunostimulants. The extracts given as feed supplement to fishes may not necessarily show the same efficacy as that of parenterally administered. Thus researchers must find ways to make sure that their extracts are efficacious via oral route also. The extraction procedure should not be very long and tedious so that farmers can supplement their feed with immunostimulant with relative ease and with less cost. For instance, crushed guava leaves were directly added to pelleted feed and given to *L. rohita* which subsequently reduced mortality upon challenge with *A. hydrophila* (Giri et al., 2015).

A large number of plants were screened for their immunostimulating efficacy based on ethnobotanical knowledge. In addition to terrestrial plants, micro- and macroalgae that are casual inhabitants of local water bodies must be investigated for their immunostimulating potential. This will reduce the labor of immunostimulant preparation. As mentioned earlier, the effect of immunostimulant is dose and time dependent (Christybapita et al., 2007). Hence, before application to aquaculture, a plant-derived immunostimulant should be studied in detail with regard to the optimal dose and duration of administration.

Finally, laboratory results should be validated or replicated in large scale farm conditions before drawing any conclusions. Further, the research personnel should know the right markers (e.g., ROS production, ceruloplasmin) that are associated and statistically correlated to disease resistance in a fish species. This can go a long way in reducing the large number of fishes killed in disease resistance experiments (Das and Sahoo, 2014).

4.3.3 Probiotics

Probiotics, derived from Greek, literally means "for life." Voluntary administration of the so-called "beneficial microorganisms" in defined numbers was shown to improve immune responses (immunostimulatory) in many tested animals ranging from crustaceans to human beings (Castex et al., 2014; Daliri and Lee, 2015). For a comprehensive review on various probiotic bacteria tested in fishes, one can read Nayak (2010). Various probiotics tested on fishes are tabulated in Table 4.8.

TABLE 4.8 Probiotics as Immunostimulants (Kesarcodi-Watson et al., 2008)

No.	Probiotic Species	Host Species	References
1	Hydrolyzate-based culture of lactic acid bacteria isolated from salmon intestines	*Salmo salar*	Gildberg et al. (1995)
2	*Bacillus amyloliquefaciens* FPTB16 and *Bacillus. subtilis* FPTB13	*Catla catla*	Kamilya et al. (2015)
3	*Kocuria* sp. SM1	*Oncorhynchus mykiss*	Sharifuzzaman and Austin (2009)
4	*Pediococcus acidilactici*	*Oncorhynchus mykiss*	Hoseinifar et al. (2015)
5	*Carnobacterium divergens*	*Gadus morhua*	Gildberg and Mikkelsen (1998)
6	*Lactobacillus plantarum*	*S. salar*	Gildberg et al. (1995)
7	*Carnobacterium* sp. (K1)	*S. salar*	Jöborn et al. (1997)
8	*Pseudomonas fluorescens*	*Oncorhynchus mykiss*	Gram et al. (1999)
9	*Cytophaga* sp.	*Sparus aurata*	Makridis et al. (2005)
10	*Shewanella putrefaciens Pdp11* and *Shewanella baltica Pdp13*	*Solea senegalensis*	Díaz-Rosales et al. (2009)
11	Dead cells of *A. hydrophila*	*Carassius auratus*	Irianto et al. (2003)
12	*Streptococcus faecium*	*Oreochromis niloticus*	Lara-Flores et al. (2003)
13	*Saccharomyces cerevisiae*	*Oreochromis nilotics*	Pinpimai et al. (2015)
14	*Aeromonas sobria*	*Oncorhynchus mykiss*	Brunt and Austin (2005)
15	*Debaryomyces hansenii*	*Dicentrarchus labrax*	Tovar et al. (2002)
16	*Enterobacter* sp.	*Oncorhynchus mykiss*	LaPatra et al. (2014)
17	*Bacillus licheniformis*	*Pangasius hypophthalmus*	Gobi et al. (2016)
18	*Bacillus subtilis, Tetraselmis chuii,* and *Phaeodactylum tricornutum*	*Sparus aurata*	Cerezuela et al. (2012)

In general, probiotics directly aid in the proliferation of various immune cells by increasing the expression of various cytokines (Irianto and Austin, 2002; Qadis et al., 2014). The effector functions of leukocyte proliferation such as ROS, RNI production, and phagocytosis in combination with various serum nonspecific parameters such as lysozyme, complement, and antiprotease eventually lead to protection of fish against harmful pathogens.

Gut is the principal organ that hosts probiotics. Studies on fish guts show that there are vast differences between fish gut and their mammalian counterpart. The immune system present in gut is referred to as gut-associated lymphoid tissue (GALT). GALT of fishes has not been shown to have well-organized lymphoid cells (called Peyer's patches in mammals), or antigen transporting tissues (M-cells), or secretory IgA. However, fish guts are not void of immune cells. Evidences are now accumulating on the effect of probiotics aiding the proliferation of cells such as T-lymphocytes, Ig^+ cells ("B-lymphocytes"), and acidophilic granulocytes. More studies aiming at the big picture of fish GALT is necessary to clearly understand the mechanism of action of probiotics in fish immunity.

Application of probiotics as a prophylactic in aquaculture industry is not without challenges. One has to consider many factors, such as optimum conditions required for probiotic microorganisms to function. Even survival of probiotics in the gut of fish is dependent upon various environmental factors (Skjermo and Vadstein, 1999). Probiotics were known to exert their effect in both host- (Madsen, 2006) and strain-specific manners (Ibnou-Zekri et al., 2003). Research involving probiotics showed that the origin and source of probiotics (Sharifuzzaman and Austin, 2009) can regulate their activities. The most dreadful fact about probiotics is that inappropriate dose and/or duration of probiotics supplementation can cause undesirable results (Dawood and Koshio, 2016; Floch et al., 2006).

4.4 CONCLUSIONS

Aquaculture is definitely the future-dependable source of food, particularly protein-rich food for the teeming human population. However, poor practices and maintenance issues cause stress to the fishes, which in turn leads to disease susceptibility, subsequently resulting in diseases and disease-outbreaks. Despite these deterring forces that play a negative role on aquaculture, the bright side is being the availability of relatively simple remedial measures that can save lots of fishes and of course, aquaculture economy. Among the various remedial measures considered in the text, prophylactic use of plant-derived immunostimulants seems better than others. This is true because source of immunostimulants are plenty, they are relatively inexpensive, biodegradable, and they confer protection against various pathogens in a single host that is currently not possible in other prophylactic measures. In conclusion, we encourage the aquaculture industries to maximally utilize the potential of immunostimulants for better fish production.

ACKNOWLEDGMENT

This work was partly supported by the Department of Biotechnology, Ministry of Science and Technology, New Delhi, India through a funded project (No. BT/PR7726/AAQ/3/634/2013).

REFERENCES

Abutbul, S., Golan-Goldhirsh, A., Barazani, O., Zilberg, D., 2004. Use of *Rosmarinus officinalis* as a treatment against *Streptococcus iniae* in tilapia (*Oreochromis* sp.). Aquaculture 238, 97–105. http://dx.doi.org/10.1016/j.aquaculture.2004.05.016.

Acar, Ü., Kesbiç, O.S., Yılmaz, S., Gültepe, N., Türker, A., 2015. Evaluation of the effects of essential oil extracted from sweet orange peel (*Citrus sinensis*) on growth rate of tilapia (*Oreochromis mossambicus*) and possible disease resistance against *Streptococcus iniae*. Aquaculture 437, 282–286. http://dx.doi.org/10.1016/j.aquaculture.2014.12.015.

Adel, M., Abedian Amiri, A., Zorriehzahra, J., Nematolahi, A., Esteban, M.Á., 2015. Effects of dietary peppermint (*Mentha piperita*) on growth performance, chemical body composition and hematological and immune parameters of fry Caspian white fish (*Rutilus frisii kutum*). Fish and Shellfish Immunology 45, 841–847. http://dx.doi.org/10.1016/j.fsi.2015.06.010.

Ahmad, H.I., Verma, A.K., Babitha, R., Rathore, G., Saharan, N., Gora, A.H., 2016. Growth, non-specific immunity and disease resistance of *Labeo rohita* against *Aeromonas hydrophila* in biofloc systems using different carbon sources. Aquaculture 457, 61–67. http://dx.doi.org/10.1016/j.aquaculture.2016.02.011.

Akrami, R., Gharaei, A., Mansour, M.R., Galeshi, A., 2015. Effects of dietary onion (*Allium cepa*) powder on growth, innate immune response and hemato–biochemical parameters of beluga (*Huso huso* Linnaeus, 1754) juvenile. Fish and Shellfish Immunology 45, 828–834. http://dx.doi.org/10.1016/j.fsi.2015.06.005.

Alarcon, J.B., Waine, G.W., McManus, D.P., 1999. DNA vaccines: technology and application as anti-parasite and anti-microbial agents. Advances in Parasitology 42, 343–410.

Albrektsen, S., Sandnes, K., Glette, J., WaagbØ, R., 1995. Influence of dietary vitamin B_6 on tissue vitamin B_6 contents and immunity in Atlantic salmon, *Salmo salar* L. Aquaculture Research 26, 331–339. http://dx.doi.org/10.1111/j.1365-2109.1995.tb00921.x.

Alexander, C.P., Kirubakaran, C.J.W., Michael, R.D., 2010. Water soluble fraction of *Tinospora cordifolia* leaves enhanced the non-specific immune mechanisms and disease resistance in *Oreochromis mossambicus*. Fish and Shellfish Immunology 29, 765–772. http://dx.doi.org/10.1016/j.fsi.2010.07.003.

Aly, S.M., Mohamed, M.F., 2010. *Echinacea purpurea* and *Allium sativum* as immunostimulants in fish culture using Nile tilapia (*Oreochromis niloticus*). Journal of Animal Physiology and Animal Nutrition 94, e31–e39. http://dx.doi.org/10.1111/j.1439-0396.2009.00971.x.

Anderson, D.P., Jeney, G., 1992. Immunostimulants added to injected *Aeromonas salmonicida* bacterin enhance the defense mechanisms and protection in rainbow trout (*Oncorhynchus mykiss*). Veterinary Immunology and Immunopathology 34, 379–389. http://dx.doi.org/10.1016/0165-2427(92)90177-R.

Anderson, D.P., 1992. Immunostimulants, adjuvants, and vaccine carriers in fish: applications to aquaculture. Annual Review of Fish Diseases 2, 281–307. http://dx.doi.org/10.1016/0959-8030(92)90067-8.

Anusha, P., Thangaviji, V., Velmurugan, S., Michaelbabu, M., Citarasu, T., 2014. Protection of ornamental gold fish *Carassius auratus* against *Aeromonas hydrophila* by treating *Ixora coccinea* active principles. Fish and Shellfish Immunology 36, 485–493. http://dx.doi.org/10.1016/j.fsi.2013.12.006.

Ardó, L., Yin, G., Xu, P., Váradi, L., Szigeti, G., Jeney, Z., Jeney, G., 2008. Chinese herbs (*Astragalus membranaceus* and *Lonicera japonica*) and boron enhance the non-specific immune response of Nile tilapia (*Oreochromis niloticus*) and resistance against *Aeromonas hydrophila*. Aquaculture 275, 26–33. http://dx.doi.org/10.1016/j.aquaculture.2007.12.022.

Arthur, J.R., McKenzie, R.C., Beckett, G.J., 2003. Selenium in the immune system. Journal of Nutrition 133, 1457S–1459S.

Awad, E., Austin, D., Lyndon, A.R., 2013. Effect of black cumin seed oil (*Nigella sativa*) and nettle extract (Quercetin) on enhancement of immunity in rainbow trout, *Oncorhynchus mykiss* (Walbaum). Aquaculture 388–391, 193–197. http://dx.doi.org/10.1016/j.aquaculture.2013.01.008.

Awad, E., Awaad, A.S., Esteban, M.A., 2015a. Effects of dihydroquercetin obtained from deodar (*Cedrus deodara*) on immune status of gilthead seabream (*Sparus aurata* L.). Fish and Shellfish Immunology 43, 43–50. http://dx.doi.org/10.1016/j.fsi.2014.12.009.

Awad, E., Cerezuela, R., Esteban, M.Á., 2015b. Effects of fenugreek (*Trigonella foenum graecum*) on gilthead seabream (*Sparus aurata* L.) immune status and growth performance. Fish and Shellfish Immunology 45, 454–464. http://dx.doi.org/10.1016/j.fsi.2015.04.035.

Baba, T., Imamura, J., Izawa, K., Iked, K., 1988. Cell-mediated protection in carp, *Cyprinus carpio* L., against *Aeromonas hydrophila*. Journal of Fish Diseases 11, 171–178. http://dx.doi.org/10.1111/j.1365-2761.1988.tb00536.x.

Bagni, M., Romano, N., Finoia, M.G., Abelli, L., Scapigliati, G., Tiscar, P.G., Sarti, M., Marino, G., 2005. Short- and long-term effects of a dietary yeast β-glucan (Macrogard) and alginic acid (Ergosan) preparation on immune response in sea bass (*Dicentrarchus labrax*). Fish and Shellfish Immunology 18, 311–325. http://dx.doi.org/10.1016/j.fsi.2004.08.003.

Balasubramanian, R., Michael, R.D., 2016. Immunostimulatory effects of N-Oxide–Quaternary alkaloid fraction of a marine Chlorophycean macroalga in the striped murrel, *Channa striata* (Bloch). Aquaculture Research 47, 591–604. http://dx.doi.org/10.1111/are.12518.

Ballesteros, N.A., Alonso, M., Saint-Jean, S.R., Perez-Prieto, S.I., 2015. An oral DNA vaccine against infectious haematopoietic necrosis virus (IHNV) encapsulated in alginate microspheres induces dose-dependent immune responses and significant protection in rainbow trout (*Oncorrhynchus mykiss*). Fish and Shellfish Immunology 45, 877–888. http://dx.doi.org/10.1016/j.fsi.2015.05.045.

Bártíková, H., Podlipná, R., Skálová, L., 2016. Veterinary drugs in the environment and their toxicity to plants. Chemosphere 144, 2290–2301. http://dx.doi.org/10.1016/j.chemosphere.2015.10.137.

Basha, K.A., Raman, R.P., Prasad, K.P., Kumar, K., Nilavan, E., Kumar, S., 2013. Effect of dietary supplemented andrographolide on growth, non-specific immune parameters and resistance against *Aeromonas hydrophila* in *Labeo rohita* (Hamilton). Fish and Shellfish Immunology 35, 1433–1441. http://dx.doi.org/10.1016/j.fsi.2013.08.005.

Baxter, D., 2007. Active and passive immunity, vaccine types, excipients and licensing. Occupational Medicine 57, 552–556. http://dx.doi.org/10.1093/occmed/kqm110.

Bebak, J., Garcia, J.C., Darwish, A., 2012. Effect of copper sulfate on *Aeromonas hydrophila* infection in channel catfish fingerlings. North American Journal of Aquaculture 74, 494–498. http://dx.doi.org/10.1080/15222055.2012.685212.

Bebak, J., Wagner, B., Burnes, B., Hanson, T., 2015. Farm size, seining practices, and salt use: risk factors for *Aeromonas hydrophila* outbreaks in farm-raised catfish, Alabama, USA. Preventive Veterinary Medicine 118, 161–168. http://dx.doi.org/10.1016/j.prevetmed.2014.11.001.

Belghit, I., Skiba-Cassy, S., Geurden, I., Dias, K., Surget, A., Kaushik, S., Panserat, S., Seiliez, I., 2014. Dietary methionine availability affects the main factors involved in muscle protein turnover in rainbow trout (*Oncorhynchus mykiss*). British Journal of Nutrition 112, 493–503. http://dx.doi.org/10.1017/S0007114514001226.

Biller-Takahashi, J.D., Urbinati, E.C., Biller-Takahashi, J.D., Urbinati, E.C., 2014. Fish immunology. The modification and manipulation of the innate immune system: Brazilian studies. Anais da Academia Brasileira de Ciências 86, 1484–1506. http://dx.doi.org/10.1590/0001-3765201420130159.

Binaii, M., Ghiasi, M., Farabi, S.M.V., Pourgholam, R., Fazli, H., Safari, R., Alavi, S.E., Taghavi, M.J., Bankehsaz, Z., 2014. Biochemical and hemato-immunological parameters in juvenile beluga (*Huso huso*) following the diet supplemented with nettle (*Urtica dioica*). Fish and Shellfish Immunology 36, 46–51. http://dx.doi.org/10.1016/j.fsi.2013.10.001.

Bizzini, B., Fattal-German, M., 1989. Potentiation by nonspecific immunostimulation of the efficacy of antibiotics in the treatment of experimental bacterial infections. Biomedicine Pharmacotheraphy 43, 753–761. http://dx.doi.org/10.1016/0753-3322(89)90164-9.

Blazer, V.S., 1992. Nutrition and disease resistance in fish. Annual Review of Fish Diseases 2, 309–323. http://dx.doi.org/10.1016/0959-8030(92)90068-9.

Boesen, H.T., Pedersen, K., Larsen, J.L., Koch, C., Ellis, A.E., 1999. *Vibrio anguillarum* resistance to rainbow trout (*Oncorhynchus mykiss*) serum: role of O-Antigen structure of lipopolysaccharide. Infection and Immunity 67, 294–301.

Bondad-Reantaso, M.G., Subasinghe, R.P., Arthur, J.R., Ogawa, K., Chinabut, S., Adlard, R., Tan, Z., Shariff, M., 2005. Disease and health management in Asian aquaculture. Veterinary Parasitology 132, 249–272. http://dx.doi.org/10.1016/j.vetpar.2005.07.005 From Science to Solutions Plenary Lectures Presented at the 20th Conference of the World Association for the Advancement of Veterinary Parasitology 20th Conference of the World Association for the Advancement of Veterinary Parasitology.

Branson, E.J., Rønsberg, S.S., Ritchie, G., 2000. Efficacy of teflubenzuron (Calicide®) for the treatment of sea lice, *Lepeophtheirus salmonis* (Krøyer 1838), infestations of farmed Atlantic salmon (*Salmo salar* L.). Aquaculture Research 31, 861–867. http://dx.doi.org/10.1046/j.1365-2109.2000.00509.x.

Bricknell, I., Dalmo, R., 2005. The use of immunostimulants in fish larval aquaculture. Fish and Shellfish Immunology 19, 457–472. http://dx.doi.org/10.1016/j.fsi.2005.03.008.

Brudeseth, B.E., Wiulsrød, R., Fredriksen, B.N., Lindmo, K., Løkling, K.-E., Bordevik, M., Steine, N., Klevan, A., Gravningen, K., 2013. Status and future perspectives of vaccines for industrialised fin-fish farming. Fish and Shellfish Immunology 35, 1759–1768. http://dx.doi.org/10.1016/j.fsi.2013.05.029.

Brunt, J., Austin, B., December 2005. Use of a probiotic to control lactococcosis and streptococcosis in rainbow trout, Oncorhynchus mykiss (Walbaum). Journal of Fish Diseases 28 (12), 693–701. http://dx.doi.org/10.1111/j.1365-2761.2005.00672.x.

Burridge, L., Weis, J.S., Cabello, F., Pizarro, J., Bostick, K., 2010. Chemical use in salmon aquaculture: a review of current practices and possible environmental effects. Aquaculture 306, 7–23. http://dx.doi.org/10.1016/j.aquaculture.2010.05.020.

Carrington, A.C., Secombes, C.J., 2006. A review of CpGs and their relevance to aquaculture. Veterinary Immunology and Immunopathology 112, 87–101. http://dx.doi.org/10.1016/j.vetimm.2006.03.015.

Castex, M., Daniels, C., Chim, L., 2014. Probiotic applications in Crustaceans. In: Merrifield, D., Ringø, E. (Eds.), Aquaculture Nutrition. John Wiley & Sons, Ltd, pp. 290–327.

Cerezuela, R., Guardiola, F.A., González, P., Meseguer, J., Esteban, M.Á., 2012. Effects of dietary *Bacillus subtilis*, *Tetraselmis chuii*, and *Phaeodactylum tricornutum*, singularly or in combination, on the immune response and disease resistance of sea bream (*Sparus aurata* L.). Fish and Shellfish Immunology 33, 342–349. http://dx.doi.org/10.1016/j.fsi.2012.05.004.

Chakrabarti, R., Srivastava, P.K., Kundu, K., Khare, R.S., Banerjee, S., 2012. Evaluation of immunostimulatory and growth promoting effect of seed fractions of *Achyranthes aspera* in common carp *Cyprinus carpio* and identification of active constituents. Fish and Shellfish Immunology 32, 839–843. http://dx.doi.org/10.1016/j.fsi.2012.02.006.

Chandran, M.R., Aruna, B.V., Logambal, S.M., Michael, R.D., 2002. Immunisation of Indian major carps against *Aeromonas hydrophila* by intraperitoneal injection. Fish and Shellfish Immunology 13, 1–9.

Choi, S.-H., Park, K.-H., Yoon, T.-J., Kim, J.-B., Jang, Y.-S., Choe, C.H., 2008. Dietary Korean mistletoe enhances cellular non-specific immune responses and survival of Japanese eel (*Anguilla japonica*). Fish and Shellfish Immunology 24, 67–73. http://dx.doi.org/10.1016/j.fsi.2007.08.007.

Christybapita, D., Divyagnaneswari, M., Michael, R.D., 2007. Oral administration of *Eclipta alba* leaf aqueous extract enhances the non-specific immune responses and disease resistance of *Oreochromis mossambicus*. Fish and Shellfish Immunology 23, 840–852. http://dx.doi.org/10.1016/j.fsi.2007.03.010.

Collins, S.A., Desai, A.R., Mansfield, G.S., Hill, J.E., Van Kessel, A.G., Drew, M.D., 2012. The effect of increasing inclusion rates of soybean, pea and canola meals and their protein concentrates on the growth of rainbow trout: concepts in diet formulation and experimental design for ingredient evaluation. Aquaculture 344–349, 90–99. http://dx.doi.org/10.1016/j.aquaculture.2012.02.018.

Collymore, C., Porelli, G., Lieggi, C., Lipman, N.S., 2014. Evaluation of 5 cleaning and disinfection methods for nets used to collect zebrafish (*Danio rerio*). Journal of the American Association for Laboratory Animal Science 53, 657–660.

Crab, R., Defoirdt, T., Bossier, P., Verstraete, W., 2012. Biofloc technology in aquaculture: beneficial effects and future challenges. Aquaculture 356–357, 351–356. http://dx.doi.org/10.1016/j.aquaculture.2012.04.046.

Cuesta, A., Chaves-Pozo, E., de las Heras, A.I., Saint-Jean, S.R., Pérez-Prieto, S., Tafalla, C., 2010. An active DNA vaccine against infectious pancreatic necrosis virus (IPNV) with a different mode of action than fish rhabdovirus DNA vaccines. Vaccine 28, 3291–3300. http://dx.doi.org/10.1016/j.vaccine.2010.02.106.

Daliri, E.B.-M., Lee, B.H., 2015. New perspectives on probiotics in health and disease. Food Science and Human Wellness 4, 56–65. http://dx.doi.org/10.1016/j.fshw.2015.06.002.

Dalmo, R.A., Bøgwald, J., 2008. ß-glucans as conductors of immune symphonies. Fish and Shellfish Immunology 25, 384–396. http://dx.doi.org/10.1016/j.fsi.2008.04.008.

Dalzell, D.J.B., Macfarlane, N.A.A., 1999. The toxicity of iron to brown trout and effects on the gills: a comparison of two grades of iron sulphate. Journal of Fish Biology 55, 301–315. http://dx.doi.org/10.1111/j.1095-8649.1999.tb00680.x.

Das, S., Sahoo, P.K., 2014. Markers for selection of disease resistance in fish: a review. Aquaculture International 22, 1793–1812. http://dx.doi.org/10.1007/s10499-014-9783-5.

Davis, J.F., Hayasaka, S.S., 1984. The enhancement of resistance of the American eel, *Anguilla rostrata* Le Sueur, to a pathogenic bacterium, *Aeromonas hydrophila*, by an extract of the tunicate, *Ecteinascidia turbinata*. Journal of Fish Diseases 7, 311–316. http://dx.doi.org/10.1111/j.1365-2761.1984.tb00936.x.

Dawood, M.A.O., Koshio, S., 2016. Recent advances in the role of probiotics and prebiotics in carp aquaculture: a review. Aquaculture 454, 243–251. http://dx.doi.org/10.1016/j.aquaculture.2015.12.033.

de Morais, M.G., Vaz, B.da S., de Morais, E.G., Costa, J.A.V., 2015. Biologically active metabolites synthesized by microalgae. BioMed Research International 2015 (2015), e835761. http://dx.doi.org/10.1155/2015/835761.

Deng, J., Kang, B., Tao, L., Rong, H., Zhang, X., 2013. Effects of dietary cholesterol on antioxidant capacity, non-specific immune response, and resistance to *Aeromonas hydrophila* in rainbow trout (*Oncorhynchus mykiss*) fed soybean meal-based diets. Fish and Shellfish Immunology 34, 324–331. http://dx.doi.org/10.1016/j.fsi.2012.11.008.

Dhayanithi, N.B., Ajith Kumar, T.T., Arockiaraj, J., Balasundaram, C., Harikrishnan, R., 2015a. Dietary supplementation of *Avicennia marina* extract on immune protection and disease resistance in *Amphiprion sebae* against *Vibrio alginolyticus*. Fish and Shellfish Immunology 45, 52–58. http://dx.doi.org/10.1016/j.fsi.2015.02.018 Special Issue Probiotics.

Dhayanithi, N.B., Ajithkumar, T.T., Arockiaraj, J., Balasundaram, C., Ramasamy, H., 2015b. Immune protection by *Rhizophora apiculata* in clownfish against *Vibrio alginolyticus*. Aquaculture 446, 1–6. http://dx.doi.org/10.1016/j.aquaculture.2015.04.013.

Díaz-Rosales, P., Arijo, S., Chabrillón, M., Alarcón, F.J., Tapia-Paniagua, S.T., Martínez-Manzanares, E., Balebona, M.C., Moriñigo, M.A., 2009. Effects of two closely related probiotics on respiratory burst activity of Senegalese sole (*Solea senegalensis*, Kaup) phagocytes, and protection against *Photobacterium damselae* subsp. *piscicida*. Aquaculture 293, 16–21. http://dx.doi.org/10.1016/j.aquaculture.2009.03.050.

Divyagnaneswari, M., Christybapita, D., Michael, R.D., 2007. Enhancement of nonspecific immunity and disease resistance in *Oreochromis mossambicus* by *Solanum trilobatum* leaf fractions. Fish and Shellfish Immunology 23, 249–259. http://dx.doi.org/10.1016/j.fsi.2006.09.015.

Dobšíková, R., Blahová, J., Mikulíková, I., Modrá, H., Prášková, E., Svobodová, Z., Škorič, M., Jarkovský, J., Siwicki, A.-K., 2013. The effect of oyster mushroom β-1.3/1.6-D-glucan and oxytetracycline antibiotic on biometrical, haematological, biochemical, and immunological indices, and histopathological changes in common carp (*Cyprinus carpio* L.). Fish and Shellfish Immunology 35, 1813–1823. http://dx.doi.org/10.1016/j.fsi.2013.09.006.

Dotta, G., de Andrade, J.I.A., Tavares Gonçalves, E.L., Brum, A., Mattos, J.J., Maraschin, M., Martins, M.L., 2014. Leukocyte phagocytosis and lysozyme activity in Nile tilapia fed supplemented diet with natural extracts of propolis and *Aloe barbadensis*. Fish and Shellfish Immunology 39, 280–284. http://dx.doi.org/10.1016/j.fsi.2014.05.020.

Dügenci, S.K., Arda, N., Candan, A., 2003. Some medicinal plants as immunostimulant for fish. Journal of Ethnopharmacology 88, 99–106. http://dx.doi.org/10.1016/S0378-8741(03)00182-X.

Duncan, T., 1978. The Use of Potassium Permanganate in Fisheries: A Literature Review. Fisheries and Wildlife Service, Fayetteville, AR. 275397.

Edwards, S.V., Hedrick, P.W., 1998. Evolution and ecology of MHC molecules: from genomics to sexual selection. Trends in Ecology and Evolution 13, 305–311. http://dx.doi.org/10.1016/S0169-5347(98)01416-5.

Eltholth, M., Fornace, K., Grace, D., Rushton, J., Häsler, B., 2015. Characterisation of production, marketing and consumption patterns of farmed tilapia in the Nile Delta of Egypt. Food Policy 51, 131–143. http://dx.doi.org/10.1016/j.foodpol.2015.01.002.

Engstad, R.E., Robertsen, B., Frivold, E., 1992. Yeast glucan induces increase in lysozyme and complement-mediated haemolytic activity in Atlantic salmon blood. Fish and Shellfish Immunology 2, 287–297. http://dx.doi.org/10.1016/S1050-4648(06)80033-1.

Eo, J., Lee, K.-J., 2008. Effect of dietary ascorbic acid on growth and non-specific immune responses of tiger puffer, *Takifugu rubripes*. Fish and Shellfish Immunology 25, 611–616. http://dx.doi.org/10.1016/j.fsi.2008.08.009.

Esteve, C., Alcaide, E., Ureña, R., 2012. The effect of metals on condition and pathologies of European eel (*Anguilla anguilla*): in situ and laboratory experiments. Aquatic Toxicology (Amsterdam, Netherlands) 109, 176–184. http://dx.doi.org/10.1016/j.aquatox.2011.10.002.

Evelyn, T.P., 1997. A historical review of fish vaccinology. Developments in Biological Standardization 90, 3–12.

Evenhuis, J.P., Wiens, G.D., Wheeler, P., Welch, T.J., LaPatra, S.E., Thorgaard, G.H., 2014. Transfer of serum and cells from *Yersinia ruckeri* vaccinated doubled-haploid hot creek rainbow trout into outcross F1 progeny elucidates mechanisms of vaccine-induced protection. Developmental and Comparative Immunology 44, 145–151. http://dx.doi.org/10.1016/j.dci.2013.12.004.

Evensen, Ø., Leong, J.-A.C., 2013. DNA vaccines against viral diseases of farmed fish. Fish and Shellfish Immunology 35, 1751–1758. http://dx.doi.org/10.1016/j.fsi.2013.10.021.

Ezeonyejiaku, C.D., Obiakor, M.O., Ezenwelu, C.O., 2011. Toxicity of copper sulphate and behavioral locomotor response of tilapia (*Oreochromis niloticus*) and catfish (*Clarias gariepinus*) species. Online Journal of Animal Feed Research 1, 130–134.

Faikoh, E.N., Hong, Y.-H., Hu, S.-Y., 2014. Liposome-encapsulated cinnamaldehyde enhances zebrafish (*Danio rerio*) immunity and survival when challenged with *Vibrio vulnificus* and *Streptococcus agalactiae*. Fish and Shellfish Immunology 38, 15–24. http://dx.doi.org/10.1016/j.fsi.2014.02.024.

FAO: Nutritional Requirements, 2016. URL http://www.fao.org/fishery/affris/species-profiles/rainbow-trout/nutritional-requirements/en/.

Ferrera, F., La Cava, A., Rizzi, M., Hahn, B.H., Indiveri, F., Filaci, G., 2007. Gene vaccination for the induction of immune tolerance. Annals of the New York Academy of Sciences 1110, 99–111. http://dx.doi.org/10.1196/annals.1423.012.

Floch, M.H., Madsen, K.K., Jenkins, D.J.A., Guandalini, S., Katz, J.A., Onderdonk, A., Walker, W.A., Fedorak, R.N., Camilleri, M., 2006. Recommendations for probiotic use. Journal of Clinical Gastroenterology 40, 275–278. http://dx.doi.org/10.1097/00004836-200603000-00022.

Forwood, J.M., Harris, J.O., Deveney, M.R., 2013. Efficacy of current and alternative bath treatments for *Lepidotrema bidyana* infecting silver perch, *Bidyanus bidyanus*. Aquaculture 416–417, 65–71. http://dx.doi.org/10.1016/j.aquaculture.2013.08.034.

Fracalossi, D.M., Lovell, R.T., 1994. Dietary lipid sources influence responses of channel catfish (*Ictalurus punctatus*) to challenge with the pathogen *Edwardsiella ictaluri*. Aquaculture 119, 287–298. http://dx.doi.org/10.1016/0044-8486(94)90183-X.

Gabriel, N.N., Qiang, J., He, J., Ma, X.Y., Kpundeh, M.D., Xu, P., 2015. Dietary *Aloe vera* supplementation on growth performance, some haemato-biochemical parameters and disease resistance against *Streptococcus iniae* in tilapia (GIFT). Fish and Shellfish Immunology 44, 504–514. http://dx.doi.org/10.1016/j.fsi.2015.03.002.

Garcia, R.L., Sanders, G.E., 2011. Efficacy of cleaning and disinfection procedures in a zebrafish (*Danio rerio*) facility. Journal of American Association for Laboratory Animal Science 50, 895–900.

GESAMP, Reports and Studies No. 65. Towards Safe and Effective Use of Chemicals in Coastal Aquaculture, 1997.

Gildberg, A., Mikkelsen, H., 1998. Effects of supplementing the feed to Atlantic cod (*Gadus morhua*) fry with lactic acid bacteria and immuno-stimulating peptides during a challenge trial with *Vibrio anguillarum*. Aquaculture 167, 103–113. http://dx.doi.org/10.1016/S0044-8486(98)00296-8.

Gildberg, A., Johansen, A., Bøgwald, J., 1995. Growth and survival of Atlantic salmon (*Salmo salar*) fry given diets supplemented with fish protein hydrolysate and lactic acid bacteria during a challenge trial with *Aeromonas salmonicida*. Aquaculture 138, 23–34. http://dx.doi.org/10.1016/0044-8486(95)01144-7.

Giri, S.S., Sen, S.S., Chi, C., Kim, H.J., Yun, S., Park, S.C., Sukumaran, V., 2015. Effect of guava leaves on the growth performance and cytokine gene expression of *Labeo rohita* and its susceptibility to *Aeromonas hydrophila* infection. Fish and Shellfish Immunology 46, 217–224. http://dx.doi.org/10.1016/j.fsi.2015.05.051.

Gobi, N., Malaikozhundan, B., Sekar, V., Shanthi, S., Vaseeharan, B., Jayakumar, R., Khudus Nazar, A., 2016. GFP tagged *Vibrio parahaemolyticus* Dahv2 infection and the protective effects of the probiotic *Bacillus licheniformis* Dahb1 on the growth, immune and antioxidant responses in *Pangasius hypophthalmus*. Fish and Shellfish Immunology 52, 230–238. http://dx.doi.org/10.1016/j.fsi.2016.03.006.

Gram, L., Melchiorsen, J., Spanggaard, B., Huber, I., Nielsen, T.F., 1999. Inhibition of *Vibrio anguillarum* by *Pseudomonas fluorescens* AH2, a possible probiotic treatment of fish. Applied and Environmental Microbiology 65, 969–973.

Gudding, R., Van Muiswinkel, W.B., 2013. A history of fish vaccination. Fish and Shellfish Immunology 35, 1683–1688. http://dx.doi.org/10.1016/j.fsi.2013.09.031.

Guimarães, I.G., Lim, C., Yildirim-Aksoy, M., Li, M.H., Klesius, P.H., 2014. Effects of dietary levels of vitamin A on growth, hematology, immune response and resistance of Nile tilapia (*Oreochromis niloticus*) to Streptococcus iniae. Animal Feed Science and Technology 188, 126–136. http://dx.doi.org/10.1016/j.anifeedsci.2013.12.003.

Guimarães, I.G., Pezzato, L.E., Santos, V.G., Orsi, R.O., Barros, M.M., June 2016. Vitamin A affects haematology, growth and immune response of Nile tilapia (*Oreochromis niloticus*, L.), but has no protective effect against bacterial challenge or cold-induced stress. Aquaculture Research 47 (6), 2004–2018. http://dx.doi.org/10.1111/are.12656.

Harikrishnan, R., Balasundaram, C., Dharaneedharan, S., Moon, Y.-G., Kim, M.-C., Kim, J.-S., Heo, M.-S., 2009a. Effect of plant active compounds on immune response and disease resistance in *Cirrhina mrigala* infected with fungal fish pathogen, *Aphanomyces invadans*. Aquaculture Research 40, 1170–1181. http://dx.doi.org/10.1111/j.1365-2109.2009.02213.x.

Harikrishnan, R., Balasundaram, C., Kim, M.-C., Kim, J.-S., Han, Y.-J., Heo, M.-S., 2009b. Innate immune response and disease resistance in *Carassius auratus* by triherbal solvent extracts. Fish and Shellfish Immunology 27, 508–515. http://dx.doi.org/10.1016/j.fsi.2009.07.004.

Harikrishnan, R., Balasundaram, C., Heo, M.-S., 2010a. Herbal supplementation diets on hematology and innate immunity in goldfish against *Aeromonas hydrophila*. Fish and Shellfish Immunology 28, 354–361. http://dx.doi.org/10.1016/j.fsi.2009.11.013.

Harikrishnan, R., Heo, J., Balasundaram, C., Kim, M.-C., Kim, J.-S., Han, Y.-J., Heo, M.-S., 2010b. Effect of *Punica granatum* solvent extracts on immune system and disease resistance in *Paralichthys olivaceus* against lymphocystis disease virus (LDV). Fish and Shellfish Immunology 29, 668–673. http://dx.doi.org/10.1016/j.fsi.2010.07.006.

Harikrishnan, R., Heo, J., Balasundaram, C., Kim, M.-C., Kim, J.-S., Han, Y.-J., Heo, M.-S., 2010c. Effect of traditional Korean medicinal (TKM) tri-herbal extract on the innate immune system and disease resistance in *Paralichthys olivaceus* against *Uronema marinum*. Veterinary Parasitology 170, 1–7. http://dx.doi.org/10.1016/j.vetpar.2010.01.046.

Harikrishnan, R., Balasundaram, C., Heo, M.-S., 2011a. Diet enriched with mushroom *Phellinus linteus* extract enhances the growth, innate immune response, and disease resistance of kelp grouper, *Epinephelus bruneus* against vibriosis. Fish and Shellfish Immunology 30, 128–134. http://dx.doi.org/10.1016/j.fsi.2010.09.013.

Harikrishnan, R., Kim, J.-S., Kim, M.-C., Balasundaram, C., Heo, M.-S., 2011b. *Kalopanax pictus* as feed additive controls bacterial and parasitic infections in kelp grouper, *Epinephelus bruneus*. Fish and Shellfish Immunology 31, 801–807. http://dx.doi.org/10.1016/j.fsi.2011.07.017.

Harikrishnan, R., Kim, J.-S., Kim, M.-C., Balasundaram, C., Heo, M.-S., 2011c. *Lactuca indica* extract as feed additive enhances immunological parameters and disease resistance in *Epinephelus bruneus* to *Streptococcus iniae*. Aquaculture 318, 43–47. http://dx.doi.org/10.1016/j.aquaculture.2011.04.049.

Harikrishnan, R., Kim, J.-S., Kim, M.-C., Balasundaram, C., Heo, M.-S., 2011d. *Hericium erinaceum* enriched diets enhance the immune response in *Paralichthys olivaceus* and protect from *Philasterides dicentrarchi* infection. Aquaculture 318, 48–53. http://dx.doi.org/10.1016/j.aquaculture.2011.04.048.

Harikrishnan, R., Balasundaram, C., Heo, M.-S., 2012a. Effect of *Inonotus obliquus* enriched diet on hematology, immune response, and disease protection in kelp grouper, *Epinephelus bruneus* against *Vibrio harveyi*. Aquaculture 344–349, 48–53. http://dx.doi.org/10.1016/j.aquaculture.2012.03.010.

Harikrishnan, R., Balasundaram, C., Heo, M.-S., 2012b. *Inonotus obliquus* containing diet enhances the innate immune mechanism and disease resistance in olive flounder *Paralichythys olivaceus* against *Uronema marinum*. Fish and Shellfish Immunology 32, 1148–1154. http://dx.doi.org/10.1016/j.fsi.2012.03.021.

Harikrishnan, R., Kim, D.-H., Hong, S.-H., Mariappan, P., Balasundaram, C., Heo, M.-S., 2012c. Non-specific immune response and disease resistance induced by *Siegesbeckia glabrescens* against *Vibrio parahaemolyticus* in *Epinephelus bruneus*. Fish and Shellfish Immunology 33, 359–364. http://dx.doi.org/10.1016/j.fsi.2012.05.018.

Harikrishnan, R., Kim, J.-S., Balasundaram, C., Heo, M.-S., 2012d. Protection of *Vibrio harveyi* infection through dietary administration of *Pueraria thunbergiana* in kelp grouper, *Epinephelus bruneus*. Aquaculture 324–325, 27–32. http://dx.doi.org/10.1016/j.aquaculture.2011.10.019.

Hernández, A.J., Romero, A., Gonzalez-Stegmaier, R., Dantagnan, P., 2016. The effects of supplemented diets with a phytopharmaceutical preparation from herbal and macroalgal origin on disease resistance in rainbow trout against *Piscirickettsia salmonis*. Aquaculture 454, 109–117. http://dx.doi.org/10.1016/j.aquaculture.2015.12.016.

Hodgkinson, J.W., Grayfer, L., Belosevic, M., 2015. Biology of bony fish macrophages. Biology 4, 881–906. http://dx.doi.org/10.3390/biology4040881.

Hoseinifar, S.H., Mirvaghefi, A., Amoozegar, M.A., Sharifian, M., Esteban, M.Á., 2015. Modulation of innate immune response, mucosal parameters and disease resistance in rainbow trout (*Oncorhynchus mykiss*) upon synbiotic feeding. Fish and Shellfish Immunology 45, 27–32. http://dx.doi.org/10.1016/j.fsi.2015.03.029. Special Issue: Probiotics.

Huising, M.O., Guichelaar, T., Hoek, C., Verburg-van Kemenade, B.M.L., Flik, G., Savelkoul, H.F.J., Rombout, J.H.W.M., 2003. Increased efficacy of immersion vaccination in fish with hyperosmotic pretreatment. Vaccine 21, 4178–4193.

Huttenhuis, H.B.T., Ribeiro, A.S.P., Bowden, T.J., Van Bavel, C., Taverne-Thiele, A.J., Rombout, J.H.W.M., 2006. The effect of oral immuno-stimulation in juvenile carp (*Cyprinus carpio* L.). Fish and Shellfish Immunology 21, 261–271. http://dx.doi.org/10.1016/j.fsi.2005.12.002.

Ibnou-Zekri, N., Blum, S., Schiffrin, E.J., von der Weid, T., 2003. Divergent patterns of colonization and immune response elicited from two intestinal lactobacillus strains that display similar properties in vitro. Infection and Immunity 71, 428–436. http://dx.doi.org/10.1128/IAI.71.1.428-436.2003.

Irianto, A., Robertson, P.A.W., Austin, B., 2003. Short communication oral administration of formalin-inactivated cells of *Aeromonas hydrophila* A3-51 controls infection by atypical *A. salmonicida* in goldfish, *Carassius auratus* (L.). Journal of Fish Diseases 26, 117–120. http://dx.doi.org/10.1046/j.1365-2761.2003.00439.x.

Irianto, A., Austin, B., 2002. Use of probiotics to control furunculosis in rainbow trout, *Oncorhynchus mykiss* (Walbaum). Journal of Fish Diseases 25, 333–342. http://dx.doi.org/10.1046/j.1365-2761.2002.00375.x.

Irvine, D.J., Swartz, M.A., Szeto, G.L., 2013. Engineering synthetic vaccines using cues from natural immunity. Nature Materials 12, 978–990. http://dx.doi.org/10.1038/nmat3775.

Jagruthi, C., Yogeshwari, G., Anbazahan, S.M., Shanthi Mari, L.S., Arockiaraj, J., Mariappan, P., Learnal Sudhakar, G.R., Balasundaram, C., Harikrishnan, R., 2014. Effect of dietary astaxanthin against *Aeromonas hydrophila* infection in common carp, *Cyprinus carpio*. Fish and Shellfish Immunology 41, 674–680. http://dx.doi.org/10.1016/j.fsi.2014.10.010.

Jantan, I., Ahmad, W., Bukhari, S.N.A., 2015. Plant-derived immunomodulators: an insight on their preclinical evaluation and clinical trials. Frontiers in Plant Science 655. http://dx.doi.org/10.3389/fpls.2015.00655.

Jeney, G., Anderson, D.P., 1993. Enhanced immune response and protection in rainbow trout to *Aeromonas salmonicida* bacterin following prior immersion in immunostimulants. Fish and Shellfish Immunology 3, 51–58. http://dx.doi.org/10.1006/fsim.1993.1005.

Jensen, I., Albuquerque, A., Sommer, A.-I., Robertsen, B., 2002a. Effect of poly I: C on the expression of Mx proteins and resistance against infection by infectious salmon anaemia virus in Atlantic salmon. Fish and Shellfish Immunology 13, 311–326. http://dx.doi.org/10.1006/fsim.2001.0406.

Jensen, I., Larsen, R., Robertsen, B., 2002b. An antiviral state induced in Chinook salmon embryo cells (CHSE-214) by transfection with the double-stranded RNA poly I: C. Fish and Shellfish Immunology 13, 367–378. http://dx.doi.org/10.1006/fsim.2002.0412.

Jian, J., Wu, Z., 2004. Influences of traditional Chinese medicine on non-specific immunity of Jian Carp (*Cyprinus carpio* var. *Jian*). Fish and Shellfish Immunology 16, 185–191. http://dx.doi.org/10.1016/S1050-4648(03)00062-7.

Jiang, C., Wu, Z.-Q., Liu, L., Liu, G.-L., Wang, G.-X., 2014. Synergy of herbal ingredients combination against *Dactylogyrus* spp. in an infected goldfish model for monogenean management. Aquaculture 433, 115–118. http://dx.doi.org/10.1016/j.aquaculture.2014.05.045.

Jiang, S.-T., 2010. The quality control status of cobia, *Rachycentron canadum*, and grouper, *Epinephelus malabaricus*, in Taiwan. Journal of World Aquaculture Society 41, 266–273. http://dx.doi.org/10.1111/j.1749-7345.2010.00354.x.

Jiao, X., Zhang, M., Hu, Y., Sun, L., 2009. Construction and evaluation of DNA vaccines encoding *Edwardsiella tarda* antigens. Vaccine 27, 5195–5202. http://dx.doi.org/10.1016/j.vaccine.2009.06.071.

Jöborn, A., Olsson, J.C., Westerdahl, A., Conway, P.L., Kjelleberg, S., 1997. Colonization in the fish intestinal tract and production of inhibitory substances in intestinal mucus and faecal extracts by *Carnobacterium* sp. strain K1. Journal of Fish Diseases 20, 383–392. http://dx.doi.org/10.1046/j.1365-2761.1997.00316.x.

Jones, S.M., Feldmann, H., Ströher, U., Geisbert, J.B., Fernando, L., Grolla, A., Klenk, H.-D., Sullivan, N.J., Volchkov, V.E., Fritz, E.A., Daddario, K.M., Hensley, L.E., Jahrling, P.B., Geisbert, T.W., 2005. Live attenuated recombinant vaccine protects nonhuman primates against Ebola and Marburg viruses. Nature Medicine 11, 786–790. http://dx.doi.org/10.1038/nm1258.

Kajita, Y., Sakai, M., Atsuta, S., Kobayashi, M., 1990. The immunomodulatory effects of levamisole on rainbow trout, *Oncorhynchus mykiss*. Fish Pathology 25, 93–98. http://dx.doi.org/10.3147/jsfp.25.93.

Kaleeswaran, B., Ilavenil, S., Ravikumar, S., 2011. Dietary supplementation with *Cynodon dactylon* (L.) enhances innate immunity and disease resistance of Indian major carp, *Catla catla* (Ham.). Fish and Shellfish Immunology 31, 953–962. http://dx.doi.org/10.1016/j.fsi.2011.08.013.

Kaleeswaran, B., Ilavenil, S., Ravikumar, S., 2012. Changes in biochemical, histological and specific immune parameters in *Catla catla* (Ham.) by *Cynodon dactylon* (L.). Journal of King Saud University—Science 24, 139–152. http://dx.doi.org/10.1016/j.jksus.2010.10.001.

Kamboj, M., Sepkowitz, K.A., 2007. Risk of transmission associated with live attenuated vaccines given to healthy persons caring for or residing with an immunocompromised patient. Infection Control and Hospital Epidemiology 28, 702–707. http://dx.doi.org/10.1086/517952.

Kamilya, D., Baruah, A., Sangma, T., Chowdhury, S., Pal, P., 2015. Inactivated probiotic bacteria stimulate cellular immune responses of catla, *Catla catla* (Hamilton) in vitro. Probiotics and Antimicrobial Proteins 7, 101–106. http://dx.doi.org/10.1007/s12602-015-9191-9.

Kaushal, D., Foreman, T.W., Gautam, U.S., Alvarez, X., Adekambi, T., Rangel-Moreno, J., Golden, N.A., Johnson, A.-M.F., Phillips, B.L., Ahsan, M.H., Russell-Lodrigue, K.E., Doyle, L.A., Roy, C.J., Didier, P.J., Blanchard, J.L., Rengarajan, J., Lackner, A.A., Khader, S.A., Mehra, S., 2015. Mucosal vaccination with attenuated *Mycobacterium tuberculosis* induces strong central memory responses and protects against tuberculosis. Nature Communications 6, 8533. http://dx.doi.org/10.1038/ncomms9533.

Kesarcodi-Watson, A., Kaspar, H., Lategan, M.J., Gibson, L., 2008. Probiotics in aquaculture: the need, principles and mechanisms of action and screening processes. Aquaculture 274, 1–14. http://dx.doi.org/10.1016/j.aquaculture.2007.11.019.

Khan, F.H., 2009. The Elements of Immunology. Pearson Education India.

Khan, K.H., 2013. DNA vaccines: roles against diseases. Germs 3, 26–35. http://dx.doi.org/10.11599/germs.2013.1034.

Khosravi, S., Bui, H.T.D., Rahimnejad, S., Herault, M., Fournier, V., Kim, S.-S., Jeong, J.-B., Lee, K.-J., 2015a. Dietary supplementation of marine protein hydrolysates in fish-meal based diets for red sea bream (*Pagrus major*) and olive flounder (*Paralichthys olivaceus*). Aquaculture 435, 371–376. http://dx.doi.org/10.1016/j.aquaculture.2014.10.019.

Khosravi, S., Rahimnejad, S., Herault, M., Fournier, V., Lee, C.-R., Dio Bui, H.T., Jeong, J.-B., Lee, K.-J., 2015b. Effects of protein hydrolysates supplementation in low fish meal diets on growth performance, innate immunity and disease resistance of red sea bream *Pagrus major*. Fish and Shellfish Immunology 45, 858–868. http://dx.doi.org/10.1016/j.fsi.2015.05.039.

Khozin-Goldberg, I., Cohen, Z., Pimenta-Leibowitz, M., Nechev, J., Zilberg, D., 2006. Feeding with arachidonic acid-rich triacylglycerols from the microalga *Parietochloris incisa* improved recovery of guppies from infection with *Tetrahymena* sp. Aquaculture 255, 142–150. http://dx.doi.org/10.1016/j.aquaculture.2005.12.017.

Kim, J.-S., Harikrishnan, R., Kim, M.-C., Jang, I.-S., Kim, D.-H., Hong, S.-H., Balasundaram, C., Heo, M.-S., 2011. Enhancement of *Eriobotrya japonica* extracts on non-specific immune response and disease resistance in kelp grouper *Epinephelus bruneus* against *Vibrio carchariae*. Fish and Shellfish Immunology 31, 1193–1200. http://dx.doi.org/10.1016/j.fsi.2011.10.015.

Kindt, T.J., Goldsby, R.A., Osborne, B.A., Kuby, J., 2007. Kuby Immunology. W. H. Freeman.

Kiron, V., Fukuda, H., Takeuchi, T., Watanabe, T., 1995a. Essential fatty acid nutrition and defence mechanisms in rainbow trout *Oncorhynchus mykiss*. Comparative Biochemistry and Physiology Part A: Physiology 111, 361–367. http://dx.doi.org/10.1016/0300-9629(95)00042-6.

Kiron, V., Watanabe, T., Fukuda, H., Okamoto, N., Takeuchi, T., 1995b. Protein nutrition and defence mechanisms in rainbow trout *Oncorhynchus mykiss*. Comparative Biochemistry and Physiology Part A: Physiology 111, 351–359. http://dx.doi.org/10.1016/0300-9629(95)00043-7.

Kirubakaran, C.J.W., Subramani, P.A., Michael, R.D., 2016. Methanol extract of *Nyctanthes arbortristis* seeds enhances non-specific immune responses and protects *Oreochromis mossambicus* (Peters) against *Aeromonas hydrophila* infection. Research in Vetrinary Science 105, 243–248. http://dx.doi.org/10.1016/j.rvsc.2016.02.013.

Kitis, M., 2004. Disinfection of wastewater with peracetic acid: a review. Environmental International 30, 47–55. http://dx.doi.org/10.1016/S0160-4120(03)00147-8.

Krogdahl, Å., Bakke-Mckellep, A.M., RØed, K.H., Baeverfjord, G., 2000. Feeding Atlantic salmon *Salmo salar* L. soybean products: effects on disease resistance (furunculosis), and lysozyme and IgM levels in the intestinal mucosa. Aquaculture Nutrition 6, 77–84.

Kum, C., Sekki, S., 2011. The immune system drugs in fish: immune function, immunoassay, drugs. In: Aral, F. (Ed.), Recent Advances in Fish Farms. InTech.

Kumar, S., Raman, R.P., Pandey, P.K., Mohanty, S., Kumar, A., Kumar, K., 2013. Effect of orally administered azadirachtin on non-specific immune parameters of goldfish *Carassius auratus* (Linn. 1758) and resistance against *Aeromonas hydrophila*. Fish and Shellfish Immunology 34, 564–573. http://dx.doi.org/10.1016/j.fsi.2012.11.038.

Kumari, J., Sahoo, P.K., 2005. High dietary vitamin C affects growth, non-specific immune responses and disease resistance in Asian catfish, *Clarias batrachus*. Molecular and Cellular Biochemistry 280, 25–33. http://dx.doi.org/10.1007/s11010-005-8011-z.

Kunttu, H.M.T., Valtonen, E.T., Jokinen, E.I., Suomalainen, L.-R., 2009. Saprophytism of a fish pathogen as a transmission strategy. Epidemics 1, 96–100. http://dx.doi.org/10.1016/j.epidem.2009.04.003.

Laghari, M.Y., Lashari, P., Zhang, Y., Sun, X., 2014. Identification of quantitative trait loci (QTLs) in aquaculture species. Reviews in Fisheries Science and Aquaculture 22, 221–238. http://dx.doi.org/10.1080/23308249.2014.931172.

Lang, Š., Mareš, J., Kopp, R., 2012. Does the water reuse affect the fish growth, welfare quality? Acta Univsitatis Agriculture et Silviculturae Mendelianae Brunensis 60, 369–374.

LaPatra, S.E., Fehringer, T.R., Cain, K.D., 2014. A probiotic *Enterobacter* sp. provides significant protection against *Flavobacterium psychrophilum* in rainbow trout (*Oncorhynchus mykiss*) after injection by two different routes. Aquaculture 433, 361–366. http://dx.doi.org/10.1016/j.aquaculture.2014.06.022.

Lara-Flores, M., Olvera-Novoa, M.A., Guzmán-Méndez, B.E., López-Madrid, W., 2003. Use of the bacteria *Streptococcus faecium* and *Lactobacillus acidophilus*, and the yeast *Saccharomyces cerevisiae* as growth promoters in Nile tilapia (*Oreochromis niloticus*). Aquaculture 216, 193–201. http://dx.doi.org/10.1016/S0044-8486(02)00277-6.

Lateef, S.S., Gupta, S., Jayathilaka, L.P., Krishnanchettiar, S., Huang, J.-S., Lee, B.-S., 2007. An improved protocol for coupling synthetic peptides to carrier proteins for antibody production using DMF to solubilize peptides. Journal of Biomolecular Techniques 18, 173–176.

Leclerc, C., Vogel, F.R., 1986. Synthetic immunomodulators and synthetic vaccines. Critical Reviews in Therapeutic Drug Carrier Systems 2, 353–406.

Lees, F., Baillie, M., Gettinby, G., Revie, C.W., 2008. The efficacy of emamectin benzoate against infestations of *Lepeophtheirus salmonis* on farmed Atlantic salmon (*Salmo salar* L.) in Scotland, 2002–2006. PLoS One 3, e1549. http://dx.doi.org/10.1371/journal.pone.0001549.

Li, M.H., Wise, D.J., Johnson, M.R., Robinson, E.H., 1994. Dietary menhaden oil reduced resistance of channel catfish (*Ictalurus punctatus*) to *Edwardsiella ictaluri*. Aquaculture 128, 335–344. http://dx.doi.org/10.1016/0044-8486(94)90321-2.

Li, P., Wang, X., Gatlin, D.M., 2004. Excessive dietary levamisole suppresses growth performance of hybrid striped bass, *Morone chrysops* × *M. saxatilis*, and elevated levamisole in vitro impairs macrophage function. Aquaculture Research 35, 1380–1383. http://dx.doi.org/10.1111/j.1365-2109.2004.01151.x.

Li, C., Wang, X., Zhang, Q., Wang, Z., Qi, J., Yi, Q., Liu, Z., Wang, Y., Yu, H., 2012. Identification of two major histocompatibility (MH) class II A genes and their association to *Vibrio anguillarum* infection in half-smooth tongue sole (*Cynoglossus semilaevis*). Journal of Ocean University of China 11, 32–44. http://dx.doi.org/10.1007/s11802-012-1802-4.

Li, M., Chen, L., Qin, J.g., Yu, N., Chen, Y., Ding, Z., Li, E., 2015. Growth, immune response and resistance to *Aeromonas hydrophila* of darkbarbel catfish *Pelteobagrus vachelli* fed diets with different linolenic acids, vitamins C and E levels. Aquaculture Nutrition. http://dx.doi.org/10.1111/anu.12287.

Li, H., Chu, X., Li, D., Zeng, Z., Peng, X., 2016. Construction and immune protection evaluation of recombinant polyvalent OmpAs derived from genetically divergent *ompA* by DNA shuffling. Fish and Shellfish Immunology 49, 230–236. http://dx.doi.org/10.1016/j.fsi.2015.12.024.

Lillehaug, A., 1989. A survey on different procedures used for vaccinating salmonids against vibriosis in Norwegian fish-farming. Aquaculture 83, 217–226. http://dx.doi.org/10.1016/0044-8486(89)90034-3.

Lim, C., Yildirim-Aksoy, M., Li, M.H., Welker, T.L., Klesius, P.H., 2009. Influence of dietary levels of lipid and vitamin E on growth and resistance of Nile tilapia to *Streptococcus iniae* challenge. Aquaculture 298, 76–82. http://dx.doi.org/10.1016/j.aquaculture.2009.09.025.

Lionel, N.D., Mirando, E.H., Nanayakkara, J.C., Soysa, P.E., 1969. Levamisole in the treatment of ascariasis in children. British Medical Journal 4, 340–341.

Liu, B., Ge, X., Xie, J., Xu, P., He, Y., Cui, Y., Ming, J., Zhou, Q., Pan, L., 2012. Effects of anthraquinone extract from *Rheum officinale* Bail on the physiological responses and HSP70 gene expression of *Megalobrama amblycephala* under *Aeromonas hydrophila* infection. Fish and Shellfish Immunology 32, 1–7. http://dx.doi.org/10.1016/j.fsi.2011.02.015.

Liu, J., Liu, Z.-Z., Zhao, X.-J., Wang, C.-H., 2014. MHC class IIα alleles associated with resistance to *Aeromonas hydrophila* in purse red common carp, *Cyprinus carpio* Linnaeus. Journal of Fish Diseases 37, 571–575. http://dx.doi.org/10.1111/jfd.12131.

Liu, F., Shi, H.-Z., Guo, Q.-S., Yu, Y.-B., Wang, A.-M., Lv, F., Shen, W.-B., 2016. Effects of astaxanthin and emodin on the growth, stress resistance and disease resistance of yellow catfish (*Pelteobagrus fulvidraco*). Fish and Shellfish Immunology 51, 125–135. http://dx.doi.org/10.1016/j.fsi.2016.02.020.

Logambal, S.M., Venkatalakshmi, S., Michael, R.D., 2000. Immunostimulatory effect of leaf extract of *Ocimum sanctum* Linn. in *Oreochromis mossambicus* (Peters). Hydrobiologia 430, 113–120. http://dx.doi.org/10.1023/A:1004029332114.

Madsen, K., 2006. Probiotics and the immune response. Journal of Clinical Gastroenterology 40, 232–234. http://dx.doi.org/10.1097/00004836-200603000-00014.

Mainous, M.E., Smith, S.A., Kuhn, D.D., 2010. Effect of common aquaculture chemicals against *Edwardsiella ictaluri* and *E. tarda*. Journal of Aquatic Animal Health 22, 224–228. http://dx.doi.org/10.1577/H10-020.1.

Makkar, H.P.S., Francis, G., Becker, K., 2007. Bioactivity of phytochemicals in some lesser-known plants and their effects and potential applications in livestock and aquaculture production systems. Animal: an International Journal of Animal Bioscience 1, 1371–1391. http://dx.doi.org/10.1017/S1751731107000298.

Makridis, P., Martins, S., Vercauteren, T., Van Driessche, K., Decamp, O., Dinis, M.T., 2005. Evaluation of candidate probiotic strains for gilthead sea bream larvae (*Sparus aurata*) using an in vivo approach. Letters in Applied Microbiology 40, 274–277. http://dx.doi.org/10.1111/j.1472-765X.2005.01676.x.

Martinon, F., Agostini, L., Meylan, E., Tschopp, J., 2004. Identification of bacterial muramyl dipeptide as activator of the NALP3/Cryopyrin inflammasome. Current Biology 14, 1929–1934. http://dx.doi.org/10.1016/j.cub.2004.10.027.

Matsuo, K., Miyazono, I., 1993. The influence of long-term administration of peptidoglycan on disease resistance and growth of juvenile rainbow trout. Nippon Suisan Gakkaishi 59, 1377–1379. http://dx.doi.org/10.2331/suisan.59.1377.

Mendoza, R., Pohlenz, C., Gatlin, D.M., 2016. Supplementation of organic acids and algae extracts in the diet of red drum *Sciaenops ocellatus*: immunological impacts. Aquaculture Research. http://dx.doi.org/10.1111/are.13015.

Michael, R.D., Quinn, C.G., Venkatalakshmi, S., 1998. Modulation of humoral immune response by ascorbic acid in *Oreochromis mossambicus* (Peters). Indian Journal of Experimental Biology 36, 1038–1040.

Midtlyng, P.J., Reitan, L.J., Speilberg, L., 1996. Experimental studies on the efficacy and side-effects of intraperitoneal vaccination of Atlantic salmon (*Salmo salar* L.) against furunculosis. Fish and Shellfish Immunology 6, 335–350. http://dx.doi.org/10.1006/fsim.1996.0034.

Midtlyng, P.J., 1996. A field study on intraperitoneal vaccination of Atlantic salmon (*Salmo salar* L.) against furunculosis. Fish and Shellfish Immunology 6, 553–565. http://dx.doi.org/10.1006/fsim.1996.0052.

Miest, J.J., Arndt, C., Adamek, M., Steinhagen, D., Reusch, T.B.H., 2016. Dietary β-glucan (MacroGard®) enhances survival of first feeding turbot (*Scophthalmus maximus*) larvae by altering immunity, metabolism and microbiota. Fish and Shellfish Immunology 48, 94–104. http://dx.doi.org/10.1016/j.fsi.2015.11.013.

Militz, T.A., Southgate, P.C., Carton, A.G., Hutson, K.S., 2013. Dietary supplementation of garlic (*Allium sativum*) to prevent monogenean infection in aquaculture. Aquaculture 408–409, 95–99. http://dx.doi.org/10.1016/j.aquaculture.2013.05.027.

Mohanty, B.R., Sahoo, P.K., Mahapatra, K.D., Saha, J.N., 2007. Innate immune responses in families of Indian major carp, *Labeo rohita*, differing in their resistance to *Edwardsiella tarda* infection. Current Science 92, 1270–1274.

Montero, D., Kalinowski, T., Obach, A., Robaina, L., Tort, L., Caballero, M.J., Izquierdo, M.S., 2003. Vegetable lipid sources for gilthead seabream (*Sparus aurata*): effects on fish health. Aquaculture, Proceedings of the Tenth International Symposium on Nutrition and Feeding in Fish (Feeding for Quality) 225, 353–370. http://dx.doi.org/10.1016/S0044-8486(03)00301-6.

Morefield, G.L., Sokolovska, A., Jiang, D., HogenEsch, H., Robinson, J.P., Hem, S.L., 2005. Role of aluminum-containing adjuvants in antigen internalization by dendritic cells in vitro. Vaccine 23, 1588–1595. http://dx.doi.org/10.1016/j.vaccine.2004.07.050.

Mourente, G., Good, J.E., Bell, J.G., 2005. Partial substitution of fish oil with rapeseed, linseed and olive oils in diets for European sea bass (*Dicentrarchus labrax* L.): effects on flesh fatty acid composition, plasma prostaglandins E_2 and $F_2\alpha$, immune function and effectiveness of a fish oil finishing diet. Aquaculture Nutrition 11, 25–40. http://dx.doi.org/10.1111/j.1365-2095.2004.00320.x.

Mulero, V., Esteban, M.A., Meseguer, J., 1998a. In vitrolevamisole fails to increase seabream (*Sparus aurata* L.) phagocyte functions. Fish and Shellfish Immunology 8, 315–318. http://dx.doi.org/10.1006/fsim.1998.0141.

Mulero, V., Esteban, M.A., Muñoz, J., Meseguer, J., 1998b. Dietary intake of levamisole enhances the immune response and disease resistance of the marine teleost gilthead seabream (*Sparus aurata* L.). Fish and Shellfish Immunology 8, 49–62. http://dx.doi.org/10.1006/fsim.1997.0119.

Musthafa, M.S., Ali, A.R.J., Mohamed, M.J., Jaleel, M.M.A., Kumar, M.S.A., Rani, K.U., Vasanth, K., Arockiaraj, J., Preetham, E., Balasundaram, C., Harikrishnan, R., 2016. Protective efficacy of Azomite enriched diet in *Oreochromis mossambicus* against *Aeromonas hydrophila*. Aquaculture 451, 310–315. http://dx.doi.org/10.1016/j.aquaculture.2015.09.006.

Muthulakshmi, M., Subramani, P.A., Michael, R.D., Summer 2016. Immunostimulatory effect of the aqueous leaf extract of *Phyllanthus niruri* on the specific and non-specific immune responses of *Oreochromis mossambicus* Peters. Iranian Journal of Veterinary Research 17 (3), 200–202. PMCID: PMC5090155c.

Nácher-Vázquez, M., Ballesteros, N., Canales, Á., Rodríguez Saint-Jean, S., Pérez-Prieto, S.I., Prieto, A., Aznar, R., López, P., 2015. Dextrans produced by lactic acid bacteria exhibit antiviral and immunomodulatory activity against salmonid viruses. Carbohydrate Polymers 124, 292–301. http://dx.doi.org/10.1016/j.carbpol.2015.02.020.

Nakajima, K., Maeno, Y., Honda, A., Yokoyama, K., Tooriyama, T., Manabe, S., 1999. Effectiveness of a vaccine against red sea bream iridoviral disease in a field trial test. Diseases of Aquatic Organisms 36, 73–75. http://dx.doi.org/10.3354/dao036073.

Nayak, S.K., 2010. Probiotics and immunity: a fish perspective. Fish and Shellfish Immunology 29, 2–14. http://dx.doi.org/10.1016/j.fsi.2010.02.017.

Newman, S.G., 1993. Bacterial vaccines for fish. Annual Review of Fish Diseases 3, 145–185. http://dx.doi.org/10.1016/0959-8030(93)90033-8.

Ngugi, C.C., Oyoo-Okoth, E., Mugo-Bundi, J., Orina, P.S., Chemoiwa, E.J., Aloo, P.A., 2015. Effects of dietary administration of stinging nettle (*Urtica dioica*) on the growth performance, biochemical, hematological and immunological parameters in juvenile and adult Victoria Labeo (*Labeo victorianus*) challenged with *Aeromonas hydrophila*. Fish and Shellfish Immunology 44, 533–541. http://dx.doi.org/10.1016/j.fsi.2015.03.025.

Nootash, S., Sheikhzadeh, N., Baradaran, B., Oushani, A.K., Maleki Moghadam, M.R., Nofouzi, K., Monfaredan, A., Aghebati, L., Zare, F., Shabanzadeh, S., 2013. Green tea (*Camellia sinensis*) administration induces expression of immune relevant genes and biochemical parameters in rainbow trout (*Oncorhynchus mykiss*). Fish and Shellfish Immunology 35, 1916–1923. http://dx.doi.org/10.1016/j.fsi.2013.09.030.

Novak, M., Vetvicka, V., 2009. Glucans as biological response modifiers. Endocrine, Metabolic and Immune Disorders—Drug Targets (Formerly Current Drugs) 9, 67–75. http://dx.doi.org/10.2174/187153009787582423.

Nya, E.J., Austin, B., 2009. Use of dietary ginger, Zingiber officinale Roscoe, as an immunostimulant to control Aeromonas hydrophila infections in rainbow trout, *Oncorhynchus mykiss* (Walbaum). Journal of Fish Diseases 32, 971–977. http://dx.doi.org/10.1111/j.1365-2761.2009.01101.x.

Ødegård, J., Baranski, M., Gjerde, B., Gjedrem, T., 2011. Methodology for genetic evaluation of disease resistance in aquaculture species: challenges and future prospects. Aquaculture Research 42, 103–114. http://dx.doi.org/10.1111/j.1365-2109.2010.02669.x.

Olivier, G., Evelyn, T.P.T., Lallier, R., 1985. Immunity to *Aeromonas salmonicida* in coho salmon (*Oncorhynchus kisutch*) inducfd by modified freund's complete adjuvant: its non-specific nature and the probable role of macrophages in the phenomenon. Developmental and Comparative Immunology 9, 419–432. http://dx.doi.org/10.1016/0145-305X(85)90005-9.

Pacitti, D., Lawan, M.M., Feldmann, J., Sweetman, J., Wang, T., Martin, S.A.M., Secombes, C.J., 2016. Impact of selenium supplementation on fish antiviral responses: a whole transcriptomic analysis in rainbow trout (*Oncorhynchus mykiss*) fed supranutritional levels of Sel-Plex®. BMC Genomics 17, 116. http://dx.doi.org/10.1186/s12864-016-2418-7.

Park, K.-H., Choi, S.-H., 2012. The effect of mistletoe, *Viscum album coloratum*, extract on innate immune response of Nile tilapia (*Oreochromis niloticus*). Fish and Shellfish Immunology 32, 1016–1021. http://dx.doi.org/10.1016/j.fsi.2012.02.023.

Park, J., Kim, P.-K., Lim, T., Daniels, H.V., 2013. Ozonation in seawater recirculating systems for black seabream *Acanthopagrus schlegelii* (Bleeker): effects on solids, bacteria, water clarity, and color. Aquacultural Engineering 55, 1–8. http://dx.doi.org/10.1016/j.aquaeng.2013.01.002.

Patterson, D., Gatlin III, D.M., 2013. Evaluation of whole and lipid-extracted algae meals in the diets of juvenile red drum (*Sciaenops ocellatus*). Aquaculture 416–417, 92–98. http://dx.doi.org/10.1016/j.aquaculture.2013.08.033.

Pedersen, L.-F., Pedersen, P.B., 2012. Hydrogen peroxide application to a commercial recirculating aquaculture system. Aquacultural Engineering 46, 40–46. http://dx.doi.org/10.1016/j.aquaeng.2011.11.001.

Pedersen, L.-F., Pedersen, P.B., Nielsen, J.L., Nielsen, P.H., 2009. Peracetic acid degradation and effects on nitrification in recirculating aquaculture systems. Aquaculture 296, 246–254. http://dx.doi.org/10.1016/j.aquaculture.2009.08.021.

Pedersen, L.-F., Meinelt, T., Straus, D.L., 2013. Peracetic acid degradation in freshwater aquaculture systems and possible practical implications. Aquacultural Engineering, Workshop on Recirculating Aquaculture Systems 53, 65–71. http://dx.doi.org/10.1016/j.aquaeng.2012.11.011.

Peterson, B.C., Peatman, E., Ourth, D.D., Waldbieser, G.C., 2015. Effects of a phytogenic feed additive on growth performance, susceptibility of channel catfish to *Edwardsiella ictaluri* and levels of mannose binding lectin. Fish and Shellfish Immunology 44, 21–25. http://dx.doi.org/10.1016/j.fsi.2015.01.027.

Pietretti, D., Vera-Jimenez, N.I., Hoole, D., Wiegertjes, G.F., 2013. Oxidative burst and nitric oxide responses in carp macrophages induced by zymosan, MacroGard® and selective dectin-1 agonists suggest recognition by multiple pattern recognition receptors. Fish and Shellfish Immunology 35, 847–857. http://dx.doi.org/10.1016/j.fsi.2013.06.022.

Pinpimai, K., Rodkhum, C., Chansue, N., Katagiri, T., Maita, M., Pirarat, N., 2015. The study on the candidate probiotic properties of encapsulated yeast, *Saccharomyces cerevisiae* JCM 7255, in Nile Tilapia (*Oreochromis niloticus*). Research in Veterinary Science 102, 103–111. http://dx.doi.org/10.1016/j.rvsc.2015.07.021.

Pulgar, R., Hödar, C., Travisany, D., Zuñiga, A., Domínguez, C., Maass, A., González, M., Cambiazo, V., 2015. Transcriptional response of Atlantic salmon families to *Piscirickettsia salmonis* infection highlights the relevance of the iron-deprivation defence system. BMC Genomics 16. http://dx.doi.org/10.1186/s12864-015-1716-9.

Pulkkinen, K., Suomalainen, L.-R., Read, A.F., Ebert, D., Rintamäki, P., Valtonen, E.T., 2010. Intensive fish farming and the evolution of pathogen virulence: the case of columnaris disease in Finland. Proceedings of the Royal Society B. Biological Sciences 277, 593–600. http://dx.doi.org/10.1098/rspb.2009.1659.

Qadis, A.Q., Goya, S., Yatsu, M., Yoshida, Y., Ichijo, T., Sato, S., 2014. Effects of a bacteria-based probiotic on subpopulations of peripheral leukocytes and their cytokine mRNA expression in calves. Journal of Veterinary Medical Science 76, 189–195. http://dx.doi.org/10.1292/jvms.13-0370.

Rajan, B.R., Omita, Y., Michael, R.D., 2013. Immunostimulation by Vitamin A in Mozambique Tilapia, *Oreochromis mossambicus* (Peters). International Journal of Frontiers in Science and Technology 1 (4), 1-12.

Ragap, H.M., Khalil, R.H., Mutawie, H.H., 2012. Immunostimulant effects of dietary *Spirulina platensis* on tilapia *Oreochromis niloticus*. Journal of Applied Pharmaceutical Science 2, 26.

Robinson, H.L., Pertmer, T.M., 2000. DNA vaccines for viral infections: basic studies and applications. Advances in Virus Research 55, 1–74.

Sahoo, P.K., Mahapatra, K.D., Saha, J.N., Barat, A., Sahoo, M., Mohanty, B.R., Gjerde, B., Ødegård, J., Rye, M., Salte, R., 2008. Family association between immune parameters and resistance to *Aeromonas hydrophila* infection in the Indian major carp, *Labeo rohita*. Fish and Shellfish Immunology 25, 163–169. http://dx.doi.org/10.1016/j.fsi.2008.04.003.

Sakai, M., Kamiya, H., Atsuta, S., Kobayashi, M., 1991. Immunomodulatory effects on rainbow trout, *Oncorlynchus mykiss*, injected with the extract of abalone, *Haliotis discus hannai*. Journal of Applied Ichthyology 7, 54–59. http://dx.doi.org/10.1111/j.1439-0426.1991.tb00594.x.

Salvesen, I., Vadstein, O., 1995. Surface disinfection of eggs from marine fish: evaluation of four chemicals. Aquaculture International 3, 155–171. http://dx.doi.org/10.1007/BF00118098.

Samad, A.P.A., Santoso, U., Lee, M.-C., Nan, F.-H., 2014. Effects of dietary katuk (*Sauropus androgynus* L. Merr.) on growth, non-specific immune and diseases resistance against *Vibrio alginolyticus* infection in grouper *Epinephelus coioides*. Fish and Shellfish Immunology 36, 582–589. http://dx.doi.org/10.1016/j.fsi.2013.11.011.

Sasaki, S., Takeshita, F., Xin, K.Q., Ishii, N., Okuda, K., 2003. Adjuvant formulations and delivery systems for DNA vaccines. Methods (San Diego, California) 31, 243–254.

Schlenk, D., Colley, W.C., El-Alfy, A., Kirby, R., Griffin, B.R., 2000. Effects of the oxidant potassium permanganate on the expression of gill metallothionein mRNA and its relationship to sublethal whole animal endpoints in channel catfish. Toxicological Sciences: an Official Journal of the Society of Toxicology 54, 177–182.

Sealey, W.M., Gatlin, I., 1999. Overview of nutritional strategies affecting health of marine fish. Journal of Applied Aquaculture 9, 11.

Selvaraj, V., Sampath, K., Sekar, V., 2009. Administration of lipopolysaccharide increases specific and non-specific immune parameters and survival in carp (*Cyprinus carpio*) infected with *Aeromonas hydrophila*. Aquaculture 286, 176–183. http://dx.doi.org/10.1016/j.aquaculture.2008.09.017.

Sen, S.S., Sukumaran, V., Giri, S.S., Park, S.C., 2015. Flavonoid fraction of guava leaf extract attenuates lipopolysaccharide-induced inflammatory response via blocking of NF-κB signalling pathway in *Labeo rohita* macrophages. Fish and Shellfish Immunology 47, 85–92. http://dx.doi.org/10.1016/j.fsi.2015.08.031.

Sharifuzzaman, S.M., Austin, B., 2009. Influence of probiotic feeding duration on disease resistance and immune parameters in rainbow trout. Fish and Shellfish Immunology 27, 440–445. http://dx.doi.org/10.1016/j.fsi.2009.06.010.

Sharma, A., Deo, A.D., Tandel Riteshkumar, S., Chanu, T.I., Das, A., 2010. Effect of *Withania somnifera* (L. Dunal) root as a feed additive on immunological parameters and disease resistance to *Aeromonas hydrophila* in *Labeo rohita* (Hamilton) fingerlings. Fish and Shellfish Immunology 29, 508–512. http://dx.doi.org/10.1016/j.fsi.2010.05.005.

Sheftel, A.D., Mason, A.B., Ponka, P., 2012. The long history of iron in the Universe and in health and disease. Biochimica et Biophysica Acta (BBA)—General Subjects, Transferrins Molecular Mechanisms of Iron Transport and Disorders 1820, 161–187. http://dx.doi.org/10.1016/j.bbagen.2011.08.002.

Shimizu, H., Thorley, B., Paladin, F.J., Brussen, K.A., Stambos, V., Yuen, L., Utama, A., Tano, Y., Arita, M., Yoshida, H., Yoneyama, T., Benegas, A., Roesel, S., Pallansch, M., Kew, O., Miyamura, T., 2004. Circulation of type 1 vaccine-derived poliovirus in the Philippines in 2001. Journal of Virology 78, 13512–13521. http://dx.doi.org/10.1128/JVI.78.24.13512-13521.2004.

Siwicki, A.K., Anderson, D.P., Dixon, O.W., 1990. In vitro immunostimulation of rainbow trout (*Oncorhynchus mykiss*) spleen cells with levamisole. Developmental and Comparative Immunology 14, 231–237. http://dx.doi.org/10.1016/0145-305X(90)90094-U.

Siwicki, A.K., 1989. Immunostimulating influence of levamisole on nonspecific immunity in carp (*Cyprinus carpio*). Developmental and Comparative Immunology 13, 87–91. http://dx.doi.org/10.1016/0145-305X(89)90021-9.

Skjermo, J., Vadstein, O., 1999. Techniques for microbial control in the intensive rearing of marine larvae. Aquaculture 177, 333–343. http://dx.doi.org/10.1016/S0044-8486(99)00096-4.

Skjermo, J., Størseth, T.R., Hansen, K., Handå, A., Øie, G., 2006. Evaluation of β-(1→3, 1→6)-glucans and High-M alginate used as immunostimulatory dietary supplement during first feeding and weaning of Atlantic cod (*Gadus morhua* L.). Aquaculture 261, 1088–1101. http://dx.doi.org/10.1016/j.aquaculture.2006.07.035.

Skov, J., Kania, P.W., Holten-Andersen, L., Fouz, B., Buchmann, K., 2012. Immunomodulatory effects of dietary β-1,3-glucan from *Euglena gracilis* in rainbow trout (*Oncorhynchus mykiss*) immersion vaccinated against *Yersinia ruckeri*. Fish and Shellfish Immunology 33, 111–120. http://dx.doi.org/10.1016/j.fsi.2012.04.009.

Snieszko, S.F., Friddle, S.B., 1949. Prophylaxis of furunculosis in Brook trout (*Salvelinus Fontinalis*) by oral immunization and Sulfamerazine. Progressive Fish-Culturist 11, 161–168. http://dx.doi.org/10.1577/1548-8640(1949)11[161:POFIBT]2.0.CO;2.

Stuart, K.R., Keller, M., Drawbridge, M., 2010. Efficacy of formalin and povidone–iodine disinfection techniques on the eggs of three marine finfish species. Aquaculture Research 41, e838–e843. http://dx.doi.org/10.1111/j.1365-2109.2010.02604.x.

Suguna, P., Binuramesh, C., Abirami, P., Saranya, V., Poornima, K., Rajeswari, V., Shenbagarathai, R., 2014. Immunostimulation by poly-β hydroxybutyrate–hydroxyvalerate (PHB–HV) from *Bacillus thuringiensis* in *Oreochromis mossambicus*. Fish and Shellfish Immunology 36, 90–97. http://dx.doi.org/10.1016/j.fsi.2013.10.012.

Sun, Y., Hu, Y.-H., Liu, C.-S., Sun, L., 2012a. Construction and comparative study of monovalent and multivalent DNA vaccines against *Streptococcus iniae*. Fish and Shellfish Immunology 33, 1303–1310. http://dx.doi.org/10.1016/j.fsi.2012.10.004.

Sun, Y., Zhang, M., Liu, C., Qiu, R., Sun, L., 2012b. A divalent DNA vaccine based on Sia10 and OmpU induces cross protection against *Streptococcus iniae* and *Vibrio anguillarum* in Japanese flounder. Fish and Shellfish Immunology 32, 1216–1222. http://dx.doi.org/10.1016/j.fsi.2012.03.024.

Swain, P., Nayak, S.K., 2009. Role of maternally derived immunity in fish. Fish and Shellfish Immunology 27, 89–99. http://dx.doi.org/10.1016/j.fsi.2009.04.008.

Swain, P., Nayak, S., Nanda, P., Dash, S., 2008. Biological effects of bacterial lipopolysaccharide (endotoxin) in fish: a review. Fish and Shellfish Immunology 25, 191–201. http://dx.doi.org/10.1016/j.fsi.2008.04.009.

Sweetman, J.W., Torrecillas, S., Dimitroglou, A., Rider, S., Davies, S.J., Izquierdo, M.S., 2010. Enhancing the natural defences and barrier protection of aquaculture species. Aquaculture Research 41, 345–355. http://dx.doi.org/10.1111/j.1365-2109.2009.02196.x.

Talpur, A.D., Ikhwanuddin, M., 2013. *Azadirachta indica* (neem) leaf dietary effects on the immunity response and disease resistance of Asian seabass, *Lates calcarifer* challenged with *Vibrio harveyi*. Fish and Shellfish Immunology 34, 254–264. http://dx.doi.org/10.1016/j.fsi.2012.11.003.

Tan, B.K.H., Vanitha, J., 2004. Immunomodulatory and antimicrobial effects of some traditional Chinese medicinal herbs: a review. Current Medicinal Chemistry 11, 1423–1430.

Tang, J., Cai, J., Liu, R., Wang, J., Lu, Y., Wu, Z., Jian, J., 2014. Immunostimulatory effects of artificial feed supplemented with a Chinese herbal mixture on *Oreochromis niloticus* against *Aeromonas hydrophila*. Fish and Shellfish Immunology 39, 401–406. http://dx.doi.org/10.1016/j.fsi.2014.05.028.

The World Bank, 2013. FISH to 2030 Prospects for Fisheries and Aquaculture (WORLD BANK REPORT NUMBER 83177-GLB). The World Bank, Washington, DC.

Thompson, I., Fletcher, T.C., Houlihan, D.F., Secombes, C.J., 1994. The effect of dietary vitamin A on the immunocompetence of Atlantic salmon (*Salmo salar* L.). Fish Physiology Biochemistry 12, 513–523. http://dx.doi.org/10.1007/BF00004453.

Thompson, M., Lochmann, R., Phillips, H., Sink, T.D., 2015. A dietary dairy/yeast prebiotic and flaxseed oil enhance growth, hematological and immunological parameters in channel catfish at a suboptimal temperature (15°C). Animal 9, 1113–1119. http://dx.doi.org/10.1017/S1751731115000300.

Torrecillas, S., Montero, D., Izquierdo, M., 2014. Improved health and growth of fish fed mannan oligosaccharides: potential mode of action. Fish and Shellfish Immunology 36, 525–544. http://dx.doi.org/10.1016/j.fsi.2013.12.029.

Tovar, D., Zambonino, J., Cahu, C., Gatesoupe, F.J., Vázquez-Juárez, R., Lésel, R., January 21, 2002. Effect of live yeast incorporation in compound diet on digestive enzyme activity in sea bass *(Dicentrarchus labrax)* larvae. Aquaculture 204 (1–2), 113–123. http://dx.doi.org/10.1016/S0044-8486(01)00650-0.

Tucker, C.S., 1984. Potassium permanganate demand of Pond waters. Progressive Fish-Culturist 46, 24–28. http://dx.doi.org/10.1577/1548-8640(1984)46<24:PPDOPW>2.0.CO;2.

Uribe, C., Folch, H., Enriquez, R., Moran, G., 2011. Innate and adaptive immunity in teleost fish: a review. Veterinarni Medicina 56 (10), 486–503.

Valenzuela, B., Obreque, J., Soto-Aguilera, S., Maisey, K., Imarai, M., Modak, B., 2016. Key cytokines of adaptive immunity are differentially induced in rainbow trout kidney by a group of structurally related geranyl aromatic derivatives. Fish and Shellfish Immunology 49, 45–53. http://dx.doi.org/10.1016/j.fsi.2015.12.008.

Van Hai, N., 2015. The use of medicinal plants as immunostimulants in aquaculture: a review. Aquaculture 446, 88–96. http://dx.doi.org/10.1016/j.aquaculture.2015.03.014.

Van Muiswinkel, W.B., Wiegertjes, G.F., Stet, R.J.M., 1999. The influence of environmental and genetic factors on the disease resistance of fish. Aquaculture 172, 103–110. http://dx.doi.org/10.1016/S0044-8486(98)00444-X.

Van Muiswinkel, W.B., 2008. A history of fish immunology and vaccination I. The early days. Fish and Shellfish Immunology NOFFI: Advances in Fish Immunology, Seventh International Symposium on Fish Immunology in Stirling, 2007 25, 397–408. http://dx.doi.org/10.1016/j.fsi.2008.02.019.

Vandenberg, G.W., 2004. Oral vaccines for finfish: academic theory or commercial reality? Animal Health Research Reviews Conference of Research Workers in Animal Diseases 5, 301–304.

Vargas, R.J., Dotta, G., Mouriño, J.L., Silva, B.C., da, Fracalossi, D.M., 2013. Dietary lipid sources affect freshwater catfish jundiá, *Rhamdia quelen*, survival, when challenged with *Aeromonas hydrophila*. Acta Scientiarum Animal Sciences 35, 349–355. http://dx.doi.org/10.4025/actascianimsci.v35i4.19617.

Verma, V.K., Rani, K.V., Sehgal, N., Prakash, O., 2012. Immunostimulatory response induced by supplementation of *Ficus benghalensis* root powder, in the artificial feed the Indian freshwater murrel, *Channa punctatus*. Fish and Shellfish Immunology 33, 590–596. http://dx.doi.org/10.1016/j.fsi.2012.06.031.

Verma, V.K., Rani, K.V., Sehgal, N., Prakash, O., 2013. Immunostimulatory effect of artificial feed supplemented with indigenous plants on *Clarias gariepinus* against *Aeromonas hydrophila*. Fish and Shellfish Immunology 35, 1924–1931. http://dx.doi.org/10.1016/j.fsi.2013.09.029.

Waagbø, R., Glette, J., Raa-Nilsen, E., Sandnes, K., 1993. Dietary vitamin C, immunity and disease resistance in Atlantic salmon (*Salmo salar*). Fish Physiology and Biochemistry 12, 61–73. http://dx.doi.org/10.1007/BF00004323.

Wang, Z., Mai, K., Liufu, Z., Ma, H., Xu, W., Ai, Q., Zhang, W., Tan, B., Wang, X., 2006. Effect of high dietary intakes of vitamin E and n-3 HUFA on immune responses and resistance to *Edwardsiella tarda* challenge in Japanese flounder (*Paralichthys olivaceus*, Temminck and Schlegel). Aquaculture Research 37, 681–692. http://dx.doi.org/10.1111/j.1365-2109.2006.01481.x.

Wang, G.-X., Jiang, D., Zhou, Z., Zhao, Y.-K., Shen, Y.-H., 2009. In vivo assessment of anthelmintic efficacy of ginkgolic acids (C13:0, C15:1) on removal of Pseudodactylogyrus in European eel. Aquaculture 297, 38–43. http://dx.doi.org/10.1016/j.aquaculture.2009.09.012.

Wang, H., Ji, D., Shao, J., Zhang, S., 2012. Maternal transfer and protective role of antibodies in zebrafish *Danio rerio*. Molecular Immunology 51, 332–336. http://dx.doi.org/10.1016/j.molimm.2012.04.003.

Wang, J.-L., Meng, X., Lu, R., Wu, C., Luo, Y.-T., Yan, X., Li, X.-J., Kong, X.-H., Nie, G.-X., 2015. Effects of *Rehmannia glutinosa* on growth performance, immunological parameters and disease resistance to *Aeromonas hydrophila* in common carp (*Cyprinus carpio* L.). Aquaculture 435, 293–300. http://dx.doi.org/10.1016/j.aquaculture.2014.10.004.

Watson, C., Yanong, R.P.E., 1989. Use of Copper in Freshwater Aquaculture and Farm Ponds.

Wedemeyer, G., 1969. Stress-induced ascorbic acid depletion and cortisol production in two salmonid fishes. Comparative Biochemistry and Physiology 29, 1247–1251. http://dx.doi.org/10.1016/0010-406X(69)91029-9.

Weimin, M., 2004. FAO Fisheries & Aquaculture—Cultured Aquatic Species Information Programme—Carassius carassius (Linnaeus, 1758).Pdf.

Welker, T.L., Lim, C., Yildirim-Aksoy, M., Klesius, P.H., 2007. Growth, immune function, and disease and stress resistance of juvenile Nile tilapia (*Oreochromis niloticus*) fed graded levels of bovine lactoferrin. Aquaculture 262, 156–162. http://dx.doi.org/10.1016/j.aquaculture.2006.09.036.

Welker, T.L., Lim, C., Yildirim-Aksoy, M., Klesius, P.H., 2010. Dietary bovine lactoferrin increases resistance of juvenile channel catfish, *Ictalurus punctatus*, to enteric septicemia. Journal of the World Aquaculture Society 41, 28–39. http://dx.doi.org/10.1111/j.1749-7345.2009.00330.x.

Wu, C.-C., Liu, C.-H., Chang, Y.-P., Hsieh, S.-L., 2010. Effects of hot-water extract of *Toona sinensis* on immune response and resistance to *Aeromonas hydrophila* in *Oreochromis mossambicus*. Fish and Shellfish Immunology 29, 258–263. http://dx.doi.org/10.1016/j.fsi.2010.04.021.

Wu, Y., Gong, Q., Fang, H., Liang, W., Chen, M., He, R., 2013. Effect of *Sophora flavescens* on non-specific immune response of tilapia (GIFT *Oreochromis niloticus*) and disease resistance against *Streptococcus agalactiae*. Fish and Shellfish Immunology 34, 220–227. http://dx.doi.org/10.1016/j.fsi.2012.10.020.

Wu, C., Ye, J., Gao, J., Chen, L., Lu, Z., 2016. The effects of dietary carbohydrate on the growth, antioxidant capacities, innate immune responses and pathogen resistance of juvenile Black carp *Mylopharyngodon piceus*. Fish and Shellfish Immunology 49, 132–142. http://dx.doi.org/10.1016/j.fsi.2015.12.030.

Xu, T., Chen, S., Ji, X., Tian, Y., 2008. MHC polymorphism and disease resistance to *Vibrio anguillarum* in 12 selective Japanese flounder (*Paralichthys olivaceus*) families. Fish and Shellfish Immunology 25, 213–221. http://dx.doi.org/10.1016/j.fsi.2008.05.007.

Xu, T., Chen, S., Zhang, Y., 2010. MHC class IIα gene polymorphism and its association with resistance/susceptibility to *Vibrio anguillarum* in Japanese flounder (*Paralichthys olivaceus*). Developmental and Comparative Immunology 34, 1042–1050. http://dx.doi.org/10.1016/j.dci.2010.05.008.

Yang, X., Guo, J.L., Ye, J.Y., Zhang, Y.X., Wang, W., 2015. The effects of *Ficus carica* polysaccharide on immune response and expression of some immune-related genes in grass carp, *Ctenopharyngodon idella*. Fish and Shellfish Immunology 42, 132–137. http://dx.doi.org/10.1016/j.fsi.2014.10.037.

Yang, J., Liu, Z., Shi, H.-N., Zhang, J.-P., Wang, J.-F., Huang, J.-Q., Kang, Y.-J., 2016. Association between MHC II beta chain gene polymorphisms and resistance to infectious haematopoietic necrosis virus in rainbow trout (*Oncorhynchus mykiss*, Walbaum, 1792). Aquaculture Research 47, 570–578. http://dx.doi.org/10.1111/are.12516.

Yeh, D.-W., Liu, Y.-L., Lo, Y.-C., Yuh, C.-H., Yu, G.-Y., Lo, J.-F., Luo, Y., Xiang, R., Chuang, T.-H., 2013. Toll-like receptor 9 and 21 have different ligand recognition profiles and cooperatively mediate activity of CpG-oligodeoxynucleotides in zebrafish. Proceedings of the National Academy of Sciences 110, 20711–20716. http://dx.doi.org/10.1073/pnas.1305273110.

Yin, G., Jeney, G., Racz, T., Xu, P., Jun, X., Jeney, Z., 2006. Effect of two Chinese herbs (*Astragalus radix* and *Scutellaria radix*) on non-specific immune response of tilapia, *Oreochromis niloticus*. Aquaculture 253, 39–47. http://dx.doi.org/10.1016/j.aquaculture.2005.06.038.

Yin, G., Ardo, L., Thompson, K.D., Adams, A., Jeney, Z., Jeney, G., 2009. Chinese herbs (*Astragalus radix* and *Ganoderma lucidum*) enhance immune response of carp, *Cyprinus carpio*, and protection against *Aeromonas hydrophila*. Fish and Shellfish Immunology 26, 140–145. http://dx.doi.org/10.1016/j.fsi.2008.08.015.

Yoshida, T., Kruger, R., Inglis, V., 1995. Augmentation of non-specific protection in African catfish, *Clarias gariepinus* (Burchell), by the long-term oral administration of immunostimulants. Journal of Fish Diseases 18, 195–198. http://dx.doi.org/10.1111/j.1365-2761.1995.tb00278.x.

Yu, X.-B., Liu, G.-L., Zhu, B., Hao, K., Ling, F., Wang, G.-X., 2014. In vitro immunocompetence of two compounds isolated from *Polygala tenuifolia* and development of resistance against grass carp reovirus (GCRV) and *Dactylogyrus intermedius* in respective host. Fish and Shellfish Immunology 41, 541–548. http://dx.doi.org/10.1016/j.fsi.2014.10.004.

Yue, G.H., 2014. Recent advances of genome mapping and marker-assisted selection in aquaculture. Fish and Fisheries 15 (3), 376–396.

Yue, Y.-R., Zhou, Q.-C., 2008. Effect of replacing soybean meal with cottonseed meal on growth, feed utilization, and hematological indexes for juvenile hybrid tilapia, *Oreochromis niloticus* × *O. aureus*. Aquaculture 284, 185–189. http://dx.doi.org/10.1016/j.aquaculture.2008.07.030.

Yunis-Aguinaga, J., Claudiano, G.S., Marcusso, P.F., Manrique, W.G., de Moraes, J.R.E., de Moraes, F.R., Fernandes, J.B.K., 2015. *Uncaria tomentosa* increases growth and immune activity in *Oreochromis niloticus* challenged with *Streptococcus agalactiae*. Fish and Shellfish Immunology 47, 630–638. http://dx.doi.org/10.1016/j.fsi.2015.09.051.

Yáñez, J.M., Houston, R.D., Newman, S., 2014. Genetics and genomics of disease resistance in salmonid species. Frontiers in Genetics 5, 415. http://dx.doi.org/10.3389/fgene.2014.00415.

Zahran, E., Risha, E., AbdelHamid, F., Mahgoub, H.A., Ibrahim, T., 2014. Effects of dietary *Astragalus* polysaccharides (APS) on growth performance, immunological parameters, digestive enzymes, and intestinal morphology of Nile tilapia (*Oreochromis niloticus*). Fish and Shellfish Immunology 38, 149–157. http://dx.doi.org/10.1016/j.fsi.2014.03.002.

Zamri-Saad, M., Amal, M.N.A., Siti-Zahrah, A., Zulkafli, A.R., 2014. Control and prevention of streptococcosis in cultured tilapia in Malaysia: a review. Pertanika Journal of Tropical Agricultural Science 37, 389–410.

Zanuzzo, F.S., Zaiden, S.F., Senhorini, J.A., Marzocchi-Machado, C.M., Urbinati, E.C., 2015. Aloe vera bathing improved physical and humoral protection in breeding stock after induced spawning in matrinxã (*Brycon amazonicus*). Fish and Shellfish Immunology 45, 132–140. http://dx.doi.org/10.1016/j.fsi.2015.02.017 Special Issue Probiotics.

Zhang, S., Wang, Z., Wang, H., 2013. Maternal immunity in fish. Developmental and Compararive Immunology 39, 72–78. http://dx.doi.org/10.1016/j.dci.2012.02.009.

Zhang, Y., Liu, B., Ge, X., Liu, W., Xie, J., Ren, M., Cui, Y., Xia, S., Chen, R., Zhou, Q., Pan, L., Yu, Y., 2014a. The influence of various feeding patterns of emodin on growth, non-specific immune responses, and disease resistance to *Aeromonas hydrophila* in juvenile Wuchang bream (*Megalobrama amblycephala*). Fish and Shellfish Immunology 36, 187–193. http://dx.doi.org/10.1016/j.fsi.2013.10.028.

Zhang, Z., Xu, L., Liu, W., Yang, Y., Du, Z., Zhou, Z., 2014b. Effects of partially replacing dietary soybean meal or cottonseed meal with completely hydrolyzed feather meal (defatted rice bran as the carrier) on production, cytokines, adhesive gut bacteria, and disease resistance in hybrid tilapia (*Oreochromis niloticus* ♀ × *Oreochromis aureus* ♂). Fish and Shellfish Immunology 41, 517–525. http://dx.doi.org/10.1016/j.fsi.2014.09.039.

Zhao, L., Seth, A., Wibowo, N., Zhao, C.-X., Mitter, N., Yu, C., Middelberg, A.P.J., 2014. Nanoparticle vaccines. Vaccine 32, 327–337. http://dx.doi.org/10.1016/j.vaccine.2013.11.069.

Zheng, Z.L., Tan, J.Y.W., Liu, H.Y., Zhou, X.H., Xiang, X., Wang, K.Y., 2009. Evaluation of oregano essential oil (*Origanum heracleoticum* L.) on growth, antioxidant effect and resistance against *Aeromonas hydrophila* in channel catfish (*Ictalurus punctatus*). Aquaculture 292, 214–218. http://dx.doi.org/10.1016/j.aquaculture.2009.04.025.

Zhou, Q., Wang, L., Wang, H., Xie, F., Wang, T., 2012. Effect of dietary vitamin C on the growth performance and innate immunity of juvenile cobia (*Rachycentron canadum*). Fish and Shellfish Immunology 32, 969–975. http://dx.doi.org/10.1016/j.fsi.2012.01.024.

Zhou, C., Ge, X., Lin, H., Niu, J., 2014. Effect of dietary carbohydrate on non-specific immune response, hepatic antioxidative abilities and disease resistance of juvenile golden pompano (*Trachinotus ovatus*). Fish and Shellfish Immunology 41, 183–190. http://dx.doi.org/10.1016/j.fsi.2014.08.024.

Zhou, Q., Jin, M., Elmada, Z.C., Liang, X., Mai, K., 2015. Growth, immune response and resistance to *Aeromonas hydrophila* of juvenile yellow catfish, *Pelteobagrus fulvidraco*, fed diets with different arginine levels. Aquaculture 437, 84–91. http://dx.doi.org/10.1016/j.aquaculture.2014.11.030.

Chapter 5

Integrated Pathogen Management Strategies in Fish Farming

Ariadna Sitjà-Bobadilla[1], Birgit Oidtmann[2]

[1]*Consejo Superior de Investigaciones Científicas (CSIC), Castellón, Spain;* [2]*Centre for Environment, Fisheries and Aquaculture Science (CEFAS), Weymouth, Dorset, United Kingdom*

5.1 THE CONCEPT: WHAT ARE INTEGRATED PATHOGEN MANAGEMENT STRATEGIES AND WHY DOES AQUACULTURE NEED THEM?

The world faces the formidable challenge of feeding the rapidly growing human population. Vegetable and animal productivities have to improve enormously to keep pace with this growth, and aquaculture is one of the most promising avenues, since it remains one of the fastest-growing food-producing sectors and now provides almost half of all fish for human food. This share is projected to rise to 62% by 2030 (FAO, 2014; Jennings et al., 2016). Pathogens are among the main limiting factors of aquaculture. When handled improperly they can evolve into unmanageable proportions and contribute to economic losses (Lafferty et al., 2015). Indiscriminate and unwise use of chemicals can result in control failure, in addition to polluting the environment and disturbing the ecological balance. The impact of pathogens on aquaculture is substantial; the financial losses are estimated to be roughly 20% of the total production value. It is estimated that the world annual grow-out loss due to parasites in finfish farming ranges from 1% to 10% of harvest size, with an annual cost of $1.05 to $9.58 billion (Shinn et al., 2015). Therefore the integration of management solutions is essential for the future increase of the aquaculture industry.

Integrated pathogen (pest) management (IPM) was initiated decades ago as a holistic way to combat pests in crops without relying solely on pesticides (Stern et al., 1959). It has now expanded to any type of pathogen and host and it is included even in legislations. As an example, the European Union (EU) requires that all EU member states develop a national action plan that ensures that IPM is implemented by all professional pesticide users in agriculture (Barzman et al., 2015). The main goal of IPM is to combine all the available preventive and curative methods to minimize the impact of pathogens in the production chain, and at the same time minimize the impact on the environment and avoid future side effects, therefore increasing sustainability, at both economic and environmental levels. IPM strategies (IPMS) should not be confused with organic practices, as IPMS do not discard the use of some chemicals when necessary, but seek to find alternative strategies to minimize their use. Its application in aquaculture is just starting. The term IPM comprises the following:

Integrated: It is a holistic approach, as it combines all available strategies to control disease, with a focus on the interactions between pathogen, host, and environment. The relationship between these three factors is complex, as the mere presence of a pathogen does not necessarily lead to the development of disease. This interaction, while complicating the epizootiology of diseases, provides opportunities to minimize the impact of the infection. Thus, integration starts with the description of factors that define the classical "disease triangle" (Stevens, 1960) (Fig. 5.1).

Pathogen: This means any organism that conflicts with plant or animal production. If an organism does not have a serious impact, it is not worth developing IPM for it. IPM works particularly well for pathogens with complex life cycles providing multiple opportunities for intervention.

Management: It is a way to keep pathogens below the levels at which they can cause serious economic damage. It does not always mean eradicating the pathogens. It means finding strategies that are effective and economical and keep environmental damage to a minimum.

The development of IPMS is a process, consisting of several steps summarized in Fig. 5.2. The process is initiated when a pathogen causes a disease outbreak, and the first step consists of gathering all the possible **knowledge** on the key pathogen(s) (life cycle, host-invasion strategies, natural enemies, vectors, etc.), as well as host and environmental risk

Fish Diseases. http://dx.doi.org/10.1016/B978-0-12-804564-0.00005-3

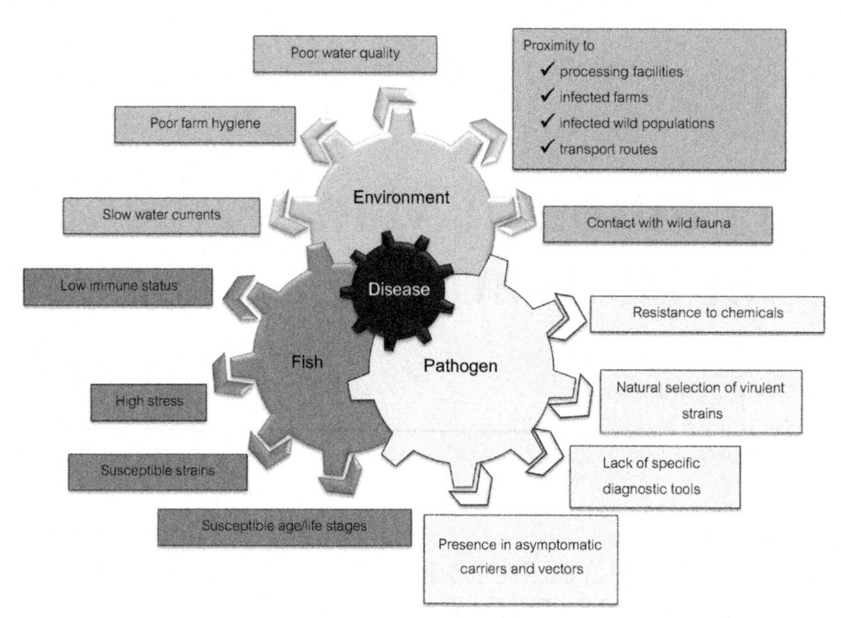

FIGURE 5.1 Venn diagram (inspired by Stevens, 1960) indicating the factors that cointervene in the occurrence of fish diseases. They are the result of the interactions between pathogens, fish (hosts), and environment.

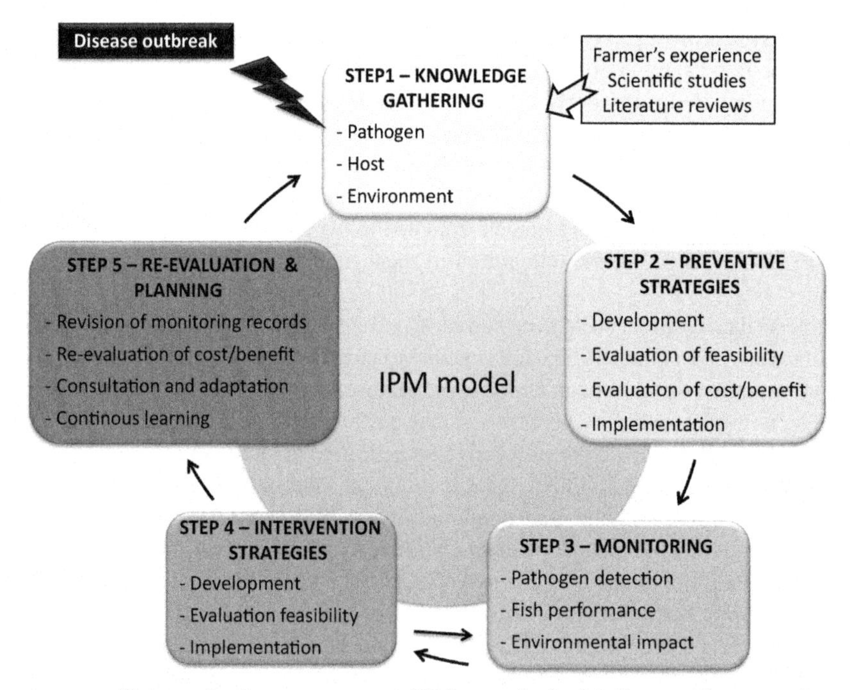

FIGURE 5.2 Development process of integrated pathogen management (IPM) strategies for fish diseases. The process is usually initiated by a disease outbreak. The first step (1) consists of gathering all the available knowledge on the pathogen and host and environmental risk factors that favor pathogen spread and impact within a fish population. The second step (2) is the development, evaluation of the feasibility and cost/benefit, and implementation of preventive strategies. The third step (3) is the monitoring of the disease. When prevention is not sufficient to stop the disease, intervention strategies have to be developed, evaluated, and applied (step 4). Monitoring also occurs after intervention. The fifth step (5) is reevaluation and planning in view of the results obtained with the various strategies, which feeds back to step 1, increasing the body of knowledge and leading to further improvement and refinement of the IPM strategy. *Adapted from integrated pest management models for plants.*

factors that favor pathogen spread and impact within a fish population. This information comes initially from farmers' experience, scientific studies, and literature reviews. The second step is **prevention**; it implies the development, evaluation of the feasibility and cost/benefit, and implementation of the best preventive strategies for each pathogen. The third step is the **monitoring** of the disease, which includes the detection of the pathogen, the surveillance of the host performance, and the possible impact on the environment. When prevention is not sufficient to stop the disease, the fourth step is **intervention**; it means the development, evaluation of the feasibility and cost/benefit, and the implementation of physical, chemical, and/or biological treatments. Monitoring also occurs after intervention. The fifth step is **reevaluation and planning** in view of the results obtained from the various strategies, as IPMS have to be constantly assessed and refined to maximize their benefits, by revising the disease records, reevaluating the costs/benefits, consulting and adapting to innovations, and continuous learning. Ideally it should allow one, in the long term, to define prediction models. Step 5 feeds back to step 1, increasing the body of knowledge. In this chapter we will describe the available choices for most of these steps in fish farming, which are the current limiting factors for their implementation in the production sites, and the future perspectives.

5.2 DIAGNOSTIC TOOLS, THE KEY FOR IDENTIFICATION AND MONITORING OF PATHOGENS

The starting point to identify a pathogen and monitor the corresponding disease is the availability of diagnostic tools. These tools are very diverse depending on the type of pathogen. According to the type of assay, they can be classified as macroscopic, microscopic, histological, microbiological, immunological, and molecular, and can be applied at different levels (Table 5.1). **Field diagnostic methods** are mainly macroscopic and include the observation of clinical signs in the external or internal appearance of the fish, and behavioral changes. They are considered only presumptive and are normally not specific for a particular disease, and some infected fish might not show any signs. **Clinical methods** (e.g., gross pathology, clinical chemistry, microscopic pathology) are applied when the field methods show a suspicion of pathogen, and may provide a presumptive or confirmatory diagnosis. For example, low hematocrit levels, blood smears with degenerate and vacuolized erythrocytes and the presence of erythroblasts with irregular nuclear shape, and differential blood cell counts with a reduction in the proportion of leukocytes relative to erythrocytes are presumptive of infectious salmon anemia (ISA) in seawater-reared Atlantic salmon.

Depending on the pathogen, a range of **direct detection methods** can be applied. Using the example of parasites, this can be done: (1) for large parasites (helminths) by naked eye or using a stereomicroscope, (2) for hemoparasites by light microscopy of Giemsa-stained blood smears, (3) for micro- and macroparasites using histology and immunohistochemistry, (4) for some protozoans and fungi by light microscopy of wet mounts, and (5) for some microparasites such as microsporidia, coccidia, or myxozoa, by transmission electron microscopy if necessary for specific classification. Various guides are available for the identification of fish protozoan and metazoan parasites in stained tissue sections (e.g., Bruno et al., 2006).

For many pathogens **the isolation of the agent** is needed for its subsequent identification. The isolation can be done by culturing the pathogen in specific cell lines, as happens with many fish viruses; by culturing in specific artificial media, as is needed for several bacteria (some are very fastidious organisms and require special media), fungi, and some microparasites; or by digestion of the tissue in which the parasite is embedded, as happens for some helminths and for the myxozoan *Myxobolus cerebralis*. The **identification of the agent** can be done conventionally by phenotypic characterization with API-ZYM systems as for several bacteria, using antibody-based antigen detection methods such as neutralization tests, immunofluorescence antibody test, or ELISA, or using molecular techniques such as PCR, RT-PCR, loop-mediated isothermal amplification, sequencing, etc. Since 2005, molecular methods have been developed for any type of pathogen and are part of a confirmatory diagnosis. Molecular tools, with the adequate automatization and standardization, allow processing of large numbers of samples, with the subsequent higher screening capacity, and render them good candidates for **early detection systems**.

Indirect detection methods are usually based on the detection of antibodies against the pathogen. In general, serological methods cannot be used to confirm the presence of a given pathogen since the antibody levels depend on the dynamics of antibody production by the fish host following pathogen exposure. Because of the current underdevelopment of serological methods in fish, these methods are not yet accepted for routine screening of the health status of fish populations. However, with further validation in the near future, the use of serological tests is expected to become more widely accepted for diagnostic purposes and they could be of value for high-scale screening of large numbers of fish as is required with the development of health control policies.

The degree of standardization of diagnostic methods depends, among other factors, on the level of obligatory reporting by the responsible authorities. As of this writing, there are only 10 notifiable fish diseases listed by the World Organisation for Animal Health (OIE, 2016 list), most of them caused by viral pathogens. The methods or tests used to classify a farming unit as either positive or negative with respect to an infection or disease can be classified as: (1) diagnostic, when applied to clinically diseased individuals; (2) screening, when applied to apparently healthy individuals; or (3) confirmatory, when applied to confirm the result of a previous positive test (OIE, 2015a). For each listed disease there is a definition for suspected cases and confirmed cases. In many cases it involves the application of several methods by at least two independent laboratories (OIE, 2015a).

The general approach to surveillance, including risk-based surveillance, for fish diseases has been summarized by Oidtmann et al. (2013).

TABLE 5.1 Diagnostic Methods for Fish Pathogens at Presumptive (P) and Confirmatory (C) Levels

Field Diagnostic Methods	Level
Clinical signs	P
Behavioral changes	P
Clinical Methods	
Gross pathology	P
Clinical chemistry	P
Microscopic pathology	P
Wet mounts	P, C
Tissue imprints and smears	P, C
Cytopathology	C
Agent Detection	
Direct Detection Methods	
Microscopic Methods	
Wet mounts	P, C
Smears: Differential stains, IF	P, C
Electron microscopy	P, C
Histology: morphology, differential stains, IHC, FISH	P, C
Agent Isolation	
Culture in cell lines	P, C
Artificial media	P, C
Tissue digestion	P, C
Agent Identification	
Phenotypic characterization	P, C
Antibody-based antigen detection methods: Neutralization test, IFAT, ELISA	C
Molecular techniques: PCR, RT-PCR, sequencing, multiplex-PCR, Q-PCR	C
Indirect Detection Methods	
Serological methods: ELISA, serum neutralization, immobilization, microagglutination, passive hemagglutination	P, C

FISH, fluorescence in situ hybridization; *IF*, immunofluorescence; *IFAT*, immunofluorescence antibody test; *IHC*, immunohistochemistry.

Regardless of the diagnostic methods applied for each pathogen, a **sampling protocol** has to be established. It should define:

- Sampling points, by the identification of critical control points (CCP). These are specific points, times, procedures, or steps in the farming chain at which control can be exercised to reduce, eliminate, or prevent the possibility of entry or action of a pathogen.
- Sample size. The required sample size is dependent on a range of factors, e.g., expected infection prevalence in the target population, diagnostic test specificity and sensitivity, and purpose of the test (e.g., estimation of prevalence to a certain level of accuracy, detection of infection in the population), to name just a few (Oidtmann et al., 2013).
- Sampling procedure. This includes how to take, preserve, store, and submit samples to the diagnostic laboratory. To avoid cross-contamination and to keep sample integrity are particularly important for molecular methods. It is important also to identify whether pooling of samples is advisable.
- Type of sample. Usually target tissues are taken from sacrificed/dead fish, but ideally, the samples could be nonlethal samples, that is to say, not killing the fish, using material such as blood, skin mucus, milt, feces, intestinal swabs, etc.

A good diagnostic method is essential in all the steps of IPM: in step 1 for helping in defining the life cycle of the pathogens and risks factors; in step 2 for selection of water sources, monitoring of host infection status, controlling quarantines, transport, etc.; in step 3 for monitoring of the disease; in step 4 for decision-making on management, treatments, population discards, etc., and in step 5 for evaluating the effects of the various intervention actions (Fig. 5.2). For this reason, the velocity of obtaining diagnostic results is critical in fish production and surveillance. Future methods are oriented toward developing **point-of-care tests (POCTs)** or field-site testing. These are defined as tests applied at or near the point at which the sample is taken, conveniently and immediately, by the fish farmer or the veterinarian at the production site. This increases the likelihood that the facilities are properly and quickly surveyed. The use of POCT devices has increased exponentially in human medicine since 2005 and in the future they will also be extensive for veterinarian purposes as they become more cost-effective.

In IPMS, the detection of a pathogen in water or other environmental samples (sediments, algae, invertebrates, etc.), before it provokes disease outbreaks, can also be very useful and has the potential to provide rapid assessment of parasite levels to inform disease management decisions. In such case, highly targeted environmental DNA tests are ideal tools for researching life-cycle diversity, host ranges, and pathogen reservoirs, and therefore can help in defining the environmental occurrence of life-cycle stages, potential transmission zones, and reservoirs and in the identification of areas to prioritize for control or eradication attempts (Longshaw et al., 2012). These assays are designed to detect individual genera, species, or subspecific strains, depending on the pathogen, and are highly specific, sensitive, and often quantitative. Many Q-PCR assays have been developed and some examples of fish parasites can be found in Bass et al. (2015).

5.3 MODELING DISEASE TRANSMISSION AND RISK ASSESSMENT

As previously indicated, the development of disease is the result of (risk) factors influencing the epidemiological triad of host, pathogen, and environment (Fig. 5.1). The development of IPMS crucially depends on the identification of such risk factors for pathogen introduction and amplification. They are part of the knowledge gathered in step 1 of the IPM cycle (Fig. 5.2). The reevaluation of the results obtained after the implementation of the various strategies will aid in the modeling of disease dispersion and therefore improving IPM.

5.3.1 Risk Factors

Whereas research into risk factors of terrestrial animal diseases has been undertaken for some time, epidemiological studies into risk factors of aquaculture diseases is relatively recent. Research into risk factors is commonly triggered when a certain disease causes relevant losses, threatens wild fish, or has other environmental impacts.

Risk factors for disease development in a farmed fish population can generally be divided into two groups: (1) risk factors for pathogen **introduction** into farms and (2) risk factors for **spread and amplification** of a pathogen within a farmed fish population. In general, risk pathways for pathogen introduction into fish farms fall into five main areas: (1) live fish and egg movements into the farm, (2) exposure via water, (3) on-site processing, (4) short-distance mechanical transmission, and (5) distance-independent mechanical transmission (Oidtmann et al., 2014). In a geographic area with low (2%) site-level prevalence of a disease, experts considered that live fish movements were the most relevant pathway for introduction of viral hemorrhagic septicemia virus (VHSV), infectious hematopoietic necrosis virus (IHNV), and koi herpesvirus (KHV) into freshwater aquaculture sites. Exposure via water, mechanical transmission pathways, and fish processing on-site were considered to be less relevant (Oidtmann et al., 2014). Such results suggest that effective control of live fish introductions and sourcing from approved disease-free sources is a very effective way of decreasing the risk of introduction of any of these pathogens (Oidtmann et al., 2011). As live fish introductions are managed through farm operations, this is a risk that can be quite easily addressed. Recognition of this risk is the foundation for the requirement of health certification when introducing fish into a declared disease-free area (e.g., Anonymous, 2006; OIE, 2015b). A far more difficult risk to control is pathogen introduction via water. One of the main measures implemented in terrestrial animal husbandry for keeping exposure to pathogens at a minimum is indoor housed farming, as is often used in poultry and pig production. This method of husbandry facilitates the implementation of **biosecurity measures**, part of which are tight controls on the introduction of animals from outside into a given farm and controlled access (e.g., of staff or people from outside, measures to prevent wild animals from accessing the rearing units). Examples of closed systems in aquaculture include hatcheries using a borehole water supply and rearing eggs and fry indoors and using indoor recirculation systems, as are used, for example, for tilapia and catfish production.

However, the majority of aquaculture systems are open systems, where sea- or river water is taken from the environment and used to supply fish rearing units, or alternatively, the fish rearing units are placed right in the aquatic environment, as is the case with sea cage farming. This leads to exposure of the fish to any pathogen contained in this water, e.g., parasitic or viral pathogens found in wild fish populations in the respective geographic area or spread from nearby fish farms. Under such conditions, the emphasis of IPMS shifts from preventing introduction of a given pathogen to limiting its impact on the fish population. Examples of risk factors for pathogen amplification on-site are stocking density, mixed fish size classes, lack of frequent removal of mortalities, and stressed fish (Fig. 5.1).

A third category is semiopen systems, where some recirculation of rearing water takes place, but the farm also replaces some water by intake of water from the environment (e.g., from a river). Here, the risk of exposure to pathogens introduced through water from the environment is somewhat reduced; however, the use of recirculation systems poses its own challenges, as pathogens might build up in such systems to considerable levels.

To focus efforts in IPM on those areas that are most effective, what needs to be considered is not only whether a certain exposure route is likely to result in pathogen introduction if it did take place, or whether a certain management practice is likely to lead to increased pathogen burden; it is also important to consider the combination of the frequency of this event occurring, the likelihood of it leading to the undesired outcome, and the impact it would have. In the case of a pathogen that is already present in the environment in or around the aquaculture site, the undesired outcome is not infection per se. It is the scale of impact caused by infection. For example, sea lice infection in salmon is normal in wild salmon species (albeit at very low levels). Therefore, it would be expected that farmed salmon are also infected. However, what is causing the damage to farmed salmon is the level at which the infection occurs. Therefore, studies into risk factors of problems such as sea lice are investigating the risk factors for increased burden of infection and not necessarily for the presence or absence of the parasite. An example of a study that investigated risk factors for increased sea lice, *Caligus rogercresseyi*, levels in farmed salmonids in southern Chile identified geographical zone, fish species, treatment against sea lice performed 1 month before sampling, stocking density, fish weight, and seawater salinity as risk factors (Yatabe et al., 2011). This is a good example of a list of risk factors that are likely to be relevant for several parasite infections in fish farmed in sea cages.

An overview of risk factors for disease problems is provided in Fig. 5.1 and specific risk factors for parasitic diseases due to management practices can be found in Fig. 5.3. It is important to note that risk factors due to management practices can be modulated, whereas there may be less scope for doing so with regard to host factors, environmental factors, and choice of location. Further examples of risk factors are mentioned in Sections 5.4 and 5.5 of this chapter.

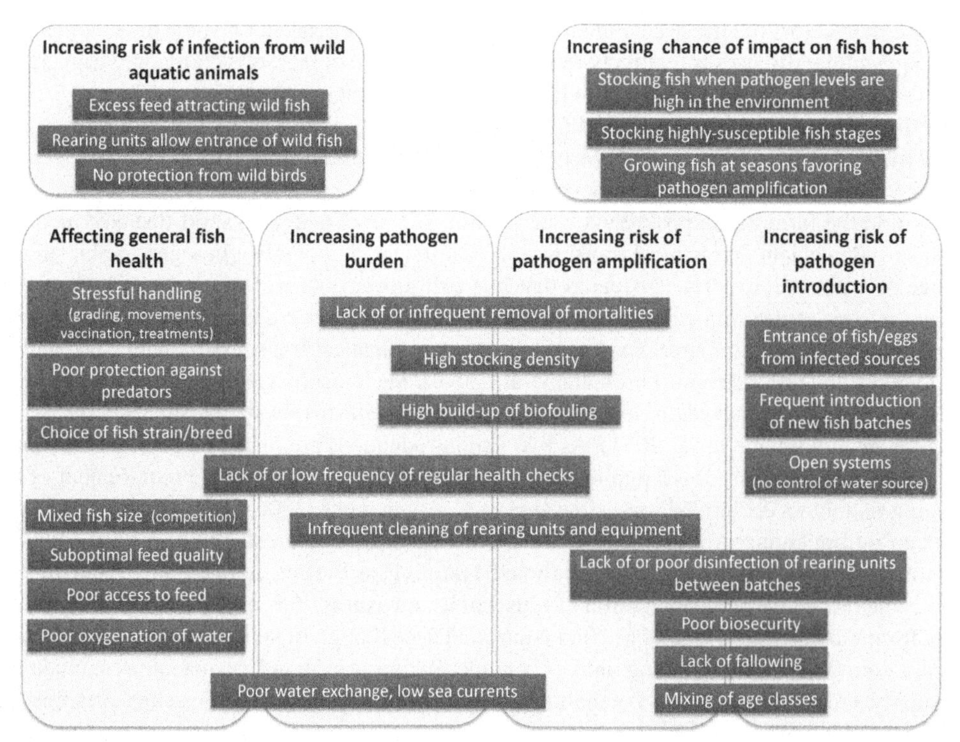

FIGURE 5.3 Management practices potentially increasing the likelihood of disease occurring and/or increasing its impact that should be considered in the design of integrated pathogen management strategies.

5.3.2 Disease Modeling

Modeling is increasingly used as a tool for predicting disease dynamics in aquaculture production. Models help us to understand the life cycle of the pathogen and the dynamics of pathogen amplification in a given population. They can simulate the effect of, e.g., temperature, host population density, and interventions on pathogen amplification (and therefore impact) in farmed fish populations. They can be applied at the individual fish level, farm level, area level, and larger scale geographic level.

Fish level models have been used to estimate the parasite burden over time and the efficacy of various treatment approaches (Revie et al., 2005; Gautam et al., 2016).

Population level models can simulate the number of susceptible (S), infected (I), and recovered (R) fish (Ogut et al., 2005; Ogut and Bishop, 2007) or—as is often done for sea lice models—simulate the average parasite burden per fish (Murray, 2011). SIR models attempt to predict the number of animals in each class over time depending on a range of factors, including environmental conditions and risk factors.

Spatial models aim to simulate the spread of disease in a defined geographic area. This can be, for example, for the purpose of predicting the likely spread of a given pathogen should it be introduced into a new geographic area. Similarly, where the pathogen has already been introduced, spatial modeling can be employed to predict further spread (Foreman et al., 2015; Salama et al., 2016). Another application of spatial modeling is for spatial planning. This has been done for setting up management areas for Atlantic salmon production (Salama et al., 2013).

Network models have been used to assess the effectiveness of fallowing strategies to control disease in salmon aquaculture (Werkman et al., 2011).

Disease models try to describe reality by explaining changes in disease dynamics with a limited number of parameters. The parameterization of models requires good estimates of how certain factors influence pathogen dynamics in the study population. Sometimes models are parameterized with very limited data. The general model structure will be based on the knowledge of the biology of host and pathogen. Scientists can then try to vary the parameter estimates to achieve a good model "fit." In this way, modeling can be applied to *generate* parameter estimates. However, this requires that some field data are available in the first place.

Economic models tend to be developed separate from dynamic disease models. Economic models have been used to undertake benefit/cost analysis for the control of notifiable diseases (Moran and Fofana, 2007). More recently, system dynamic models have been applied to the evaluation of the most cost-beneficial approach of controlling sea lice (Hamza et al., 2014).

Disease models allow the exploration of control options for diseases—at fish, cage, site, or large geographic level—in silico. These have been used with success in the control of sea lice, from which it emerged that treatments of Atlantic salmon early, when parasite burden is still low, are more efficient at controlling sea lice populations, compared with delaying treatments until a higher count of sea lice is found per individual fish (Hamza et al., 2014). This leads to an overall reduction of economic losses due to sea lice.

Where risk factors can be controlled, it is crucial to evaluate the impact that the control is likely to have in terms of reduction of disease or mortality and the costs. The most desirable control strategies are those requiring the lowest costs with highest impact. IPM is about the identification and implementation of such strategies. Examples of control strategies that require a relatively high upfront investment are land-based saltwater aquaculture with disinfection of water intake, or new types of seawater rearing systems with a stronger separation of farmed fish from the environment (and pathogens present within). Comparatively cheaper strategies are farming of synchronized fish year classes and synchronized fallowing.

The translation of the cheaper strategies for disease reduction may not always be accessible to all farming systems. For example, synchronized fallowing requires the coordination of stocking of multiple sites in the respective management area. The fish in a given management area need to be harvested at the same time, which reduces the window during which the farmer can supply to the market. Smaller farms, which may require continuous income, may struggle to participate in synchronized fallowing schemes as they are unlikely to have multiple sites located in other management areas, where the time of harvest takes place at a different time. In this respect, some management strategies favor large companies. In some cases, the implementation of IPMS may require regulation by a government body (e.g., where multiple companies have sites in the same management area). Both the cheap and the expensive IPMS favor large companies, which are able to afford higher investment costs upfront.

A range of risk factors for development of disease or mortality in farmed fish are common to several fish pathogens. Therefore, where a disease problem starts to occur and there is no clear identification of the pathogen, a range of measures can be implemented that are aimed at restricting introduction and amplification of the pathogen in the farmed fish population. These measures tend to be associated with farm biosecurity and good animal husbandry practices that keep fish health at high levels and aim at limiting the spread of pathogens within fish farms.

Aquaculture itself creates a range of risks for disease emergence, notably the rearing of fish at increased density (compared to conditions in the wild), which facilitates fish-to-fish transmission of pathogens. One example where this is evident is the farming of Atlantic salmon, which has expanded rapidly since 1995 and has experienced the emergence of several diseases caused mainly by viral (e.g., pancreas disease, ISA) or parasite [e.g., sea lice and amoebic gill disease (AGD)] infections.

5.4 PREVENTION STRATEGIES IN FISH FARMING

"Prevention is better than cure" applies to human and veterinary medicine. Preventive measures are especially important in aquaculture, in which the number of animals at risk is high and treatment is a challenge from practical, technical, and economic points of view. In the past decades, a notable decrease in the use of antibiotics has been noted in salmonid aquaculture because of the development and application of vaccines (see Section 5.5.1.1). Prevention is the main focus for health management here, as once a given pathogen becomes established, the required measures for disease control can require considerable effort and costs, particularly for viral and some parasitic infections. Fig. 5.4 summarizes the possible prevention strategies in IPMS.

5.4.1 Management of Environmental Risks

5.4.1.1 Management of Water

Water quality is critical for minimizing the impact of pathogens. For inland facilities, the **treatment of incoming water** can be very effective in controlling some parasitic stages, such as actinospores of myxozoa by ozonation, chlorination, and ultraviolet light (UV) (Bedell, 1971; Sanders et al., 1972; Bower and Margolis, 1985; Hedrick et al., 2000). Passive sand filtration and sand–charcoal filters have also shown some efficacy in the removal of *M. cerebralis* actinospores (Arndt and Wagner, 2003; Nehring et al., 2003). However, these options can be expensive to implement or unaffordable for all the production cycle, and choosing the critical vulnerable fish life stages to rear on clean water supplies may prevent overt disease, if not infection, when these fish are eventually exposed to the pathogen. This practice has been used to decrease the severity of whirling disease, as fish quickly become more resistant to clinical disease as cartilage ossifies to bone (Bartholomew and Reno, 2002).

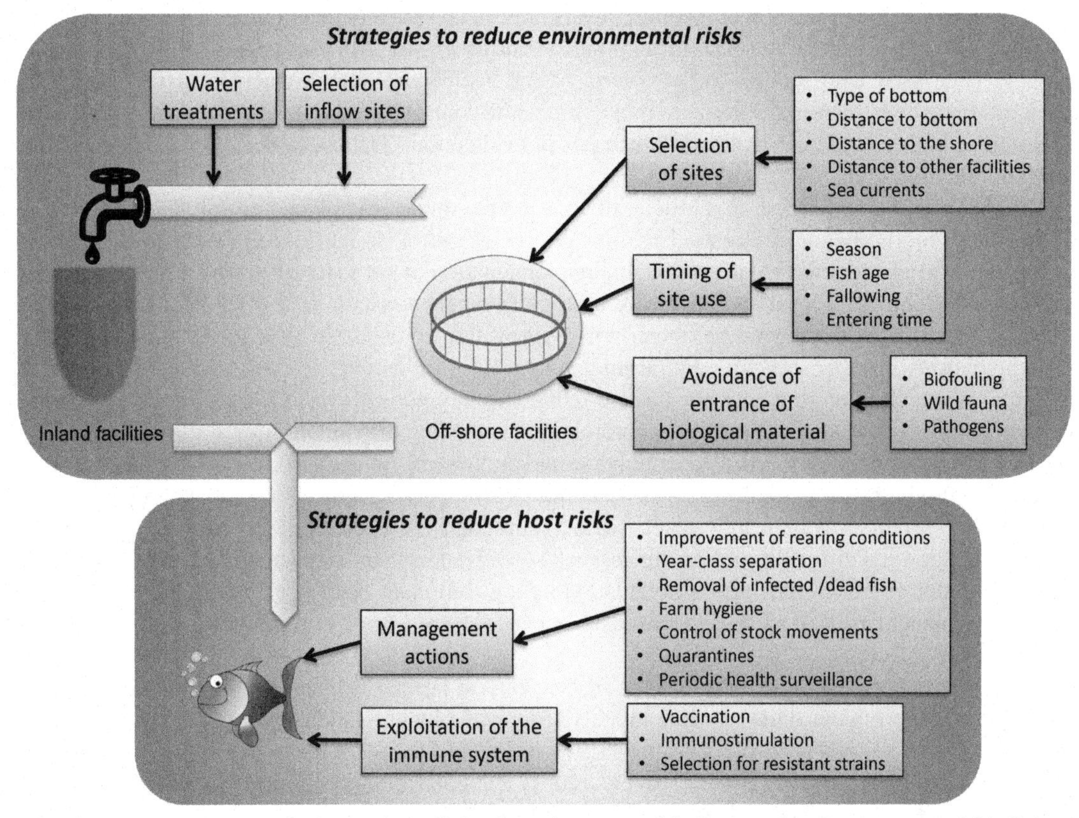

FIGURE 5.4 Preventive strategies that can be applied to the environment and the host to reduce the occurrence of fish diseases.

The control of virus and bacteria in inland facilities (either open-flow or recirculating systems) is mainly done by UV radiation and ozonation of water. UV disinfection systems are chemical free, are automatic, and cannot be overdosed. This ensures the maintenance of the physical characteristics of the water, such as pH levels and temperature, while avoiding the introduction of harmful toxins, such as chlorine, into the environment. UV disinfection is nonselective in destruction and almost all waterborne bacteria, viruses, and fungi can be destroyed. However, the required dose is very variable. The minimum UV dose for a 3-log reduction in mJ/cm^2 can range from 2 to 6 for bacteria, from 1 to more than 300 for viruses, and from 150 to 205 for fungi (Yoshimizu et al., 1990, 2005; Liltved et al., 2006). Ozone is an extremely reactive oxidant and a very effective bactericide and viricide and plays an important role in maintenance of water quality by microflocculating fine particulate matter and oxidizing nonbiodegradable organic molecules, nitrite, and refractory organic molecules. However, ozone is dangerous to humans and fish. Epithelial surfaces (skin and gills) can be damaged, and fish mortality may result. Thus, prior to installation of an ozone system, professional advice should be sought to determine if it is appropriate for the system in question.

Some water treatments are not meant to destroy or filter pathogens, but to maintain **water quality** to a level that limits further development of other microorganisms. This means to reduce organic material in suspension that would otherwise facilitate growth of algae and many protozoans, and reduce the oxygen content of the water. The use of **sonic wave technology** has been introduced in pond cultures to destroy algae, most bacteria, and other unicellular organisms and significantly reduces nitrates in the water in an environmentally friendly way. It consists of the transmission of inaudible directed pulses of ultrasound at precise levels for set durations into the water. These ultrasonic waves create microscopic bubbles that implode (cavitation), producing an intense cleaning effect on any surface, without affecting fish. They have been used in barramundi farms with success. More information on the water quality–disease relationship can be found in Chapters 8 and 9 of this book.

When the treatment of the water is not possible, the **selection of the inflow site** can be effective at least in restricting the possibility of the entrance of some pathogens. This is particularly relevant for some parasites that have an intermediate host that lives in specific types of environments and therefore its avoidance will be effective. It is also important that the inflow site is far from the outlet site to avoid the reentry of pathogens leaving the facility.

Water cannot be controlled directly in most large aquaculture settings, such as sea cages or large continental ponds. Even then, some actions can be adopted to avoid low water exchange and poor bottom hygiene, which are critical in the transmission of most parasites. For example, in monogenean infections, as the hatched oncomiracidia have a limited viability time, any actions that reduce their infective ability can reduce prevalence of infection. In ponds, frequent seining to remove fry (every 3 weeks), and the flow of freshwater through the system every fortnight, can interrupt parasite transmission as well as wash out some of the oncomiracidia in tilapia farming (Akoll et al., 2012).

Manipulation of water temperature in a facility can also be used to manage some diseases. There are often differences between ideal temperatures and ranges for hosts and pathogens, such that by raising or lowering one may be selected for. In an epidemiological study performed in Finland salmonid farms from 1986 to 2006, a significant annual mean water temperature increase was associated with a higher prevalence of infection (i.e., proportion of fish tanks infected each year) for some diseases (*Ichthyophthirius multifiliis*, *Flavobacterium columnare*), whereas for other diseases, the pattern was the opposite (*Ichthyobodo necator*) or absent (*Chilodonella* spp.) (Karvonen et al., 2010). Water temperature is also a critical risk factor in the transmission and onset of enteromyxosis (Sitjà-Bobadilla and Palenzuela, 2012). For the myxozoan *Enteromyxum leei*, the optimal development is between 20 and 25°C (Yanagida et al., 2006). The onset of the disease is largely delayed or even suppressed at low temperatures (below 15°C). However, a high water temperature seems to have preventive/curative effects in some fish species. Anemone fish reared at 30°C have significantly lower prevalence of infection than fish reared at 23°C. The prevalence of infection of anemone fish at 23°C for 20 days is decreased when transferred to 30°C. Infected Malabar grouper when reared at 30°C cleared the parasite within 6 days (China et al., 2014).

A study performed in Malaysian farms revealed a strong positive correlation between water temperature, the presence of *Vibrio*, and fish mortality. The study concluded that water temperature can enhance the susceptibility of cultured marine fish species to vibriosis. Although in practice it is difficult to control temperature in an open water system, the frequent removal of biofilm on net cages and floating structures and strict adherence to good aquaculture practices may help minimize vibriosis outbreaks and hence reduce fish mortality in net cage aquaculture facilities (Albert and Ransangan, 2013).

For some parasites, mathematical models have been obtained to estimate parasite age and time to sexual maturity at different water temperatures. For the skin monogenean *Benedenia seriolae* the equations provide a simple tool that accurately determines sexual maturity. This is especially important for parasite management on commercial farms because it allows scheduling of treatments to break the parasite's life cycle by killing parasites before they can mature and lay eggs (which cannot be controlled or treated) (Lackenby et al., 2007).

5.4.1.2 Management of Farming Conditions

This section includes other strategies that can be applied at the farm level to modulate the environmental risk factors involved in disease introduction and spread. They range from the proper selection of the farm site and the timing of its use to the reduction and prevention of the entrance of biological material carrying or transmitting pathogen stages, such as the removal or reduction of biofouling and the avoidance of contact with wild fauna, either fish or birds.

5.4.1.2.1 Site Location and Timing of Site Use

Site location and timing of site use may play a crucial role in controlling disease. In sea cages and inland ponds the avoidance of the seabed bottom and earth bottom that favor the growth of intermediate hosts can be helpful. In sea cages, the distance to the seabed is also critical. For sea cages, the distance from the shore and the water bed can influence considerably the transmission of monogenean oncomiracidia in tilapia culture (Akoll et al., 2012). In addition, seawater and tide currents may play a role in the dispersal of a pathogen, and thus the precise alignment, orientation, and location of the farming units can avoid or favor pathogen transmission. The dispersal of *B. seriolae* is considerable in sea-caged *Seriola lalandi*, and distances greater than 8 km between farming units may be required for effective parasite management (Chambers and Ernst, 2005).

In the ranching of southern bluefin tuna (SBT) (*Thunnus maccoyii*) in Australia, the temporary translocation of cages to a zone at a greater distance from the shore reduced the loads of sea lice and blood flukes, but also increased fish performance and health condition. The offshore cohort (at 25 nautical miles) had no *Cardicola forsteri* and a 5% prevalence of *Caligus* spp. compared to a prevalence of 85% for *C. forsteri* and 55% for *Caligus* spp. near shore (at 16 nautical miles) at 6 weeks posttransfer (Kirchhoff et al., 2011). These authors argued that the effects were probably due to the different environmental conditions (higher flow rate, higher distance to the bottom, and different type of bottom). The greater depth of the sea bottom and current velocity may decrease the incidence of cercariae of the blood fluke within the cages, as the intermediate host is known to be a benthic polychaete, *Longicarpus modestus*. It may also be possible that the intermediate host is absent from the offshore site, as its distribution is not known.

Fallowing is the practice of not restocking fish at a particular farming site for months or years. It was initially meant to allow the sediment below sea cages or in earth ponds to undergo natural recovery, both geochemically and ecologically, from the impacts of nutrient loading. Currently, this practice is widely and successfully implemented around the world as a method for preventing long-lasting damage to the benthic environment. It is also used for disease management purposes as it breaks pathogen life cycles. The World Organisation for Animal Health aquatic code (OIE, 2015b) defines fallowing as an operation in which an aquaculture establishment is emptied of aquatic animals susceptible to a disease of concern or known to be capable of transferring the pathogenic agent and, where feasible, of the carrying water. For aquatic animals of unknown susceptibility and those agreed not to be capable of acting as carriers of a disease of concern, decisions on fallowing should be based on a risk assessment. This practice sometimes consists of switching to another farmed fish species, as different species have different susceptibilities to pathogens or have different target pathogens.

The best example of the systematic application of fallowing to control fish diseases is sea lice. After salmon harvesting from a particular farm site, the cages are left empty for a time, breaking the cycle of the disease and also interrupting the breeding cycle of sea lice. More than 2 decades ago, Bron et al. (1993) showed that fallowing between harvesting and restocking led to lower numbers of *Lepeophtheirus salmonis* on newly introduced fish compared to fish in nonfallowed sites. Currently, the Scottish salmon farmers' Code of Good Practice specifies a minimum fallow period of 4 weeks, although many salmon farming companies choose to allow much longer fallow breaks. The process works most effectively if an entire sea loch/bay is fallowed at the same time, so it is an essential part of good single-bay management (AST, 2010). As an example, in the Hardangerfjord system on the western coast of Norway, the Norwegian Sea Food Agency has implemented a special regulation for the control of sea lice that implies the synchronized fallowing of geographically restricted production zones. Thus, communication, coordination, and cooperation between farmers are essential. The results show that the spatial–temporal infection risk on wild sea trout and in sentinel cages clearly has shifted from high risk in outer fjord areas before fallowing to low risk after fallowing 1 year later (Bjørn et al., 2012).

Fallowing is also the most powerful management option for the control of monogenean parasites in sea cage–farmed *Seriola* spp. It not only greatly reduces the requirements for treatment, but also improves fish health and performance. Fallowing is particularly effective where the main source of infection is the caged fish. The fallow period may be short, between 10 and 30 days. It requires a minimum of two farming sites, preferably three per farm (depending on the production cycle). Sites may need to be separated by large distances. Costs and logistical issues need to be considered (Ernst, 2010).

Fallowing is also applied after a disease outbreak to avoid reemergence. It is compulsory following clearance of a site under official suspicion or confirmation of a List I or List II disease in the EU. According to the EU diagnostic manual, the fallow period for farms infected with IHNV, VHSV, and KHV should be at least 6 weeks after the farm was emptied, cleansed, and disinfected. When all farms officially declared infected within the same protection zone are emptied, at least 3 weeks of synchronized fallowing shall be carried out. For farms officially declared infected with HPR-deleted ISA virus (ISAV) or confirmed ISA, the fallow period is at least 3 months (after emptying, cleansing, and disinfection), with a requirement for a synchronized fallowing period of at least 6 weeks in the protection and surveillance zones (EU Commission, 2015). Coordinated fallowing within management areas combined with risk-based surveillance lessens the likelihood that virulent ISAV will persist and reestablish (Murray et al., 2010). A 2011 modeling study of the effects of different fallowing strategies on disease transmission of two viruses (infectious pancreatic necrosis virus and pancreas disease virus) showed that synchronized fallowing is highly effective at eradicating epidemics when the transmission rate of the virus is low ($\beta=0.10$), even when long-distance contacts (directed movements both between and within management areas) are fairly common (1.5 contacts/farm and month). For higher transmission rates ($\beta=0.25$), if synchronized fallowing is applied, long-distance contacts have to be kept at lower levels (0.15 contacts/month) for eradication to be effective. These results demonstrate the potential benefits of having epidemiologically isolated management areas and applying synchronized fallowing (Werkman et al., 2011).

Fallowing is a normal procedure in northern European countries and even mandatory in salmon farming. However, in the Mediterranean region (producing mainly sea bass, sea bream, and meagre), fallowing of cage farms, year separation, and all-in all-out management practices are not utilized. This is due to a lack of common sector-wide health management principles and epidemiological studies in the spatial planning of Mediterranean mariculture. The lack of EU-notifiable diseases among the marine cultured species in this region has delayed the implementation of detailed epidemiological studies and the application of zoning policies to prevent and contain disease outbreaks. Owing to increased competition with other coastal zone users, the current trend to allocate aquaculture zones that concentrate large numbers of fish farms in limited areas should consider the epidemiology of the main fish pathogens to avoid disease "hot spots" in the near future. Coordinated area health management is required to break parasite life cycles, as well as notification systems to report on disease outbreaks. Licensing of new sites with appropriate distances between them (minimum of two sites, preferably three per farm depending on production cycle) to allow fallowing and site rotation requires legislative measures and stakeholder consultations that will require at least 8–10 years in some countries (AE, 2014).

The timing of moving stock into a given site can have a substantial impact on disease development. Introduction of naïve (usually juvenile) fish into sites close to other sites with infected fish (e.g., of other age classes) presents a disease risk to the naïve fish. Therefore, the introduction of new vulnerable fish ages close to rearing units with previous pathogen records should be avoided. Time of year (season) of stock introduction is also relevant, as the susceptibility and exposure of fish to various pathogens vary with season, because of a lower immunocompetence of the fish, worse environmental conditions (storms, etc.), higher presence of the pathogen in the water, etc. For diseases like proliferative kidney disease, for which an effective immune response has been demonstrated, stocking fish in the late fall, when parasite release is declining, may result in immunity when exposure to higher parasite numbers occurs the following spring (Foott and Hedrick, 1987; Longshaw et al., 2002). Similarly, Canadian farmers' data on *Kudoa thyrsites* infection in Atlantic salmon have shown that fish coming from some nursery sites with a low level of the parasite in the waters, when grown up at enzootic sites, have lower prevalence and severity of the infection than fish coming from clean waters. Therefore, the exposure to low levels of the parasite in some sites was acting as a vaccination. These empirical data have been experimentally demonstrated, suggesting an acquired immunity and the possibility of vaccination as a management strategy (Jones et al., 2016).

5.4.1.2.2 Removal/Reduction of Biofouling

Aquaculture surfaces (net cages, ropes, pipes, platforms, buoys, etc.) placed in the aquatic environment are rapidly colonized by a variety of organisms. Biofouling starts with a biofilm produced by microalgae and bacteria, which is followed by macrofouling species (green filamentous algae, bryozoans, serpulid tube worms, barnacles, seaweeds, sponges, sea anemones, sea squirts, and corals) and finally by mobile benthic and epibenthic animals (errant polychaete worms, skeleton shrimps, amphipods, isopods, crabs, nudibranchs, whelks, crinoids, and even fish). A range of commensals, parasites, and pathogens may intimately accompany these biota. Biofouling inflicts production losses as a consequence of: (1) reduced water flow through nets, resulting in reduced food supply and oxygen; (2) competition of biofouling communities with cultured species for resources; and (3) introduction of diseases and parasites (CRAB, 2006, 2007). In caged fish, fouling leads to a need for more frequent net replacement and application of antifouling products. This increases stress on cultured fish, reducing growth rates and, hence, productivity. Current estimates based on figures from the industry and the Food and Agriculture Organization suggest that biofouling impacts on fish cages and shellfish cost the European industry between 5% and 10% of the industry value (up to €260 million/year). The cost of changing nets for medium-sized salmon farmers is €60,000 per year (GISP, 2008). Furthermore, unintentional introductions of pathogens can occur with the movement of fouled aquaculture equipment such as settlement lines, ropes, nets, well boats, etc., from one area to another.

Relatively little attention, however, has been paid to the role of biofouling in the spreading of fish diseases. Most of the available data refer to parasites, and few cases have been documented concerning virus and bacteria. The loads of *Neoparamoeba pemaquidensis* in biofouling communities, when combined with amoeba loads in the water column, may contribute to AGD outbreaks in cultured salmon (Tan et al., 2002). Infection by *Gilquinia squali* metacestodes has been implicated in the deaths of farmed Chinook salmon smolts, whereby an unidentified crustacean that lives within the cage biofouling community probably acts as an intermediate host, and transfer occurs directly through ingestion (Kent et al., 1991). The life cycle of *C. forsteri*, a major blood fluke pathogen of SBT in Australian aquaculture cages, has an intermediate life history stage within polychaete biofoulers attached to the net pens, with other biofouling species acting as a reservoir for this parasite (Cribb et al., 2011).

The presence of zoonotic nematodes in farmed salmon runts is thought to be due to the fact that these fish (with clear signs of poor performance and abnormal appearance) feed mainly on the macrofouling of the nets and do not eat the commercial pellets, and therefore the chances of getting infected by preying infected crustaceans is enhanced (Mo et al., 2014). Previously, Sepúlveda et al. (2004) found crustaceans in the stomach of farmed Atlantic salmon in Chile. Therefore, the European Food Safety Authority (EFSA, 2010) conclusion that risk of farmed salmon infection by anisakids is negligible, because they almost exclusively feed on dry pellets, should be revised.

Parasitic gill and skin blood flukes (monogeneans) produce hundreds of eggs that are released into the water, sometimes forming visible entangled masses. These eggs have a rugged shell and generally long tendrils ending in hooks that allow them to attach directly to netting or to fouling on the netting. Bathing infected fish in various chemical compounds is a usual practice in fish farms. However, this removes only the infective stages of the parasite from the fish, leaving unhatched eggs attached to the netting unharmed. These eggs will hatch and reinfect the fish within the cage again. Thus, synchronization of the fish baths with cleaning/removal of fouled nets is essential for prolonging the periods in which the fish are not infected (Sitjà-Bobadilla et al., 2006). Another alternative preventive strategy involves the use of netting material that significantly reduces the ability of parasite eggs to efficiently colonize it. This has been explored for *Neobenedenia* in Hawaiian offshore mariculture. Small crevices formed by the strands of a braided rope enable egg attachment, whereas smooth surface netting materials tend to facilitate easier removal of eggs and fouling with active cleaning methods such as pressure washing. An antifouling compound coated on the netting, or made of a compound that has antifouling properties, such as copper, reduced egg attachment significantly (Lowell, 2012).

Viral pathogens of finfish accumulate and persist for long periods within shellfish. Viruses isolated from bivalves and identified as finfish pathogens include at least 13p2 reovirus, the chum salmon virus, infectious pancreatic necrosis strains, and IHNV (Leong and Turner, 1979; Meyers, 1984). In addition, a number of bacterial agents that cause disease in finfish are common to bivalve tissues (e.g., *Vibrio* spp.). These bivalves are part of the biofouling. The occurrence of netpen liver disease (caused by the toxin microcystin-LR) in caged fish was linked to the consumption of fouling organisms by the cultured species (Kent, 1990; Andersen et al., 1993; Fitridge et al., 2012). A probable role of biofouling was also suggested for ISA in salmon outbreaks (Stagg et al., 2001).

5.4.1.2.3 Minimizing Contact With Wild Fauna

Wild fish, either surrounding or entering cages, are a source of pathogens for farmed fish. Connections among farms and other marine areas through wild fish movements have been demonstrated both in Norway (Uglem et al., 2009) and in Mediterranean fish farms (Arechavala-Lopez et al., 2010, 2011). Wild fish are attracted to farming areas by waste food, especially beneath cages. Therefore, efficient management of the feed will reduce the attracted wild population. There are several examples of the transmission of parasites from wild to farmed fish, either directly or through the involvement of intermediate hosts. Groupers fed commercial pellets exclusively can become infected with helminths that have a complex life cycle (involving several hosts), because parasite transfer occurs via organisms that naturally live in, on, and in the surroundings of the net cages (Rückert et al., 2009). The ingestion of small fish that enter salmon cages has also been suggested as the infection route for some zoonotic nematodes (Mo et al., 2014). For sea lice, a high prevalence of infection has been found in many studies of wild salmonids for decades. An autumn rise in sea lice abundance in mariculture has been associated with the seasonal return of infected adult Pacific salmon during their spawning migrations, as they harbor sea lice mainly in their motile (preadults or adult) stages (Jones and Hargreaves, 2007; Saksida et al., 2007; Marty et al., 2010).

In contrast to *Caligus* infections in other farmed fish, larval stages are rarely detected on ranched SBT. It is believed that Degen's leatherjacket (*Thamnaconus degeni*) acts as a reservoir of mobile adult *Caligus* infections. These fish are benthic scavengers that are commonly attracted to the SBT grow-out cages during feeding. It has been suggested that moving SBT into deeper water may reduce interactions between SBT and this source of *Caligus*, therefore reducing infection rates. Enhanced feeding protocols may also reduce the attractiveness of the cages to opportunistic feeding by demersal fish (Kirchhoff et al., 2011).

It seems that parasite transfer is more likely to occur if fish are in close vicinity, as well as if the hosts (e.g., farmed and wild) are of the same species. Farming conditions (oceanography of the site, density on both sides of the net pen) further contribute to the transfer chances. In any case, it should be carefully assessed case by case before drawing conclusions.

Unlike parasitic pathogens, bacteria seem to have a higher potential to spread between wild and farmed fish, probably because the ecological barriers that exist for parasite transfer do not represent a great obstacle for them. This is due to the fact that bacteria are almost always present on fish surfaces and are often generalists; thus they do not need wild conspecifics to spread to or from farmed fish. Transmission of bacterial diseases among wild gray mullets and sparids and farmed fish is well documented (Arechavala-Lopez et al., 2013). Concerning viral infections, relatively little research has been conducted on the prevalence and maintenance of infections in wild fish populations including those adjacent to affected farming sites. Furthermore, the knowledge of the transfer pathways from wild to cultured fish is very sparse. For ISAV and salmon alpha virus several wild fish have been considered candidate reservoir species (Jones et al., 2015). Panzarin et al. (2012) showed that transfer of betanodavirus between wild and farmed fish in Mediterranean waters was possible, since viruses isolated from feral and farmed fish in some cases were found to be similar.

Other members of the wild fauna that can also introduce pathogens into the farming site are birds. Fish-eating birds not only cause significant economic losses due to their predatory activity, but also can contribute to disease transmission. They may transfer diseases by carrying the pathogen on their body or feet, by dropping fish or fish parts at other locations, or by discharging their guano. Birds are the final hosts of the complicated life cycle of many digenean parasites, with several larval stages that infect one or more hosts. With rare exceptions, the first intermediate host is a mollusk, without which the life cycle generally cannot be completed. Guano discharged by infected birds releases the parasitic stages that infect the mollusk and the parasite is transmitted to fish when they ingest infected mollusks. As an example, pond-reared, juvenile, tropical fish may develop severe gill disease from metacercarial cysts of *Centrocestus formosanus* in gill tissue. Prevention of the disease by elimination of the intermediate host in the farm, the freshwater snail *Melanoides tuberculata*, can be effective. However, this task might be difficult since this snail exhibits resistance to desiccation, molluscicides, and disinfectants and has been documented to outcompete established mollusks and continues to spread in various countries (Mitchell et al., 2005). The American white pelican, a host of the trematode *Bolbophorus damnificus*, spreads this parasite to commercial catfish farms, resulting in severe economic losses due to mortalities and unmarketability.

There is also evidence that birds can transmit bacteria and viruses through their droppings. Observations confirm their role at least in the transmission of viral hemorrhagic septicemia, infectious pancreatic necrosis, spring viremia of carp, and bacterial diseases from one place to another in freshwater pond aquaculture (FAO, 1989). Birnaviruses taken up by birds can be viable when excreted and can infect fish experimentally, but the principal method of translocation remains live fish and eggs (Reno, 1999). Concerning aquatic bacteria, the chance of being passed to birds feeding on scraps, discarded products, or live fish, and then to fish again through guano, depends on the bacterium's capacity to survive outside the fish host (Evelyn, 2001).

Another pathogen that can be potentially transmitted by bird droppings is the mesomycetozoan *Ichthyophonus hoferi*. This parasite infects many wild and farmed fish species (McVicar, 1999) and even birds (Chauvier and Mortier-Gabet, 1984). The spores can resist passage through the digestive tract of the fish-eating birds and therefore birds are potential transmitters.

The methods to minimize the presence of birds at aquaculture facilities often include scaring and exclusionary techniques such as noisemakers and netting. The use of netting, wire grids, and fencing, which offer fish farmers long-term protection, is generally recommended, but the cost associated with the installation of these physical barriers makes them impractical for some farmers. In addition, some farmers report that the barriers interfere with normal fish-rearing operations and require substantial monitoring and repairs. Noisemakers, such as propane cannons and pyrotechnics, and visual devices (effigies), like "eye-spot" balloons, predator-like remote-control boats and airplanes, and scarecrows, are also used. Dispersal techniques should be used before, during, and immediately after feeding, while the fish remain at the surface of the water (Barras and Godwin, 2005). Unfortunately, many birds quickly adapt to the sight and sound of such devices (WS, 2010).

In conclusion, the distribution of fish pathogens and their pathogenicity, prevalence, and incidence in wild populations must be taken into account when developing proper IPMS.

5.4.1.2.4 Reducing Contact With the Pathogen

Here we provide some examples of promising technological solutions thus far developed to avoid the contact of farmed salmon with sea lice. They are gathered in the toolbox of IPMS for sea lice shown in Fig. 5.5. They are based on the fact that salmon lice copepodits are typically found near the surface during daylight. The salmon lice copepodit positions itself in the upper water column to increase encounter probability with a potential host. It is therefore hypothesized that decreasing the chances of contact between fish and sea lice in the sea cages will reduce lice infection rates on the farmed salmon.

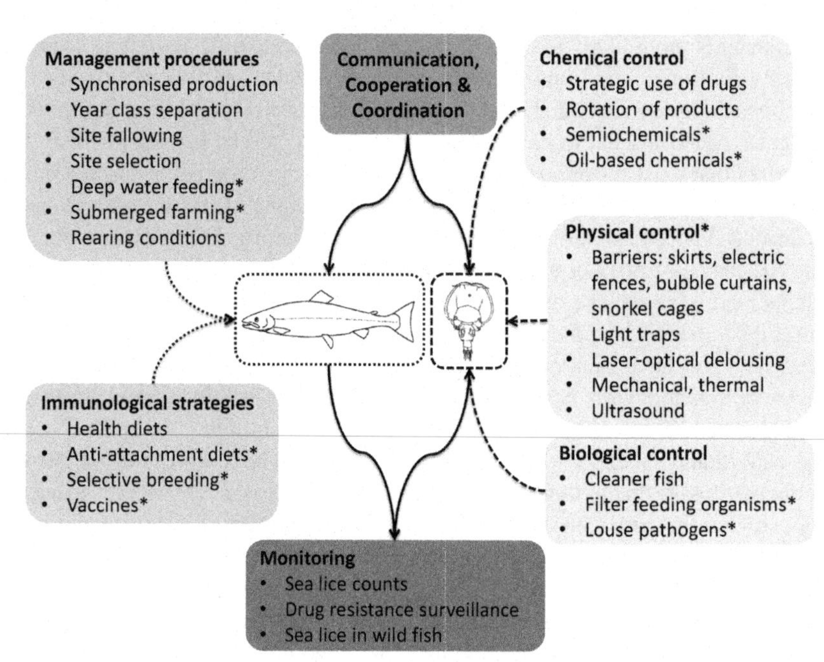

FIGURE 5.5 Toolbox for integrated pathogen management of sea lice in Atlantic salmon farming. *Strategies for future development. *Drawings:* Salmo salar *from Linnaeus, 1758. https://commons.wikimedia.org/wiki/File:Salmo_salar_Linnaeus_1758_Fig_123_(Matschie_et_al._1909).svg. Fig. 123;* Lepeophtheirus salmonis *redrawn by Dr. F.E. Montero (UV) from Whelan, K., 2010. A review of the impacts of the salmon louse,* Lepeophtheirus salmonis *(Krøyer, 1837) on wild Salmonids. Atlantic Salmon Trust, 1–27, Fig. 5.1.*

Snorkel cages: Hevrøy et al. (2003) found that salmon held at 0–4 m depth developed higher infestation than salmon held at 4–8 and 8–12 m depth. Therefore, it may be possible to keep the parasite at bay by placing net cages below the "louse zone." The problem with this solution is that salmon need access to air to fill their swim bladder. One innovative approach to the problem is the snorkel cage, developed by researchers from the Institute of Marine Research and the University of Melbourne in cooperation with Egersund Net. At a depth of 3–4 m inside the cage, netting is used to form a ceiling and a central cylindrical passage that extends above the surface of the water. This snorkel is covered by a tarpaulin or plankton cloth, securing a louse-free water passage through which the salmon can swim up to the surface to gulp some air. This method is extremely promising, as the first tested prototypes in the summer had decreased incidence of salmon lice by 80–84%.

Nets and skirts: Plankton nets (net mesh of 100 μm) surrounding sea cages reduce significantly sea lice numbers. They act as shields for sea lice larvae, breaking the life cycle. One 2011 trial showed a 75% reduction in sea lice. Preventing the water from the upper column from passing through the salmon cages has been proposed as a way to reduce salmon lice infestation. One simple implementation proposed by the industry is to place a permanent tarpaulin skirt around the upper part of the sea cages. It is said in the industry that the skirt method has been tried with great success (Tveit, 2012). The most relevant publications on the topic are a report on the use of very deep skirts (10 m deep) to reduce infestation of salmon lice (Næs et al., 2012). However, pilot studies have shown that putting a tarpaulin skirt around a full-scale commercial sea cage may seriously decrease the oxygen saturation levels available for the fish inside the skirt (Stien et al., 2012).

Deep water feeding: Results from commercial scale trials confirmed that salmon swimming behavior is altered under submerged feeding conditions, with fish attracted to the feeding corridor during the feeding process. Submerged lighting attracted the fish to the illuminated water depths during the night. The number of *L. salmonis* on fish exposed to deep submerged lighting was significantly lower than the number of lice found on salmon in cages with surface lighting during the summer months. The results of the study suggest that swimming depth manipulation can be used at a commercial scale to reduce salmon lice burdens on Atlantic salmon stocks (Frenzl et al., 2014).

5.4.2 Management of Host Risks

5.4.2.1 Management of Farming Conditions

Management options should consider **reducing risk factors** (see Section 5.3) that aggravate infections especially (Figs. 5.3 and 5.4). High stocking density is a critical factor not only because it increases the chances of pathogen transmission by fish-to-fish contact, but also because it potentially increases stress and the occurrence of skin lesions. Skin condition plays an important role in the onset of some bacterial, protozoan, and fungal diseases, since skin injuries act as sites of pathogen entry. Therefore, farmers should consider decreasing density and **reducing stress-related procedures** to improve the immune status of fish and subsequently avoid the emergence of opportunistic pathogens. This is particularly important if we consider that, for example, the most frequent Mediterranean pathogens belong to secondary microorganisms that are firmly ingrained at the farm site and emerge periodically, triggered by shifts in environmental conditions or anthropogenic stress (OIE, 2015b). Shortened production cycles (shorter time the fish are in the water) will reduce exposure times to pathogens.

Particulate matter (feces, uneaten food, etc.) should be removed on a regular basis. **Dead or dying fish should be removed** promptly as they can serve as potential disease sources to the remaining stock and a breeding ground for others, as well as fouling the water when decomposing. As an example, rapid removal of sick or dead stock is effective in limiting ISAV transmission (Jarp and Karlsen, 1997). Frequent monitoring for disease should be done by employees. This early identification of fish diseases can help minimize their spread and impact on the farm. Dead fish should be disposed of immediately in accordance with local and state laws and in a manner such that predators or wild birds cannot access carcasses and further spread the disease.

Control of stock movements is a key element in disease prevention. The movement of fish onto, within, or off the farm is the greatest risk factor for disease introduction and spread in aquaculture. This includes new fish brought to the farm for breeding, grow out, or restocking or those returned to the farm. These "new" additions can introduce pathogens to resident fish, and for some diseases, these fish may be infected without showing signs of illness, serving as carriers or reservoirs to other fish. To prevent disease introduction onto farms, fish should be purchased from known and trusted suppliers. These fish should be inspected and found free of important diseases. The number of sources fish are purchased from should be minimized as well as the frequency and number of new introductions onto the farm. When possible, eggs or fish from certified disease-free brood stock should be purchased. Eggs should be disinfected upon arrival.

Farm hygiene: It is important that siphon hoses, brushes, nets, and other equipment used to clean tanks be stored off the floor and chemically disinfected between uses. Equipment should be thoroughly rinsed in clean water following disinfection and before being placed into a tank containing live fish. Footbaths and areas to wash hands with a disinfecting soap should be placed at the entrance to buildings and between rooms within buildings. These steps directly decrease the potential for the spread of pathogens.

Quarantine: This should be applied to new arrivals. Once fish are brought to the farm they should be isolated before adding them to the current stocks. The length of quarantine is variable and depends on the disease of concern; 4–6 weeks is commonly suggested. The quarantine area should be away from resident fish and should have a separate water source or flow circuit. Any water effluent on or off the farm from the quarantine area should be managed to avoid contaminating water sources or fish production areas. Dedicated equipment for the quarantine area should be used to avoid fomite transfer to resident fish. Fish in the quarantine area should be cared for or worked with after resident fish, or an employee should be assigned solely to the care and handling of the quarantined fish.

5.4.2.2 Exploitation of the Immune System

This section reviews the three main strategies for enhancement of the fish immune system. The reader can find more information in Chapters 1, 2, and 6 of this book.

Vaccination: Vaccination is a preventive measure that protects fish against a future disease and the associated costs due to morbidity, mortality, and therapeutic treatment. It plays an important role in large-scale commercial fish farming and has been a key reason for the success of salmon cultivation. Nowadays commercial vaccines are available for salmonids, channel catfish, carp, European seabass, gilt-head sea bream, Japanese amberjack, yellowtail, tilapia, turbot, and Atlantic cod. Vaccines are applied by immersion, by injection, and orally depending upon the natural route of infection, the life stage of the fish, the production techniques, and other logistical considerations depending on the fish size, species, pathogen, etc. The number of available vaccines against bacteria is higher than for viruses, and as of this writing, there are no fish vaccines for diseases caused by fungi and parasites (Sommerset et al., 2005; Gudding, 2014). Vaccination for bacterial/viral diseases can also reduce susceptibility to parasitic diseases, such as sea lice, by reducing stress. Inactivated vaccines were the first ones produced and DNA vaccines are the most recent ones, though only a few of them work at the commercial level (IHNV) and experimental small-scale level (VHSV) (Hølvold et al., 2014). In any case, proper fish management with good hygiene and limited stress are key factors in the prophylaxis of infectious diseases and are also a necessity for the optimal effect of vaccines.

Selection for resistant strains: This approach is based on the selection of specific fish strains/lines that have innate immunity and are more resistant to a particular pathogen. It is a key element of disease management in intensive animal production and provides a cheap and effective alternative in situations in which resistance to treatments is a concern and no vaccines are at hand. The feasibility of selective breeding depends in part upon the heritability of the trait or traits used to measure disease resistance, as well as the amount of variation among animals. Selective breeding for enhanced disease resistance is used now against a number of diseases in aquaculture or has potential for application in the future (Das and Sahoo, 2014; Henryon et al., 2005; LaFrenz et al., 2016). For some pathogens even double-pathogen-resistant lines are feasible (e.g., columnaris disease and bacterial cold water disease) (Evenhuis et al., 2015), but for others enhanced resistance to one pathogen increases susceptibility to another; for example, rainbow trout strains resistant to *Ceratomyxa shasta* are susceptible to *M. cerebralis*, suggesting that different mechanisms might be specifically involved in the resistance to each myxozoan (Hedrick et al., 2001). For Atlantic salmon, substantially fewer chemical treatments would be needed to control sea lice outbreaks in selected populations and chemical treatment could be unnecessary after 10 generations of selection (Gharbi et al., 2015).

Immunostimulation, functional diets, health diets: Functional diets are those that lead to physiological benefits beyond basic nutritional requirements. They include diets that reduce organic load in recirculation systems, diets that facilitate the acclimatization of salmonids to seawater, and diets that improve health status and reduce the impact of diseases. The last are named health diets and are becoming a key element of the preventative strategies employed in fish farms (Jensen et al., 2015). Health diets contain various ingredients that promote health-related functions and are therefore often termed immunostimulant diets. An immunostimulant is a substance that enhances a host's defense mechanisms or immune response (both specific and nonspecific), thus rendering the animal more resistant to diseases and other external influences (Anderson, 1992). The range of products and molecules having such properties is wide, including plant extracts, organic acids, essential oils, prebiotics, probiotics, symbiotics, nucleotides, vitamins, etc. (Meena et al., 2013; Tafalla et al., 2013; Hoseinifar et al., 2014; Hai, 2015; Koshio, 2016). For each pathogen and host the specific dose, product, and feeding time have to be established. Health diets for Atlantic salmon have become an important component of the IPMS targeting sea lice. Salmon fed various immunostimulant formulations prior to infection by the two main species of sea lice, *L. salmonis* or *C. rogercresseyi*, showed significant reductions in the number of lice compared to untreated groups (Covello et al., 2012; Igboeli et al., 2012; Jensen et al., 2015; Núñez-Acuña et al., 2015). In some trials, selected antiattachment compounds, based on natural plant extracts with a repellent effect on sea lice, have been added to immunostimulants, which has also provided a significant reduction in the number of parasites (Nuñez-Acuña et al., 2015). For these ectoparasites differences in a number of specific immune gene responses correlated closely with the differences in infection levels. However, for many other pathogens the specific pathways involved in the improvement are still unknown.

5.5 TREATMENT STRATEGIES IN FISH FARMING

5.5.1 Chemical Control

5.5.1.1 Antibiotics

Antibiotic treatment in aquaculture is achieved by medicated baths and medicated food. In both cases, they are likely to pass into the environment, affecting wildlife, remaining in the environment for extended periods of time and exerting their antibiotic effects. Some compounds are persistent in sediments and can therefore affect the marine microbiota near aquaculture sites. One of the major concerns with the use of antibiotics is the potential for bacteria to develop resistance to the compounds and for the resistance traits to be manifested in other bacteria, including human pathogens. The reader can find more information on their environmental impact in Chapter 10.

Strong regulation of antibiotic use in aquaculture has led to a drastic reduction in the classes and volumes of antibiotics used for this purpose in many countries. However, large quantities of antibiotics have been applied elsewhere (reviewed in Romero et al., 2012a). Norway, the world leader in salmon production, is today an example of how the implementation of IPMS has reduced the use of antibiotics in salmon farming. The Norwegian aquaculture industry has reduced the use of antibiotics by 99.8% per ton of trout and salmon produced, compared to the 1987 level, thanks to the introduction of vaccines. As of this writing, only 1 ton of antibiotics is used annually altogether to treat diseases. By contrast, the development of aquaculture in Chile has been accompanied by the use of antibiotics. An example, the use of flumequine alone increased from 30 to 100 tons between 1998 and 2002 (Bravo et al., 2005; Cabello, 2004).

The first consideration for reducing the use of antibiotics is to avoid their prophylactic use. The second is to try using vaccines (if available) before antibiotic treatment. The third is to use them only when there is a bacterial diagnosis sensitive to a specific compound. On top of all these considerations is the implementation of the other preventive strategies that have been exposed in Section 5.4.

5.5.1.2 *Parasiticides*

Most of the chemical products used to combat parasites in aquaculture are directed against ectoparasites located on or in the skin, gills, or buccal cavity. Parasiticides are used globally, and the quantity of these compounds being applied is considerable. The principal parasitic pathogen affecting Atlantic salmon aquaculture is the sea louse, and the number of efficacious products licensed for application is getting smaller. In the Northern Hemisphere, emamectin benzoate has, until recently, been the main treatment of choice. However, reliance on one or two key therapeutants is poor practice, as resistance development is accelerated and, once resistance is present, treatment options become severely restricted. Evidence of resistance has been reported for all classes of compounds used to date (Aaen et al., 2015; Helgesen et al., 2015). Therefore, the strategic use of parasiticides is crucial to maintain efficacy and provide minimal environmental impact. Best practice can include the deployment of early spring treatments before water temperatures rise (in the Northern Hemisphere), the rotation of product classes, and the minimization of treatment concentrations. Bath treatments using aqueous formulations normally involve surrounding the cages with tarpaulins, with drugs released into the sea after treatment. More controlled use of bath treatments may be achieved by the use of wellboats, with fish pumped into tanks, treated with uniform concentrations of compounds, and returned to cages, with the possibility of retention of the treatment chemicals and removed lice. There is a promising experimental method that consists of using oil-based treatments that form an oil film on the surface of the cage. Salmon must pass through the film to obtain contact with the product. To increase the natural leaping behavior of salmon, fish are prevented from reaching the water surface for a day using a net barrier. In the 2h following the barrier's removal, salmon leap more than normal, probably because they feel the urge to swallow air to refill their swim bladders (Dempster et al., 2011). Thus, the development of oil-based treatments will be more environmentally friendly, as it will be possible to remove the chemicals after the treatment, e.g., using practices designed for dealing with oil leaks in the offshore industry.

5.5.1.3 *Other Chemicals*

There are many other chemicals used worldwide for the treatment of ectoparasites and fungi, mainly in baths, with a subsequent impact on the environment. Some of them are banned in several countries. Formalin is used extensively throughout the world and for some pathogens is the only available commercial chemical. Its use in aquaculture is banned in some countries because of the problems associated with its handling, as it can cause respiratory and pulmonary problems in humans and is believed to be a carcinogen. Formalin can reduce oxygen concentration in water, so constant aeration is required during treatment. Some species (i.e., eels) and/or very young fish are considered to be highly sensitive to formalin. A white precipitate (paraformaldehyde) may form in formalin, which must not be used, as it is toxic to fish.

Various research lines are trying to find alternative products and methods to reduce their impact on the environment and the toxicity for fish and fish products. They include the use of plant extracts with antibacterial, antiviral, and antiparasitic properties (Reverter et al., 2014). Salt can control some fungal and ectoparasitic diseases in freshwater fish. It is also an effective therapeutant, as it promotes the production of mucus and the repair of damaged skin and gill tissues, and reduces stress and lessens the work fish have to do for osmoregulation. Salt is safe, cheap, and easy to handle. On the other hand, freshwater bathing is a potential control option against numerous sea lice species. A study has shown 96–100% mortality of the first attached (copepodid) stage of *L. salmonis* after 1h in freshwater. Thus, regular freshwater bathing methods targeting the more susceptible attached copepodid stage may successfully treat *L. salmonis* and potentially other sea lice on fish cultured in marine and brackish waters (Wright et al., 2016). Semiochemicals are chemicals that mediate interactions between organisms. Semiochemicals are subdivided into allelochemicals and pheromones depending on whether the interactions are interspecific or intraspecific, respectively. Pheromones are commonly used in many insect IPM programs for crops, but their use in aquaculture is just starting, but with a promising future. Some semiochemicals have been tested to combat sea lice. They hide the "scent" of the fish host or repel the sea lice. The delivery of these compounds from a controlled release system has been shown to be significantly more attractive than salmon conditioned water for *L. salmonis*, setting the basis for the development of odor-baited monitoring traps for the attraction and retention of lice (Ingvarsdóttir et al., 2002; Pino-Marambio et al., 2008).

5.5.2 Biological Control

Biological control is a bioeffector method of controlling pathogens using other living organisms. It relies on predation, commensalism, or other natural mechanisms, but typically also involves an active human management role. Biocontrol as a complement to other management tools has the potential to increase the likelihood of success of early intervention and eradication efforts.

5.5.2.1 Cleaner Fish

Certain organisms feed on ectoparasites or intermediate hosts of fish, therefore reducing the number of parasites on the target fish or the chances that an intermediate parasitic stage can reach the farmed fish. The most successful application of biocontrol in fish farming is the use of cleaner fish to combat sea lice. In Norwegian farms, three different species of cleaning wrasse (Labridae) are cocultured with salmon (up to 5% of the total pen density) to remove sea lice (Skiftesvik et al., 2013). This is feasible because wrasses feed on ectoparasites on the skin of farmed salmon and do not eat salmon food (Treasurer, 2005). Other candidate cleaner fish species are being studied, such as lump fish and three-spined stickleback. Their cleaning behavior differs from those of wrasses (Losos et al., 2010).

Other cleaner fish eat parasites located in the oral cavity or gills of other fish. This is the case of some fish that eat gnathiid isopods from cooperating fish. It has been shown that these cleaner fish have a significant effect on the short-term (12 h) abundance (Grutter, 1999) and long-term (2 years) size–frequency distribution (Gorlick et al., 1987) of parasites. These cleaner fish are known to benefit from cleaning because of the nutritional value of gnathiids (energetic value and UV-absorbing compounds) and mucus from client fish (Eckes et al., 2015). This is consistent with the hypothesis that cleaning behavior is mutually beneficial to both participants, and paves the way to using these cleaner fish also to fight some isopods that affect farmed fish, such as *Ceratothoa oestroides* in Mediterranean farmed fish (Mladineo, 2002).

The role of this type of biocontrol has become increasingly important not only in terms of economic productivity and fish welfare in salmon farming, but also by reducing the spread of parasites into the wild fish population. A key issue to address is the ability to obtain biocontrol agents in sufficient quantities for effective control at the scales of interest. As of this writing, it is not possible to obtain enough cleaner fish from fisheries, and thus farming of cleaner fish is becoming a parallel increasing industry. At the same time, the control of the health status of cleaner fish is also becoming of concern, since they might be carriers of pathogens for salmon (Hall et al., 2013; Ecofish Project).

5.5.2.2 Microorganisms

There are several microorganisms that can be used to combat pathogens, either as probiotics in feed or in the water.

Phage therapy: This is based on the use of bacteriophages (viruses that infect bacteria) to inactivate pathogenic bacteria. This approach presents several advantages: (1) phages are target specific; (2) nearly no serious or irreversible side effects of phage addition have been described to date, although phages may carry virulence or toxin genes; (3) phages can be chosen to have no environmental impact, infecting only pathogenic bacteria; (4) phages have high resistance to different conditions; (5) while phage-resistant cells regrow after treatment, these mutants can recover phage sensitivity after being grown in culture medium; and (6) phage therapy is a highly flexible, fast, and inexpensive technology (Silva et al., 2016). It has been explored to combat pathogenic bacteria that affect farmed fish only recently. A promising example is the use of the AS-A phage to fight furunculosis through direct administration of phages to the culture water. This practice appears to promote a limited regrowth of resistant cells and has no significant impact on natural bacterial communities present in aquaculture water. Under lab conditions, AS-A phage protected Senegalese sole juveniles against *Aeromonas salmonicida* (Silva et al., 2016) or reduced mortality of Atlantic salmon (Imbeault et al., 2006) and rainbow trout (Kim et al., 2015). A major concern of bacterial inactivation by phages is the emergence of phage-resistant bacteria. However, previous studies suggest that phage-resistant bacteria tend to be less fit or lose their virulence properties (Castillo et al., 2015), so they can be expected to be eliminated from the environment faster than their wild-type relatives.

Biosurfactants: Some bacteria are capable of inhibiting the growth of other bacteria or even other multicellular organisms. This is the case of some *Pseudomonas* species isolated from salmon eggs; in particular, the *Pseudomonas* isolate H6 significantly reduced salmon egg mortality caused by the oomycete *Saprolegnia diclina*. Live colony mass spectrometry showed that this strain produces a viscosin-like lipopeptide surfactant. This biosurfactant inhibited growth of *Saprolegnia* in vitro, but did not protect salmon eggs against saprolegniosis. These results indicate that a live inoculum of aquatic *Pseudomonas* strains, instead of their bioactive compound, can provide new (micro)biological and sustainable means to mitigate oomycete diseases in aquaculture (Liu et al., 2015).

Quorum quenching (QQ): This is one of the mechanisms to inhibit growth or the production of virulence factors of bacteria through the disruption of quorum sensing (QS). QS is a process that involves bacterial cell-to-cell communication with the participation of low-molecular-weight signaling molecules that elicit population density–dependent responses. These molecules regulate and coordinate the expression of certain bacterial genes. Bacteria that are able to degrade QS molecules might be useful as biocontrol agents in aquaculture. The signal molecules involved in the regulation of virulence factors in many pathogenic bacteria, including fish pathogens, are acyl-homoserine lactones (AHLs), autoinducing oligopeptides, and autoinducer 2. Defoirdt et al. (2008) suggested that some *Bacillus* strains that are currently used as probiotics for other properties in aquaculture would be worthwhile testing with respect to their QQ ability, since various *Bacillus* spp. have been reported to degrade AHLs. If so, these bacteria could be used as a new kind of probiotic.

Various bacteria isolated from water or fish have been shown to quench the QS system or various fish pathogenic bacteria in either in vitro or in vivo trials. There are some promising candidates for aquaculture, either used in feed or injected, that produced a higher survival in fish after bacterial challenge, delayed the time of death, attenuated biofilm formation, inhibited protease production, or repressed the expression of virulent genes of pathogenic bacteria (Romero et al., 2012b, 2014; Ibacache-Quiroga et al., 2013; Chu et al., 2014). Genome mining is helping to find bacterial candidates for their use as QQ producers. The extensive genomic analysis of a *Flaviramulus ichthyoenteri* Th78T isolated from the intestine of a healthy flounder (*Paralichthys olivaceus*) identified a QQ enzyme, *N*-acyl-homoserine lactonase FiaL, whose QQ activity against *Aeromonas hydrophila*, *Edwardsiella tarda*, *Vibrio salmonicida*, and *Vibrio anguillarum* was confirmed in vitro (Zhang et al., 2015).

QQ can also be induced by microorganisms other than bacteria. The microalgae *Chlorella saccharophila* CCAP211/48, which is commonly used in aquaculture, exhibited stable inhibitory activity on *Vibrio harveyi* with the production of QS antagonistic metabolites, which had not been reported previously in microalgae (Natrah et al., 2011).

5.5.2.3 Filter-Feeding Organisms

Filter-feeding organisms can be used for reducing the amount of pathogens in the water, and therefore the chances of disease in farming facilities. Most of them are just starting to be explored and applied in aquaculture. The marine predator equivalent for sea lice, the blue mussel, is being investigated as a biofilter. Initial laboratory experiments have shown that mussels are capable of consuming sea lice nauplii at a rate of approximately 0.5 lice per mussel per hour. Very rough calculations suggest that 12 rafts of mussels placed strategically on a site could theoretically consume 8.4 million lice per hour. As there are indications that the early life stages of sea lice, including the eggs, may have a benthic component, filter feeders should be placed below a site (Robinson, 2010). As of this writing, several projects are studying the most appropriate bivalve species and sizes for their use as biological agents in the control of sea lice at salmon farms as a benefit of integrated multitrophic aquaculture (Webb et al., 2013).

There are several examples of the potential use of other filtering organisms against bacteria. The freshwater mussel *Pilsbryoconcha exilis* seems to control the population of *Streptococcus agalactiae* in a laboratory-scale tilapia culture system. Future studies should focus on the dynamic interactions among fish, mussels, and bacteria as well as on how input such as feed and other organic materials affects these interactions (Othman et al., 2015). Stabili et al. (2010) provided an interesting example in which large numbers of the Mediterranean fanworm *Sabella spallanzanii* were transferred to and cultured within the vicinity of a finfish aquaculture farm as a means of filtering out harmful bacteria.

Some filter-feeding fish, such as *Polyodon* spp., have been suggested as a way to combat some protozoan infections in pond cultures. Empirical data in carp polyculture shows that the prevalence and intensity of *Ichthyophthirius multifilis* infection are lower than in monocultures, suggesting that this fish removes from the water the infecting stages of the ciliate (Dr. C. Szekely, Hungarian Academy of Sciences, personal communication). Further studies are needed to demonstrate and explore the potential of this biological control.

5.5.3 Physical Control

Drying in sunlight: This has been since early times of aquaculture an easy and cheap way to inactivate aquatic pathogens. This procedure is done in drainable freshwater and marine earth ponds. The exposure time to sunlight needed will vary depending upon intensity, temperature, and other factors, and little research has been done on finding effective exposure times for each pathogen. Some parasitic stages with resistant cystic stages, spores, or eggs, and even some viruses (Kumar et al., 2013), can be resistant to drying.

Light traps: Another area of development is the use of sea lice traps using light. An LED-based light trap caught ~70% of salmon lice larval stages and ~24% of the adults in the water of tanks. It also acted as a delousing agent by removing ~8% of adult salmon lice from Chinook salmon smolts in tank experiments. In field studies, the light trap caught 21 sea lice, comprising free-swimming and attached stages, while plankton net tows failed to capture any (Novales-Flamarique et al., 2009). However, in sea cages, it could be used only at nighttime, as the sun will usually be brighter than the light source. A possible further caveat is that fish could be drawn to the lights to feed on other invertebrates also attracted by the light and therefore be at an increased risk of infection by other parasites, creating in theory an additional risk of infection (Professor J.E. Bron, personal communication). With further improvements, light traps have the potential to become in the future an effective, noninvasive, environmentally friendly method to monitor sea lice.

Electric fences: A Seafarm Pulse Guard (SPG) system has been developed to combat sea lice at aquaculture sites. SPG was developed in Norway and is a patented system whereby a skirt of electrical cables forms an electric field in the sea in close proximity to the sea pen (Ingvarsdóttir et al., 2010).

Laser-optical delousing: There are several companies working on this approach. One of them, Stingray Marine Solutions AS (Norway), has developed a submersible device, which employs a green laser to remove the parasites noninvasively. It consists of a barrel-shaped device that submerges directly into a salmon farm's net pens and uses an onboard camera vision system with real-time image recognition software to capture images of the lice as the salmon swim past the device. Whenever the outer shape of the sea lice is recognized on the salmon's skin, the system's green laser (which is also used in human medicine) releases a continuous-wave beam that destroys the tissue of the parasite without damaging the fish. Optical delousing takes place while the salmon is swimming in its natural habitat, and does not require forcing or moving the fish. A number of fish welfare experiments indicate that optical delousing causes less damage than human handling of the fish (Dubay and Goode, 2014).

Mechanical and thermal delousing: There are several methods in use for mechanical removal of sea lice from salmon. The one developed by SkaMik AS consists of a water flush plus an additional but undisclosed element. Full-scale tests have demonstrated that the technology kills up to 90% of lice. The salmon are transported from the cage and treated individually with soft brushes, a water flush, and an unrevealed component. A filter collects and destroys all lice and the process water is filtered. The company claims that passage through the system does not result in ulcerations, loss of fish scales, or reduction of the skin mucus cover. Another device, the OCEA Delouser, flushes fish with slightly warm seawater (28–35°C) for ~20–30 s. It removes and kills more than 98% of adult lice from the fish and approximately 40% of juvenile lice. Thus far, it has been used in Chilean salmon farming against *C. rogercresseyi* with success. However, further studies on the impact of these devices on animal health and welfare are required, before they can be used at large scale in fish farms.

Ultrasound: It has been shown that the use of ultrasound, applied underwater in fish pens, does not affect salmon, wild fish, or marine mammals, but affects some sea lice stages. This is due to the low power and frequencies used: only 20 W per transmitter and 20 kHz. Experimental application of ultrasound has been shown to destroy the first nauplius stage of *C. rogercresseyi* in Chile; an effect was also seen in full-scale trials in sea cages. However, the ultrasound had almost no effect against adult sea lice, which means that they will continue to infect fish, but that the cycle might be broken in the nauplius phase alone, thus having a limited effect (CORFO report). In general, the application of ultrasound reduces the frequency of antiparasitic treatments, but this effect will be modulated by environmental parameters that can exacerbate reinfection levels.

5.6 CONCLUSIONS AND FUTURE PERSPECTIVES

This review has shown that there is a substantial scientific and empirical base for the implementation of IPMS in fish farming and that nowadays integration of all the available preventive and treatment strategies is indispensable to fighting fish diseases. The strict control of the use of antibiotics and other chemicals in the modern aquaculture industry and the appearance of drug resistance have spurred the development of alternative control measures based on technological development and biological tools. For some pathogens, such as sea lice, there is no single magic bullet solution, and IPMS is clearly the only way to keep this parasite under control in salmon farming (Fig. 5.5).

The degree of development of IPMS is very variable depending on the farmed species, the pathogen, and the region. In general, salmonid farming is ahead of that of other fish species in the implementation of IPMS, and bacteria and viruses are better covered than parasites, since preventive strategies based on vaccine development and selective breeding are already established. This is partly due to the complexity of the life cycles of some parasites and their antigenic diversity, so that the basic knowledge (step 1 of the IPM model) is not as advanced. Furthermore, IPMS are guided by the identification of CCP in the fish production system. This approach has not been generally adopted in aquaculture for many parasites, owing to the lack of adequate knowledge of measures that can prevent infection of fish. The remedy for this is a greater understanding of the epidemiology in aquaculture systems and the identification of CCP in transmission. Although risk assessment studies are limited, several significant risk factors have been identified and are essential for developing adequate biosecurity measures and strategic health plans.

IPMS are key for a successful future of aquaculture production. Chemical treatments are likely to remain the exception or only part of disease control measures. Identifying strategies that reduce the impact of pathogens is crucial. These are in almost all cases likely to require a combination of methods. Increasing our knowledge of how best to manage pathogens and associated diseases is indispensable to reducing their impact. Novel methods and approaches (e.g., submerged sea cages, ultrasonic treatment of sea lice) are valuable additions to the IPMS toolbox. There are several good ideas for pathogen control. Bringing them to the level of application at which they are cost efficient and practical is one of the major challenges ahead.

ABBREVIATIONS

AGD Amoebic gill disease
CCP Critical control points
ELISA Enzyme-linked immunosorbent assay
FISH Fluorescence in situ hybridization
IF Immunofluorescence
IFAT Immunofluorescence antibody test
IHC Immunohistochemistry
IHNV Infectious hematopoietic necrosis virus
IPNV Infectious pancreatic necrosis virus
ISAV Infectious salmon anemia virus
KHV Koi herpesvirus
LAMP Loop-mediated isothermal amplification
PCR Polymerase chain reaction
PDV Pancreas disease virus
PKD Proliferative kidney disease
Q-PCR Quantitative PCR
RT-PCR Reverse transcription PCR
SAV Salmon alpha virus
VHSV Viral hemorrhagic septicemia virus

ACKNOWLEDGMENT

This work has received funding from the European Union's Horizon 2020 research and innovation program to the project ParafishControl, under Grant Agreement 634429. This publication reflects the views only of the authors, and the European Commission cannot be held responsible for any use that may be made of the information contained herein. Additional funds have been obtained from Defra Contract FB002 and Spanish AGL2013-48560 and PROMETEO FASEII-2014/085. The authors thank Professor J.E. Bron from the University of Stirling (UK) and P. Christofilogiannis (AQUARK, Greece) for constructive comments on some management strategies, and Dr. F.E. Montero from the University of Valencia (Spain) for kindly providing a drawing of sea lice.

REFERENCES

Aaen, S.M., Helgesen, K.O., Bakke, M.J., Kaur, K., Horsberg, T.E., 2015. Drug resistance in sea lice: a threat to salmonid aquaculture. Trends in Parasitology 31, 72–81.

AE, 2014. Performance of the sea bass and sea bream sector in the Mediterranean. Minutes of a workshop held within Aquaculture Europe (AE) Conference. In: 16th October 2014, San Sebastián, Spain, Organised by European Aquaculture Society and the European Aquaculture Technology & Innovation Platform.

Akoll, P., Konecny, R., Mwanja, W.W., Schiemer, F., 2012. Risk assessment of parasitic helminths on cultured Nile tilapia (*Oreochromis niloticus*, L.). Aquaculture 356–357, 123–127.

Albert, V., Ransangan, J., 2013. Effect of water temperature on susceptibility of culture marine fish species to vibriosis. International Journal of Research in Pure and Applied Microbiology 3, 48–52.

Andersen, R.J., Luu, H.A., Chen, D.Z.X., Homes, C.F.B., Kent, M.L., Le Blanc, F., Taylor, F.J.R., Williams, D.E., 1993. Chemical and biological evidence links microcystins to salmon 'netpen liver disease'. Toxicon 31, 1315–1325.

Anderson, D.P., 1992. Immunostimulants, adjuvants, and vaccine carriers in fish: applications to aquaculture. Annual Review of Fish Diseases 2, 281–307.

Anonymous, 2006. In: Union, C.o.t.E. (Ed.), Council Directive 2006/88/EC of 24 October 2006 on Animal Health Requirements for Aquaculture Animals and Products Thereof, and on the Prevention and Control of Certain Diseases in Aquatic Animals.

Arndt, R.E., Wagner, E.J., 2003. Filtering *Myxobolus cerebralis* Triactinomyxons from contaminated water using rapid sand filtration. Aquacultural Engineering 29, 77–91.

Arechavala-Lopez, P., Uglem, I., Sanchez-Jerez, P., Fernandez-Jover, D., Bayle-Sempere, J.T., Nilsen, R., 2010. Movements of grey mullet *Liza aurata* and *Chelon labrosus* associated with coastal fish farms in the western Mediterranean Sea. Aquaculture Environment Interactions 1, 127–136.

Arechavala-Lopez, P., Sanchez-Jerez, P., Bayle-Sempere, J., Fernandez-Jover, D., Martinez-Rubio, L., Lopez-Jimenez, J.A., Martinez-Lopez, F.J., 2011. Direct interaction between wild fish aggregations at fish farms and fisheries activity at fishing grounds: a case study with *Boops boops*. Aquaculture Research 42, 996–1010.

Arechavala-Lopez, P., Sanchez-Jerez, P., Bayle-Sempere, J., Uglem, I., Mladineo, I., 2013. Reared fish, farmed escapees and wild fish stocks-a triangle of pathogen transmission of concern to Mediterranean aquaculture management. Aquaculture Environment Interactions 3, 153–161.

AST, 2010. Atlantic Salmon Trust News. Available from: www.atlanticsalmontrust.org/latest-news/ast-launches-new-sea-lice-policy-78.html.

Barras, S.C., Godwin, K.C., 2005. Controlling Bird Predation at Aquaculture Facilities: Frightening Techniques. Southern Regional Aquaculture Center, Publication No. 401.

Bartholomew, J.L., Reno, P., 2002. The history and dissemination of whirling disease. In: Bartholomew, J.L., Wilson, J.C. (Eds.), Whirling Disease: Reviews and Current Topics. American Fisheries Society Symposium 29. Bethesda, Maryland, pp. 3–24.

Barzman, M., Barberi, P., Birch, A., Boonekamp, P., Dachbrodt-Saaydeh, S., Graf, B., Hommel, B., Jensen, J.E., Kiss, J., Kudsk, P., Lamichhane, J.R., Messean, A., Moonen, A.C., Ratnadass, A., Ricci, P., Sarah, J.L., Sattin, M., 2015. Eight principles of integrated pest management. Agronomy for Sustainable Development 35, 1199–1215.

Bass, D., Stentiford, G.D., Littlewood, D.T.J., Hartikainen, H., 2015. Diverse applications of environmental DNA methods in parasitology. Trends in Parasitology 31, 499–513.

Bedell, G.W., 1971. Eradicating *Ceratomyxa shasta* from infected water by chlorination and ultraviolet irradiation. The Progressive Fish-Culturist 33, 51–54.

Bjørn, P.A., Johnsen, I.A., Nilsen, R., Serra Llinares, R.M., Heuch, P.A., Finstad, B., Boxaspen, K., Asplin, L., 2012. The effects of fallowing zones on the distribution and risks of salmon lice infection in wild salmonids in an intensively farmed Norwegian fjord system. In: Book of Abstracts 9th International Sea Lice Conference, Bergen, Norway, May 2012, pp. 35–36.

Bower, S.M., Margolis, L., 1985. Microfiltration and ultraviolet irradiation to eliminate *Ceratomyxa shasta* (Myxozoa: Myxosporea), a salmonid pathogen, from Fraser River water, British Columbia. Canadian Technical Report of Fisheries and Aquatic Sciences 1364.

Bravo, S., Dolz, H., Silva, M.T., Lagos, C., Millanao, A., Urbina, M., 2005. Final Report. Diagnosis on the Use of Pharmaceuticals and Other Chemicals in Aquaculture. Austral University of Chile. Faculty of Fishery and Oceanography, Aquaculture Institute, Port Montt, Chile. Project No. 2003-28.

Bron, J.E., Sommerville, C., Wooten, R., Rae, G.H., 1993. Fallowing of marine Atlantic salmon, *Salmo salar* L. farms as a method for the control of sea lice, *Lepeoptherius salmonis* (Kroyer, 1837). Journal of Fish Diseases 16, 487–493.

Bruno, D.W., Nowak, B., Elliott, D.G., 2006. Guide to the identification of fish protozoan and metazoan parasites in stained tissue sections. Diseases of Aquatic Organisms 70, 1–36.

Cabello, F.C., 2004. Antibiotics and aquaculture in Chile: implications for human and animal health. Medical Journal of Chile 132, 1001–1006.

Castillo, D., Christiansen, R.H., Dalsgaard, I., Madsen, L., Middelboe, M., 2015. Bacteriophage resistance mechanisms in the fish pathogen *Flavobacterium psychrophilum*: linking genomic mutations to changes in bacterial virulence factors. Applied and Environmental Microbiology 81, 1157–1167.

Chambers, C.B., Ernst, I., 2005. Dispersal of the skin fluke *Benedenia seriolae* (Monogenea: Capsalidae) by tidal currents and implications for sea-cage farming of *Seriola* spp. Aquaculture 250, 60–69.

Chauvier, G., Mortier-Gabet, J., 1984. First observations of the pathogenicity of *Ichthyophonus* in birds – two cases of natural infection. Annales de Parasitologie Humaine et Comparee 59, 427–431.

China, M., Nakamura, H., Hamakawa, K., Tamaki, E., Yokoyama, H., Masuoka, S., Ogawa, K., 2014. Efficacy of high water temperature treatment of myxosporean emaciation disease caused by *Enteromyxum leei* (Myxozoa). Fish Pathology 49, 137–140.

Chu, W., Zhou, S., Zhu, W., Zhuang, X., 2014. Quorum quenching bacteria *Bacillus* sp. QSI-1 protect zebrafish (*Danio rerio*) from *Aeromonas hydrophila* infection. Scientific Reports 4, 5446.

CORFO report. Use of Ultrasound to Control Chilean sea lice (*Caligus rogercresseyi*). Available from: https://gaalliance.org/wp-content/uploads/2016/02/Day2_RodrigoPrado.pdf.

Covello, J.M., Purcell, S.L., Pino, J., González Vecino, J.L., González, J., Troncoso, J., Fast, M., Wadsworth, S.L., 2012. Effects of orally administered immune stimulants on Atlantic salmon (*Salmo salar*) transcriptional responses and subsequent sea lice (*Lepeophtheirus salmonis*). In: Book of Abstracts 9th International Sea Lice Conference Bergen, Norway, May 2012, pp. 32–33.

CRAB, 2007. European Best Practice in Aquaculture Biofouling. 60 pp. Available from: www.crabproject.com.

CRAB, 2006. Towards European best practice in aquaculture biofouling. In: A Workshop Organized by the EU-Funded Collective Research on Aquaculture Biofouling and Held during AQUA 2006, Firenze, Italy Available from: www.crabproject.com.

Cribb, T.H., Adlard, R.D., Hayward, C.J., Bott, N.J., Ellis, D., Evans, D., Nowak, B.F., 2011. The life cycle of *Cardicola forsteri* (Trematoda: Aporocotylidae), a pathogen of ranched southern bluefin tuna, *Thunnus maccoyi*. International Journal for Parasitology 41, 861–870.

Das, S., Sahoo, P.K., 2014. Markers for selection of disease resistance in fish: a review. Aquaculture International 22, 1793–1812.

Defoirdt, T., Boon, N., Sorgeloos, P., Verstraete, W., Bossier, P., 2008. Quorum sensing and quorum quenching in *Vibrio harveyi*: lessons learned from in vivo work. ISME Journal 2, 19–26.

Dempster, T., Kristiansen, T.S., Korsøen, Ø.J., Fosseidengen, J.E., Oppedal, F., 2011. Technical note: modifying Atlantic salmon (*Salmo salar*) jumping behavior to facilitate innovation of parasitic sea lice control techniques. Journal of Animal Science 89, 4281–4285.

Dubay, L., Goode, B., July–August 2014. Laser device can slay sea lice on farmed salmon noninvasively. BioOptics 44.

Eckes, M., Dove, S., Siebeck, U.E., Grutter, A.S., 2015. Fish mucus versus parasitic gnathiid isopods as sources of energy and sunscreens for a cleaner fish. Coral Reefs 34, 823–833.

Ecofish Project. The Health of Farmed Wrasse. Ecofish Ballan Wrasse Project. Technical Leaflet 06. Available from: www.eco-fish.org.

EFSA, 2010. European Food Safety Agency (EFSA) panel on Biological Hazards (BIOHAZ); scientific opinion on risk assessment of parasites in fishery products. EFSA Journal 8, 1543. 91 pp. Available from: www.efsa.europa.eu.

Ernst, I., 2010. Management of monogenean parasites in sea-cage aquaculture. Tahiti Aquaculture. In: Book of Abstracts of Sustainable Aquaculture on Tropical Islands, Arue, Tahiti, December 6–11, 2010, p. 48.

EU Commission, 2015. Commission Implementing Decision (EU) 2015/1554 of 11 September 2015 laying Down rules for the application of Directive 2006/88/EC as regards requirements for surveillance and diagnostic methods. In: Europeon Union (Ed.), Official Journal of the European Union, L247 241–262.

Evelyn, T.P.T., 2001. The effects of chilling, freezing and cold-smoking on the infectious titre of certain microbial fish pathogens that may occasionally be present in marketed salmonid flesh. In: Rodgers, C.J. (Ed.), Risk Analysis in Aquatic Animal Health. Proceedings of an International Conference, Paris, France, February 8–10, 2000, pp. 215–229.

Evenhuis, J.P., Leeds, T.D., Marancik, D.P., LaPatra, S.E., Wiens, G.D., 2015. Rainbow trout (*Oncorhynchus mykiss*) resistance to columnaris disease is heritable and favorably correlated with bacterial cold water disease resistance. Journal of Animal Science 93, 1546–1554.

FAO, 1989. European Inland Fisheries Advisory Commission, 1988 Report of the EIFAC Working Party on Prevention and Control of Bird Predation in Aquaculture and Fisheries Operations. EIFAC Tech. Pap., 51 , p. 79 Available from: www.fao.org/docrep/009/t0054e/t0054e00.htm.

FAO, 2014. The State of World Fisheries and Aquaculture Opportunities and Challenges. Food and Agriculture Organization of the United Nations, Rome. 243 pp.

Fitridge, I., Dempster, T., Guenther, J., de Nys, R., 2012. The impact and control of biofouling in marine aquaculture: a review. Biofouling 28, 649–669.

Foott, J.S., Hedrick, R.P., 1987. Seasonal occurrence of the infectious stage of proliferative kidney disease (PKD) and resistance of rainbow trout, *Salmo gairdneri* Richardson, to reinfection. Journal of Fish Biology 30, 477–483.

Foreman, M.G.G., Guo, M., Garver, K.A., Stucchi, D., Chandler, P., Wan, D., Morrison, J., Tuele, D., 2015. Modelling infectious hematopoietic necrosis virus dispersion from marine salmon farms in the Discovery Islands, British Columbia, Canada. PLoS One 10, e0130951.

Frenzl, B., Stien, L.H., Cockerill, D., Oppedal, F., Richards, R.H., Shinn, A.P., Bron, J.E., Migaud, H., 2014. Manipulation of farmed Atlantic salmon swimming behaviour through the adjustment of lighting and feeding regimes as a tool for salmon lice control. Aquaculture 424–425, 183–188.

Gautam, R., Boerlage, A.S., Vanderstichel, R., Revie, C.W., Hammell, K.L., 2016. Variation in pre-treatment count lead time and its effect on baseline estimates of cage-level sea lice abundance. Journal of Fish Diseases 39, 1297–1303. http://dx.doi.org/10.1111/jfd.12460.

Gharbi, K., Matthews, L., Bron, J., Roberts, R., Tinch, A., Stear, M., 2015. The control of sea lice in Atlantic salmon by selective breeding. Journal of the Royal Society Interface 12, 20150574.

GISP, 2008. Global Invasive Species Programme (GISP). Marine Biofouling: An Assessment of Risks and Management Initiatives. Compiled by Lynn Jackson on behalf of the GISP and the UNEP Regional Seas Programme. 68 pp.

Gorlick, D.L., Atkins, P.D., Losey, G.S., 1987. Effect of cleaning by *Labroides dimidiatus* (Labridae) on an ectoparasite population infecting *Pomacentrus vaiuli* (Pomacentridae) at Enewetak Atoll. Copeia 1, 41–45.

Grutter, A.S., 1999. Cleaner fish really do clean. Nature 398, 672–673.

Gudding, R., 2014. Vaccination as a preventive measure. In: Gudding, R., Lillehaug, A., Evensen, Ø. (Eds.), Fish Vaccination. Wiley & Sons, Ltd, Oxford, pp. 12–20.

Hai, N.V., 2015. The use of probiotics in aquaculture. Journal of Applied Microbiology 119, 917–935.

Hall, L.M., Smith, R.J., Munro, E.S., Matejusova, I., Allan, C.E.T., Murray, A.G., Duguid, S.J., Salama, N.K.G., McBeath, A.J.A., Wallace, I.S., Bain, N., Marcos-Lopez, M., Raynard, R.S., 2013. Epidemiology and control of an outbreak of viral haemorrhagic septicaemia in wrasse around Shetland commencing 2012. Scottish Marine and Freshwater Science 4, 1–46.

Hamza, K., Rich, K.M., Wheat, I.D., 2014. A system dynamics approach to sea lice control in Norway. Aquaculture Economics & Management 18, 344–368.

Hedrick, R.P., McDowell, T.S., Marty, G.D., Mukkatira, K., Antonio, D.B., Andree, K.B., Bukhari, Z., Clancy, T., 2000. Ultraviolet irradiation inactivates the waterborne infective stages of *Myxobolus cerebralis*: a treatment for hatchery water supplies. Diseases of Aquatic Organisms 42, 53–59.

Hedrick, R.P., McDowell, T.S., Mukkatira, K., Georgiadis, M.P., MacConnell, E., 2001. Salmonids resistant to *Ceratomyxa shasta* are susceptible to experimentally induced infections with *Myxobolus cerebralis*. Journal of Aquatic Animal Health 13, 35–42.

Helgesen, K.O., Aaen, S.M., Romstad, H., Horsberg, T.E., 2015. First report of reduced sensitivity towards hydrogen peroxide found in the salmon louse *Lepeophtheirus salmonis* in Norway. Aquaculture Reports 1, 37–42.

Henryon, M., Berg, P., Olesen, N.J., Kjær, T.E., Slierendrecht, W.J., Jokumsen, A., Lund, I., 2005. Selective breeding provides an approach to increase resistance of rainbow trout (*Onchorhynchus mykiss*) to the diseases, enteric redmouth disease, rainbow trout fry syndrome, and viral haemorrhagic septicaemia. Aquaculture 250, 621–636.

Hevrøy, E.M., Boxaspen, K., Oppedal, F., Taranger, G.L., Hol, J.C., 2003. The effect of artificial light treatment and depth on the infestation of the sea louse *Lepeophtheirus salmonis* on Atlantic salmon (*Salmo salar* L.) culture. Aquaculture 220, 1–14.

Hølvold, L.B., Myhr, A.I., Dalmo, R.A., 2014. Strategies and hurdles using DNA vaccines to fish. Veterinary Research 45, 1–11.

Hoseinifar, S.H., Esteban, M.A., Cuesta, A., Sun, Y.Z., 2014. Prebiotics and fish immune response: a review of current knowledge and future perspectives. Reviews in Fisheries Science & Aquaculture 23, 315–328.

Ibacache-Quiroga, C., Ojeda, J., Espinoza-Vergara, G., Olivero, P., Cuellar, M., Dinamarca, M.A., 2013. The hydrocarbon-degrading marine bacterium *Cobetia* sp. strain MM1IDA2H-1 produces a biosurfactant that interferes with quorum sensing of fish pathogens by signal hijacking. Microbial Biotechnology 6, 394–405.

Igboeli, O.O., Purcell, S., Wotton, H., Poley, J., Burka, J.F., Fast, M.D., 2012. Host immunostimulation and its effects on *Lepeophtheirus salmonis* P-glycoprotein expression and subsequent emamectin benzoate exposure. In: Book of Abstracts 9th International Sea Lice Conference, Bergen, Norway, May 2012, p. 32.

Imbeault, S., Parent, S., Lagacé, M., Uhland, C.F., Blais, J.-F., 2006. Using bacteriophages to prevent furunculosis caused by *Aeromonas salmonicida* in farmed brook trout. Journal of Aquatic Animal Health 18, 203–214.

Ingvarsdóttir, A., Birkett, M.A., Duce, I., Genna, R.L., Mordue, W., Pickett, J.A., Wadhams, L.J., Mordue, A.J., 2002. Semiochemical strategies for sea louse control: host location cues. Pest Management Science 58, 537–545.

Ingvarsdóttir, A., Provan, F., Bredal, H., 2010. Seafarm Pulse Guard (SPG): protecting farmed salmon from sealice. In: Book of Abstracts 9th International Sea Lice Conference Bergen, Norway, May 2012, p. 50.

Jarp, J., Karlsen, E., 1997. Infectious salmon anaemia (ISA) risk factors in sea-cultured Atlantic salmon *Salmo salar*. Diseases of Aquatic Organisms 28, 79–86.

Jennings, S., Stentiford, G.D., Leocadio, A.M., Jeffery, K.R., Metcalfe, J.D., Katsiadaki, I., Auchterlonie, N.A., Mangi, S.C., Pinnegar, J.K., Ellis, T., Peeler, E.J., Luisetti, T., Baker-Austin, C., Brown, M., Catchpole, T.L., Clyne, F.J., Dye, S.R., Edmonds, N.J., Hyder, K., Lee, J., Lees, D.N., Morgan, O.C., O'Brien, C.M., Oidtmann, B., Posen, P.E., Ribeiro Santos, A., Taylor, N.G.H., Turner, A.D., Townhill, B.L., Verner-Jeffreys, D.W., 2016. Aquatic food security: insights into challenges and solutions from an analysis of interactions between fisheries, aquaculture, food safety, human health, fish and human welfare, economy and environment. Fish & Fisheries 17, 893–938. http://dx.doi.org/10.1111/faf.12152.

Jensen, L.B., Provan, F., Larssen, E., Bron, J.E., Obach, A., 2015. Reducing sea lice (*Lepeophtheirus salmonis*) infestation of farmed Atlantic salmon (*Salmo salar* L.) through functional feeds. Aquaculture Nutrition 21, 983–993.

Jones, S.R.M., Hargreaves, N., 2007. The abundance and distribution of *Lepeophtheirus salmonis* (Copepoda: Caligidae) on pink (*Oncorhynchus gorbuscha*) and chum (*O. keta*) salmon in coastal British Columbia. Journal of Parasitology 93, 1324–1331.

Jones, S.R.M., Bruno, D.W., Madsen, L., Peeler, E.J., 2015. Disease management mitigates risk of pathogen transmission from maricultured salmonids. Aquaculture Environment Interactions 6, 119–134.

Jones, S.R.M., Cho, S., Nguyen, J., Mahony, A., 2016. Acquired resistance to *Kudoa thyrsites* in Atlantic salmon *Salmo salar* following recovery from a primary infection with the parasite. Aquaculture 451, 457–462.

Karvonen, A., Rintamäki, P., Jokela, J., Tellervo Valtonen, E., 2010. Increasing water temperature and disease risks in aquatic systems: climate change increases the risk of some, but not all, diseases. International Journal for Parasitology 40, 1483–1488.

Kent, M.L., 1990. Netpen Liver Disease (NLD) of salmonid fishes reared in sea water: species susceptibility, recovery and probable cause. Diseases of Aquatic Organisms 8, 21–28.

Kent, M., Margolis, L., Fournie, J., 1991. A new eye disease in pen-reared chinook salmon caused by metacestodes of *Gilquinia squali* (Trypanorhyncha). Journal of Aquatic Animal Health 3, 134–140.

Kim, J., Choresca, C., Shin, S., Han, J., Jun, J., Park, S., 2015. Biological control of *Aeromonas salmonicida* subsp. *salmonicida* infection in rainbow trout (*Oncorhynchus mykiss*) using *Aeromonas* phage PAS-1. Transboundary and Emerging Diseases 62, 81–86.

Kirchhoff, N.T., Rough, K.M., Nowak, B.F., 2011. Moving cages further offshore: effects on Southern bluefin tuna, *T. maccoyii*, parasites, health and performance. PLoS One 6, e23705.

Koshio, S., 2016. Immunotherapies targeting fish mucosal immunity – current knowledge and future perspectives. Frontiers in Immunology 6, 1–4.

Kumar, S.S., Bharathi, R.A., Rajan, J.J.S., Alavandi, S.V., Poornima, M., Balasubramanian, C.P., Ponniah, A.G., 2013. Viability of White Spot Syndrome Virus (WSSV) in sediment during sun-drying (drainable pond) and under non-drainable pond conditions indicated by infectivity to shrimp. Aquaculture 402, 119–126.

Lackenby, J.A., Chambers, C.B., Ernst, I., Whittington, I.D., 2007. Effect of water temperature on reproductive development of *Benedenia seriolae* (Monogenea: Capsalidae) from *Seriola lalandi* in Australia. Diseases of Aquatic Organisms 74, 235–242.

Lafferty, K.D., Harvell, C., Conrad, J.M., Friedman, C.S., Kent, M.L., Kuris, A.M., Powell, E.N., Rondeau, D., Saksida, S.M., 2015. Infectious diseases affect marine fisheries and aquaculture economics. Annual Review of Marine Science 7, 471–496.

LaFrentz, B.R., Lozano, C.A., Shoemaker, C.A., García, J.C., Xu, D.-H., Løvoll, M., Rye, M., 2016. Controlled challenge experiment demonstrates substantial additive genetic variation in resistance of Nile tilapia (*Oreochromis niloticus*) to *Streptococcus iniae*. Aquaculture 458, 134–139.

Leong, J., Turner, S., 1979. Isolation of Waterborne Infectious Hematopoietic Necrosis Virus, vol. 8. Fish News, pp. vi–viii.

Liltved, H., Vogelsang, C., Modahl, I., Dannevig, B., 2006. High resistance of fish pathogenic viruses to UV irradiation and ozonated seawater. Aquacultural Engineering 34, 72–82.

Liu, Y., Rzeszutek, E., van der Voort, M., Wu, C.H., Thoen, E., Skaar, I., Bulone, V., Dorrestein, P.C., Raaijmakers, J.M., de Bruijn, I., 2015. Diversity of aquatic *Pseudomonas* species and their activity against the fish pathogenic *Oomycete saprolegnia*. PLoS One 10, e0136241.

Longshaw, M., Le Deuff, R.-M., Harris, A.F., Feist, S.W., 2002. Development of proliferative kidney disease in rainbow trout, *Oncorhynchus mykiss* (Walbaum), following short-term exposure to *Tetracapsula bryosalmonae* infected bryozoans. Journal of Fish Diseases 25, 443–449.

Longshaw, M., Feist, S.W., Oidtmann, B., Stone, D., 2012. Applicability of sampling environmental DNA for aquatic diseases. Bulletin of the European Association of Fish Pathologists 32, 69–76.

Losos, C.J.C., Reynolds, J.D., Dill, L.M., 2010. Sex-selective predation by three spine sticklebacks on sea lice: a novel cleaning behaviour. Ethology 116, 981–989.

Lowell, J., 2012. Effect of Netting Materials on Fouling and Parasite Egg Loading on Offshore Net Pens in Hawaii. Blue Ocean Mariculture. ICA Study Number: TEK 1049-7.

Marty, G.D., Saksida, S.M., Quinn II, T.J., 2010. Relationship of farm salmon, sea lice and wild salmon populations. Proceedings of the National Academy of Sciences 107, 22599–22604.

McVicar, A.H., 1999. *Ichthyophonus* and related organisms. In: Woo, P.T.K., Bruno, D.W. (Eds.), Fish Diseases and Disorders, Volume 3: Viral, Bacterial and Fungal Infections. CAB International, Wallingford, Oxfordshire, pp. 661–688.

Meena, D.K., Das, P., Kumar, S., Mandal, S.C., Prusty, A.K., Singh, S.K., Akhtar, M.S., Behera, B.K., Kumar, K., Pal, A.K., Mukherjeeet, S.C., 2013. Beta-glucan: an ideal immunostimulant in aquaculture (a review). Fish Physiology and Biochemistry 39, 431–457.

Meyers, T.R., 1984. Marine bivalve molluscs as reservoirs of viral finfish pathogens: significance to marine and anadromous finfish culture. Marine Fisheries Review 46, 14–17.

Mitchell, A.J., Overstreet, R.M., Goodwin, A.E., Brandt, T.M., 2005. Spread of an exotic fish-gill trematode: a far-reaching and complex problem. Fisheries 30, 11–16.

Mladineo, I., 2002. Prevalence of *Ceratothoa oestroides* (Risso, 1826), a cymothoid isopode parasite, in cultured sea bass *Dicentrarchus labrax* L. on two farms in middle Adriatic Sea. Acta Adriatica 43, 97–102.

Mo, T.A., Gahr, A., Hansen, H., Hoel, E., Oaland, Ø., Poppe, T.T., 2014. Presence of *Anisakis simplex* (Rudolphi, 1809 det. Krabbe, 1878) and *Hysterothylacium aduncum* (Rudolphi, 1802) (Nematoda; Anisakidae) in runts of farmed Atlantic salmon, *Salmo salar* L. Journal Of Fish Diseases 37, 135–140.

Moran, D., Fofana, A., 2007. An economic evaluation of the control of three notifiable fish diseases in the United Kingdom. Preventive Veterinary Medicine 80, 193–208.

Murray, A.G., 2011. A simple model to assess selection for treatment-resistant sea lice. Ecological Modelling 222, 1854–1862.

Murray, A.G., Munro, L.A., Wallace, I.S., Berx, B., Pendrey, D., Fraser, D., Raynard, R.S., 2010. Epidemiological investigation into the re-emergence and control of an outbreak of infectious salmon anaemia in the Shetland Islands, Scotland. Diseases of Aquatic Organisms 91, 189–200.

Natrah, F.M.I., Kenmegne, M.M., Wiyoto, W., Sorgeloos, P., Bossier, P., Defoirdt, T., 2011. Effects of micro-algae commonly used in aquaculture on acyl-homoserine lactone quorum sensing. Aquaculture 317, 53–57.

Nehring, R.B., Thompson, K.G., Taurman, K., Atkinson, W., 2003. Efficacy of passive sand filtration in reducing exposure of salmonids to the actinospore of *Myxobolus cerebralis*. Diseases of Aquatic Organisms 57, 77–83.

Novales-Flamarique, I.N., Gulbransen, C., Galbraith, M., Stucchi, D., 2009. Monitoring and potential control of sea lice using a LED-based light trap. Canadian Journal of Fisheries and Aquatic Sciences 66, 1371–1382.

Núñez-Acuña, G., Gonçalves, A.T., Valenzuela-Muñoz, V., Pino-Marambio, J., Wadsworth, S., Gallardo-Escárate, C., 2015. Transcriptome immunomodulation of in-feed additives in Atlantic salmon *Salmo salar* infested with sea lice *Caligus rogercresseyi*. Fish & Shellfish Immunology 47, 450–460.

Næs, M., Heuch, P.A., Mathisen, R., 2012. Bruk av "luseskjørt" for å redusere påslag av lakselus *Lepeophtheirus salmonis* (Krøyer) på oppdrettslaks (Use of "sea lice skirt" to reduce infestation of salmon lice on farmed salmon). NCE Aquaculture (in Norwegian).

Ogut, H., Bishop, S.C., 2007. A stochastic modelling approach to describing the dynamics of an experimental furunculosis epidemic in Chinook salmon, *Oncorhynchus tshawytscha* (Walbaum). Journal of Fish Diseases 30, 93–100.

Ogut, H., LaPatra, S.E., Reno, P.W., 2005. Effects of host density on furunculosis epidemics determined by the simple SIR model. Preventive Veterinary Medicine 71, 83–90.

Oidtmann, B., Thrush, M., Denham, K., Peeler, E., 2011. International and national and biosecurity strategies in aquatic animal health. Aquaculture 320, 22–33.

Oidtmann, B., Peeler, E., Lyngstad, T., Brun, E., Bang, J.B., Stärk, K.D., 2013. Risk-based methods for fish and terrestrial animal disease surveillance. Preventive Veterinary Medicine 112, 13–26.

Oidtmann, B.C., Peeler, E.J., Thrush, M.A., Cameron, A.R., Reese, R.A., Pearce, F.M., Dunn, P., Lyngstad, T.M., Tavornpanich, S., Brun, E., Stark, K.D., 2014. Expert consultation on risk factors for introduction of infectious pathogens into fish farms. Preventive Veterinary Medicine 115, 238–254.

OIE. World Organisation for Animal Health (OIE), 2015a. Manual of Diagnostic Tests for Aquatic Animals. Available from: http://www.oie.int/en/international-standard-setting/aquatic-manual/access-online.

OIE. World Organisation for Animal Health (OIE), 2015b. Aquatic Animal Health Code 2015. Available from: www.oie.int/international-standard-setting/aquatic-code/access-online.

OIE. World Organisation for Animal Health (OIE), 2016. OIE-Listed Diseases, Infections and Infestations in Force in 2016. Available from: http://www.oie.int/en/animal-health-in-the-world/oie-listed-diseases-2016/.

Othman, F., Islam, M.S., Sharifah, E.N., Shahrom-Harrison, F., Hassan, A., 2015. Biological control of streptococcal infection in Nile tilapia *Oreochromis niloticus* (Linnaeus, 1758) using filter-feeding bivalve mussel *Pilsbryoconcha exilis* (Lea, 1838). Journal of Applied Ichthyology 31, 724–728.

Panzarin, V., Fusaro, F., Monne, I., Cappellozza, E., Patarnello, P., Bovo, G., Capua, I., Holmes, E.C., Cattoli, G., 2012. Molecular epidemiology and evolutionary dynamics of betanodavirus in southern Europe. Infection, Genetics and Evolution 12, 63–70.

Pino-Marambio, J., Mordue, A.J., Birkett, M., Carvajal, J., Asencio, G., Pickett, J.A., Quiroz, A., 2008. Parasite-host relationship between the salmon lice, *Lepeophtheirus salmonis* and *Caligus rogercresseyi* (Copepoda: Caligidae) and salmonid fish mediated by semiochemicals. In: Book of Abstracts of the 7th International Conference Sea Lice, Puerto Varas, Chile, 31 March–1 April 2008, pp. 1–2.

Reno, P.W., 1999. Infectious pancreatic necrosis and associated aquatic birnaviruses. In: Woo, P.T.K., Bruno, D.W. (Eds.), Fish Diseases and Disorders, Volume 3: Viral, Bacterial and Fungal Infections. CAB International, Wallingford, Oxfordshire, pp. 1–55.

Reverter, M., Bontemps, N., Lecchini, D., Banaigs, B., Sasal, P., 2014. Use of plant extracts in fish aquaculture as an alternative to chemotherapy: current status and future perspectives. Aquaculture 433, 50–61.

Revie, C.W., Robbins, C., Gettinby, G., Kelly, L., Treasurer, J.W., 2005. A mathematical model of the growth of sea lice, *Lepeophtheirus salmonis*, populations on farmed Atlantic salmon, *Salmo salar* L., in Scotland and its use in the assessment of treatment strategies. Journal of Fish Diseases 28, 603–613.

Robinson, S., 2010. Evaluation of the efficiency of non-chemical methods to reduce the impact of sea lice associated with salmon aquaculture sites using the principles of bio-filtration and trapping. In: Final Report of the 2010 Technical Review and 2011 Sea Lice Management and Program Development Workshop, Algonquin, Canada, pp. 14–15. 30.

Romero, J., Feijoo, C.G., Navarrete, P., 2012a. Antibiotics in aquaculture – use, abuse and alternatives. In: Carvalho, E. (Ed.), Health and Environment in Aquaculture. InTech. Available from: www.intechopen.com/books/health-and-environment-in-aquaculture/antibioticsin-aquaculture-use-abuse-and-alternatives.

Romero, M., Acuña, L., Otero, A., 2012b. Patents on quorum quenching: interfering with bacterial communication as a strategy to fight infections. Recent Patents on Biotechnology 6, 2–12.

Romero, M., Muras, A., Mayer, C., Bujan, N., Magariños, B., Otero, A., 2014. In vitro quenching of fish pathogen *Edwardsiella tarda* AHL production using marine bacterium *Tenacibaculum* sp. strain 20J cell extracts. Diseases of Aquatic Organisms 108, 217–225.

Rückert, S., Klimpel, S., Al-Quraishy, S., Mehlhorn, H., Palm, H.W., 2009. Transmission of fish parasites into grouper mariculture (Serranidae: *Epinephelus coioides* (Hamilton, 1822)) in Lampung Bay, Indonesia. Parasitology Research 104, 523–532.

Saksida, S., Karreman, G.A., Constantine, J., Donald, A., 2007. Differences in *Lepeophtheirus salmonis* abundance levels on Atlantic salmon farms in the Broughton Archipelago, British Columbia, Canada. Journal of Fish Diseases 30, 357–366.

Salama, N.K., Collins, C.M., Fraser, J.G., Dunn, J., Pert, C.C., Murray, A.G., Rabe, B., 2013. Development and assessment of a biophysical dispersal model for sea lice. Journal of Fish Diseases 36, 323–337.

Salama, N.K., Murray, A.G., Rabe, B., 2016. Simulated environmental transport distances of *Lepeophtheirus salmonis* in Loch Linnhe, Scotland, for informing aquaculture area management structures. Journal of Fish Diseases 39, 419–428.

Sanders, J.E., Fryer, J.L.D., Leith, A., Moore, K.D., 1972. Control of the infectious protozoan *Ceratomyxa shasta* by treating hatchery water supplies. Progressive Fish-Culturist 34, 13–17.

Sepúlveda, F., Marín, S.L., Carvajal, J., 2004. Metazoan parasites in wild fish and farmed salmon from aquaculture sites in southern Chile. Aquaculture 235, 89–100.

Shinn, A.P., Pratoomyot, J., Bron, J.E., Paladini, J., Brooker, E.E., Brooker, A.J., 2015. Economic impacts of aquatic parasites on global finfish production. Global Aquaculture Advocate 18, 58–61. Available from: http://advocate.gaalliance.org/economic-impacts-of-aquatic-parasites-on-global-finfish-production/.

Silva, Y.J., Moreirinha, C., Pereira, C., Costa, L., Rocha, R.J.M., Cunha, A., Gomes, N.C.M., Calado, R., Almeida, A., 2016. Biological control of *Aeromonas salmonicida* infection in juvenile Senegalese sole (*Solea senegalensis*) with Phage AS-A. Aquaculture 450, 225–233.

Sitjà-Bobadilla, A., Palenzuela, O., 2012. Enteromyxum species. In: Woo, P.T.K., Buchmann, K. (Eds.), Parasites: Pathobiology and Protection. CAB International., Oxfordshire, pp. 163–176.

Sitjà-Bobadilla, A., Conde de Felipe, M., Alvarez-Pellitero, P., 2006. In vivo and in vitro treatments against *Sparicotyle chrysophrii* (Monogenea: Microcotylidae) parasitizing the gills of gilthead sea bream (*Sparus aurata* L.). Aquaculture 261, 856–864.

Skiftesvik, A.B., Bjelland, R.M., Durif, C.M.F., Johansen, I.S., Browman, H.I., 2013. Delousing of Atlantic salmon (*Salmo salar*) by cultured vs. wild ballan wrasse (*Labrus bergylta*). Aquaculture 402, 113–118.

Sommerset, I., Krossøy, B., Biering, E., Frost, P., 2005. Vaccines for fish in aquaculture. Expert Review of Vaccines 4, 89–101.

Stabili, L., Schirosi, R., Licciano, M., Mola, E., Giangrande, A., 2010. Bioremediation of bacteria in aquaculture waste using the polychaete *Sabella spallanzanii*. New Biotechnology 27, 774–781.

Stagg, R.M., Bruno, D.W., Cunningham, C.O., Raynard, R.S., Munro, P.D., Murray, A.G., Allan, C.E.T., Smail, D.A., McVicar, A.H., Hastings, T.S., 2001. Epizootiological Investigations into an Outbreak of Infectious Salmon Anaemia (ISA) in Scotland. FRS Marine Laboratory Report No 13/01.

Stern, V.M., Smith, R.F., van den Bosch, R., Hagen, K.S., 1959. The integrated control concept. Hilgardia 29, 81–101.

Stevens, R.B., 1960. Cultural practices in disease control. In: Horsfall, J.G., Dimond, A.E. (Eds.). Horsfall, J.G., Dimond, A.E. (Eds.), Plant Pathology, an Advanced Treatise, vol. 3. Academic Press, NY, pp. 357–429.

Stien, L.H., Nilsson, J., Hevrøy, E.M., Oppedal, F., Kristiansen, T.S., Lien, A.M., Folkedal, O., 2012. Skirt around a salmon sea cage to reduce infestation of salmon lice resulted in low oxygen levels. Aquacultural Engineering 51, 21–25.

Tafalla, C., Bogwald, J., Dalmo, R.A., 2013. Adjuvants and immunostimulants in fish vaccines: current knowledge and future perspectives. Fish & Shellfish Immunology 35, 1740–1750.

Tan, C.K.F., Nowak, B.F., Hodson, S.L., 2002. Biofouling as a reservoir of *Neoparamoeba pemaquidensis* (Page, 1970), the causative agent of amoebic gill disease in Atlantic salmon. Aquaculture 210, 49–58.

Treasurer, J.W., 2005. Cleaner fish: a natural approach to the control of sea lice on farmed fish. Veterinary Bulletin 75, 17–29.

Tveit, K.J., 2012. Nytt "luseskjørt" stoppar lusa (New "Sea Lice Skirt" Stopped the Lice). Kyst. no. 30.04.2012 (in Norwegian).

Uglem, I., Dempster, T., Bjørn, P.A., Sanchez-Jerez, P., Økland, F., 2009. High connectivity of salmon farms revealed by aggregation, residence and repeated movements of wild fish among farms. Marine Ecology Progress Series 384, 251–260.

Webb, J.L., Vandenbor, J., Pirie, B., Robinson, S.M.C., Cross, S.F., Jones, S.R.M., Pearce, C.M., 2013. Effects of temperature, diet, and bivalve size on the ingestion of sea lice (*Lepeophtheirus salmonis*) larvae by various filter-feeding shellfish. Aquaculture 406–407, 9–17.

Werkman, M., Green, D.M., Murray, A.G., Turnbull, J.F., 2011. The effectiveness of fallowing strategies in disease control in salmon aquaculture assessed with an SIS model. Preventive Veterinary Medicine 98, 64–73.

Whelan, K., 2010. A review of the impacts of the salmon louse, *Lepeophtheirus salmonis* (Krøyer, 1837) on wild Salmonids. Atlantic Salmon Trust 1–27.

Wright, D.W., Oppedal, F., Dempster, T., 2016. Early-stage sea lice recruits on Atlantic salmon are freshwater sensitive. Journal of Fish Diseases 39, 1179–1186. http://dx.doi.org/10.1111/jfd.12452.

WS, 2010. Wildlife Problems at Fish Farms. Wildlife Services (WS) Factsheet, November 2010 Assisting American Aquaculture.

Yanagida, T., Sameshima, M., Nasu, H., Yokoyama, H., Ogawa, K., 2006. Temperature effects on the development of *Enteromyxum* spp. (Myxozoa) in experimentally infected tiger puffer, *Takifugu rubripes* (Temminck & Schlegel). Journal of Fish Diseases 29, 561–567.

Yatabe, T., Arriagada, G., Hamilton-West, C., Urcelay, S., 2011. Risk factor analysis for sea lice, *Caligus rogercresseyi*, levels in farmed salmonids in southern Chile. Journal of Fish Diseases 34, 345–354.

Yoshimizu, M., Takizawa, H., Manabu, S., Kataoka, H., Kugo, T., Kimura, T., 1990. Disinfectant effects of ultraviolet irradiation in hatchery water supply. In: Hirano, R., Haoyu, I. (Eds.), The Second Asian Fisheries Forum 1990. Asian Fisheries Society, Manila, pp. 643–646.

Yoshimizu, M., Yoshinaka, T., Hatori, S., Kasai, H., 2005. Survivability of fish pathogenic viruses in environmental water, and inactivation of fish viruses. Bulletin of Fisheries Research Agency Supplement 2, 47–54.

Zhang, Y., Liu, J., Tang, K., Yu, M., Coenye, T., Zhang, X.-H., 2015. Genome analysis of *Flaviramulus ichthyoenteri* Th78T in the family Flavobacteriaceae: insights into its quorum quenching property and potential roles in fish intestine. BMC Genomics 16, 38.

Part III

Environment

Chapter 6

General Relationship Between Water Quality and Aquaculture Performance in Ponds

Claude E. Boyd
Auburn University, Auburn, AL, United States

6.1 INTRODUCTION

Any physical, chemical, or biological property that affects the use of water for any purpose is a water-quality variable (Boyd and Lichtkoppler, 1979). Fish and other aquatic animals live in water and are especially sensitive to many substances, both natural and anthropogenic, found in water. Water quality is of paramount importance in aquaculture, because it affects the health, survival, and production of aquaculture species (Boyd and Tucker, 1998).

In the most primitive type of aquaculture, aquatic animals are stocked, their food derives from natural sources, and the fertility of the ecosystem determines the amount of natural food organisms available and the level of production that can be obtained, e.g., ponds and net pen culture without feeding and open-water culture of molluskan species. Fertilizers (animal dung, other agricultural waste, and commercial chemical fertilizers) are applied as sources of nitrogen, phosphorus, and other nutrients to increase natural productivity in aquaculture ponds (Mortimer, 1954; Boyd and Tucker, 1998; Mischke, 2012). The effectiveness of fertilizers is influenced by water quality (Hickling, 1962; Boyd and Tucker, 1998). But, fertilizers also influence water quality, especially the density of plankton and the ranges of pH and dissolved oxygen and carbon dioxide concentrations.

In feed-based aquaculture, feeds allow much greater production than possible in fertilized ponds, and they also permit high production in flow-through systems and cages. Nevertheless, wastes from feeding—uneaten feed, feces, and metabolic excretions by the culture species—greatly impair the water quality in culture systems. The culture species also impose an oxygen demand and excretes carbon dioxide, ammonia, and other metabolites.

Water quality should be considered at all stages of aquaculture production, i.e., for the water source, in culture systems before stocking, and in the culture system during grow-out. These tasks are facilitated by measurements of key variables, interpretation of the measured values, and implementation of interventions to improve water quality when necessary. Proper water-quality management in aquaculture is essential for promoting good aquatic animal health required for efficient production.

The environmental health of water bodies receiving aquaculture effluents must also be considered, because aquaculture wastewater may pollute the receiving water bodies (Tucker and Hargreaves, 2008). However, when water quality is properly managed within culture systems, the potential negative impacts of aquaculture effluents on the biota and biodiversity of receiving water bodies are greatly diminished.

The purpose of this chapter is to provide a simple overview of water quality as it relates to the health of aquatic animals in aquaculture.

6.2 WATER QUALITY–AQUATIC ANIMAL HEALTH INTERACTIONS

6.2.1 Stress

The ultimate effect of water quality on aquatic animals can be expressed in one of two ways. A variable can be either at a high concentration or at a low concentration such that direct mortality of the culture species results. The most familiar example is low dissolved oxygen concentration that can lead to high mortality of otherwise completely healthy aquatic

Fish Diseases. http://dx.doi.org/10.1016/B978-0-12-804564-0.00006-5

animals. High concentrations of carbon dioxide, ammonia nitrogen, nitrite, and sulfide produced within culture systems can also result in high mortality, but a low dissolved oxygen concentration will usually occur before the other variables become toxic factors. Of course, a toxic substance, such as high concentration of a pesticide in the water supply can have a lethal effect, but aquaculture facilities are usually selected to minimize the possibility of toxic pollutants entering culture systems.

The more common effect of water quality on aquatic animals is the stress caused as a result of chronic exposure to a lower or higher than optimal concentration of one or more water-quality variables. Effects of stress on aquatic animals have been studied more in fish than in other taxa. According to Bly et al. (1997), the fish immune system and other physiological functions can be stressed by natural factors (season, temperature, and salinity) and social factors such as crowding and hierarchy. These authors also emphasized that stress could result from artificial (human-induced) factors and especially from pollution. Of course, in aquaculture, crowding is a human-induced factor. Stress suppresses immune functions in both the innate (or nonspecific) and adaptive arms of the immune system of fish (and presumably of other aquatic animals). Bly et al. (1997) concluded that a suppression of the immune system made fish more susceptible to infectious diseases. Although the mechanisms responsible for immunosuppression are not well understood, elevated serum adrenocorticotrophic hormone and cortisol levels may be indicators of stress.

Conte (2004) emphasized that aquatic animals utilize a three-dimensional culture medium, and density of culture incorporates the number (or weight) of animals per unit volume of static water or per unit volume of flowing water per unit of time. He concluded that a combination of these factors influence water quality and animal to animal interaction to affect animal welfare. The carrying capacity in aquaculture systems is reported to be determined by the metabolic loading density and food availability in relation to the dissolved oxygen supply (Ellis et al., 2001). These authors basically ignored the spatial needs and referred to carrying capacity as the maximum biomass of animals based upon physiological requirements. Ellis et al. (2002) reviewed literature on rainbow trout to include food consumption, feed-conversion ratio (FCR), nutritional status, growth rate and size variation, health and condition profile, blood profiles, condition of organs, and plasma cortisol. The conclusion from this review is that the degradation of water quality was the most important factor affecting the health negatively in high-density culture. However, in the case of arctic charr (*Salvelinus alpinus*) stress-related behavior increased with greater density. Thus, all species may not respond in the same way, but adverse environments created by pollution alter the susceptibility of the host to pathogens in the environment. This increases the chance of the host contacting the pathogens (Arkoosh et al., 1998). Aquaculture environments are polluted by nutrient inputs, and it is a widely held premise that stress from adverse water quality is the major factor predisposing aquaculture animals to disease (Rottmann et al., 1992; Boyd and Tucker, 1998).

An important caveat related to stress is that, as Barton (2002) pointed out, stress is not necessarily detrimental to fish (and presumably other aquatic animals). Stress is a state of threatened homeostasis, and the animal can respond to this stress and reestablish homeostasis. However, Barton also emphasized that an intense stressor may compromise the physiological response mechanisms and be detrimental to fish health. An excellent example of the influence of an environmental stressor on common carp *Cyprinus carpio* was provided by Wang et al. (1997). In freshwater, carp had a daily weight gain of 2.32%, which decreased with greater salinity, becoming negative at a salinity of 10.5 g/L (Table 6.1). The rate of ammonia excretion by the fish increased progressively from 24.8 μg N g/day in freshwater to 131.0 μg N g/day at 10.5 g/L salinity. The carp diverted more and more food energy to maintain homeostasis and less to growth as the salinity increased. Although the study did not measure indicators of immunosuppression, it seems reasonable to assume that immune response could have been compromised somewhat in proportion to growth rate as it also requires energy.

TABLE 6.1 Effects of Salinity on Food Energy Use for Growth in Common Carp (*Cyprinus carpio*)

Salinity (g/L)	Total Food Energy Recovered as Growth (%)
0.5	33.4
2.5	31.8
4.5	22.2
6.5	20.1
8.5	10.4
10.5	−1.0

Modified from Wang, J.Q., Lui, H., Po, H., Fan, L., 1997. Influence of salinity on feed consumption, growth, and energy conversion efficiency of common carp (*Cyprinus carpio*) fingerlings. Aquaculture 148, 115–124.

6.2.2 Toxicity

It is important to know the relative toxicities of various water-quality variables in order to assess the concentrations of toxins measured in culture systems. Toxicity of substances to aquatic animals has traditionally been expressed as the lethal concentration to 50% of the test animals (LC50) exposed under well-defined conditions for a specific time period as illustrated in Fig. 6.1. The 96-h LC50 of 8.3 mg/L shown in Fig. 6.1 refers to a concentration of the toxicant of 8.3 mg/L killing 50% of the test animals in 96 h. As a rule, the LC50 value decreases (greater toxicity) in response to higher temperature and longer exposure time. Water quality, especially pH, alkalinity, and hardness, affect the toxicity of most variables. Increased pH lessens the amount of total ammonia nitrogen necessary for toxicity, while increased pH, alkalinity, and hardness also increase the concentration of metals necessary to cause toxicity (Boyd et al., 2016).

In aquaculture, the LC50 is not of direct interest because culture animals should not even be stressed by a potential toxicant much less succumb to it. Moreover, the exposure time of animals in aquaculture may be much greater than 96 h, and water may differ from the conditions of the LC50 test. The acquisition of data on the safe concentration of toxicants for continuous exposure require data from long-term toxicity tests, ideally for the entire life cycle of a species. Few such tests were conducted, and when they were, the safe concentration for continuous exposure was estimated by taking the geometric mean of the lowest concentration that had no observed effect (NOEC) and the lowest concentration of observed effect (LOEC) in behavior, physiological function, or other endpoint. According to Boyd (2015), the safe concentration is often estimated as the geometric mean of the NOEC and LOEC:

$$\text{Safe concentration} = \sqrt{\text{NOEC} + \text{LOEC}}$$

In many instances, the safe concentration was around 0.05 to 0.01 times the 96 h LC50. Thus, when the NOEC and LOEC are not known, it is a common practice to divide the 96 h LC50 by a factor ranging from 20 to 100. Usually, the lower the 96-h LC50, the greater the factor applied. Nevertheless, it is extremely difficult to establish a safe concentration of a toxicant, because conditions in culture systems differ from those in toxicity tests. This is especially true for temperature and pH that tend to be constant in toxicity tests but may vary hourly in culture systems.

6.2.3 Optimum Ranges

Aquaculture species, like all other plants and animals, have an optimum range for environmental variables as expressed by Liebig's Law of the Minimum illustrated in Fig. 6.2 and Shelford's Law of Tolerance illustrated in Fig. 6.3. According to Liebig's Law, growth will be limited by the growth factor in shortest supply. Shelford's Law shows that a species has an optimum range for each environmental factor in which it grows at the best rate, while outside this range, growth decreases with progressively lower or higher levels of the variable until a lethal level is reached. Although not implicitly mentioned in Shelford's Law, one would presume that the animal or plant is increasingly stressed as the level of a factor deviates more from the optimum range.

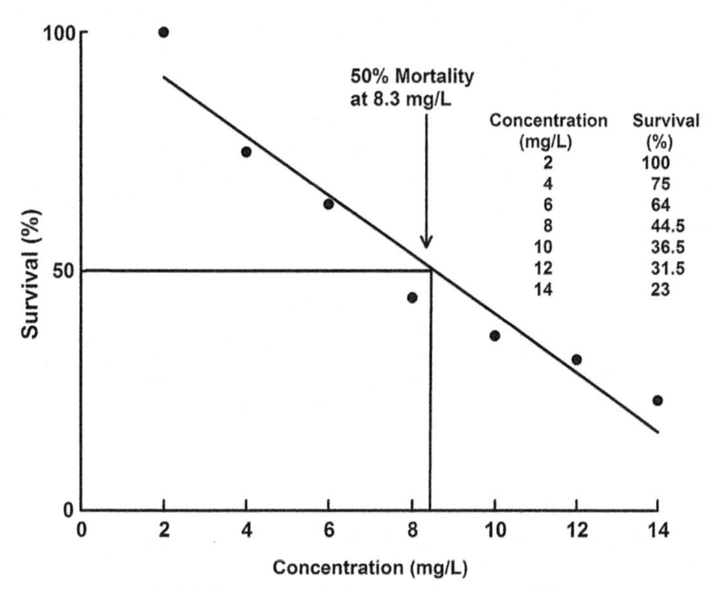

FIGURE 6.1 Graphical estimation of the LC50.

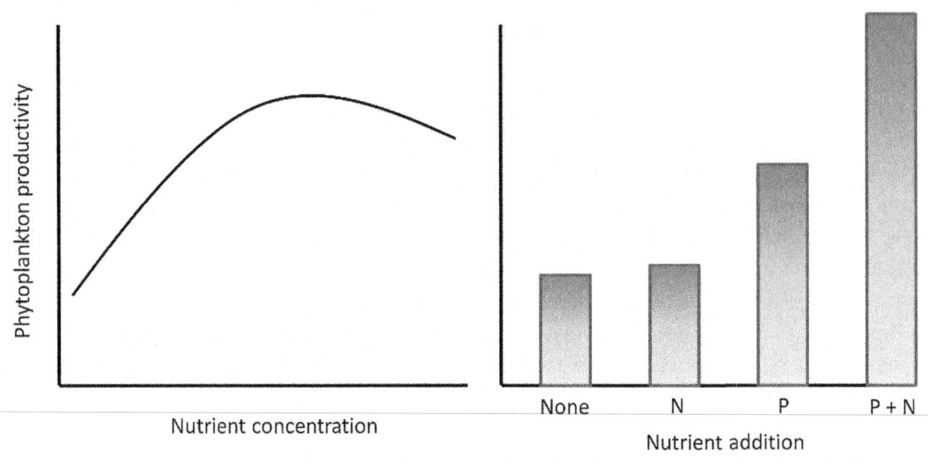

FIGURE 6.2 (Left) Phytoplankton productivity in response to additions of a single limiting nutrient. (Right) Phytoplankton productivity in response to additions of nitrogen alone, phosphorus alone, and phosphorus plus nitrogen fertilization. Productivity in this hypothetical system was limited first by phosphorus availability. Adding nitrogen alone did not increase productivity until the requirement for the first growth-limiting nutrient—phosphorus—was met.

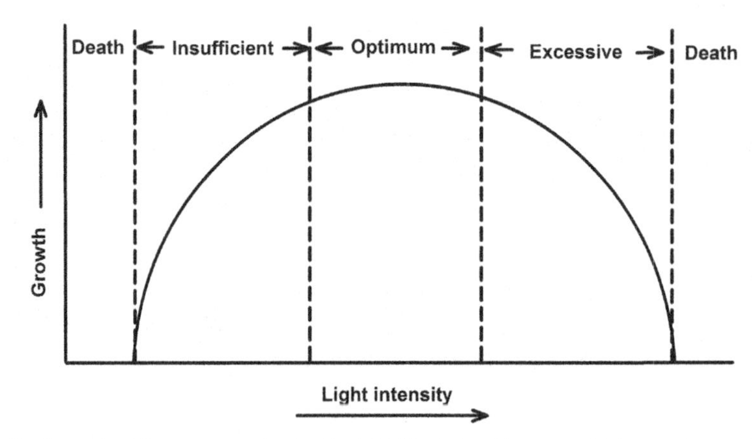

FIGURE 6.3 Example of the response of a plant to light, illustrating the range of tolerance to a particular factor.

In aquaculture it is important to maintain all environmental factors within optimal ranges. Not only does this avoid stress by an individual factor, but a combination of suboptimal concentrations of two or more factors is often more stressful than any one of the individual factors alone. For example, shrimp were more adversely affected by low temperature and low salinity than by either factors acting alone (Lester and Pante, 1992). Walters and Plumb (1980) injected channel catfish intraperitoneally with sublethal doses of *Aeromonas hydrophilia* and subjected them to suboptimal concentrations of dissolved oxygen, dissolved carbon dioxide, and ammonia. There was a decrease in survival compared to the injected control held in water with optimal concentrations of the variables. It was also found that a combination of suboptimal concentrations of the combination of the three variables result in lower survival than do suboptimal concentrations of a single variable while other variables are optimal.

6.2.4 Avoiding Diseases

All instances of disease in aquaculture species cannot be avoided by preventing stress as some might infer from the discussion above. Improving water quality will lessen stress and allow animals to fend off diseases better. But, even the healthiest animals may contact a contagious disease present in the culture system. An analogy is flu in humans. When a new strain of flu emerges, it may infect people who have a well-functioning immune system. Of course, some people may avoid the flu by chance or through natural immunity to the new flu strain. For example, when I was a first-year college student in 1957, a new strain of Asian flu struck the United States. There were around 40 of us, all seeming healthy boys, living in a dormitory. All but one in this group contacted the flu within a period of 1 week—that fellow never contacted the flu, although he did nothing different from the rest of us to avoid the sickness. There was, however, a considerable degree of difference in the severity of the flu among us which I could observe as we were quarantined. This difference was likely related to the robustness of our individual immune systems and other physiological processes.

Most common diseases are present at the majority of aquaculture facilities, and maintaining good water quality is likely the best way of avoiding outbreaks of these diseases. However, biosecurity measures to prevent diseases, especially new ones, from entering facilities via brood stock, water, or other avenues is critical.

6.3 MANAGEMENT AND WATER QUALITY

Aquaculturists have no control over source water quality aside from site selection. However, once aquaculture is started in a pond or other culture system, the main factor that affects water quality is management. Some management interventions are solely for the purpose of improving water quality, while others, especially feeding, cause water-quality deterioration. However, aquaculture is usually not economically feasible without inputs of fertilizer and feed.

In pond culture, one of the most widely used interventions is liming, because freshwater ponds are often located in high rainfall areas where soils are highly leached and acidic. Liming materials increase the background pH by neutralizing soil acidity and increase the alkalinity. Greater alkalinity buffers water against wide daily fluctuations in pH (Fig. 6.4) resulting from a net removal of carbon dioxide by aquatic plants in the daytime for use in photosynthesis and a return of carbon dioxide to the water by respiration at night when there is no photosynthesis. Greater alkalinity also provides more inorganic carbon for photosynthesis. Liming is practiced in both fertilized ponds and ponds receiving feed where alkalinity is less than 50–60mg/L.

Fertilizers—either manure or other agricultural wastes, commercial nitrogen and phosphorus fertilizers, or both—are used to increase phytoplankton productivity that serves as the base of the food web culminating in aquaculture biomass. Photosynthesis and respiration cause higher dissolved oxygen concentrations during the day, but lower concentrations at night in proportion with phytoplankton abundance (Fig. 6.5).

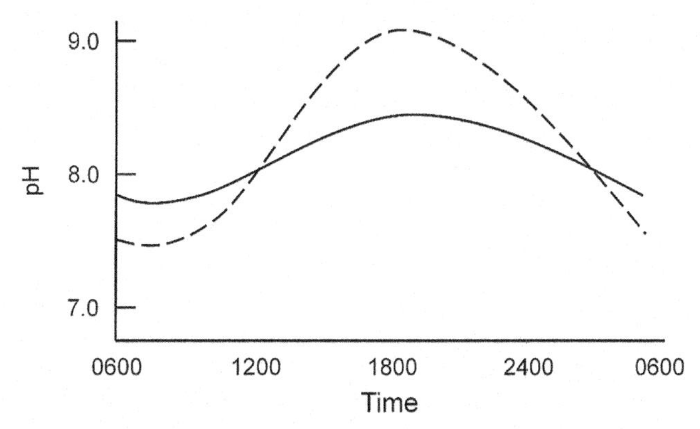

FIGURE 6.4 Daily pH cycle in ponds with sparse (*solid line*) and abundant (*dashed line*) phytoplankton.

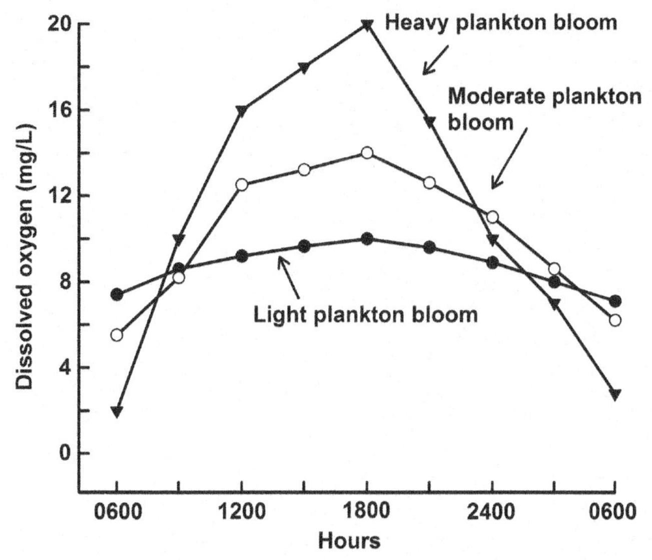

FIGURE 6.5 Changes in dissolved oxygen concentration in the surface water during a 24-h period in ponds with different densities of phytoplankton.

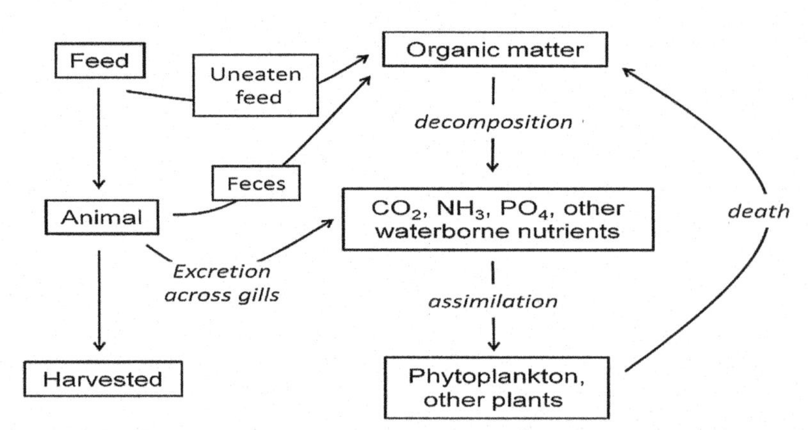

FIGURE 6.6 Fate of feed offered to fish or crustaceans in aquaculture ponds. Oxygen-consuming processes include animal respiration; oxidation of organic matter derived from uneaten feed, animal feces, and dead plankton; and oxidation of ammonia to nitrate by nitrifying bacteria.

In order to increase production above that possible in fertilized ponds, manufactured feeds are increasingly applied. The fate of feed in a pond is illustrated in Fig. 6.6. The culture animals and the microorganisms decomposing uneaten feed and feces use oxygen in respiration and excrete carbon dioxide, ammonia nitrogen, phosphate, and other inorganic substances into the water. These nutrients stimulate phytoplankton growth, and ponds with feeding tend to have dense phytoplankton blooms. Thus, both photosynthesis and respiration occur at rapid rates in ponds with feeding.

The oxygen demand of feed is typically around 1.1–1.2 kg O_2 kg/feed (Boyd and Tucker, 2014), while net oxygen production by photosynthesis is about 2–5 g $O_2 \cdot m^2$/day or 20–40 g O_2/ha/day in a pond 1 m deep (Boyd, 1973). Thus, it should not be surprising that feeding rates above 30 kg/ha/day often lead to dangerously low nighttime dissolved oxygen concentrations in spite of a 10–20 kg ha/day gain in dissolved oxygen by diffusion from the air in the pond at night when water is undersaturated with dissolved oxygen.

Mechanical aeration can be used to allow much higher feeding rates—daily feeding rates often exceed 100 kg/day in fertilized ponds, but phytoplankton abundance tends to increase with greater feed input, and aeration must be carefully sized to avoid low dissolved oxygen. An aeration rate of 1 hp/ha for each 10 kg feed/ha/day has been recommended (Boyd and Tucker, 2014).

Ammonia is the major nitrogenous excretory product of aquatic animals. Thus, in feed-based aquaculture, high-standing stocks of aquatic animals release large amounts of ammonia nitrogen into the culture system. Unionized ammonia diffuses into the air, phytoplankton take up ammonia nitrogen for use in growth, and biological nitrification oxidizes ammonia nitrogen to nitrate (Gross et al., 2000). Nevertheless, ammonia nitrogen concentration often is elevated in ponds resulting in the possibility of toxicity, especially when pH is high.

Nitrite can also accumulate in ponds to toxic concentrations. In aquaculture ponds, elevated nitrite concentration is derived from the nitrification process when ammonia is oxidized to nitrite faster than nitrite is oxidized to nitrate. This phenomenon usually occurs when total ammonia nitrogen concentrations suddenly increase and stimulate ammonia-oxidizing bacteria. Nitrite may accumulate at a greater rate than nitrite-oxidizing bacteria can convert it to nitrate. In the temperate zone, in the fall, day length shortens and temperature decreases resulting in less ammonia nitrogen uptake by phytoplankton. This event often coincides with high nitrite concentration in ponds (Boyd and Tucker, 2014).

Sulfide can be produced in anaerobic sediment by sulfate-reducing bacteria and enter the water column. Hydrogen sulfide is extremely toxic to aquatic animals.

Phytoplankton blooms themselves may be a health hazard to the culture animals. Several kinds of algae can produce and excrete compounds toxic to fish and other aquatic animals. These include certain species of cyanobacteria (often called blue-green algae by aquaculturists), euglenoids, golden-brown algae, raphidophytes, dinoflagellates, and diatoms (Boyd and Tucker, 2014).

Cage culture is feed-based, and all of the components of feed not becoming fish biomass enter the water surrounding cages. This can cause water quality in the water body containing the cages to diminish and negatively impact fish in cages.

Water passing through raceways and other flow-through systems deteriorates in quality as fish remove oxygen and excrete carbon dioxide and ammonia (Soderberg, 1995). If not removed, uneaten feed and feces that accumulate in flow-through systems decompose to degrade water quality. However, water passes through these systems too rapidly for phytoplankton blooms to occur.

In water-recirculating systems, water is used over and over, and treatment equipment is installed to maintain adequate water quality. However, if this equipment fails or is inadequately sized, water quality may deteriorate and cause stress or mortality of culture animals.

In molluskan shellfish culture, fertilizers and feeds are not used, and the culture organisms filter their food from the water. However, mollusks are also sensitive to water quality, and they can be negatively impacted at sites with degraded water quality. Also, in polluted water mollusks may filter harmful bacteria and virus particles from the water or accumulate algal toxins. Such mollusks, when consumed by humans, represent a health hazard by causing serious illnesses and even death.

6.4 WATER QUALITY–RELATED STRESSORS

The most common water-quality variables causing stress in aquaculture animals are: temperature, salinity, and pH that may be too low or too high; low dissolved oxygen concentration; high concentrations of carbon dioxide, ammonia nitrogen, nitrite, and hydrogen sulfide; gas supersaturation (which has not been discussed); and algal toxins. Heavy metals, pesticides, and industrial chemicals are common toxins and stressors affecting fish and other aquatic life in streams, lakes, reservoirs, and coastal waters. However, aquaculture facilities are usually located in areas where such pollution can be largely avoided. The main exception would be cage culture in some lakes and molluskan shellfish culture in polluted waters.

6.4.1 Temperature and Salinity

The effects of temperature and salinity on aquaculture animals have been a subject of much interest. Fish are usually classified as warmwater with optimum growth above 20°C or coldwater with optimum growth below 20°C. Some species that grow best at temperatures slightly higher than optimum for coldwater species, but below the optimum for warmwater species, are called coolwater species. Of course, warmwater species can be separated into tropical and nontropical species—the tropical species will die if temperature falls much below the optimum, but nontropical species can survive during the winter in the temperate zone.

It is equally difficult to classify fish and other aquatic species according to salinity. There are freshwater and marine species, species adapted to estuarine life where salinity varies greatly, and anadromous species that spend part of their lives in freshwater and part of it in saltwater.

Most aquaculture species have a fairly wide range in both temperature and salinity tolerance, but changes in either variables outside the optimal range leads to stress, and, if extreme enough, death. The temperature and salinity ranges of most aquaculture species have been investigated and can be found in the literature. The topic will not be discussed further here. But, in spite of our knowledge of temperature and salinity preferences, some aquaculturists insist on attempting to culture certain species in areas where temperature and salinity are outside acceptable ranges.

6.4.2 pH

The pH can be either too low or too high for an aquaculture species (Table 6.2). Gills are the primary target of elevated hydrogen ion concentration (low pH). The major effects of low pH are (1) changes in gill structure and function leading to a decreased ability to maintain internal ion balance, and (2) inhibition of gas exchange as a result of excessive mucous formation, changes in the structure of the gill epithelium, and blood acidosis (McDonald, 1983). Under mild acid stress, animals expend extra metabolic energy for maintenance of gill function at the expense of growth, immune function, and other processes. If stress becomes more severe, the animal will die. High environmental pH also impairs ion-exchange processes by the gill leading to a decrease in osmoregulation capacity, an increase in blood alkalosis, and a decrease in the gradient for ammonia excretion across the gill into the surrounding water (Wilkie and Wood, 1994).

TABLE 6.2 Effects of pH on Fish and Other Aquatic Life

pH	Effects
4	Acid death point
4–5	No reproduction[a]
4–6.5	Slow growth of many species[a]
6.5–9	Optimum range
9–11	Slow growth and adverse reproductive effects
11	Alkaline death point

[a] *Some fish in rivers flowing from jungles do very well at low pH.*

Tolerance to pH extremes is influenced by environmental factors. Fish and crustaceans tolerate low pH better when salinity is near optimum. High calcium concentration affects the permeability and stability of biological membranes thereby making the gill epithelium less "leaky" to electrolytes in the blood when pH is high (Boyd and Tucker, 2014).

6.4.3 Dissolved Oxygen

The responses of warmwater (including tropical) species to dissolved oxygen concentration are reported by many authors, and Table 6.3 provides a summary assessment. Of course, a few species can tolerate lower dissolved oxygen concentrations, e.g., tilapia and *Clarius* catfish. Coldwater species have a higher dissolved oxygen requirement than warmwater species; the lethal concentration is greater for coldwater species than for warmwater ones.

Most aquatic animals are healthiest and grow fastest when dissolved oxygen concentrations are near air-saturation (Collins, 1984). The hemoglobin (cyanoglobin in crustaceans) becomes completely saturated when dissolved oxygen concentration reaches near saturation, and higher dissolved oxygen concentration provides no benefit. As dissolved oxygen concentrations fall below 60–70% of saturation (4.7–5.5 mg/L at 28°C and 6.0–7.0 mg/L at 15°C in freshwater), animals compensate by changing behavior and physiology. The adaptions and responses allow healthy animals to survive for long periods even when dissolved oxygen levels are as low as 20–30% of saturation, but they eat less feed, grow more slowly, and may become more susceptible to infectious diseases (Torrans, 2008; Andrews et al., 1973).

6.4.4 Carbon Dioxide

An increase in carbon dioxide in the water reduces the efficiency of carbon dioxide excretion from the blood across the gill to the environment. Thus, as carbon dioxide increases in the water, the greater accumulation of carbon dioxide in blood (hypercapnia) results in a decrease in hemolymph pH (acidosis). These conditions decrease the affinity of respiratory pigments for oxygen, and reduce oxygen uptake at the gills. A moderate environmental carbon dioxide concentration increases the dissolved oxygen tension necessary to completely saturate hemoglobin with oxygen, while a higher environmental concentration may prevent complete oxygenation of hemoglobin even at oxygen saturation in the environment. This phenomenon is known as the Bohr effect that results in lower respiratory efficiency and decreases the tolerance of aquatic animals to low dissolved oxygen concentration. It is important to note that high dissolved carbon dioxide concentration usually occurs when dissolved oxygen concentration is low. This results in an elevation of the minimum dissolved oxygen concentration necessary to lessen the possibility of stress or lethality.

Tolerance of fish and crustaceans to aqueous carbon dioxide concentrations vary. Marine fish are particularly intolerant to carbon dioxide. Growth of Atlantic cod was reduced at 8 mg/L of carbon dioxide (Moran and Stöttrup, 2011). Colt and Orwicz (1991) suggested 20 mg/L dissolved carbon dioxide as the upper limit for salmonids, but Good et al. (2010) reported no ill effects of up to 24 mg/L aqueous carbon dioxide in trout culture. Warmwater fish evolved in environments where fluctuations in aqueous carbon dioxide were common and wide, and they can tolerate higher concentrations—60–80 mg/L has a narcotic effect, while higher concentrations can be lethal (Boyd and Tucker, 2014).

Prolonged exposure to elevated aqueous carbon dioxide can cause calcareous deposits within fish kidneys or nephrocalcinosis (Landolt, 1975). While this condition does not usually cause death, it impairs growth. High carbon dioxide concentration is also noted to cause cataracts in fish (Moran et al., 2012).

Effects of carbon dioxide, similar to those of other stressors, have typically been studied at constant concentration. In culture systems and especially in ponds, carbon dioxide concentration varies daily, often ranging from 0 mg/L in the

TABLE 6.3 Effects of Dissolved Oxygen Concentration on Warmwater Fish

Dissolved Oxygen (mg/L)	Effects
0–0.3	Small fish survive short exposure.
0.3–1.5	Lethal if exposure is prolonged for several hours.
1.5–5.0	Fish survive, but growth will be slow and fish will be More susceptible to disease.
5.0–saturation	Desirable range
Above saturation	Possible gas bubble trauma if exposure prolonged.

afternoon to 5–20 mg/L or occasionally even more in the early morning. Thus, it is difficult to predict the effect of carbon dioxide exposure.

6.4.5 Ammonia

Ammonia nitrogen exists in water in a pH temperature–dependent equilibrium:

$$NH_3 + H^+ = NH_4^+$$

The toxicity of ammonia nitrogen is mostly from the unionized form (NH_3), and the concentration of this form increases in proportion to the total ammonia nitrogen ($NH_3 + NH_4^+$) concentration as both temperature and pH rise, especially as pH rises (Table 6.4).

The main nitrogenous waste produced by aquatic animals is ammonia. This waste is transported in blood mainly as ammonium ion but excreted as unionized ammonia. Excretion rate is facilitated by a gradient of high unionized ammonia concentration in the blood to a low concentration in water outside the gills (Wilkie, 1997). This gradient is enhanced by acidification of water at the gill surface by carbon dioxide, because it quickly converts ammonia to ammonium. However, if the external unionized ammonia concentration increases, it inhibits the movement of unionized ammonia from blood across the gill to the water. This allows ammonia to accumulate in the blood.

Acute ammonia toxicity affects the central nervous system leading to effects ranging from hyperactivity and convulsions to lethargy. Loss of equilibrium will occur, and ultimately coma and death. Long-term exposure to elevated but nonlethal concentrations of ammonia causes osmoregulatory disturbances, blood acidosis, and diminished respiratory efficiency—the results of which are reduced growth and greater susceptibility to infectious diseases.

At pH above 7.5, ammonia toxicity is primarily related to unionized ammonia. However, ammonium ion also contributes to toxicity at lower pHs.

The 96-h LC50 of unionized ammonia to aquaculture species range from about 0.3 mg/L to 2 mg/L, with coldwater species being the most sensitive (Boyd and Tucker, 1998). Studies in the laboratory revealed, however, that any concentration of unionized ammonia above that of the water supply (0.07 mg/L) diminished growth of channel catfish, and no growth occurred above 1.07 mg/L (Colt and Tchobanoglous, 1978). When unionized ammonia concentrations in laboratory toxicity tests were caused to fluctuate daily as they do in ponds, channel catfish growth was not affected at unionized concentrations below mean daily values of 1.81 mg/L (Hargreaves and Kucuk, 2001).

The unionized ammonia concentration must be calculated from total ammonia nitrogen concentration, pH, and temperature. It is more convenient in practical application to assess the likelihood of ammonia toxicity from total ammonia nitrogen concentration. The US Environmental Protection Agency developed general equations for estimating maximum safe concentrations of total ammonia nitrogen at different pH and temperature values for 1-h exposures for non-salmonids and salmonids, and for 3-day exposures to ammonia in freshwater (with or without early life stages present). The equations

TABLE 6.4 Decimal Fractions of Total Ammonia-Nitrogen Existing as Un-ionized Ammonia-Nitrogen at Various pH Values and Water Temperatures

	Temperature (°C)						
pH	5	10	15	20	25	30	35
7.0	0.001	0.002	0.003	0.004	0.005	0.007	0.010
7.5	0.004	0.005	0.008	0.011	0.016	0.023	0.032
8.0	0.011	0.017	0.024	0.035	0.050	0.069	0.093
8.5	0.035	0.051	0.073	0.103	0.141	0.189	0.246
9.0	0.103	0.146	0.200	0.267	0.342	0.424	0.507
9.5	0.266	0.350	0.442	0.535	0.622	0.699	0.765
10.0	0.534	0.630	0.715	0.784	0.839	0.880	0.911

Values were calculated using the "ammonia Calculator" at http://fisheries.org/hatchery for a typical, dilute fresh water with a total dissolved Solids concentration of 250 Mg/L.
Modified from Boyd, C.E., Tucker, C.S., 2014. Handbook for Aquaculture Water Quality. Craftmaster Printer, Inc., Auburn, Alabama.

were used by Boyd and Tucker (2014) to calculate safe total ammonia nitrogen concentrations (Tables 6.5–6.7). Marine species usually are slightly less susceptible to ammonia than are freshwater ones.

6.4.6 Hydrogen Sulfide

Hydrogen sulfide inhibits oxidative phosphorylation in aquatic animals by blocking the reoxidation of reduced cytochrome a_3 by molecular oxygen. The overall effect is inhibition of energy metabolism by cells, similar to the effect of hypoxia. Fish exposed to acutely lethal concentrations of sulfide have increased ventilation rates, but ventilation soon stops and fish die. Work by Dombkowski et al. (2004) suggests that hydrogen sulfide is an endogenous vasoactive "gasotransmitter" in trout and that the cardiovascular system might also be a focal point of toxicity.

TABLE 6.5 Criterion Maximum Concentration (CMC, mg N/L) for Non-salmonid and Salmonid Fish

	CMC (mg N/L)	
pH	Non-salmonids	Salmonids
6.6	46.8	31.3
6.8	28.1	42.0
7.0	24.1	36.1
7.2	19.7	29.5
7.4	15.4	23.0
7.6	11.4	17.0
7.8	8.1	12.1
8.0	5.6	8.4
8.2	3.8	5.7
8.4	2.6	3.9
8.6	1.8	2.7
8.8	1.2	1.8

The CMC is the maximum allowable total ammonia-nitrogen concentration for 1-h exposure at different pH values.
Modified from United States Environmental Protection Agency (USEPA), 1999. 1999 Update of Ambient Water Quality Criteria for Ammonia. EPA-822-R-99-014. USEPA, Washington, DC.

TABLE 6.6 Criterion Chronic Concentration (CCC, mg N/L) for Early Life Stages of Fish

	CCC (mg N/L) for Fish When Early Life Stages Are Present					
	Temperature (°C)					
pH	0	14	18	22	26	30
6.5	6.7	6.7	5.3	4.1	3.2	2.5
7.0	5.9	5.9	4.7	3.7	2.8	2.2
7.5	4.4	4.4	3.5	2.7	2.1	1.6
8.0	2.4	2.4	1.9	1.5	1.2	0.9
8.5	1.1	1.1	0.9	0.7	0.5	0.4
9.0	0.5	0.5	0.4	0.3	0.2	0.2

The CCC is the maximum allowable total ammonia-nitrogen concentration for a 30-day exposure at different pH and temperature values. At water temperatures of 15°C and above, CCC values when fish early life stages are absent are the same for when fish early life stages are present.
Modified from United States Environmental Protection Agency (USEPA), 2009. 2009 Update Aquatic Life Ambient Water Quality Criteria for Ammonia – Freshwater. EPA-822-D-09-001. USEPA, Washington, DC.

TABLE 6.7 Criterion Chronic Concentration (CCC, mg N/L) for Fish Production When Early Life Stages Are Absent

	CCC (mg N/L) for Fish When Early Life Stages Are Absent					
	Temperature (°C)					
pH	0–7	8	10	12	14	16
6.5	10.7	10.1	8.8	7.7	6.8	5.9
7.0	9.6	9.0	7.9	7.0	6.1	5.4
7.5	7.1	6.6	5.8	5.1	4.5	4.0
8.0	4.0	3.7	3.5	2.9	2.5	2.2
8.5	1.8	1.7	1.5	1.3	1.1	1.0
9.0	0.8	0.7	0.7	0.6	0.5	0.4

The CCC is the maximum allowable total ammonia-nitrogen concentration for a 30-day exposure at different pH and temperature values. At water temperatures ≥15°C, CCC values when fish early life stages are absent are the same as those for fish early life stages (Table 6.6).
Modified from United States Environmental Protection Agency (USEPA), 2009. 2009 Update Aquatic Life Ambient Water Quality Criteria for Ammonia – Freshwater. EPA-822-D-09-001. USEPA, Washington, DC.

Hydrogen sulfide dissociates in two steps:

$$H_2S = HS^- + H^+$$

$$HS^- = S^- + H^+$$

It is usually assumed that hydrogen sulfide is the primary toxic form because the unionized molecule readily crosses the cell membranes while ionized sulfide is largely excluded by membranes. The proportion of the unionized form relative to the ionized form increases with decreasing pH (Table 6.8).

Hydrogen sulfide is extremely toxic to freshwater species with 96-h LC50s of 4.2–34.8 µg/L (Adelman and Smith, 1970; Gray et al., 2002). The USEPA (1977) considered 2 µg/L to be the maximum safe exposure. Marine species are usually more tolerant than freshwater species. The 96-h LC50 values are usually between 50 and 200 mg/L (Bagarinao and Lantin-Olaguer, 1999; Gopakumar and Kuttyamma, 1996; Chien, 1992). Exposure of kuruma shrimp to 88 µg/L hydrogen sulfide for 144 h did not cause mortality. However, hydrogen sulfide weakened the animals' immune system because 80% of animals injected with *Vibrio alginalyticus* and exposed to the same hydrogen sulfide concentration died within 144 h (Cheng et al., 2007). Similar results were obtained with whiteleg shrimp (Hsu and Chen, 2007).

6.4.7 Gas Supersaturation

Water containing more dissolved gas than it can accommodate at equilibrium with air is said to be supersaturated with gas. Supersaturation is an unstable state; the supersaturating gas will transfer by diffusion or via bubbles to the air. Aquatic animals living in gas-supersaturated water may develop a stressful or lethal condition known as gas bubble disease when bubbles come out of solution in body fluids inside tissues. The degree of gas supersaturation often is expressed as the delta pressure (ΔP) that is calculated by subtracting barometric pressure from the total gas pressure. The ΔP represents the degree of gas supersaturation (in mmHg) at the surface, but at a greater depth, the value of ΔP is less for the same total gas pressure by an amount equal to the hydrostatic pressure. Corrected for hydrostatic pressure at any depth below the surface ΔP is known as the uncompensated ΔP and calculated as follows:

$$\Delta P_{uncompensated} = \Delta P_{surface} - \rho g z$$

where ρ is the specific weight of mercury; g, gravitational constant (9.81 m/s^2); and z, water depth (m). On an average, the term ρg has a value of about 73 mm Hg/m.

Acute gas bubble disease typically occurs at uncompensated ΔP values of over 100 mmHg and is often fatal. Chronic gas bubble disease occurs at uncompensated ΔP values of 25–100 mmHg. Symptoms of chronic gas bubble disease include bubbles in the gut and mouth, hyperinflation of the swim bladder, and a low-level mortality usually as a result of secondary, stress-related infections (Colt, 1986; Shrimpton et al., 1990).

TABLE 6.8 Decimal Fractions (Proportions) of Total Sulfide-Sulfur Existing as Un-ionized Hydrogen Sulfide in Freshwater at Different pH Values and Temperatures

	Temperature (°C)								
pH	16	18	20	22	24	26	28	30	32
5.0	0.993	0.992	0.992	0.991	0.991	0.990	0.989	0.989	0.989
5.5	0.977	0.976	0.974	0.973	0.971	0.969	0.967	0.965	0.963
6.0	0.932	0.928	0.923	0.920	0.914	0.908	0.903	0.897	0.891
6.5	0.812	0.802	0.792	0.781	0.770	0.758	0.746	0.734	0.721
7.0	0.577	0.562	0.546	0.530	0.514	0.497	0.482	0.466	0.450
7.5	0.301	0.289	0.275	0.263	0.250	0.238	0.227	0.216	0.206
8.0	0.120	0.114	0.107	0.101	0.096	0.090	0.085	0.080	0.076
8.5	0.041	0.039	0.037	0.034	0.032	0.030	0.029	0.027	0.025
9.0	0.013	0.013	0.012	0.011	0.010	0.010	0.009	0.009	0.008

Multiply values by 0.9 for seawater.
Modified from Boyd, C.E., Tucker, C.S., 2014. Handbook for Aquaculture Water Quality. Craftmaster Printer, Inc., Auburn, Alabama.

The most common causes of gas supersaturation in aquaculture are photosynthesis causing supersaturation in dissolved oxygen concentration, a decline in gas solubility when water warms rapidly, air leakage on suction sides of pumps, and bubbles released by submerged aerators. Also, animals at facilities that use water that has fallen over high dams may be at high risk of gas bubble disease. Water entrains air while falling over spillways and becomes supersaturated with air in the plunge pool area below the spillway. Gas supersaturation is rare in ponds and cage culture, but more common in hatcheries, flow-through systems, and water-recirculating systems. In ponds, a high ΔP is normally from a high dissolved oxygen concentration. Aquatic animals can move to deeper water where lower dissolved oxygen concentration and hydrostatic pressure both act to reduce the ΔP.

6.4.8 Toxic Algae

Aquaculture ponds often have an abundance of cyanobacteria (Tucker, 1996, 2000). Toxic species produce hepatotoxins, neurotoxins, or both. The most common hepatotoxins are microsystins produced by species of *Microcystis*, *Anabaena*, and *Planktothrix*. Microcystins cause liver damage and gill damage in fish. Fish injected with microcystin become lethargic, and ventilate rapidly and irregularly—death occurs within 1–2 days (Madsen et al., 2012). Neurotoxins include anatoxins and saxitoxins that are produced mainly by species of *Anabaena* and *Aphanizomenon*, respectively. Anatoxins inhibit acetylcholine resulting in overstimulation of muscle cells. Symptoms include convulsions, tetany, and disoriented swimming with death by respiratory failure.

Some strains of *Euglena sanguinea* produce the neurotoxin euglenophycin that has killed fish in freshwater ponds (Zimba et al., 2010). The symptoms of poisoning by euglenophycin are loss of appetite and disoriented swimming, not very specific symptoms, but death can follow.

Haptophyceae are mostly marine, flagellated organisms commonly known as golden-brown algae. Species of *Prymnesium* and *Chrysochromulina* are well known for causing fish kills in estuaries and along the coast, and *Prymnesium* has been found in saline, inland waters. The species *P. parvum* produces toxins called prymnesins that act to disrupt gill-membrane permeability. Once inside the fish or shrimp, this toxin lyses red blood cells and produces neurotoxic effects. Gills bleed easily and hemorrhagic areas develop around the mouth, eyes, and bases of fins (Barkoh et al., 2003). Fish may quickly recover if removed from the presence of the toxin, but they often die upon prolonged exposure.

Raphidophytes are flagellated algae that have caused fish kill in net pens (Smayda, 2006). Toxic forms of this algal group produce both neurotoxins and hemolytic toxins. Dinoflagellates can be toxic to fish, shrimp, and shellfish; these organisms are responsible for red tides. A species of the marine diatom *Chaetoceros* has spines that break off, penetrate, and damage the epithelium of fish gills (Bruno et al., 1989).

Of course, toxic algae in coastal waters can lead to contamination of molluskan shellfish with algal toxins that can cause serious illness and even death in humans (Shumway, 1990). The dinoflagellates are usually associated with shellfish poisoning in humans. However, the saxitoxins produced by some blue-green algae are the same as those produced by dinoflagellates (Malbrouck and Kestemont, 2006). Toxins produced by diatoms of the genus *Pseudo-nitzschia* can also be responsible for shellfish poisoning in humans.

6.4.9 Ionic Imbalance

There are many regions where saline water is available from aquifers, and in arid regions saline surface water is common. This water represents an unused resource, and there is growing interest in cultivating marine or estuarine species in such waters (Roy et al., 2010). Inland saline water is often quite different in the proportions of major cations as compared to the proportions of these ions in oceanic or estuarine water. Ionic imbalance, especially low concentrations of potassium and magnesium, has been associated with low survival and production of shrimp and several other species when cultured in low-salinity, inland waters (McNevin et al., 2004; Partridge et al., 2008)

6.5 WATER-QUALITY MANAGEMENT

The selection of aquaculture sites with unfavorable environmental features is the cause of many water-quality problems. One feature that should be avoided in all types of aquaculture is a polluted water source. In pond aquaculture, areas with unfavorable soil characteristics should also be avoided. Some examples are mangrove habitats, organic soils, potential acid-sulfate soils, and very sandy or rocky soils. Cage culture should be installed in areas with good water circulation, and with a low likelihood of low nighttime dissolved oxygen and of sudden thermal destratification. Thermal destratification of lakes and reservoirs with cage culture has caused many massive fish kills as a result of low dissolved oxygen concentration that often follows the event. There are many other factors to consider in site selection, but the above are the critical water quality–related ones. Of course, many times the prospective aquaculturist has a site available and decides on a species without adequate consideration of the site. The site may not be suitable for the species that has been selected, especially with respect to water temperature and salinity. Thus, the site should be considered from the perspective of the species to be cultured, or conversely, the species should be chosen based on the characteristics of the available site.

6.5.1 Alkalinity and pH

An acceptable alkalinity concentration in all types of culture is critical. Waters of acceptable alkalinity will not have low pH and there will be sufficient buffering capacity to avoid wide daily fluctuations in pH, especially high afternoon pH in ponds. Avoiding high pH is especially beneficial by limiting the proportion of potentially toxic, unionized ammonia in systems with elevated total ammonia nitrogen concentration. Greater alkalinity also increases the availability of inorganic carbon for photosynthesis. This is particularly important in fertilized ponds where production is based upon the availability of natural food organisms.

Freshwater ponds with total alkalinity below 50 mg/L usually receive some benefit from liming, but the benefits are greater at lower alkalinity. Liming materials do not dissolve well at higher alkalinities. The liming rate for ponds should be based upon a lime requirement analysis of sediment, because the acidity in sediment will neutralize alkalinity in pond water over time. A buffer method for lime requirement (Pillai and Boyd, 1985) can be used by soil-testing laboratories for assessing the liming rate for ponds. Han et al. (2014) developed a version of the buffer method for lime requirement that could be conducted at many aquaculture farms.

There are some waters in which total alkalinity concentration is high, but total hardness is low. Such waters are particularly subject to high afternoon pH, because there is inadequate dissolved calcium ion to react with carbonate that increases when phytoplankton use bicarbonate as a carbon source at pH above 8.3. In waters where alkalinity and calcium hardness have similar concentrations, there is sufficient calcium to bind the carbonate to form calcium carbonate before it hydrolyzes to release hydroxide ion that elevates the pH. Calcium carbonate does not precipitate quickly and often remains suspended in the water to re-dissolve at night when carbon dioxide is returned to the water by respiration.

The usual treatment for increasing the calcium hardness in waters of moderate to high alkalinity is to apply gypsum (calcium sulfate). Around 2 mg/L of gypsum are necessary to cause a 1 mg/L increase in calcium hardness.

In highly intensive culture with high daily feed inputs, the nitrification of ammonia nitrogen excreted by the culture species and microorganisms of decay can cause alkalinity and pH to fall very rapidly. Each milligram per liter of ammonia nitrogen oxidized can potentially neutralize 7.14 mg/L of alkalinity (Boyd and Tucker, 2014). Thus, in highly intensive systems, it is prudent to check alkalinity concentration frequently and apply liming material to maintain the target alkalinity. Because agricultural limestone and lime (burnt limestone or hydrated lime) are not quickly soluble, it is more efficient to apply sodium bicarbonate. Each milligram per liter of sodium bicarbonate is equal to 0.56 mg/L of alkalinity.

6.5.2 Dissolved Oxygen

The concentration of dissolved oxygen is of great concern in all types of aquaculture. In manured ponds, excessive inputs of manure can lead to dissolved oxygen depletion through the oxygen demand imposed by bacteria decomposing manure. As a general rule, application rates of more than 50 kg/ha/day (dry weight basis) should be avoided in unaerated ponds. However, manure varies greatly in moisture content, and most farmers have no way of determining its dry matter content. In fertilized ponds, excessive nutrient inputs can result in dense phytoplankton blooms that can cause low dissolved oxygen concentration at night. Usually, if the Secchi disk visibility is measured frequently and fertilization is adjusted to avoid Secchi disk visibilities less than 30–40 cm, nighttime dissolved oxygen concentration should be acceptable in ponds.

In ponds with feeding, stocking and feeding rates should be balanced with the availability of dissolved oxygen. The carrying capacity of most feed-based ponds is similar, and feeding rates above 30 kg/ha/day are likely to lead to low dissolved oxygen concentrations, especially in the early morning. Aeration applied at about 1 hp for each 10 kg/ha/day increment of feed input should prevent dissolved oxygen concentrations lower than 4 mg/L in most ponds. Of course, some species such as tilapia can do well even when dissolved oxygen concentration is as low as 2 mg/L for short periods at night. Less aeration can be used with such species. Water exchange is sometimes used in ponds as an alternative to mechanical aeration but it is not nearly as effective in improving dissolved oxygen concentration unless exchange rates are 25% culture volume or greater.

In flow-through systems, the dissolved oxygen concentration of incoming water provides a way of estimating safe feeding rates (Soderberg, 1995). For coldwater species, such as rainbow trout, the dissolved oxygen concentration should not fall below 5 mg/L. If the flow rate into the system and the dissolved oxygen concentration of inflow are known, the available dissolved oxygen can be estimated:

$$DO_{available} = (DO_{in} - 5)\,(Q)\,(1440\ \text{min/day})\,(0.001\ \text{kg/g})$$

where $DO_{available}$ is the available dissolved oxygen (kg/day); DO_{in}, DO concentration in inflow ($g/m^3 = mg/L$); and Q, inflow (m^3/min). The amount of feed (kg/day) that can be applied may be calculated as $5 \times DO_{available}$. Of course, if aeration is applied, the feed input can be increased. For example, if the aerator adds 1.5 kg O_2 hp/h and a 5-hp aerator is installed, the available oxygen would increase by 180 kg/day.

Cages are not usually aerated, and it is critical to place cages in an area where dissolved oxygen concentration entering the cages is high. Moreover, water flowing through the cages should be sufficient to maintain adequate dissolved oxygen concentrations within the cages as fish will remove considerable dissolved oxygen.

In water-recirculating systems, dissolved oxygen is usually supplied by aeration. The biochemical oxygen demand of feed is about 1.1–1.2 kg/feed, and the aeration rate should be sized to supply enough oxygen to overcome the oxygen demand of the feed and maintain an acceptable oxygen concentration in the culture vessels. Of course, the biological filter in water-recirculating systems also consumes oxygen for nitrification, but this oxygen demand was included in the calculation of feed oxygen demand.

Although experience and research on dissolved oxygen have allowed the estimates discussed above, events do not always follow the predicted trend, especially with respect to biological processes. Therefore, it is prudent to measure dissolved oxygen concentrations in culture systems and adjust fertilization and feeding rates, aeration rates, or both as necessary to maintain adequate dissolved oxygen concentration and minimize fish stress.

6.5.3 Nitrogenous Metabolites

There are no effective treatments for removing ammonia nitrogen from pond water, despite the wide use of microbial products and zeolite for this purpose (Boyd and Tucker, 2014; Zhou and Boyd, 2014; Li and Boyd, 2016). Ammonia nitrogen concentration cannot be controlled in cage culture, but in flow-through systems, removal of uneaten feed and feces can lessen the amount of ammonia nitrogen, and where possible, increasing flow rate can increase ammonia discharge from culture units. Of course, in water-recirculatory systems, solid waste can be removed before it decays, and biofilter capacity can be conserved. The best approach to ammonia management is to operate systems so that the input of ammonia nitrogen does not exceed a system's capacity to eliminate ammonia nitrogen or to transform it to a less harmful form of nitrogen. In ponds, ammonia nitrogen is lost by diffusion to air, uptake by phytoplankton and other plants, biological nitrification, and outflow of water from ponds. The main benefit is usually from phytoplankton uptake (Tucker et al., 1984), but diffusion and nitrification are also important. Nitrification depends upon plenty of dissolved oxygen, and attention to aeration rate has already been discussed as it relates to preventing dissolved oxygen stress to the culture species. Aeration obviously increases the contact of water with air and enhances diffusion of ammonia from the water. However, aeration often is not applied when dissolved oxygen concentration is high, the same time the pH is highest. Thus, the benefits of aeration to ammonia removal are probably not great in most ponds, and the increased removal rate of ammonia by aeration when pH is likely the highest would not be cost-effective. Water exchange at rates less than 20–30% per day are not effective in reducing ammonia concentration appreciably. Hargreaves and Tucker (2004) provide excellent guidelines for ammonia management in ponds.

In flow-through systems, the feeding rate often is adjusted based on available dissolved oxygen concentration. However, in a system where the pH of the incoming water is relatively high, say pH 8.0–8.5, the ammonia nitrogen concentration may become excessive within an acceptable feed input range with respect to oxygen. Obviously, the concentration of ammonia nitrogen should be measured at intervals in flow-through systems to assure that it is not excessive. There are basically two options if ammonia nitrogen is excessive, lessen feed input or increase flow rate. Increasing flow rate may not be an option at many facilities.

In cage culture, water flowing through the cages removes ammonia nitrogen. Thus, as already mentioned, cages should be in areas where water movement is good, and the netting material for cages should be as large as possible to facilitate flow. The netting also should be cleaned periodically to avoid clogging of the opening in the mesh by encrusting particles and organisms.

The biofilter is the main control on ammonia nitrogen in water-recirculating systems, and the capacity of the biofilter must be sufficient or ammonia nitrogen concentration will rise. The dissolved oxygen concentration is critical for the biofilter, because nitrification requires 4.57 mg/L oxygen for each 1-mg/L increment of ammonia nitrogen nitrified.

One of the most important steps that can be taken to minimize the demand for oxygen and release of carbon dioxide, ammonia nitrogen, and phosphate into culture systems is to minimize the FCR. The FCR is the ratio of feed applied to net live weight production. Typical FCR values range from around 1.2 to 2.2, but higher FCR values are not uncommon. A decrease in FCR by 0.1 unit, e.g., 1.5 to 1.4 lowers the feed input for 1 ton of production by 100 kg and reduces oxygen demand by around 110–120 kg, and nitrogen and phosphorus input by the amounts in 100 kg feed, around 4–6 kg and 0.75–1.5 kg, respectively. Carbon dioxide release would be lessened by about 120–140 kg per 100 kg feed.

Practices for lessening FCR are: use high quality feed; apply feed conservatively to avoid uneaten feed; and maintain good water quality in ponds to avoid stress and possible loss of appetite by the animals. It is surprising that aquaculturists do not place more emphasis on FCR, because the 0.1unit reduction mentioned above lessens the feed cost by an amount equal to the price of 100 kg feed.

6.5.4 Hydrogen Sulfide

Hydrogen sulfide toxicity would be an extremely rare occurrence in cage culture, flow-through systems, and water-recirculating systems. However, there have been documented instances of hydrogen sulfide toxicity in ponds. These situations usually resulted from diffusion of hydrogen sulfide from anaerobic zones in sediment into the water column at a greater rate than the rate of oxidation of hydrogen sulfide to sulfate by chemical oxidation or oxidation by sulfate-oxidizing bacteria. This scenario results in a residual concentration of hydrogen sulfide in the water column until the rate of hydrogen sulfide diffusion decreases, the rate of its oxidation increases, or both. Toxicity is most likely when this phenomenon occurs in ponds with water of $pH<7$, because the proportion of sulfide in the form of toxic hydrogen sulfide is greatest at lower pH. Maintaining plenty of dissolved oxygen in the water column will maximize the rate of hydrogen sulfide oxidation.

When hydrogen sulfide toxicity is suspected, measurement of its concentration can be attempted. This is a difficult undertaking for several reasons: methodology is rather complex and tedious (except with kits, the accuracy of which is questionable); hydrogen sulfide concentrations usually are quite low; and contamination of samples with oxygen between collection and analysis may result in oxidation and greatly lessen the measured concentration of hydrogen sulfide. Often, it is unnecessary to make analyses to reveal the presence of hydrogen sulfide, because the extremely strong, rotten-egg odor of this substance is detectable even at very low concentration.

Sediment analyses of hydrogen sulfide are not recommended, because many anaerobic sediments contain hydrogen sulfide but seldom contaminate the pond water. There is a common belief that black sediment contains hydrogen sulfide. This observation is not always correct, because the black color is from ferrous iron. The redox potential must decline considerably below that necessary for the appearance of ferrous iron for occurrence of hydrogen sulfide. Nevertheless, a sediment that contains hydrogen sulfide will be black or at least darker than its color when it is oxidized.

6.5.5 Toxic Algae

Toxic algal blooms can develop in ponds. When this occurs, copper sulfate should not be applied because it will lyse algal cells releasing the toxin into the water. Potassium permanganate will kill a portion of algae and also oxidize algal toxins to nontoxic substances (Rodriguez et al., 2007). If used early enough, potassium permanganate can prevent or lessen fish or shrimp mortality. The effective dose rate is 2–3 mg/L, but some of the applied potassium permanganate will be expended in oxidizing organic matter in water. Thus, a potassium permanganate demand test should be conducted. Samples of the water should be treated at 1.0 mg/L increments of potassium permanganate to provide a range of 0.0–10.0 mg/L. The lowest concentration that results in the pink color of permanganate after 15 min can be taken as the potassium permanganate demand. An additional 2–3 mg/L above the demand should be the dose. The demand may range from 1 mg/L or less in clear water to more than 5 mg/L in water with abundant phytoplankton. The dose should be applied in several increments to avoid overtreatment.

Water bodies that are water suppliers for ponds and flow-through systems, or contain cages may develop toxic algal blooms. Such blooms may spread over areas of many square kilometers, and treatment to avoid the blooms is impossible. Because the blooms affect shellfish, both wild and cultured, governments often send alerts of toxic algal blooms. However, blooms are unpredictable and can occur suddenly, and it may or may not be possible to suspend water intake to a culture facility before the onset of a toxic algae episode. Nevertheless, if it is known that a toxic algae event is occurring, water intake should be suspended if possible. Of course, in cage culture, there is no possibility to isolate the fish from algal toxins in the surrounding water.

Intensive culture facilities, even for pond culture, often detain intake water in a reservoir for a few days before using it in culture ponds or other grow-out units. These reservoirs provide an opportunity for treatment with potassium permanganate to oxidize algal toxins. There are other means of denaturing algal toxins, e.g., ozone and ultraviolet light, but only a few facilities would be equipped for using these treatments.

6.5.6 Gas Supersaturation

Oxygen supersaturation from mid-morning until late afternoon is a daily occurrence in most aquaculture ponds, but gas bubble trauma is rare. Fish apparently sound in deeper waters where ΔP is less, and crustaceans usually live on the bottom where there may not be a ΔP. Cases of gas bubble trauma generally are limited to early life stages that may not escape surface waters of high ΔP. In water bodies with cage culture, phytoplankton blooms usually are not as great as in ponds, and oxygen supersaturation is also less. Most cages are at least 1–2-m deep, allowing for the benefit of sounding if ΔP is high.

The likelihood of gas bubble disease is greater in flow-through systems and water-recirculating systems. In flow-through systems, springs often are the water supply, especially for coldwater species. Water that seeps into aquifers in areas for coldwater fish culture is often cold and saturated with air. When it is introduced into culture facilities, it often gets warmed up causing supersaturation. The best way to avoid this problem is to pass the water through a degassing device. These devices may consist of a series of screens through which the water falls. The water is broken into drops that increase the surface area for diffusion of supersaturating gases to the atmosphere. Packed columns that contain media with a large void volume can also be used very effectively to increase surface area and facilitate degassing.

In raceways for warmwater fish culture, the water supply often is from streams or lakes, and it may be supersaturated with oxygen. Cases of gas bubble disease have been reported in shallow raceways supplied with such water (Supplee and Lightner, 1976). In shallow raceways, sounding to lessen the ΔP is of limited benefit in preventing gas bubble trauma.

In water-recirculating systems or in any system where water is transferred by pumps, there is a risk of air leaks on suction sides of pumps. These air leaks can result in supersaturation with air, and care should be taken to prevent them.

6.5.7 Ionic Imbalance

In low-salinity culture of marine or estuarine species, the water can be supplemented with ions to avoid stress and mortality related to ionic imbalance. The most well-known examples are for inland, low-salinity shrimp culture in which low potassium and magnesium concentrations have been increased by application of muriate of potash (potassium chloride) fertilizer Kmag (potassium magnesium sulfate). At an inland shrimp farm in Alabama, during the first years of operation, shrimp grew to a few grams and then began to die exhibiting cramps and white tails. Analysis of the water revealed very low potassium concentration, and muriate of potash was applied to increase potassium from around 5 mg/L to 50 mg/L. The shrimp improved greatly, but because of the delay in the use of potassium fertilizer until the problem was manifest, survival for the crop in 16 ponds averaged 19% and production was only 750 kg/ha. In the next crop, potassium was applied before stocking, and the survival was around 70%. Moreover, production at the same stocking rate as the previous crop averaged 3500 kg/ha.

Analyses of water sources for inland low-salinity ponds should be done to determine if potassium, magnesium, or other cations appear low. If so, mineral amendments should be added before stocking, also additional analyses should be done during the crop at intervals to ascertain if additional mineral supplementation is necessary. The interpretation of the mineral analyses is fraught with difficulty because the optimal proportions of ions unknown. The proportions of ions that would be expected in seawater diluted to the salinity of the culture system water often is used as a guideline.

6.6 CONCLUSION

This chapter briefly summarized the interactions among water quality, use of feeds, and aquatic animal health. The major conclusions that can be drawn from the discussion are listed below:

1. Impaired water quality can kill aquatic animals directly, or it may cause such stressful conditions that the fish divert energy that would normally be used for growth to other physiological processes in order to maintain homeostasis. The effect is often a lowered immune response and greater susceptibility to infectious diseases.
2. In aquaculture, the most common water-quality stressors result from the natural characteristics of the water supply or from alterations in water quality that occur within culture systems as a result of inputs of fertilizer and feed. Pollution usually is less likely to affect animals in aquaculture systems than in other water bodies, because aquaculture facilities, at least ideally, are located at sites where water pollution is minimal.
3. Water-quality variables most likely to be outside optimal concentration ranges causing excessive stress or mortality in aquaculture animals are as follows: water temperature, salinity, pH, dissolved oxygen, ammonia nitrogen, nitrite, carbon dioxide, hydrogen sulfide, presence and abundance of toxic algae, gas supersaturation, and proportions of cations in low-salinity inland water.
4. Water temperature and salinity issues usually are the results of site selection, species selection, or both. Although these two variables can be controlled in water-recirculating systems, they cannot be controlled in most other systems that are responsible for the vast majority of aquaculture production worldwide.
5. Dissolved oxygen concentration is generally the most critical variable in aquaculture. Dissolved oxygen concentration should be near saturation for best performance of aquaculture animals. This is seldom possible during nighttime, but dissolved oxygen concentrations for most warmwater species should not fall below 4 mg/L, while those for coldwater species should remain above 5–6 mg/L through each 24-h period.
6. Dissolved oxygen management should focus on matching the stocking and feeding rate with the carrying capacity of the system. Mechanical aeration can be used to increase system carrying capacity.
7. Good feed management includes selecting high quality feed, applying no more feed than animals will eat, and maintaining an adequate culture environment that minimizes the FCR that is the key to efficient feed-based aquaculture. Of course, in ponds without feeding, it is important not to overfertilize.
8. Improving the efficiency of fertilizer and feed use lessens the waste load in pond, minimizes oxygen demand, reduces the quantities of potentially toxic metabolites, and lowers the nutrient inputs that stimulate excessive phytoplankton blooms.
9. Feed management, water-quality management, and aquatic animal health management are intricately intertwined.
10. Toxic algae are a potential problem in many aquaculture facilities. Episodes of toxic algae are rather rare in comparison to many other water-quality problems, but there are relatively few effective control interventions when they do occur.
11. Gas supersaturation is rarely a problem in ponds and cage culture. It is more common in flow-through systems and water-recirculating systems where methods are available for its control in most instances.
12. Ionic imbalances in low-salinity inland culture of estuarine or marine species can be controlled through application of mineral amendments.

REFERENCES

Adelman, I.R., Smith Jr., L.L., 1970. Effect of hydrogen sulfide on northern pike eggs and sac fry. Transactions of the American Fisheries Society 99, 501–509.

Andrews, J.W., Murai, T., Gibbons, G., 1973. The influence of dissolved oxygen on the growth of channel catfish. Transactions of the American Fisheries Society 102, 835–838.

Arkoosh, M., Casillas, E., Clemons, E., Kagley, A.N., Olson, R., Reno, P., Stein, J.W., 1998. Effect of pollution on fish disease: potential impacts on salmonid populations. Journal of Aquatic Animal Health 10, 182–190.

Bagarinao, T., Lantin-Olaguer, I., 1999. The sulfide tolerance of milkfish and tilapia in relation to fish kills in farms and natural waters in the Philippines. Hydrobiologia 382, 137–150.

Barkoh, A., Smith, D.G., Schlechte, J.W., 2003. An effective minimum concentration of un-ionized ammonia nitrogen for controlling *Prymnesium parvum*. North American Journal of Aquaculture 65, 220–225.

Barton, B.A., 2002. Stress in fishes: a diversity of responses with particular reference to changes in circulating corticosteroids. Integrated and Comparative Biology 42, 517–525.

Bly, J.E., Quiniou, S.M., Clem, L.W., 1997. Environmental effects on fish immune mechanisms. Developmental Biology Standard 90, 33–43.

Boyd, C.E., 1973. Summer algal communities and primary productivity in fish ponds. Hydrobiologia 41, 357–390.

Boyd, C.E., 2015. Water Quality, an Introduction, second ed. Springer, New York.

Boyd, C.E., Lichtkoppler, F., 1979. Water Quality in Fish Culture. Research and Development Series No. 22. Alabama Agricultural Experiment Station, Auburn University, Auburn, Alabama.

Boyd, C.E., Tucker, C.S., 1998. Pond Aquaculture Water Quality Management. Kluwer Academic Publishers, Boston.

Boyd, C.E., Tucker, C.S., 2014. Handbook for Aquaculture Water Quality. Craftmaster Printer, Inc., Auburn, Alabama.

Boyd, C.E., Tucker, C.S., Somridhivej, B., 2016. Alkalinity and hardness: critical but elusive concepts in aquaculture. Journal of the World Aquaculture Society 47 (1), 6–41.

Bruno, D.W., Dear, G., Seaton, D.D., 1989. Mortality associated with phytoplankton blooms among farmed Atlantic salmon, *Salmo salar* L., in Scotland. Aquaculture 78, 217–222.

Cheng, S., Hsu, S., Chen, J., 2007. Effect of sulfide on the immune response and susceptibility to *Vibrio alginolyticus* in the kuruma shrimp *Marsupenaeus japonicas*. Fish and Shellfish Immunology 22, 16–26.

Chien, Y., 1992. Water quality requirements and management for marine shrimp culture. In: Proceedings of the Special Session on Shrimp Farming. World Aquaculture Society, Baton Rouge, pp. 144–156.

Collins, G., 1984. Fish growth and lethality versus dissolved oxygen. In: Proceedings Specialty Conference on Environmental Engineering, June 25–27. ASCE, Los Angeles, California.

Colt, J., 1986. Gas supersaturation – impact on the design and operation of aquatic systems. Aquacultural Engineering 5, 49–85.

Colt, J., Orwicz, K., 1991. Modeling production capacity of aquatic culture systems under freshwater conditions. Aquacultural Engineering 10, 1–29.

Colt, J., Tchobanoglous, G., 1978. Chronic exposure of channel catfish, *Ictalurus punctatus*, to ammonia: effects on growth and survival. Aquaculture 15, 353–372.

Conte, F.S., 2004. Stress and the welfare of cultural fish. Applied Animal Behaviour Science 86, 205–233.

Dombkowski, R.A., Russell, R.A., Olson, K.R., 2004. Hydrogen sulfide as an endogenous regulator of vascular smooth muscle tone in trout. American Journal of Physiology: Regulatory, Integrative and Comparative Physiology 286, R678–R685.

Ellis, T., Scott, A.P., Bromage, N., North, B., Porter, M., 2001. What is stocking density? Trout News 32, 35–37.

Ellis, T., North, B., Scott, A.P., Bromage, N.R., Porter, M., Gadd, D., 2002. The relationship between stocking density and welfare in farmed rainbow trout. Journal of Fish Biology 61, 493–531.

Good, C., Davidson, J., Welsh, C., Snekvik, K., Summerfelt, S., 2010. The effects of carbon dioxide on performance and histopathology of rainbow trout *Oncorhynchus mykiss* in water recirculation aquaculture systems. Aquacultural Engineering 42, 51–56.

Gopakumar, G., Kuttyamma, V.H., 1996. Effect of hydrogen sulphide on two species of penaeid prawns *Penaeus indicus* (H. Milne Edwards) and *Metapenaeus dobsoni* (Miers). Environmental Contamination and Toxicology 57, 824–828.

Gray, J.S., Wu, R.S., Or, Y.Y., 2002. Effects of hypoxia and organic enrichment on the coastal marine environment. Marine Ecology Progress Series 238, 249–279.

Gross, A., Boyd, C.E., Wood, C.W., 2000. Nitrogen transformations and balance in channel catfish ponds. Aquacultural Engineering 24, 1–14.

Han, Y., Boyd, C.E., Viriyatum, R., 2014. A bicarbonate titration method for lime requirement to neutralize exchangeable acidity of pond bottom soils. Aquaculture 434, 282–287.

Hargreaves, J.A., Kucuk, S., 2001. Effects of diel un-ionized ammonia fluctuation on juvenile hybrid striped bass, channel catfish, and blue tilapia. Aquaculture 195, 163–181.

Hargreaves, J.A., Tucker, C.S., 2004. Management of Ammonia in Fish Ponds. Publication 4603. Southern Regional Aquaculture Center, Stoneville, Mississippi.

Hickling, C.F., 1962. Fish Cultures. Faber and Faber, London.

Hsu, S., Chen, J., 2007. The immune response of white shrimp *Penaeus vannamei* and its susceptibility to *Vibrio alginolyticus* under sulfide stress. Aquaculture 27, 61–69.

Landolt, M.L., 1975. Visceral granuloma and nephrocalcinosis in trout. In: Ribelin, W.E., Migaki, G. (Eds.), The Pathology of Fishes. University of Wisconsin Press, Madison, pp. 793–801.

Lester, L.J., Pante, M.J.R., 1992. Penaeid temperature and salinity responses. In: Fast, A.W., Lester, L.J. (Eds.), Marine Shrimp Culture: Principles and Practices. Elsevier, Amsterdam, pp. 515–534.

Li, Y., Boyd, C.E., 2016. Laboratory tests of bacterial amendments for accelerating oxidation rates of ammonia, nitrate and organic matter in aquaculture pond waters. Aquaculture 460, 45–58.

Madsen, J.D., Richardson, R.J., Wersal, R.M., 2012. Managing aquatic vegetation. In: Neal, J.W., Willis, D.W. (Eds.), Small Impoundment Management in North America. American Fisheries Society, Bethesda, pp. 275–305.

Malbrouck, C., Kestemont, P., 2006. Effects of microcystins on fish. Environmental Toxicology and Chemistry 25, 72–86.

McDonald, D.G., 1983. The effects of H^+ upon the gills of freshwater fish. Canadian Journal of Zoology 61, 691–703.

McNevin, A.A., Boyd, C.E., Silapajarn, O., Silapajarn, K., 2004. Ionic supplementation of pond waters for inland culture of marine shrimp. Journal of the World Aquaculture Society 35, 460–467.

Mischke, C.C. (Ed.), 2012. Aquaculture Pond Fertilization Impacts of Nutrient Input on Production. Wiley-Blackwell, Ames.

Moran, D., Stöttrup, J.G., 2011. The effect of carbon dioxide on growth of juvenile Atlantic cod *Gadus morhua* L. Aquatic Toxicology 102, 24–30.

Moran, D., Tubbs, L., Stöttrup, J.G., 2012. Chronic CO_2 exposure markedly increases the incidence of cataracts in juvenile Atlantic cod *Gadus morhua* L. Aquaculture 364–365, 212–216.

Mortimer, C.H., 1954. Fertilizers in Fish Ponds. Her Majesty's Stationery Office, London.

Partridge, G.J., Lymbery, A.J., George, R.J., 2008. Finfish mariculture in inland Australia: a review of potential water sources, species, and production systems. Journal of the World Aquaculture Society 39, 291–310.

Pillai, V.K., Boyd, C.E., 1985. A simple method for calculating liming rates for fish ponds. Aquaculture 46, 157–162.

Rodriguez, E., Onstad, G.D., Kull, T.P.J., Metcalf, J., Acero, J.L., Von Gunton, U., 2007. Oxidative elimination of cyanotoxins: comparison of ozone, chlorine, chlorine dioxide and permanganate. Water Research 41, 3381–3393.

Rottmann, R.W., Francis-Floyd, R., Durborow, R., 1992. The Role of Stress in Fish Diseases. Publication 474. Southern Regional Aquaculture Center, Stoneville, Mississippi.

Roy, L.A., Davis, D.A., Saoud, I.P., Boyd, C.A., Pine, H.J., Boyd, C.E., 2010. Shrimp culture in inland low salinity waters. Reviews in Aquaculture 2, 191–208.

Shrimpton, J.M., Randall, D.J., Fidler, L.E., 1990. Factors affecting swim bladder volume in rainbow trout (*Oncorhynchus mykiss*) help in gas supersaturated water. Canadian Journal of Zoology 68 (5), 962–968.

Shumway, S.E., 1990. A review of the effects of algal blooms on shellfish and aquaculture. Journal of the World Aquaculture Society 21, 65–104.

Smayda, T., 2006. Harmful Algal Bloom Communities in Scottish Coastal Waters: Relationship to Fish Farming and Regional Comparison – a Review. Paper 2006/3 prepared for the Scottish Executive Environmental and Rural Affairs Department. http://www.scotland.gov.uk/Publications/2006/02/03095327/0.

Soderberg, R.W., 1995. Flowing Water Fish Culture. Lewis Publishers, Boca Raton.

Supplee, V.C., Lightner, D.B., 1976. Gas bubble disease due to oxygen supersaturation in raceway-reared California brown shrimp. Progressive Fish-Culturist 38, 158–159.

Torrans, E.L., 2008. Production responses of channel catfish to minimum daily dissolved oxygen concentrations in earthen ponds. North American Journal of Aquaculture 50, 371–381.

Tucker, C.S., 1996. The ecology of channel catfish ponds in northwest Mississippi. Reviews in Fisheries Science 4, 1–55.

Tucker, C.S., 2000. Off-flavor problems in aquaculture. Reviews in Fisheries Science 8, 45–88.

Tucker, C.S., Hargreaves, J.A., 2008. Environmental Best Management Practices for Aquaculture. Wiley-Blackwell, Oxford.

Tucker, C.S., Lloyd, S.W., Busch, R.l, 1984. Relationships between phytoplankton periodicity and the concentrations of total and un-ionized ammonia in channel catfish ponds. Hydrobiologia 111, 75–79.

United States Environmental Protection Agency (USEPA), 1977. Quality Criteria for Water. Office of Water and Hazardous Materials, USEPA, Washington, DC.

United States Environmental Protection Agency (USEPA), 1999. 1999 Update of Ambient Water Quality Criteria for Ammonia. EPA-822-R-99-014. USEPA, Washington, DC.

United States Environmental Protection Agency (USEPA), 2009. 2009 Update Aquatic Life Ambient Water Quality Criteria for Ammonia – Freshwater. EPA-822-D-09-001. USEPA, Washington, DC.

Walters, G.R., Plumb, J.A., 1980. Environmental stress and bacteria infection in channel catfish, *Ictalurus punctatus* Rafinesque. Journal of Fish Biology 17, 177–185.

Wang, J.Q., Lui, H., Po, H., Fan, L., 1997. Influence of salinity on feed consumption, growth, and energy conversion efficiency of common carp (*Cyprinus carpio*) fingerlings. Aquaculture 148, 115–124.

Wilkie, M.P., 1997. Mechanisms of ammonia excretion across fish gills. Comparative Biochemistry and Physiology 118, 39–50.

Wilkie, M.P., Wood, C.M., 1994. The effects of extremely alkaline water (pH 9.5) on rainbow trout gill function and morphology. Journal of Fish Biology 45, 87–98.

Zhou, L., Boyd, C.E., 2014. Total ammonia nitrogen removal from aqueous solutions by the natural zeolite, mordenite: a laboratory test and experimental study. Aquaculture 432, 252–257.

Zimba, P.V., Moeller, P.D., Beauchesne, K., Lane, H.E., Triemer, R.E., 2010. Identification of euglenophycin – a toxin found in certain euglenoids. Toxicon 55, 100–104.

Chapter 7

Water Quality–Disease Relationship on Commercial Fish Farms

Zdenka Svobodova, Jana Machova, Hana Kocour Kroupova, Josef Velisek
University of South Bohemia in Ceske Budejovice FFPW CENAKVA, Vodnany, Czech Republic

There are significant interrelationships between host organisms, their pathogens, and environmental factors (Snieszko, 1974). In an unpolluted environment with only the normal fluctuations in ambient conditions, there will be a natural balance between host organisms, pathogens, and environmental factors, leading to sporadic outbreaks of disease. However, a reduction in the quality of the environment will lead to immune system damage and reduced resistance of the host organisms to diseases and, consequently, marked increases in the frequency and severity of diseases. Also, an increase in the population density of fish will usually increase the risk of disease, because of impairment of the water quality (Snieszko, 1974).

It is possible that adverse environmental conditions may decrease the ability of organisms to maintain an effective immunological response system, so that an increased susceptibility to various diseases might be expected to occur. This certainly happens in aquatic organisms, particularly in fish, in which acute and/or chronic pollution of surface waters can cause a reduction in the level of unspecific immunity to disease.

Any marked change in surface water quality is reflected both directly and indirectly in the structure of the fish population. Indirect effects can occur from damage to the food web, which consists of lower organisms in the aquatic environment. There is a wide range of susceptibility of individual species of aquatic organisms to various pollutants. In most cases, the lower aquatic organisms (i.e., components of the zooplankton and zoobenthos) comprise the more susceptible species.

Thus, at low concentrations of pollutants damage and mortality of sensitive food organisms can occur. In consequence, although fish are not affected primarily, they suffer from secondary effects; the reduction and/or complete absence of natural food leads to poorer condition of the fish and this may be accompanied by a decrease of antibody production (Noga, 2010). In this way, the disease resistance of fish may be decreased (Noga, 2010; Andrews et al., 2010).

The aim of this chapter is to evaluate the effects of changes in the basic indicators of water quality on the fish organism. From the viewpoint of commercial fish farming the basic indicators are water temperature, pH, oxygen, ammonia and nitrites, and chlorine, which is often used in aquaculture. An important part of this chapter is about prevention of damage to the fish due to changes in water quality.

7.1 WATER TEMPERATURE

7.1.1 Etiology

Water temperature ranks among factors significantly affecting not only water quality, but, most importantly, fish metabolism intensity (Alabaster and Lloyd, 1980). Water temperature has an impact on the form in which some substances (e.g., ammonia, chlorine, and others) occur in water, as well as on their solubility and therefore their bioavailability to fish. Water temperature also significantly affects chemical and biochemical reactivity. The rate of most biochemical processes slows down at temperatures close to zero, and some processes are stopped completely. The nitrification process, for example, slows down considerably at temperatures lower than 5°C. In biological wastewater treatment, nitrification is typically inhibited at temperatures below 12°C, with the highest rate in the range of 20–30°C (Pitter, 2009).

Generally, higher water temperature accelerates the rate of organic matter decomposition. Consequently, the risk of oxygen deficiency increases with higher water temperatures combined with the presence of organic substances in water. Oxygen dissolved in water is depleted at a faster rate than it is replenished from the atmosphere. In addition, oxygen solubility in water drops with growing temperature.

Fish Diseases. http://dx.doi.org/10.1016/B978-0-12-804564-0.00007-7

7.1.2 Mechanism of Action

Fish are poikilothermic animals. This means that their body temperature is equal to that of the aquatic environment, or higher by just 0.5–1°C. Water temperature is very closely interrelated with metabolism intensity. Metabolic rate increases with rising temperature (up to the optimal value). A general rule states that metabolism intensity is doubled with a temperature rise of 10°C. This generalization applies to thermophilic fishes. The metabolism of psychrophilic fishes (e.g., salmonids, whitefish, and burbot) allows chemical transformations at relatively low temperatures, while higher temperatures, typically above 20°C, reduce their activity and food intake (Svobodova et al., 1993).

All organisms, including fish, show an increase in heat shock protein (HSP) synthesis under thermal stress conditions. These proteins provide universal protection of cells from shock by preserving the spatial arrangement of proteins within the cells, protect membrane lipids and nucleic acids, and reduce the production of toxic radicals. In addition, they have an important thermoregulatory function (Feige et al., 1996). As stated by Lewis et al. (2012), the induction of HSP70 guarantees protection and blocks the cell death cascade induced by heat shock. HSPs also provide protection from other stressors, e.g., pollutants and pathogenic microorganisms (Iwama et al., 2005). For these reasons, HSPs are widely used as stress markers in environment monitoring (Madeira et al., 2012). HSPs are classified on the basis of their relative molecular weight. The best-explored ones include HSP60, HSP70, and HSP90. Compared to other cells, a particularly prominent response to heat shock is manifested by erythrocytes in fish (Currie et al., 2000).

When assessing the impact of temperature on particular fishes, a distinction must be made between the optimal temperature (i.e., a temperature guaranteeing optimal metabolism) and temperature tolerance (i.e., temperatures that can be tolerated by fish on a long-term basis, but which significantly affect their metabolic rate).

The other temperature-related concepts include a critical thermal minimum and maximum. These are temperatures causing significant damage or even death to fish. The actual values vary across fishes, and also within a single species, depending on previous thermal acclimatization.

The optimal temperature for growth and development is 18–28°C in cyprinids and 8–18°C in salmonids. In the moderate climate of central Europe, fish, depending on the species, can easily adapt to seasonal changes in water temperature in their natural habitats, i.e., to temperatures dropping to 0°C in winter and rising to 20–30°C in summer (Svobodova et al., 1993). The optimal temperatures vary in the course of fish development, with different optima in the course of spawning, the embryonic and the larval stages, and subsequently the juvenile period and the adult stages of life. In rainbow trout (*Oncorhynchus mykiss*), for example, Reinchenbach-Klinke (1975) suggests 6–14°C as the embryonic and larval optimum, 6–19°C for the juvenile stage, and 10–18°C for the adult stage.

Simcic et al. (2015) state that psychrophilic fishes (particularly their early developmental stages) may be significantly affected by future climatic changes as they are highly sensitive to higher temperatures. An increase in temperature (from 2 to 8°C) in the course of lake whitefish (*Coregonus clupeaformis*) egg incubation has been demonstrated to result in hatching time reduction by as much as 50%, while larval mortality rose from 45% to 83%, and the individual weight dropped from 1.27 to 0.61 mg (Mueller et al., 2015).

Although fish generally manifest a high degree of temperature tolerance, they may be harmed by sudden changes in water temperature. A sudden change in temperature of 12°C or more induces a thermal shock in fish, potentially resulting in death. Early developmental stages of fish are more vulnerable and should not be exposed to sudden temperature changes of more than 3°C. This must be borne in mind, for example, when fish in early larval stages are removed from hatcheries, with temperatures around 20°C, and stocked into ponds that may be substantially colder. Stocking usually takes place in May, when water temperature in ponds tends to be around 15°C, and may be even lower in cold weather (Svobodova et al., 1987).

Stocking fish that have been fed into water colder by 5–8°C or more causes digestion disorders, or digestion may be stopped completely. Undigested or partially digested feed in the digestive system produces gases, which accumulate and extend the internal cavity, leading to loss of balance and death. Increased pressure in the body cavity owing to gas accumulation can be expected to damage internal organs. A drop in the metabolic rate reduces the amount of ammonia excreted by the gills and leads to high ammonia concentrations in the blood plasma, causing autointoxication and eventually death (Svobodova et al., 1987). A particularly high risk of autointoxication occurs in carp consuming nitrogen-rich feed (natural food, some feed mixtures). This happens when marketable carp are harvested in summer, from ponds with water temperatures of up to 25°C, and subsequently stocked into tanks supplied with colder water (15–18°C) (Svobodova et al., 2007a,b).

7.1.3 Clinical Symptoms

When fish are exposed to water colder than the temperature optimum for a given species, feed intake stops, their growth slows down, and the function of their immune system is inhibited. Conversely, exposure to high water temperatures, which reduce oxygen solubility in water, may induce clinical symptoms of suffocation. Early symptoms of response to a sudden temperature change include loss of balance, and when suffering a thermal shock, fish die with symptoms of breathing and heart muscle paralysis. As stated above, loss of balance may also be caused by expansion of the body cavity resulting from accumulation of gases produced by undigested food (Svobodova et al., 2008).

7.1.4 Pathological–Morphologic Symptoms

Thermal shock may produce convulsive opening of the mouth and the opercula. However, these symptoms may be inconspicuous in some species, and may also be observed in predatory fishes as a result of oxygen deficiency (Svobodova et al., 1987). In some cases hemolysis may occur (Svobodova et al., 2008). If digestion was disturbed or stopped as a result of a sudden temperature change, autopsy reveals a large accumulation of gases and the body cavity is conspicuously extended. In cases of ammonia autointoxication, gills suffer hyperemia and edema (they bleed considerably when mechanically damaged).

7.1.5 Diagnosis

Determination of diagnosis must take into account water temperature, possible changes in water temperature, and clinical and pathological changes. When ammonia autointoxication is suspected, ammonia content is measured in the blood plasma of the fish showing symptoms of poisoning, as its levels exceed the usual ones by 5–10 times. In such cases, water quality must be assessed (in particular pH levels and ammonia concentrations) to exclude the possibility of poisoning by ammonia from the external environment—water.

7.1.6 Prevention

Sudden temperature changes must be avoided, particularly in the larval stage. When stocking fry transported in polyethylene bags filled with oxygen, the bags should first be placed onto the surface of the pond, and the fry stocked only when the temperature of transport water matches that of the water in the pond. If water temperature in the hatchery is substantially different from that in the pond where fry are to be stocked, the water temperature in the hatchery must be lowered gradually before stocking. An alternative (if conditions in the hatchery allow this) is postponing the stocking until the weather becomes warmer and water temperature in the ponds rises accordingly.

In summer harvests, all possible care must be taken to prevent the water temperature in the transport containers and in the hatchery from being substantially lower than the water temperature in the pond from which the fish are harvested. The risk of autointoxication may also be reduced by excluding high-nitrogen feed before the harvest. Similarly, the supply of such feed must be reduced to fish fry reared in ponds at the end of the feeding season, when sudden weather changes involving a substantial drop in temperature can be expected.

7.2 WATER pH

7.2.1 Etiology

With the exception of water from peat moors, the pH of surface waters ranges between 6.0 and 8.5. A shift toward the alkaline part of the scale may be due to photoautotrophic organisms carrying out intensive photosynthetic assimilation, depleting free carbon dioxide. Conversely, relatively low pH values of water are characteristic of the spring season, when snow is melting, particularly in waters from peat moors (Pitter, 2009).

Water pH may be affected by human activity. High pH values may be due to concrete mixtures running off into watercourses during construction work conducted close to or right in the recipient water body (e.g., during the construction of bridges or weirs, or channeling work). On the other hand, low pH values (3.5–4.5) are found in silage juices containing organic acids—lactic, acetic, and butyric (Pitter, 2009). Sudden changes in pH values are also recorded in accidental acid and hydroxide spills (Svobodova et al., 1987).

7.2.2 Mechanism of Action and Toxicity

The pH value of water is important to fish homeostasis, development, and survival. Exposure to strongly acidic (pH 3.5) or alkaline (pH 11) water is lethal to fish after just a few hours, owing to disturbed ionic balance (Parra and Baldisserotto, 2007).

The surface of the whole fish body (e.g., the skin and gills) exposed to extreme pH values suffers damage (chemical burn), generally resulting in gas exchange and osmoregulation disturbances.

The optimum pH value of water for fish ranges between 6.5 and 8.5. Damage and death occur at pH values below 4.8 or above 9.2, in salmonids, and below 5.0 or above 10.8 in cyprinids (Svobodova et al., 2008). This means that salmonids are more sensitive to high pH values and more resistant to low pH values (Svobodova et al., 1987). Fish protect themselves against the effects of low or high pH by dropping the frequency of respiratory movements, reducing the volume of water flowing through the gills (Svobodova et al., 2008).

Alkaline pH values (9.5–10) slow down the conversion of ammonia into ammonia ions on the surface of the gills and so inhibit ammonia diffusion. Alkaline pH also inhibits sodium cation absorption (Wilkie and Wood, 1996). Bolner et al. (2014) found that, in South American catfish (*Rhamdia quelen*), an increase in water pH values resulted in a drop in the levels of ammonia, glucose, and glycogen in the liver; a drop in glucose and lactate levels in the kidneys; and also an increase in ammonia levels and a drop in glucose and glycogen values in muscle tissue. Higher ammonia levels in muscle appear to be related to reduced ammonia excretion by fish caused by higher pH values of water (Golombieski et al., 2013). Similar results were produced in experiments involving rainbow trout (*O. mykiss*) exposed to higher pH values of water over 15 days (Wilson et al., 1998).

The mechanism of harmful effects of a strongly acidic (<pH 5) aquatic environment on physiological processes in fish is described by Wright and Welbourn (2002). They point out that the presence of H^+ ions increases gill epithelium permeability and leads to important electrolyte loss by:

- H^+ ions moving through the gills into the fish body owing to the concentration gradient,
- high H^+ ion concentrations in the external environment decreasing the rate of active exchange of sodium (Na^+) for H^+ (or ammonia ions) from fish bodies,
- higher branchial epithelium passive permeability for Na^+, potassium (K^+), and chloride (Cl^-) ions, and possibly also for other ions, facilitating the transfer of these ions from fish bodies into the external environment.

It was found by Bolner et al. (2014) that in South American catfish (*R. quelen*) a drop in water pH values produced an increase in ammonia, glucose, and glycogen levels in the liver; an increase in the glucose and lactate levels in the kidneys; and a drop in ammonia levels and an increase in glucose and glycogen levels in muscle tissue. Kidney ammonia and glycogen levels were lowest in fish reared in water of neutral pH values, and an increase occurred following both a drop and a rise in pH values. A similar pattern was observed in liver and muscle lactate. Increased lactate levels in the liver and muscle suggest that acidic or alkaline pH values cause lactate to be used as a substrate for anaerobic gluconeogenesis, a process that was observed in other fishes.

In South American catfish (*R. quelen*), acidic waters (pH of 5.5) slowed down growth; however, increased NaCl supply in the feed reduced the loss of sodium ions, thus moderating the negative impact of acidic pH on fish growth (Copatti et al., 2011). A very high degree of low pH tolerance is manifested by brook trout (*Salvelinus fontinalis*), which can tolerate values as low as 4.5 (Svobodova et al., 1987). Conversely, a much lower tolerance of low pH values is characteristic of early developmental stages of common carp (*Cyprinus carpio*) (Machova et al., 1983). These authors compared the growth and ontogenetic development rates of common carp age 4–22 days reared in water of pH values ranging between 5.7 and 6.3, with common carp reared in pH ranging from 8.75 to 9.16. The fry exposed to lower pH values showed a worse feed conversion ratio, leading to a reduction of individual weight (by c. 50%), and a mortality rate higher by nearly 20%, and a slightly slower ontogenetic development rate than the group reared in water of higher pH values.

The effects of extreme pH values on early developmental stages of northern pike (*Esox lucius*) were studied by Le Louarn and Webb (1998), who exposed fertilized eggs and embryos to water pH values ranging from 4.5 to 10.5, on both a short- and a long-term basis. The results of the tests suggest that pH values lethal for embryos are 6 and 10, respectively. Just a small proportion of eggs exposed to pH values of 4.5 and 10.5 actually hatched, and the hatchlings manifested a high degree of abnormalities in the following development.

Araujo et al. (2008) studied the impact of low pH values on the survival of young guppies (*Poecilia reticulata*), and determined the median lethal time (LT_{50}) (i.e., the time after which 50% of the organisms tested die) for different pH values (from 3 to 6.5). The lowest LT_{50} value of ca. 1.5 h was identified for pH 3. From there LT_{50} values grew along with rising pH, up to a pH of 5, at which the LT_{50} was ca. 75 h. At a pH of 6–6.5, the fry of this fish species survived without explicit signs of damage for over 96 h. Further results come from Dunson et al. (1977), who identified a 100% mortality rate in guppies exposed for 11 days to water at pH 4.75, and, simultaneously, recorded a drop in sodium ions and a rise in hydrogen ions in these organisms.

Tolerance of sudden pH rises was confirmed to increase with the age and size of fish in experiments involving channel catfish (*Ictalurus punctatus*) and their hybrids (Mischke and Chatakondi, 2012).

Sensitivity to extreme pH values depends not only on the species and age of the fish (increased pH sensitivity was observed in early developmental stages of fish), but also on the aquatic environment oxygen levels. So, for instance, in highly eutrophic ponds, where the water is saturated with oxygen, pH levels of over 10 may be tolerated by fish on a long-term basis. Lloyd (1961) suggests that increased levels of dissolved oxygen in water may intensify CO_2 excretion, resulting in a local drop in pH values close to the surface of the gills.

7.2.3 The Impact of Water pH on the Toxicity of Some Substances

The pH value significantly affects the forms in which some substances occur in water, which in turn affects their solubility and toxicity. The substances important to fish farming include ammonia, hydrogen sulfide, cyanides, and toxic metals. The effects of these substances in water depend so closely on the actual pH that without the knowledge of its value, the potential dangers to fish cannot be estimated (Svobodova et al., 2008).

The most striking pH dependence is found in ammonia. Growing pH values along with rising water temperature sharply increase ammonia toxicity to fish, as the proportion of free (toxic) ammonia grows. Conversely, cyanide and hydrogen sulfide toxicity drops with rising water pH (Svobodova et al., 2007a).

7.2.4 Clinical Symptoms

Extreme pH values are manifested as initial excitement in fish, followed by lethargy. A drop in ventilation frequency may be observed, reducing the volume of water flowing through the gills and thus protecting them from extreme pH values (Svobodova et al., 2008).

7.2.5 Pathological–Morphologic Symptoms

Typical symptoms in fish include increased mucus secretion on the skin, on the inside of the opercula and the gills, often bloody. The mucus is glassy and watery. Hemorrhage may occur on the gills and the underside of the body. Extreme pH values cause corneal opacity, frayed fins, and damage to the skin (Svobodova et al., 2008).

7.2.6 Diagnosis

The diagnosis is based on the medical history and takes into account water pH, as well as clinical and pathological changes. To record the highest pH values, which tend to be caused by photosynthetic activity of algae and cyanobacteria in highly eutrophic water reservoirs, pH measurements must be taken in the afternoon, when photosynthesis is at its peak, owing to intense sunlight (Svobodova et al., 2008).

7.2.7 Therapy

When fish are reared in ponds and reservoirs, a sufficient supply of safe water must be provided. In tanks and aquaria, the water must be replaced or the organisms must be moved to clean water.

7.2.8 Prevention

An important preventive measure is pH monitoring of the incoming water in trout farms. If low pH values are recorded, immediate steps must be taken (e.g., the application of sodium carbonate). Decreased pH values typically occur in spring, when snow is melting, particularly in water coming from peat moors (Svobodova et al., 2008). Low pH values are also characteristic of recirculating aquaculture systems, where they result from nitrification processes. All these systems therefore require consistent pH monitoring, and, when necessary, pH value adjustment by addition of limestone or sodium carbonate.

Increased attention must be given to construction work taking place close to water recipients, to prevent runoff of concrete mixtures. In highly eutrophic reservoirs, with high pH values occurring in spring and summer as a result of photosynthetic activity of green organisms, the fish stock must be manipulated in terms of number and species structure to prevent excessive cyanobacterial and algal growth (Svobodova et al., 2008).

7.3 OXYGEN

7.3.1 Etiology and Mechanism of Action

Water gets oxygen from the atmosphere by diffusion and from photosynthetic assimilation of aquatic plants, algae, and cyanobacteria.

The principal cause of low oxygen levels in water (oxygen deficiency) is surface water pollution by easily degradable organic substances. These are decomposed in water by the activity of microorganisms in a process that uses oxygen dissolved in water. In flow-through fish farms, oxygen deficiency may be caused by a drop in the volume or a complete disruption of incoming water. Oxygen deficiency may affect ponds where organic matter was not sufficiently decomposed at the end of the vegetation season, and the surface froze over because of a sudden drop in temperature. The risk increases when the ice is completely covered in snow that blocks sunrays and prevents photosynthesis. In highly eutrophic ponds, oxygen deficiency often develops on early summer mornings as a result of increased oxygen consumption at night by photoautotrophic organisms in the process of dissimilation, and owing to bacterial decomposition of organic matter in water. In spring and summer, oxygen deficiency may be due to excessive growth of large zooplankton, especially in eutrophic and hypertrophic reservoirs with unsuitable fish stocks. This happens when photosynthesis raises pH values, which in turn cause damage to the stock. Weakened fish (often also suffering from the presence of nondissociated/toxic ammonia) consume less zooplankton and cannot therefore reduce its growth, which becomes excessive. At the same time, the amount of oxygen-producing phytoplankton drops. If such a condition is not identified in time and no steps are taken, oxygen deficiency grows worse, possibly causing death of fish on a mass scale. When oxygen levels in pond water drop suddenly, ammonia autointoxication may occur in carp, followed by toxic gill necrosis (Jeney et al., 1992a,b).

Like temperature, oxygen is another significant factor affecting metabolic rate. Insufficient oxygen supply reduces and disturbs metabolic rate, and consequently causes death by suffocation. Sensitivity to oxygen deficiency is common to all species and categories of fish, albeit to different degrees, depending on the species, the developmental stage (fertilized egg, larva, adult fish), and the fish's life processes (feeding, growing period, reproduction period). Salmonids are very sensitive, with optimal oxygen levels in water in the range of 8–10 mg/L (Lloyd, 1961; Svobodova et al., 1993). They develop symptoms of suffocation when the oxygen level drops below 3 mg/L. Cyprinids are less demanding with respect to oxygen, with an oxygen optimum of 6–8 mg/L, and develop suffocation symptoms at levels below 2 mg/L (Svobodova et al., 1987, 1993; Dwyer et al., 2014).

So, for instance, in sturgeon (*Acipenser sturio* L.), no effect of oxygen saturation of water (30%, 50%, or 90%) on embryonic survival rate was identified. On the other hand, a high hatch rate of embryos (nearly 80%) was achieved only at a level of oxygen saturation of 90%, while hypoxic conditions effectively stopped hatching (Delage et al., 2014).

Under certain conditions water may become oversaturated with oxygen (hyperoxia) by up to tens of percentage points. This happens naturally in highly turbulent water (rapids, weirs, waterfalls) or as a result of intense assimilation of autotrophic organisms (Pitter, 2009). Oxygen oversaturation may also occur in closed transport systems for fish, e.g., in polyethylene bags filled with oxygen. In such cases, oxygen oversaturation may be as high as 300%, and the fish may be seriously damaged in the transport. The critical oxygen saturation level in water is considered to be between 250% and 300% for cyprinids (Svobodova et al., 2008).

7.3.2 The Impact of Dissolved Oxygen Level on the Toxicity of Selected Substances

The presence of oxygen is a necessary precondition for aerobic processes occurring in surface water self-cleaning processes and in biological wastewater treatment. The concentration of dissolved oxygen may affect fish sensitivity to high pH values. It has been shown, e.g., by Wiebe (1931), that bluegill sunfish (*Lepomis macrochirus*) showed stress symptoms and some fish even died at a pH value of 9.6 and dissolved oxygen level of 5 mg/L, while no signs of damage were observed at 10 mg/L of dissolved oxygen and a pH of 9.5. Lloyd (1961) suggests that at higher dissolved oxygen concentrations, fish excrete more carbon dioxide, which reduces pH values on the surface of the gills.

7.3.3 Clinical Symptoms

Under oxygen-deficient conditions (hypoxia) symptoms of suffocation and dying occur gradually in different species and age categories, depending on their oxygen deficiency sensitivity. The fish stop feeding and move close to the surface, and cyprinids employ emergency breathing. In ponds, fish gather near the source of incoming water, become lethargic, stop responding to stimuli, lose their escape reflex, and eventually die (Svobodova et al., 2008).

Oxygen oversaturation of water (hyperoxia) results in damage to fish transported in polythene bags with an oxygen-filled air space, whose principal symptoms include gas accumulation in the eyes, fins, skin, and gills (Gultepe et al., 2011). Large necrotic patches appear on the skin, disturbed by bubbles (Vatsos and Angelidis, 2010). According to Svobodova et al. (2008), the color of the gills of fish reared in oxygen-oversaturated water is strikingly light red, and the gill filament edges are frayed. After fish damaged in this way are stocked, their gill epithelium becomes necrotic, followed by secondary dermatomycosis, and by death.

7.3.4 Pathological–Morphologic Symptoms

The skin of fish exposed to hypoxic environments is conspicuously pale. The gills are hyperemic or cyanotic (blue to purplish), gill filaments are stuck together, and slight hemorrhage occurs in the anterior chamber of the eyeball and also on the gill cover. In most predatory fishes the mouth is convulsively open, with the opercula set conspicuously apart. This symptom is particularly prominent while the fish are still in the state of rigor mortis (Svobodova et al., 2008).

7.3.5 Diagnosis

The diagnosis is based on the anamnesis and on the assessment of oxygen concentration in the water where fish death occurred, taking into account the requirements of the fish species and categories. If oxygen levels were not tested on the spot, the diagnosis may be supported by data on chemical oxygen demand (COD) and biological oxygen demand (BOD_5) obtained from water samples sent for laboratory tests. While these analyses do not diagnose oxygen deficiency itself, they indicate its probability (the values obtained suggest the presence of degradable organic substances, whose decomposition processes the use of oxygen). Cyprinid farming usually requires COD values determined by the Kubel method (COD_{Mn}) up to 20–30 mg/L and BOD_5 values up to 8–15 mg/L (depending on farming intensity). Water of better quality is required in salmonid farming, where the values should be up to 10 mg/L COD_{Mn} and 5 mg/L BOD_5 (Svobodova et al., 2008). Water quality required for a land-based catfish farm should be up to 50 mg/L BOD_5 (Svobodova et al., 1987).

7.3.6 Therapy

A supply of water with optimum oxygen levels must be provided, and, possibly, additional aeration must be applied to the water in the reservoir (Svobodova et al., 2008).

7.3.7 Prevention

An essential preventive measure is consistent care of the aquatic environment, in particular protection from possible contamination of the water with organic substances, as well as strict observance of technological procedures (a suitable stock density, regular dredging and cleaning of fish-rearing reservoirs, oxygen level testing). Eutrophic and hypertrophic ponds require regular monitoring of dissolved oxygen levels and water transparency and the development of hydrobiological parameters. A fast-increasing water transparency is usually due to a drop in the amount of vegetation causing water opacity. This happens when excessively growing zooplankton reduces the volume of oxygen-producing phytoplankton. Zooplankton dissimilation, along with a drop in oxygen production, leads to oxygen deficiency. If oxygen deficiency is imminent, the supply of clean water must be restored or increased, and aeration systems can be installed.

In the case of threatening oxygen deficiency induced by a temporary drop in phytoplankton photosynthetic activity caused by disposable phosphorus deficiency, a simple and cheap solution is application of small amounts of dissolved phosphorus fertilizers (Faina et al., 2011).

7.4 AMMONIA

7.4.1 Etiology

Ammonia nitrogen is the primary product of decomposition of nitrogen-containing organic substances of both animal and plant origin. The anthropogenic sources of organic ammonia nitrogen primarily include wastewaters and agricultural wastes. Ammonia nitrogen is a significant component of atmospheric water, where it is often the dominant cation, particularly in regions plagued by air pollution. Significant amounts may come from ammonia emissions produced by and around animal farms (e.g., the mean annual ammonia production of a single dairy cow is 60 kg). Nitrogen fertilizers are another important source of ammonia nitrogen; they contaminate both surface and underground waters by infiltration and runoff from fields. Large amounts of ammonia nitrogen are present in industrial wastewaters from thermal processing of coal and electroplating plants. Ammonia is also used as a cooling medium in large cooling systems. Ammonia-containing compounds are sometimes added to water for hygienic purposes in the process of chloramination (Pitter, 2009).

Ammonia is also the principal final product of nitrogen (protein) metabolism in bony fishes. Freshwater bony fishes excrete about 90% of waste nitrogen substances in the form of ammonia, with the remaining 10% excreted as urea. Seawater bony fishes excrete 70% of waste nitrogen substances in the form of ammonia, and 30% as urea. The proportions of excreted ammonia and urea may vary slightly in bony fishes under the influence of a variety of internal and external factors, such as water salinity and pH, food quality, etc. (Altinok and Grizzle, 2004). Conversely, freshwater cartilaginous fishes (sturgeons) excrete waste nitrogen-containing substances solely in the form of urea (Schreckenbach and Spangenberg, 1978; Lloyd, 1992; Wood, 1993).

7.4.2 Mechanism of Action

Water and biological fluids contain ammonia in two forms: the molecular NH_3 form (sometimes referred to as nondissociated or free) and the ammonium NH_4^+ ion (sometimes referred to as dissociated or bound). These two forms coexist in water, their relative proportions depending on the pH value and water temperature (Pitter, 1981). The proportion of free NH_3 ammonia in total ammonia ($NH_3 + NH_4^+$) is given in Table 7.1.

TABLE 7.1 NH_3 Proportion (in Percentage of Total Ammonia) Dependence on Water pH and Temperature

	Water Temperature (°C)					
pH	0	5	10	15	20	25
7.0	0.082	0.12	0.175	0.26	0.37	0.55
7.2	0.13	0.19	0.28	0.41	0.59	0.86
7.4	0.21	0.30	0.44	0.64	0.94	1.36
7.6	0.33	0.48	0.69	1.01	1.47	2.14
7.8	0.52	0.75	1.09	1.60	2.32	3.35
8.0	0.82	1.19	1.73	2.51	3.62	5.21
8.2	1.29	1.87	2.71	3.91	5.62	8.01
8.4	2.02	2.93	4.23	6.06	8.63	12.13
8.6	3.17	4.57	6.54	9.28	13.02	17.95
8.8	4.93	7.05	9.98	13.95	19.17	25.75
9.0	7.60	10.73	14.95	20.45	27.32	35.46
9.2	11.53	16.00	21.79	28.95	37.33	46.55
9.4	17.12	23.19	30.36	39.23	48.56	57.99
9.6	24.66	32.37	41.17	50.58	59.94	68.62
9.8	34.16	43.14	52.59	61.86	70.34	77.62
10.0	45.12	54.59	63.74	71.99	78.98	84.60
10.2	56.58	65.58	73.59	80.29	85.63	89.70
10.4	67.38	75.12	81.54	86.59	90.42	93.24
11.0	89.16	92.32	94.62	96.26	97.41	98.21

From Pitter, P., 1981. Hydrochemie, SNTL, Praha.

Nearly all exchange of essential ions between the internal environment of fish and the external environment (water) is conducted through the gill epithelium (Wilkie and Wood, 1996). This process is demonstrated in Fig. 7.1. Free NH_3 ammonia passes from water through the gill epithelium into the blood, and the other way round. The transport is passive, regulated solely by the free-ammonia concentration gradient (Wilkie, 2002). At neutral or slightly alkaline pH values of water (≤ 8), nearly all excreted free ammonia is transformed into the ionized NH_4^+ form. As the gill cell membrane is almost nonpermeable to the ammonium ion, it cannot transfer from the aquatic environment into the blood. It can, however, pass in the opposite direction, from the blood into the water, in exchange for a sodium Na^+ ion (Evans et al., 1999). When breathing, fish excrete carbon dioxide, which reduces the pH value of water on the gill surface, i.e., on the boundary between the gill epithelium and the surrounding water (Wilkie, 2002). This shifts the balance of the two forms of ammonia in favor of the ionized one. Consequently, under identical conditions, the free ammonia concentration is lower on the gill surface than in the surrounding water. This helps to preserve the concentration gradient of free ammonia and enables its excretion by fish even at higher pH values in water. The degree to which this effect applies is related to water pH, its buffering capacity, and the current fish fitness (Di Giulio and Hinton, 2008).

Ammonia shows a special affinity for the brain and the nervous system. This is why nervous disorders are very prominent symptoms of ammonia poisoning in fish. Ammonia disturbs energy metabolism in nerve cells, binds to α-ketoglutaric

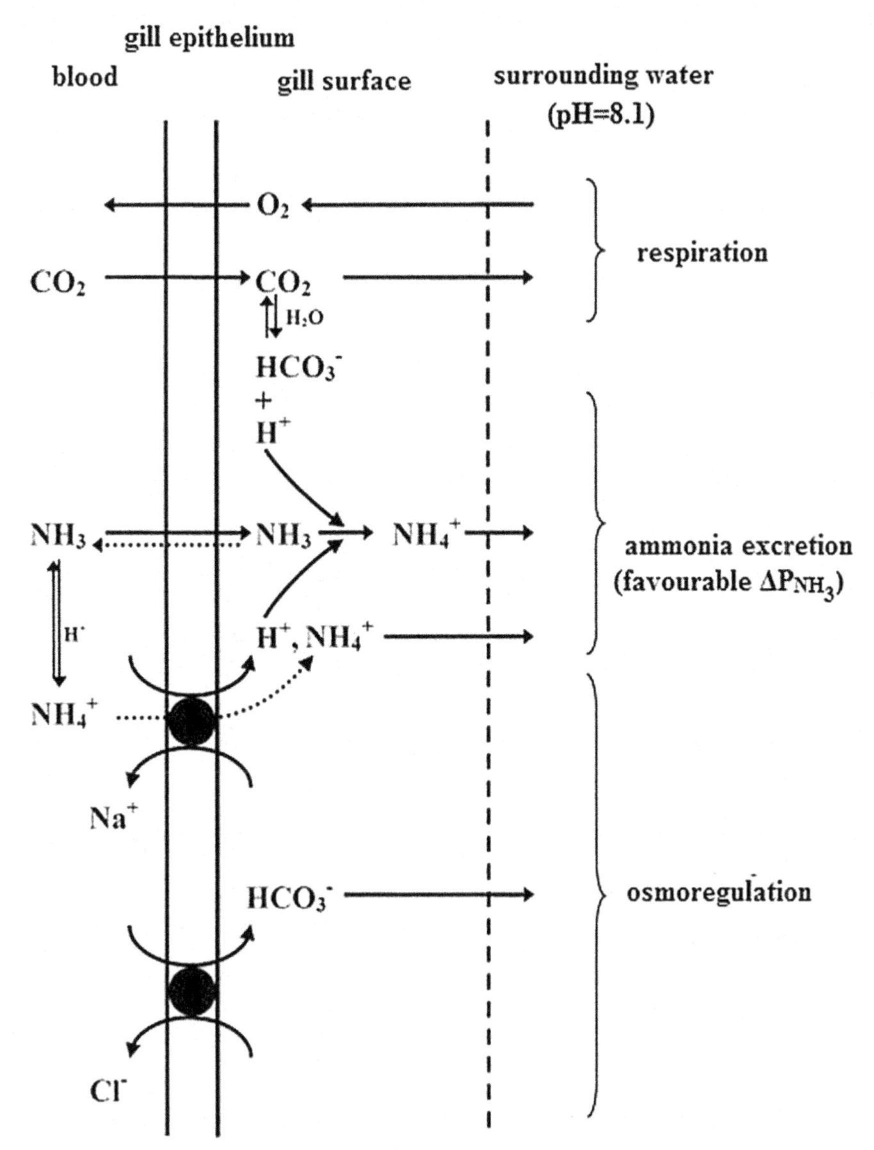

FIGURE 7.1 Essential ion exchange between the internal environment of fish and the surrounding water. *Adapted from Wilkie, M.P., Wood, C.M., 1996. The adaptations of fish to extremely alkaline environments. Comparative Biochemisty and Physiology 113, 665–673.*

acid from the Krebs citric acid cycle, and consequently reduces ATP production (Smutna et al., 2002). Additionally, the process produces glutamine, which is toxic to nerve cells and which represents another source of toxic ammonia, as well as being the precursor of γ-aminobutyric acid (GABA), which acts as an inhibitory neurotransmitter. High GABA concentrations may lead to neurotransmission inhibition, with symptoms of "sleeping sickness" in common carp (*C. carpio*). The toxic effect of ammonia is supposed to result from neuron depolarization occurring when potassium (K^+) is replaced by the NH_4^+ ion. Neuron depolarization subsequently causes increased activation of *N*-methyl-D-aspartate receptors and eventually nerve cell necrosis (Smutna et al., 2002; Svobodova et al., 2008).

Ammonia occupies a special position among the substances toxic to fish. This is because it is transmitted into the fish body from the external environment and at the same time is a product of fish metabolism. The overall ammonia concentration in fish blood is therefore the sum of the ammonia received from the external environment (exogenous ammonia) and the ammonia produced by protein metabolism (endogenous ammonia). In this way, endogenous ammonia contributes to ammonia accumulation within the fish body, potentially reaching toxic levels in the organism. It is therefore of primary importance for fish to retain the ability to excrete ammonia. If the balance between ammonia production and excretion is disturbed, ammonia concentrations may rise significantly in fish blood, leading to autointoxication (Svobodova et al., 2007a,b). This balance can be disturbed, e.g., by high pH values in the water (Fig. 7.2), since the ammonia released remains in its nonionized NH_3 form, which reduces the nonionized ammonia gradient at the blood–water phase interface (Russo et al., 1988). Disruption of this balance may also occur as a result of a variety of internal and external factors. Ammonia

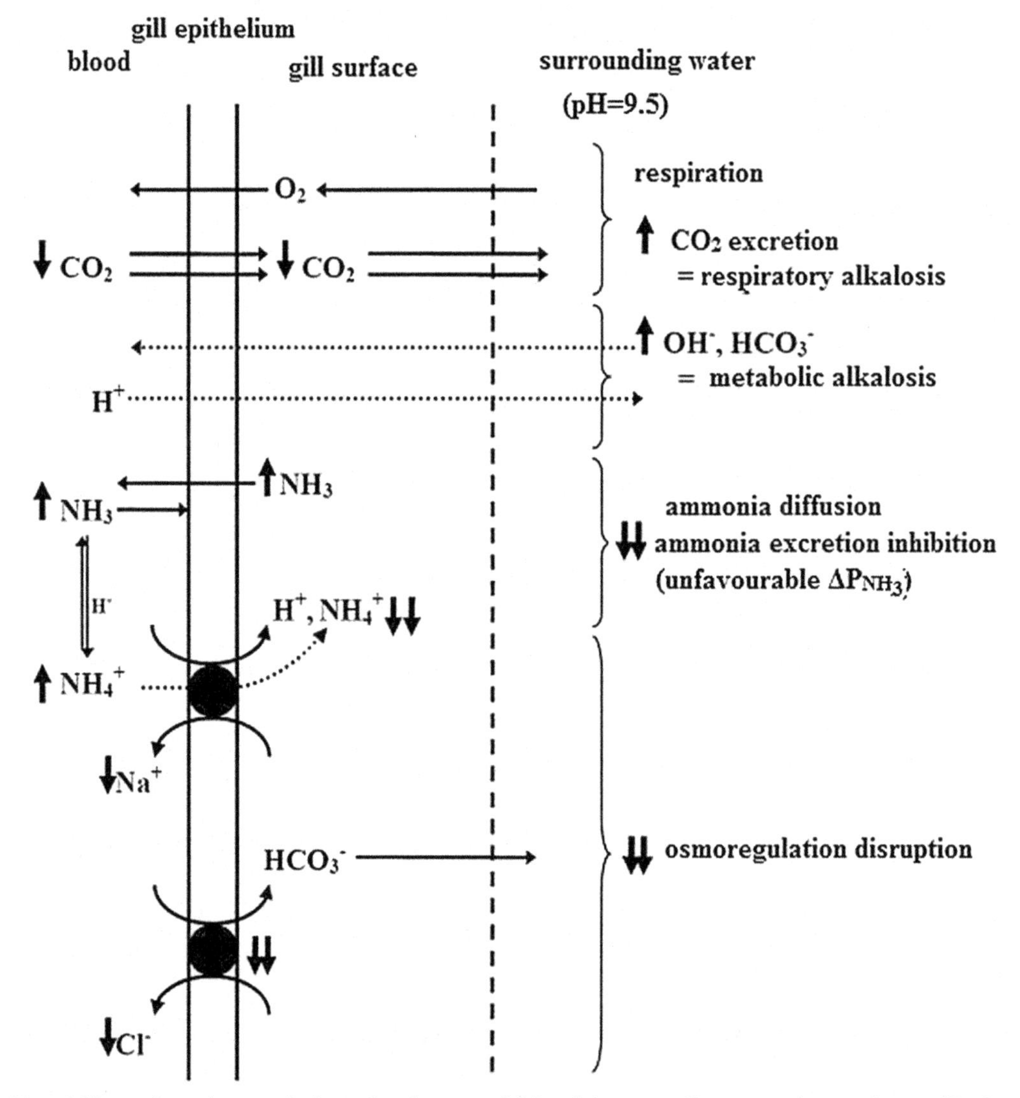

FIGURE 7.2 Essential ion exchange between the internal environment of fish and the surrounding water at increased water pH values. *Adapted from Wilkie, M.P., Wood, C.M., 1996. The adaptations of fish to extremely alkaline environments. Comparative Biochemisty and Physiology 113, 665–673.*

autointoxication cases usually coincide with sudden temperature drops (of 5–8°C), or drops in oxygen levels (from over-saturation to 20–40% saturation) taking place at times when the digestive tract of the fish is full (Svobodova et al., 1993; Olah et al., 1985). Cases of complete ammonia excretion cessation were described by Russo et al. (2006) in South American catfish (*R. quelen*) after exposure to hypoxic conditions (2.5 mg/L O_2) over 24 h, while reduced ammonia excretion was recorded in fish exposed to oxygen concentrations of 3.5 mg/L over 120 h.

A drop in water temperature or in oxygen levels significantly reduces the metabolic rate, as well as the efficiency of the Krebs citric acid cycle. This triggers mediator deficiency (α-ketoglutaric acid), disrupting ammonia transport into the gills, and leads to a subsequent rise in ammonia concentrations in the blood and to autointoxication.

7.4.3 Toxicity

Ammonia in its molecular NH_3 form ranks among substances that are highly toxic to fish. According to Svobodova et al. (1987), the 48-h lethal concentration, 50% (LC_{50}), value obtained for salmonids is 0.5–0.8 mg/L, and the highest allowable concentration is 0.0125 mg/L NH_3.

In cyprinids, the 48-h LC_{50} values for ammonia range within 1–1.5 mg/L NH_3. Exposure of common carp (*C. carpio*) fry to ammonia at a concentration of 0.2 mg/L over 150 h produced serious histopathological changes, especially in the respiratory epithelium of the gills (Svobodova et al., 1987). Common carp fry exposed between days 4 and 26 after hatching to ammonia concentrations of 0.14–0.40 mg/L NH_3 showed a conspicuous slowing down of development, a disruption of fin development (particularly in the pelvic fins), and a drop in the level of skin pigmentation (caused by loss of branches and reduction in melanophore size) (Machova et al., 1983).

Fish sensitivity to ammonia varies with age. The sensitivity of chub (*Squalius cephalus*) to free ammonia in its early development has been found by Gomulka et al. (2011) to drop with age and the degree of ontogenetic development. The 48-h LC_{50} value was 0.62 mg/L NH_3 for larvae tested 1 day after the onset of exogenous feeding, while for those tested 30 days after the onset of exogenous feeding it rose to 1.73 mg/L NH_3.

Free ammonia toxicity is related to water temperature, since the latter affects the metabolic rate, i.e., the intake and excretion of ammonia and the intensity of detoxification processes. Another important factor is the concentration of oxygen dissolved in water. Its deficiency reduces the metabolic rate and the intensity of detoxification processes, increasing in turn the fish's sensitivity to ammonia (Magaut et al., 1997).

7.4.4 Clinical Symptoms

Acute ammonia intoxication initially causes slight agitation in fish, an increase in opercular movement frequency, and the movement of fish toward the water surface. In cyprinids, emergency breathing and growing restlessness can be observed. Fish movements become progressively faster and opercular movement irregular, and the excitation phase sets in. Fish respond strongly to external stimuli, lose balance, and jump above the water surface and tonic–clonic muscle spasms occur. After that the fish turn onto their side, with their mouth and opercula convulsively open, and then, seemingly, a brief period of convalescence follows. The fish take up their normal position and show slight restlessness. This phase is followed by another strong excitation, the skin turning pale, and eventually death (Svobodova et al., 2008).

Sublethal ammonia concentrations may reduce or even stop food intake (Wicks and Randall, 2002).

When autointoxicated by ammonia, fish gather near the walls of reservoirs or in shaded sections of ponds and do not feed. Sharp jerky movements can be observed in them at times, with the fish jumping out of the water while showing muscle spasms. In advanced stages of autointoxication, the body surface becomes dark, the escape reflex is weakened or lost altogether, and death follows (Svobodova et al., 2008). In some cases, particularly in marketable common carp kept in tanks in autumn, autointoxication need not be acute, but may instead manifest itself as carp sleeping sickness, with the fish turning onto the side and showing reduced response to stimuli, but not dying.

7.4.5 Pathological–Morphologic Symptoms

The skin of fish intoxicated by ammonia is pale and covered in a strong or even very strong layer of mucus. In some cases small hemorrhages occur, especially at the base of pectoral fins and in the anterior chamber of the eyeball. The gills are strongly hyperemic and covered in mucus, and, in cases of exposure to high ammonia concentrations, gill hemorrhage occurs. The mucus on the skin and the gills is matted. Organs in the body cavity are hyperemic, and parenchymatous organs show dystrophic changes (Svobodova et al., 2008).

The surface of the body of fish suffering from ammonia autointoxication is darker. The gills show characteristic changes, i.e., strong hyperemia, dark red color, and strong edema. An injury to the gills causes strong bleeding. Respiratory epithelium necrosis follows, the epithelium comes off the gills, and sometimes the gill filament pillars are left completely bare (Svobodova et al., 2008). Gill hyperemia and strong edema are also characteristic of sleeping sickness in common carp (*C. carpio*) (Svobodova et al., 2007a,b).

7.4.6 Diagnosis

Ammonia poisoning in fish is diagnosed by physical–chemical water analysis. The concentration of total ammonia ($NH_3 + NH_4^+$) is established, and water temperature and pH are measured. The toxic effect depends on the nondissociated toxic ammonia (NH_3) concentration, which is calculated from the values obtained by measuring total ammonia ($NH_3 + NH_4^+$), water temperature, and pH, in accordance with Table 7.1.

In living fish with symptoms of poisoning, the diagnosis may be stated by blood plasma analysis for ammonia content. However, interpretation of the results must take into account other factors affecting ammonia concentration in blood plasma, e.g., the season of the year, the water temperature, the degree to which the digestive tract is full, and the quality of the feed supplied—particularly the concentration of nitrogen compounds in it. So, for instance, the physiological ammonia concentration in common carp is around 50–100 µmol/L in winter and 500–800 µmol/L in summer (Svobodova et al., 2008). The impact of food intake on the ammonia level in blood serum is demonstrated by results obtained by Wicks and Randall (2002), who state that ammonia levels in blood plasma of rainbow trout exposed to sublethal free ammonia concentrations (0.03 mg/L NH_3) were, on average, twice as high 6 h after feeding, compared to the average concentration in fish that have not been fed (14.43 and 6.78 µg/mL, i.e., 850 and 400 µmol/L, respectively).

In cases of poisoning, ammonia concentrations in blood plasma rise 5–10 times, and the same levels occur in autointoxication. Blood plasma analysis is therefore an essential instrument in diagnosing cases of ammonia intoxication and autointoxication.

7.4.7 Therapy

Suitable therapy is chosen individually, depending on particular circumstances. When fish are reared in ponds and reservoirs, a sufficient supply of clean water must be provided. If fish are reared in smaller tanks (aquaria), the water has to be exchanged, or the fish must be moved into clean water.

7.4.8 Prevention

Prevention consists in avoiding contamination of the aquatic environment with ammonia. To prevent autointoxication, sudden drops in water temperature and oxygen concentrations must be avoided (Svobodova et al., 2008).

7.5 NITRITE

7.5.1 Etiology

Increased nitrite levels (tenths to tens of mg/L NO_2^-) may occur in water in intensive farming units producing commercial and aquarium species, especially in recirculation systems. This may happen immediately after the operation is started or later, because of insufficient efficiency of biological filtering (Avnimelech et al., 1986; Kamstra et al., 1996; Dvorak, 2004; Svobodova et al., 2005). Biological filters employ a process of nitrification to reduce the concentration of ammonia, the principal waste product of fish nitrogen metabolism (Wood, 1993). In the course of nitrification, ammonia nitrogen is first oxidized biochemically to nitrites, and then to nitrates, which are less toxic to fish than both ammonia and nitrites. If the second nitrification phase is completely or even partly inhibited, nitrites accumulate in the system, causing deterioration of fish health and sometimes mass death (Svobodova et al., 2005).

7.5.2 Mechanism of Action

Exposed to nitrites, most freshwater fishes actively accumulate the nitrite ion in their blood plasma (Jensen, 2003). Nitrite concentrations in fish blood plasma therefore may reach levels over 60 times higher than those in the surrounding water (Fontenot et al., 1999). Although to a lesser degree, nitrites also accumulate in some tissues, namely in the gills, liver, brain, and muscle (Margiocco et al., 1983). They make their way into fish bodies through the eosinophilic chloride cells of the gills, which, among other processes, carry out Cl^-/HCO_3^- ion exchange (uptake/excretion) between the organism and the surrounding aquatic environment. Nitrites act as competitive inhibitors of chloride uptake, and the same is true about the reverse process (Crawford and Allen, 1977; Harris and Coley, 1991).

Fish sensitivity to nitrites is therefore closely interrelated with the rate of their chloride uptake. Fishes demonstrating a high rate of chloride uptake through gills, e.g., rainbow trout (*O. mykiss*), European perch (*Perca fluviatilis*), channel catfish (*I. punctatus*), and northern pike (*E. lucius*), have been proved very sensitive to nitrites. Conversely, fishes with a low or zero chloride uptake rate, e.g., European eel (*Anguilla anguilla*), common carp (*C. carpio*), tench (*Tinca tinca*), and bluegill (*L. macrochirus*), are more nitrite resistant (Williams and Eddy, 1986, 1988; Tomasso and Grosell, 2005). The mechanism of nitrite uptake also explains why increased Cl^- concentrations in water protect fish from absorbing nitrites and from their toxic effects (Jensen, 2003).

In fish, nitrites move from blood plasma further into erythrocytes, where they oxidize bivalent iron in hemoglobin into trivalent iron, turning it into methemoglobin, which lacks the oxygen-carrying capacity (Bodansky, 1951). Since methemoglobin is brown, its increased concentrations in blood are manifested by brown coloring of the blood and gills. Even in the absence of nitrites, fish blood contains a certain proportion of methemoglobin naturally, at concentrations possibly reaching 10% (Cameron, 1971). This is because hemoglobin in fish is more prone to autoxidation by oxygen than in mammals (Beutler, 1968; Kiese, 1974). When fish are moved into clean water after nitrite poisoning, methemoglobin content usually drops to physiological levels within 24–72 h (Huey et al., 1980; Knudsen and Jensen, 1997), unless the poisoning has progressed to an advanced stage. In addition, nitrites affect processes regulating ion levels in the body (Jensen et al., 1987; Knudsen and Jensen, 1997; Gisbert et al., 2004), respiratory processes (Cameron, 1971; Aggergaard and Jensen, 2001), cardiovascular functions (Aggergaard and Jensen, 2001), functions of endocrine glands, and excretory functions (Jensen, 2003). Moreover, fish acutely or chronically exposed to sublethal nitrite concentrations develop infectious diseases more easily, which suggests that nitrites negatively affect the fish immune system (Hanson and Grizzle, 1985; Carballo et al., 1995).

7.5.3 Toxicity

Nitrite sensitivity varies substantially across fishes, with acute lethal concentrations ranging from ones to hundreds of mg/L NO_2^-. Among the most sensitive fishes are salmonids. Conversely, largemouth bass (*Micropterus salmoides*) (Palachek and Tomasso, 1984) and bluegill (Tomasso and Grosell, 2005) are highly nitrite-resistant species. It is to be noted, though, that concentrations lethal to different species can be compared only if they were obtained from tests using dilution water of comparable chloride concentrations. Studies that did not include measuring of the chloride concentration in the water have very low validity, because this parameter affects nitrite toxicity more significantly than all the other factors explored.

Nitrite toxicity to fish depends substantially on a number of internal and external factors. The most important external factors affecting nitrite toxicity include water quality, in particular, temperature, saturation with oxygen, and cation and anion concentrations (especially chlorides), and, obviously, length of exposure (Crawford and Allen, 1977; Palachek and Tomasso, 1984; Kroupova et al., 2006). Among the significant internal factors are the species, age, and size of the fish, as well as their individual sensitivity (Palachek and Tomasso, 1984; Bartlett and Neumann, 1998; Aggergaard and Jensen, 2001).

Tests of several fish species have confirmed a linear relationship between nitrite toxicity and chloride concentrations in water (Palachek and Tomasso, 1984; Machova and Svobodova, 2001; Kroupova et al., 2004). The European Inland Fisheries Advisory Commission recommends keeping the $Cl^-/N\text{-}NO_2^-$ weight ratio above 17, in salmonid farming, and above 8 in the farming of secondary fish species (EIFAC, 1984). However, our experience shows that death or damage to fish may occur even at higher $Cl^-/N\text{-}NO_2^-$ ratios. Deaths of wels catfish (*Silurus glanis*) and tench (*T. tinca*) reared in intensive recirculating aquaculture systems were recorded at $Cl^-/N\text{-}NO_2^-$ weight ratio values in the range of 13–28 and 11–19, respectively (Svobodova et al., 2005).

Among other water quality parameters, the presence of bromides has been proved to reduce nitrite toxicity to fish (Eddy et al., 1983), due to their chemical properties that are similar to chlorides. Some authors observed a positive effect of increased calcium concentrations in reducing nitrite toxicity to fish, although their effect was never as prominent as that of chlorides (Crawford and Allen, 1977; Mazik et al., 1991; Fontenot et al., 1999). Other anions and cations are insignificant with respect to nitrite toxicity, and similarly negligible is the effect of water pH, particularly at values commonly occurring in natural waters (Lewis and Morris, 1986). On the other hand, factors potentially increasing fish sensitivity to nitrites include low oxygen concentrations in water (Bowser et al., 1983) and higher water temperature (Huey et al., 1984; Kroupova et al., 2006). At the same time, higher water temperature can accelerate the process of fish regeneration following nitrite poisoning (Huey et al., 1984).

Fish sensitivity to nitrites has also been found to vary with age. Repeated tests have shown that older fish are more sensitive to nitrites than in their earlier developmental stages (Russo et al., 1974; Bartlett and Neumann, 1998; Kroupova et al., 2010).

Reduced sensitivity of early developmental stages of fish may be due to methemoglobin reductase (an enzyme converting methemoglobin back to hemoglobin) activity, which is higher in young individuals compared to older ones. Another explanation may be that, unlike juveniles and adults, early developmental stages do not possess a fully developed gill system, which means that they primarily employ skin respiration and transport of oxygen in the blood is less important (Kroupova et al., 2005).

7.5.4 Clinical Symptoms

The fish are lethargic and lose their escape reflex, and tonic–clonic muscle spasms occur. These symptoms are followed by agony and death in the lateral or dorsal position.

7.5.5 Pathological–Morphologic Symptoms

The macroscopic indication of nitrite poisoning is brown color of the blood and gills. In gills, the brown coloring is apparent at a methemoglobin content of c. 25% (Svobodova et al., 2005). Microscopically, nitrite poisoning manifests itself as edema and gill epithelium hyperplasia, or an increased presence of eosinophilic chloride cells.

7.5.6 Diagnosis

The first step in diagnosing cases of nitrite poisoning in fish is water analysis to determine nitrite and chloride concentrations. The diagnosis can then be corroborated if high methemoglobin levels in blood are detected. In nitrite-poisoned fish, methemoglobin concentration values are ≥20–25% of total hemoglobin, and death usually occurs at methemoglobin concentrations of over 70% (Svobodova et al., 2008). The fish affected also have high nitrite concentrations in their blood plasma and in tissues, often exceeding those in the surrounding water.

7.5.7 Therapy

Fish suffering nitrite-induced damage must be moved into clean water as soon as possible.

7.5.8 Prevention

The essential preventive measures taken in fish-rearing recirculating systems include the following (Svobodova et al., 2005):

1. When operation starts, fish should be stocked gradually, and feed portions should be optimized in accordance with the current biofilter efficiency, to guarantee standards of water quality required for fish farming. Full stocking is possible only after the biofilters become fully functional. The facility should be monitored continually for pH values and for concentrations of oxygen, ammonia, nitrites, nitrates, and chlorides, as well as for the chloride–nitrite weight or molar ratio.
2. Treatment of fish in recirculating systems should not use antibiotics or disinfectants in the form of bath. Therapy must be conducted outside the recirculating section to prevent biofilter damage (Svobodova et al., 2008).
3. Chloride concentrations should be kept above 100 mg/L Cl^- in fish-rearing facilities where nitrite accumulation is a threat. Increased chloride concentrations in the water may be achieved by addition of common salt (sodium chloride). A dose of 165 mg of salt per liter raises chloride concentrations by 100 mg/L.
4. The feed supplied to fish should be rich in ascorbic acid (c. 8000 mg/kg) to prevent nitrite-induced methemoglobinemia (Wise et al., 1988).

7.6 CHLORINE

7.6.1 Etiology

Chlorine and active chlorine-releasing compounds have long been used as disinfectants in both human and veterinary medicine. However, it was found as early as the mid-1970s that chlorination might produce substances potentially dangerous to fish and humans (Ferraris et al., 2005). This danger was pointed out, e.g., by Gehrs et al. (1974), who stated that chlorine reacted with a wide range of organic substances in the course of chlorination, producing stable organochlorine compounds, potentially dangerous to fish even at very low concentrations of about 1 μg/L. As the insight into the effects of chlorination on the aquatic ecosystem grew deeper in the following years, more risks were identified, not only to fish, but also to humans. It is now common knowledge that chemical disinfectants react with organic compounds, both those naturally occurring in water and contaminants. A range of disinfection by-products arise in the process, some of them stable, others unstable. So, for instance, a reaction with phenols produces unpleasantly smelling chlorophenols, while contact of chlorine with some organic substances (e.g., humic substances) triggers the haloform reaction, in the course of which various trihalomethanes are formed, as well as halogen organic compounds and hydrofurans, which are insalubrious. In a number of tests conducted on animals, many chlorination products were later proved to possess carcinogenic, mutagenic, and/or teratogenic effects. Some epidemiological studies published in the United States, Finland, and Taiwan describe a positive relationship between chlorinated drinking water mutagenicity and the prevalence of tumor diseases of the urinary and digestive tracts (Bull, 1985; Monarca et al., 2004; Pitter, 2009).

To cut these potential health risks, new methods of water disinfection are being explored, such as oxidation with ozone or peracetic acid, inactivation by UV radiation, and microfiltration (Ferraris et al., 2005).

The effects of active chlorine have aquacultural applications. Sodium hypochlorite (chlorine powder) is used for a complete disinfection of pond bottoms (500–600 kg/ha), storage tanks and other facilities used in fish farming and transport. Chlorine powder is recommended in cases of gill disease in fish (e.g., branchiomycosis), applied to the pond surface in doses of 10–15 kg/ha, with the average depth of 1 m. Similarly, chloramine B and chloramine T are very common disinfectants in fish farming. They are used in antibacterial fish treatment, and they have proved to be very effective in gill flavobacteriosis treatment in salmonids (Svobodova et al., 2007a; Noga, 2010).

An overdose or improper application of chlorine and active chlorine-releasing compounds causes damage to fish or even death. Damage to fish from chlorine also occurs when marketable fish are kept in fish shops in tanks supplied with drinking water from the tap, which may contain 0.05–0.3 mg/L of active chlorine (Svobodova et al., 2007a).

7.6.2 Mechanism of Action

Active chlorine affects fish bodies both locally and globally. Local chlorine effects are particularly apparent on the skin and the gills. Typical symptoms of skin damage include pale color, anemia, and increased mucus production. Gill epithelium cells are edemic and swollen, and histological examination reveals vacuolization and swelling of gill epithelium cells, as well as an increase in the number of chloride cells (Mahjoor and Loh, 2008). More serious damage causes respiratory cell dystrophy or even necrosis, followed by desquamation (Svobodova et al., 1987, 1993; Powell et al., 1995).

Overall effects (after chlorine has been absorbed into blood) are, in particular, nervous system disorders. Free chlorine binds to the SH groups in amino acids, which are constituent parts of enzymes; forms firm covalent bonds; and irreversibly impairs the activity of these enzymes. This explains why chlorine-poisoned fish that have lost their balance do not recover after being moved into clean water (Alabaster and Lloyd, 1980). Fish exposed to active chlorine, particularly in chlorite form, show mild symptoms of methemoglobinemia. After long-term exposure, symptoms of hemolytic anemia combined with increased potassium concentration and a drop in sodium concentrations in blood plasma were identified (Buckley, 1977).

Genotoxic effects of sodium hypochlorite (NaClO) at a concentration of 0.5 mg/L after 36 h of exposure have been shown in Angora loach (*Nemacheilus angorae*), on the basis of increased erythrocyte micronucleus frequency (Gül et al., 2008). The effects of chloramine T and chlorine dioxide have been studied on the basis of oxidative stress biomarkers [thiobarbituric acid-reactive substances (TBARS), carbonyl derivate concentration] in the cardiac muscle tissue of rainbow trout (*O. mykiss*). The findings suggest that treatment of fish with chlorine dioxide led to a significant rise in TBARS levels and in carbonyl derivatives, while no change in oxidative stress markers was recorded after chloramine T application (Tkachenko et al., 2015).

7.6.3 Toxicity

Active chlorine is strongly toxic to fish. The intensity of chlorine's effects on fish depends on the water quality, especially on the level of dissolved oxygen (with higher oxygen concentrations reducing the toxic effect of chlorine). Water temperature and pH also affect the forms in which chlorine occurs, hence also active chlorine toxicity (Alabaster and Lloyd, 1980). Active chlorine concentrations of 3.5 mg/L at temperatures of 3–7°C are sublethal to common carp; however, at 15–20°C the same concentrations cause death in 1–2 h. Active chlorine concentrations of 0.6–0.7 mg/L at 18–20°C caused death in 24 h, and concentrations of 0.4 mg/L at 18–20°C led to common carp dying in 6–7 days. Generally, active chlorine concentrations of 0.04–0.2 mg/L can be considered lethal to most fishes (Fisher et al., 1999; Svobodova et al., 1987, 2008).

An interesting observation concerning adaptability to increased active chlorine concentrations has been made by Alabaster and Lloyd (1980), who demonstrated increasing resistance in fathead minnow (*Pimephales promelas*). Fish exposed for 3 weeks to concentrations roughly corresponding to 10%, 20%, and 50% 4-day LC_{50} active chlorine concentrations were able to survive the following week at concentrations of roughly 1.5 times 4-day LC_{50}.

In addition to direct effects on fish, chlorine may have an impact on the effects of other substances present in water (e.g., the rise of highly toxic substances as a consequence of chlorination). A possibility has been described of abiotic chlorpyrifos (the active component of insecticides from the organophosphate group) transformation into chlorpyrifos oxon, whose anticholinesterase effect is, by up to three orders, more potent than that of the original substance (Wu and Laird, 2003). Conversely, the chlorine oxidation capacity is used to reduce cyanide toxicity by oxidizing it into cyanates, which are up to 1000 times less toxic than free cyanides (Alabaster and Lloyd, 1980).

7.6.4 Clinical Symptoms

Clinical symptoms initially include intense restlessness, with fish jumping out of water, and muscle spasms. After that, the fish turn on their side and convulsive movements are observed in the mouth, fins, and tail. Mouth spasms disturb the breathing rhythm, the fish suffocate, become lethargic, and die (Alabaster and Lloyd, 1980; Svobodova et al., 1987, 2008).

7.6.5 Pathological–Morphologic Symptoms

The skin and gills of poisoned fish are covered in a thick layer of mucus, and high active chlorine concentrations cause gill hyperemia and possibly hemorrhage. The body becomes pale, and the edges of the gill filaments and fins bear a distinct gray-white coating. Histopathological examination reveals dystrophy or even necrosis and desquamation of gill epithelium and skin epidermis (Powell et al., 1995; Svobodova et al., 1987, 2008).

7.6.6 Diagnosis

Diagnosis is based on the assessment of clinical symptoms and pathological changes, supported by anamnestic data, and active chlorine concentrations detected in water. If the laboratory is not equipped to measure active chlorine content in water, or if samples cannot be tested immediately after they were collected (recommended time is within 5 min of collecting), sensory evaluation of the water is recommended as a minimum. This evaluation employs threshold odor concentration, which is very low in active chlorine. It is about 0.075 mg/L at pH 5, c. 0.156 mg/L at pH 7, and c. 0.450 mg/L at pH 9 (Pitter, 2009).

7.6.7 Therapy

If possible, the fish should be moved into clean unchlorinated water.

7.6.8 Prevention

Prevention of chlorine poisoning in fish consists in strict observance of technological procedures of disinfection and treatment of fish. Before fish are stocked in an aquatic environment potentially containing active chlorine, the water must be tested for chlorine, or a biological test of water toxicity must be conducted. Special attention is needed if drinking water is used for transporting or storing fish. It has to be borne in mind that residual active chlorine concentrations in drinking water may be up to 0.3 mg/L. In some cases, reduction of active chlorine levels may be achieved by allowing the water to stand, and by aeration, but these measures may not be sufficient.

ACKNOWLEDGMENTS

This work was supported by the projects CENAKVA (No. CZ.1.05/2.1.00/01.0024) and CENAKVA II (No. LO1205 under the NPU I program).

REFERENCES

Aggergaard, S., Jensen, F.B., 2001. Cardiovascular changes and physiological response during nitrite exposure in rainbow trout. Journal of Fish Biology 59, 13–27.

Alabaster, J.S., Lloyd, R., 1980. Water Quality Criteria for Freshwater Fish. Butterworths, Toronto.

Altinok, I., Grizzle, J.M., 2004. Excretion of ammonia and urea by phylogenetically diverse fish species in low salinities. Aquaculture 238, 499–507.

Andrews, C., Exell, A., Carrington, N., 2010. Manual of Fish Health: Everything You Need to Know about Aquarium Fish, Their Environment and Disease Prevention. Firefly Books, Tonawanda.

Araujo, C.V., Cohin-De-Pinho, S.J., Chastinet, C.B., Santos, J.S., Da Silva, E.M., 2008. Discriminating the pH toxicity to *Poecilia reticulata* Peters, 1859 in the Dunas Lake (Camacari, BA, Brazil). Chemosphere 73, 365–370.

Avnimelech, Y., Weber, B., Hepher, B., Milstein, A., Zorn, M., 1986. Studies in circulated fish ponds: organic matter recycling and nitrogen transformation. Aquaculture and Fisheries Management 17, 231–242.

Bartlett, F., Neumann, D., 1998. Sensitivity of brown trout alevins (*Salmo trutta* L.) to nitrite at different chloride concentrations. Bulletin of Environmental Contamination and Toxicology 60, 340–346.

Beutler, E., 1968. Hereditary Disorders of Erythrocyte Metabolism. Grune and Stratton, New York.

Bodansky, O., 1951. Methemoglobinemia and methemoglobin producing compounds. Pharmacological Review 3, 144–196.

Bolner, K.C.S., Copatti, C.E., Rosso, F.L., Loro, V.L., Baldisserotto, B., 2014. Water pH and metabolic parameters in silver catfish (*Rhamdia quelen*). Biochemical Systematics and Ecology 56, 202–208.

Bowser, P.R., Falls, W.W., Van Zandt, J., Collier, N., Phillips, J.D., 1983. Methemoglobinemia in channel catfish: methods of prevention. Progressive Fish Culturist 45, 154–158.

Buckley, J.A., 1977. Heinz body hemolytic anemia in coho salmon (*Oncorhynchus kisutch*) exposed to chlorinated wastewater. Journal of the Fisheries Research Board of Canada 34, 215–224.

Bull, R.J., 1985. Carcinogenic and mutagenic properties of chemicals in drinking water. Science of the Total Environment 47, 385–413.

Cameron, J.N., 1971. Methemoglobin in erythrocytes of rainbow trout. Comparative Biochemistry and Physiology A 40, 743–749.

Carballo, M., Munoz, M.J., Cuellar, M., Tarazona, J.V., 1995. Effect of waterborne copper, cyanide, ammonia, and nitrite on stress parameters and changes in susceptibility to saprolegniosis in rainbow trout (*Oncorhynchus mykiss*). Applied and Environmental Microbiology 61, 2108–2112.

Copatti, C.E., Garcia, L.O., Kochhann, D., Cunha, M.A., Baldisserotto, B., 2011. Dietary salt and water pH effects on growth and Na^+ fluxes of silver catfish juveniles. Acta Scientiarum. Animal Sciences 33, 261–266.

Crawford, R.E., Allen, G.H., 1977. Seawater inhibition of nitrite toxicity to Chinook salmon. Transactions of the American Fisheries Society 106, 105–109.

Currie, S., Moyes, C.D., Tufts, B.L., 2000. The effect of head shock and acclimation temperature on hsp70 and hsp30 mRNA expression in rainbow trout: in vivo and vitro comparisons. Journal of Experimental Biology 56, 398–408.

Delage, N., Cachot, J., Rochard, E., Fraty, R., Jatteau, P., 2014. Hypoxia tolerance of European sturgeon (*Acipenser sturio* L., 1758) young stages at two temperatures. Journal of Applied Ichthyology 30, 1195–1202.

Di Giulio, R.T., Hinton, D.E., 2008. The Toxicology of Fishes. CRC Press Taylor & Francis Group, Boca Raton.

Dunson, W.A., Swarts, F., Silvestri, M., 1977. Exceptional tolerance to low pH of some tropical blackwater fish. Journal of Experimental Zoology 201, 157–162.

Dvorak, P., 2004. Vybraná specifika onemocnění akvarijních ryb. Bulletin VÚRH Vodňany 40, 101–108.

Dwyer, G.K., Stoffels, R.J., Pridmore, P.A., 2014. Morphology, metabolism and behaviour: responses of three fishes with different lifestyles to acute hypoxia. Freshwater Biology 59, 819–831.

Eddy, F.B., Kunzlik, P.A., Bath, R.N., 1983. Uptake and loss of nitrite from the blood of rainbow trout, *Salmo gairdneri* Richardson, and Atlantic salmon, *Salmo salar* L. in fresh water and in dilute sea water. Journal of Fish Biology 23, 105–116.

EIFAC (European Inland Fisheries Advisory Commission), 1984. Water Quality Criteria for European Freshwater Fish: Report on Nitrite and Freshwater Fish. EIFAC Technical Paper, No. 46.

Evans, D.H., Piermarini, P.M., Potts, W.T.W., 1999. Ionic transport in the fish gill epithelium. Journal of Experimental Zoology 283, 641–652.

Faina, R., Machova, J., Valentova, O., 2011. Možnost řešení kritických deficitů kyslíku v rybničním chovu ryb pomocí aplikace nízké dávky superfosfátu. Edice metodik, FROV JU, Vodňany. no. 116.

Feige, U., Morimoto, R.J., Yahara, J., Polla, B.S., 1996. Stress – Inducible Cellular Responses. Birkhauser, Basel.

Ferraris, M., Chiesara, E., Radice, S., Giovara, A., Fregeiro, S., Marabini, L., 2005. Study of potential toxic effects on rainbow trout hepatocytes of surface water treated with chlorine or alternative disinfectants. Chemosphere 60, 65–73.

Fisher, D.J., Burton, D.T., Yonkos, L.T., Turley, S.D., Ziegler, G.P., 1999. The relative acute toxicity of continuous and intermittent exposures of chlorine and bromine to aquatic organisms in the presence and absence of ammonia. Water Research 33, 760–768.

Fontenot, Q.C., Isely, J.J., Tomasso, J.R., 1999. Characterization and inhibition of nitrite uptake in shortnose sturgeon fingerlings. Journal of Aquatic Animal Health 11, 76–80.

Gehrs, C.W., Eyman, L.D., Jolley, R.L., Thompson, J.E., 1974. Effects of stable chlorine-containing organics on aquatic environments. Nature 249, 675–676.

Gisbert, E., Rodriguez, A., Cardona, L., Huertas, M., Gallardo, M.A., Sarasquete, C., Sala-Rabanal, M., Ibarz, A., Sanchez, J., Castello-Orvay, F., 2004. Recovery of Siberian sturgeon yearlings after an acute exposure to environmental nitrite: changes in the plasmatic ionic balance, Na^+–K^+ ATPase activity, and gill histology. Aquaculture 239, 141–154.

Golombieski, J.I., Koakoski, G., Becker, A.J., Almeida, A.P.G., Toni, C., Finamor, I.A., Pavanato, M.A., De Almeida, T.M., Baldisserotto, B., 2013. Nitrogenous and phosphorus excretions in juvenile silver catfish (*Rhamdia quelen*) exposed to different water hardness, humic acid, and pH levels. Fish Physiology and Biochemistry 39, 837–849.

Gomulka, P., Zarski, D., Kucharszyk, D., Kupren, K., Krejczeff, S., Targonska, K., 2011. Acute ammonia toxicity during early ontogeny of chub, *Leuciscus cephalus* (Cyprinidae). Aquatic Living Resources 24, 211–218.

Gül, S., Ozkan, O., Nur, G., Aksu, P., 2008. Genotoxic effects and LC50 value of NaOCl on *Orthrias angorae* (Steindachner 1897). Bulletin of Environmental Contamination and Toxicology 80, 544–548.

Gultepe, N., Ates, O., Hisar, O., Beydemir, S., 2011. Carbonic anhydrase activities from the rainbow trout correspond to the development of acute gas bubble disease. Journal of Aquatic Animal Health 23, 134–139.

Hanson, L.A., Grizzle, J.M., 1985. Nitrite-induced predisposition of channel catfish to bacterial diseases. The Progressive Fish Culturist 47, 98–101.

Harris, R.R., Coley, S., 1991. The effects of nitrite on chloride regulation in the crayfish *Pacifastacus leniusculus* Dana (Crustacea: Decapoda). Journal of Comparative Physiology 161, 199–206.

Huey, D.W., Simco, B.A., Criswell, D.W., 1980. Nitrite-induced methemoglobin formation in channel catfish. Transactions of the American Fisheries Society 109, 558–562.

Huey, D.W., Beitinger, T.L., Wooten, M.C., 1984. Nitrite-induced methemoglobin formation and recovery in channel catfish (*Ictalurus punctatus*) at three acclimation temperatures. Bulletin of Environmental Contamination and Toxicology 32, 674–681.

Iwama, G., Vijayan, M., Morgan, J., 2005. Stress and ion regulation in fish. Comparative Biochemistry and Physiology 141, 181–182.

Jensen, F.B., 2003. Nitrite disrupts multiple physiological functions in aquatic animals. Comparative Biochemistry and Physiology 135, 9–24.

Jeney, G., Nemcsók, J., Jeney, Z., Oláh, J., 1992a. Acute effect of sublethal ammonia concentrations on common carp (*Cyprinus carpio* L.). II Effect of ammonia on blood plasma transaminases (GOT, GPT), G1DH enzyme activity, and ATP value. Aquaculture 104, 149–156.

Jeney, Z., Nemcsok, J., Jeney, G., Olah, J., 1992b. Acute effect of sublethal ammonia concentrations on common carp (*Cyprinus carpio* L.) II. Effect of ammonia on adrenalin and noradrenaline levels in different organs. Aquaculture 104, 139–156.

Jensen, F.B., Andersen, N.A., Heisler, N., 1987. Effects of nitrite exposure on blood respiratory properties, acid–base and electrolyte regulation in the carp (*Cyprinus carpio*). Journal of Comparative Physiology 157, 533–541.

Kamstra, A., Span, J.A., Van Weerd, J.H., 1996. The acute toxicity and sublethal effects of nitrite on growth and feed utilization of European eel. Aquaculture Research 27, 903–911.

Kiese, M., 1974. Methemoglobinemia: A Comprehensive Treatise. CRC Press, Cleveland.

Knudsen, P.K., Jensen, F.B., 1997. Recovery from nitrite-induced methaemoglobinaemia and potassium balance disturbances in carp. Fish Physiology and Biochemistry 16, 1–10.

Kroupova, H., Machova, J., Piackova, V., Flajshans, M., Svobodova, Z., Poleszczuk, G., 2006. Nitrite intoxication of common carp (*Cyprinus carpio* L.) at different water temperatures. Acta Veterinaria Brno 75, 561–569.

Kroupova, H., Machova, J., Svobodova, Z., Valentova, O., 2004. Akutní toxicita dusitanů pro *Poecilia reticulata*. In: Spurný, P. (Ed.), 55 let výuky rybářské specializace na MZLU v Brně, 2004, Brno, pp. 237–245.

Kroupova, H., Machova, J., Svobodova, Z., 2005. Dusitany ve vodním prostředí a jejich účinky na ryby – přehled. Bulletin VÚRH Vodňany 41, 154–170.

Kroupova, H., Prokes, M., Macova, S., Penaz, M., Barus, V., Novotny, L., Machova, J., 2010. Effect of nitrite on early-life stages of common carp (*Cyprinus carpio* L.). Environmental Toxicology and Chemistry 29, 535–540.

Le Louarn, H., Webb, D.J., 1998. Negative effects of extreme environmental pH on embryonic and larval development of the pike *Esox lucius* L. Bulletin Français de la Pêche et de la Pisciculture 350, 325–336.

Lewis, J.M., Klein, G., Walsh, P.J., Currie, S., 2012. Rainbow trout (*Oncorhynchus mykiss*) shift the age composition of circulating red blood cells towards a younger cohort when exposed to thermal stress. Journal of Comparative Physiology 182, 663–671.

Lewis, W.M., Morris, D.P., 1986. Toxicity of nitrite to fish: a review. Transactions of the American Fisheries Society 115, 183–195.

Lloyd, R., 1961. Effect of dissolved oxygen concentrations on the toxicity of several poisons to rainbow trout (*Salmo gairdneri*, Richardson). Journal of Experimental Biology 38, 447–455.

Lloyd, R., 1992. Pollution and Freshwater Fish. Fishing News Books, Oxford.

Madeira, D., Narciso, L., Cobral, H.N., Vinagre, C., Diniz, M.S., 2012. HSP70 production patterns in costal and estuarine organisms facing increasing temperatures. Journal of Sea Research 73, 137–147.

Magaut, H., Migeon, B., Morfin, P., Garric, J., Vindimian, E., 1997. Modelling fish mortality due to urban storm run-off: interacting effects of hypoxia and un-ionized ammonia. Water Research 31, 211–218.

Mahjoor, A.A., Loh, R., 2008. Some histopathological aspects of chlorine toxicity in rainbow trout. Asian Journal of Animal and Veterinary Advances 3, 303–306.

Machova, J., Penaz, M., Kouril, J., Hamackova, J., Machacek, J., Groch, L., 1983. Vliv různých hodnot pH a zvýšené koncentrace amoniaku na růst a ontogenetický vývoj plůdku kapra. Buletin VÚRH Vodňany 19, 3–14.

Machova, J., Svobodova, Z., 2001. Nitrite toxicity for fish under experimental – and farming conditions – poster. In: 10th International Conference of the EAFP Diseases of Fish and Shellfish 2001, Dublin, Irsko.

Margiocco, C., Arillo, A., Mensi, P., Shenone, G., 1983. Nitrite bioaccumulation in *Salmo gairdneri* and haematological consequences. Aquatic Toxicology 3, 261–270.

Mazik, P.M., Hinman, M.L., Winkelmann, D.A., Klaine, S.J., Simco, B.A., 1991. Influence of nitrite and chloride concentrations on survival and hematological profiles of striped bass. Transactions of the American Fisheries Society 120, 247–254.

Mischke, C.H.C., Chatakondi, N., 2012. Effects of abrupt pH increases on survival of different ages of young channel catfish and hybrid catfish. North American Journal of Aquaculture 74, 160–163.

Monarca, S., Zani, C., Richardson, S.D., Thurston, A.D.J.R., Moretti, M., Feretti, D., Villarini, M., 2004. A new approach to evaluating the toxicity and genotoxicity of disinfected drinking water. Water Research 38, 3809–3816.

Mueller, C.A., Eme, J., Manzon, R.G., Somers, C.M., Boreham, D.R., Wilson, J.Y., 2015. Embryonic critical windows: changes in incubation temperature alter survival, hatchling phenotype, and cost of development in lake whitefish (*Coregonus clupeaformis*). Journal of Comparative Physiology B 185, 315–331.

Noga, E.J., 2010. Fish Disease Diagnosis and Treatment. Wiley-Blackwell.

Olah, J., Papp, J., Meszaros-Kiss, A., Mucsy, G., Kallo, D., 1985. Simultaneous separation of suspended solids, ammonium and phosphate ion from wastewater by modified clinoptilolite. Studies in Surface Science and Catalysis 46, 711–719.

Palachek, R.M., Tomasso, J.R., 1984. Toxicity of nitrite to channel catfish (*Ictalurus punctatus*), tilapia (*Tilapia aurea*), and largemouth bass (*Micropterus salmoides*): evidence for a nitrite exclusion mechanism. Canadian Journal of Fisheries and Aquatic Sciences 41, 1739–1744.

Parra, J.E.G., Baldisserotto, B., 2007. Effect of water pH and hardness on survival and growth of freshwater teleosts. In: Baldisserotto, B., Mancera, J.M., Kapoor, B.C. (Eds.), Fish Osmoregulation. Enfiled: Science Publishers.

Pitter, P., 1981. Hydrochemie. SNTL, Praha.

Pitter, P., 2009. Hydrochemie. VŠCHT, Praha.

Powell, M.D., Wright, G.M., Speare, D.J., 1995. Morphological changes in rainbow trout (*Oncorhynchus mykiss*) gill epithelia following repeated intermittent exposure to Chloramine-T. Canadian Journal of Zoology 73, 154–165.

Reinchenbach-Klinke, H.H., 1975. Fisch und Umwelt. Georg Fischer, München.

Rosso, F.L., Bolner, K.C.S., Baldisserotto, B., 2006. Ion fluxes in silver catfish (*Rhamdia quelen*) juveniles exposed to different dissolved oxygen levels. Neotropical Ichthyology 4, 435–440.

Russo, R.C., Randall, D.J., Thurston, R.V., 1988. Ammonia toxicity and metabolism in fishes. In: Ryans, R.C. (Ed.), Protection of River Basins, Lakes, and Estuaries. American Fisheries Society, Bethesda.

Russo, R.C., Smith, C.E., Thurston, R.V., 1974. Acute toxicity of nitrite to rainbow trout (*Salmo gairdneri*). Journal of the Fisheries Research Board of Canada 31, 1653–1655.

Schreckenbach, K., Spangenberg, R., 1978. pH-Wert-abhängige Ammoniakvergiftungen bei Fischen und Möglichkeiten ihrer Beeinflussung. Zeitschrift Binnenfischerei 22, 299–314.

Simcic, T., Jesensek, D., Brancelj, A., 2015. Effects of increased temperature on metabolic activity and oxidative stress in the first life stages of marble trout (*Salmo marmoratus*). Fish Physiology and Biochemistry 41, 1005–1014.

Smutna, M., Vorlova, L., Svobodova, Z., 2002. Pathobiochemistry of ammonia in the internal environment of fish (Review). Acta Veterinaria Brno 71, 169–181.

Snieszko, S.F., 1974. The effects of environmental stress on outbreaks of infectious diseases of fishes. Journal of Fish Biology 6, 197–208.

Svobodova, Z., Gelnarova, J., Justyn, J., Krupauer, V., Machova, J., Simanov, L., Valentova, V., Vykusova, B., Wolgemuth, E., 1987. Toxikologie vodních živočichů. SZN, Praha.

Svobodova, Z., Kolarova, J., Navratil, S., Vesely, T., Chloupek, P., Tesarcik, J., Citek, J., 2007a. Nemoci sladkovodních a akvarijních ryb. Informatorium, Praha.

Svobodova, Z., Lloyd, R., Machova, J., Vykusova, B., 1993. Water Quality and Fish Health. EIFAC Technical Paper 54. FAO, Rome.

Svobodova, Z., Machova, J., Hamackova, J., Huda, J., Kroupova, H., 2005. Nitrite poisoning of fish in aquaculture facilities with water-recirculating systems: three case studies. Acta Veterinaria Brno 74, 129–137.

Svobodova, Z., Machova, J., Kroupova, H., 2008. Otravy ryb. In: Svobodová, Z. (Ed.), Veterinární toxikologie v klinické praxi. Profi Press, Praha.

Svobodova, Z., Machova, J., Kroupova, H., Smutna, M., Groch, L., 2007b. Ammonia autointoxication of common carp: case studies. Aquaculture International 15, 277–286.

Tkachenko, H., Kurhaluk, N., Grudniewska, J., 2015. Biomarkers of oxidative stress and antioxidant defences as indicators of different disinfectants exposure in the heart of rainbow trout (*Oncorhynchus mykiss* Walbaum). Aquaculture Research 46, 679–689.

Tomasso, J.R., Grosell, M., 2005. Physiological basis for large differences in resistance to nitrite among freshwater and freshwater-acclimated euryhaline fishes. Environmental Science and Technology 39, 98–102.

Vatsos, I.N., Angelidis, P., 2010. Water quality and fish diseases. Journal of the Hellenic Veterinary Medical Society 61, 40–48.

Wicks, B.J., Randall, D.J., 2002. The effect of sub-lethal ammonia exposure on fed and unfed rainbow trout: the role of glutamine in regulation of ammonia. Comparative Biochemistry and Physiology A 132, 275–285.

Wiebe, A.H., 1931. Note on the exposure of several species of pond fishes to sudden changes in pH. Transactions of the American Microscopical Society 50, 380–383.

Williams, E.M., Eddy, F.B., 1986. Chloride uptake in freshwater teleosts and its relationship to nitrite uptake and toxicity. Journal of Comparative Physiology 156, 867–872.

Williams, E.M., Eddy, F.B., 1988. Anion transport, chloride cell number and nitrite-induced methaemoglobinaemia in rainbow trout (*Salmo gairdneri*) and carp (*Cyprinus carpio*). Aquatic Toxicology 13, 29–42.

Wilkie, M.P., 2002. Ammonia excretion and urea handling by gills: present understanding and future research challenges. Journal of Experimental Zoology 293, 284–301.

Wilkie, M.P., Wood, C.M., 1996. The adaptations of fish to extremely alkaline environments. Comparative Biochemisty and Physiology 113, 665–673.

Wilson, J.M., Iwata, K., Iwama, G.K., Randall, D.J., 1998. Inhibition of ammonia excretion and production in rainbow trout during severe alkaline exposure. Comparative Biochemistry and Physiology B 121, 99–109.

Wise, D.J., Tomasso, J.R., Brandt, T.M., 1988. Ascorbic-acid inhibition of nitrite-induced methemoglobinemia in channel catfish. Progressive Fish Culturist 50, 77–80.

Wood, C.M., 1993. Ammonia and urea metabolism and excretion. In: Ewans, D.H. (Ed.), Physiology of Fishes. CRC Press, Boca Raton.

Wright, D.A., Welbourn, P., 2002. Environmental Toxicology. Cambridge University Press, Cambridge.

Wu, J.G., Laird, D.A., 2003. Abiotic transformation of chlorpyrifos to chorpyrifos oxon in chlorinated water. Environmental Toxicology and Chemistry 22, 261–264.

Chapter 8

Stress and Disease in Fish

Ana Patrícia Mateus, Deborah M. Power, Adelino V.M. Canário
University of Algarve, Faro, Portugal

8.1 STRESS AND DISEASE

Aquaculture is by definition cultivation in water. Water as a solvent can remove most metabolites, provided there is enough renewal with clean and oxygenated water, but it can also be a source of contaminants and may lack enough oxygen to supply the demands of the organisms (fishes in this case) under cultivation. These conditions and many others are physical and chemical sources of discomfort and ultimately stress. While scientists have argued for decades the definition of stress, they all agree on the progressive steps of the syndrome. What is special about stress is that it triggers physiological and behavioral responses that can compromise bodily functions, growth, and resistance to disease and ultimately can cause death. While immunosuppression is a well-established effect of stress, the aquatic environment makes it riskier for pathogens to take hold, with devastating consequences.

Many reviews have been produced on the stress response and the relationship to disease. This chapter is directed at those less familiar with stress physiology and interactions with the immune system. It will start by reviewing the historical concepts of stress and the progressive nature of the stress response, including the energetic demands, the effects on ion balance, and the role of the hypothalamus–pituitary–interrenal axis as well as other interacting endocrine factors. It will then explore the links between the immune system and disease resistance, in particular, markers that have been identified by transcriptomic and other studies linking innate and acquired immunity to particular kinds of pathogens. Finally, some considerations are made about managing stress in the fish farm using behavioral and physiological markers to monitor stress.

8.1.1 What Is Stress? The Impact of Stress on Aquaculture

Stress is a general and widely accepted concept but there has been much debate about its scientific definition (for reviews of stress in fishes see Barton, 2011; Ellis et al., 2012; Miller and O'Callaghan, 2002; Pottinger, 2008; Sapolsky, 2002; Wendelaar Bonga, 1997; Schreck, 2010). The term "stress" was originally coined by Selye (1936, 1973) and defined as "the non-specific response of the body to any demand for change" although he also suggested "Everyone knows what stress is, but nobody really knows." Selye (1936) also included such notions in his attempts to describe the stress response and proposed it to be a "generalized effort of the organism to adapt itself to new conditions, termed the general adaptation syndrome" and compared it to other general defense reactions. One definition that seems to fit aquaculture is "a condition in which the dynamic equilibrium of the organism, called homeostasis, is threatened or disturbed as a result of the actions of intrinsic or extrinsic stimuli that act as stressors" (Wendelaar Bonga, 1997).

A key element common to most definitions of stress is the "disruption of homeostasis" by the threatening stimulus (Chrousos, 1998; Miller and O'Callaghan, 2002). Such definitions refer to the stress response as an attempt to reestablish the disrupted homeostasis through physiological and behavioral actions, a reaction that seems to be beneficial for the survival of the organism. However, it is clear that the adaptive response of an animal may be inadequate, excessive, or prolonged, and this may lead to disease. This reality was already contemplated by Brett (1958), who described stress "as a state produced by an environmental or other factor which extends the adaptive responses of an animal beyond the normal range or which disturbs the normal functioning to such an extent that the chances of survival are significantly reduced."

The general adaptation syndrome in response to a stressor develops in three stages. The first phase, the alarm stage or reaction, initiates the "fight or flight" response with the rapid release of hormones. The second phase, the resistance stage, begins within 48 h of contact with the stressor and is characterized by the resolution of the fight or flight response with

Fish Diseases. http://dx.doi.org/10.1016/B978-0-12-804564-0.00008-9

reestablishment of the normal state. The third stage, the phase of exhaustion, is a result of continued stimulus/threat by the stressor, when the animal loses its resistance and succumbs because of the reappearance of the alarm stage manifestations. In fact, later definitions of stress include elements of the diseases that result from overstimulation of the physiological and psychological responses (the so-called chronic stress effects) (McEwen, 2000; Sapolsky, 2002; Tort and Teles, 2011; Barton, 2011).

The ultimate outcome of a stressor depends on the response and may underpin either the survival of the animal or the onset of a pathological condition or even death. The magnitude of physiological responses and its duration determines the outcome of the stress response, and is largely dependent on the nature of the stressor, acute or chronic (Davis, 2006). Acute stressors are characterized by high severity, short duration, and abrupt onset of the stress response, which is based on the fight or flight reaction to ensure survival. Chronic stressors induce a less intense and slower onset stress response but have a higher energetic cost because of their duration and can lead to distress and maladaptation, seriously compromising survival (Ellis et al., 2012; Iwama et al., 2011; Martínez-Porchas et al., 2009; Tort, 2011).

Since acute stress generally triggers a beneficial stress response, it is chronic stress that requires special attention when considering farmed fish (Davis, 2006). The effects of the stress response are generally regarded as detrimental because farmed fish cannot escape from the continuous exposure to the stressors associated with aquaculture (Davis, 2006; Tort et al., 2011). In particular, farmed fish, unlike wild fish, are exposed at all stages of the production cycle to multiple stressors, including environmental (temperature, salinity, ammonia, dissolved oxygen, pH, chemicals, pathogens), physical (cleaning, grading, handling, crowding, confinement, feeding, vaccination), and social (competition, crowding, aggressiveness) (Iwama et al., 2005; Pavlidis and Mylonas, 2011; Tort, 2011; Braithwaite and Ebbesson, 2014).

Establishing a definition for fish welfare is not simple (Huntingford et al., 2006; Segner et al., 2012), and identifying indicators of disturbed welfare is difficult, particularly since they may be mistaken for natural and adaptive responses to stress (Ellis et al., 2012). Indicators of chronic stress linked to a compromised welfare status include behavioral signs (reduced appetite, erratic swimming, and repeated actions), physiological indicators (decreased growth, reproductive incapacity, reduced nutritional status), and compromised health (increased number of injuries and immunosuppression) (reviewed in Ashley, 2007; Braithwaite and Ebbesson, 2014; Huntingford and Kadri, 2014; Martins et al., 2012; Segner et al., 2012).

Controlling stress in farmed fish is a key goal of aquaculture to optimize production by improving fish growth, disease resistance, reproductive success, and quality. This major concern in aquaculture has led to diverse experiments directed toward understanding the physiology of stress and its consequences for fish well-being, with many reviews being published in the past decades (Ashley, 2007; Barton and Iwama, 1991; Huntingford and Kadri, 2014; Iwama et al., 2011; Pavlidis and Mylonas, 2011; Pickering, 1998; Schreck, 2010). In addition, the search for stress markers has led to improvements in husbandry, as fish well-being has gained increasing attention in recent years (Vijayan et al., 2010).

8.1.2 Physiological Effects and Relationship to Disease

8.1.2.1 Stress: A Three-Stage Model

The stress response in fish has been broadly categorized into primary, secondary, and tertiary responses (Barton, 2011; Ellis et al., 2012; Miller and O'Callaghan, 2002; Pottinger, 2008; Sapolsky, 2002; Wendelaar Bonga, 1997). The stress response initiates rapidly with the activation of the adrenergic system, which causes an increase in plasma adrenaline and noradrenaline and is followed by the activation of the hypothalamus–pituitary–interrenal (HPI) axis and an increase in plasma cortisol. The secondary response encompasses physiological modifications and includes metabolic, hydromineral, hematological, immunological, and structural changes. The tertiary response describes whole animal changes and is associated with reductions in growth, condition factor, resistance to disease, metabolic scope, behavior, and ultimately survival (Barton, 2011; Ellis et al., 2012; Miller and O'Callaghan, 2002; Pottinger, 2008; Sapolsky, 2002; Wendelaar Bonga, 1997).

The organization of the stress axis in fish is similar to that of other vertebrates, with the head kidney interrenal tissue functionally analogous to the adrenal gland (Avella et al., 1991; Barton, 2002; Brown et al., 1978; Iwama et al., 2011; Rand-Weaver et al., 1993; Tort and Teles, 2011). On perception of a stressor the sympathetic component of the central nervous system stimulates the release of corticotropin-releasing factor (CRF) from the hypothalamus and catecholamines from the chromaffin cells (Reid et al., 1998). CRF reaches the pituitary corticotropic cells via direct neural contact and stimulates secretion of adrenocorticotropic hormone (ACTH), α-melanocyte-stimulating hormone (α-MSH), and β-endorphins (end-products of the prohormone proopiomelanocortin, POMC). ACTH binds to the melanocortin-2 receptor on target cells in the interrenal tissue and stimulates release of cortisol, the main circulating corticosteroid in teleosts, which has both glucocorticoid and mineralocorticoid actions (Mommsen et al., 1999). In steroidogenic cells of the interrenal tissue, steroidogenic acute regulatory protein transports cholesterol from the outer to the inner mitochondrial membrane, the rate-limiting step for cortisol production (Mommsen et al., 1999; Martínez-Porchas et al., 2009). Cortisol binding to the glucocorticoid receptor (GR) mediates the negative feedback regulation of the HPI axis at all levels (Alderman et al., 2012; Schaaf et al., 2009; Milla et al., 2009).

Selye (1973) defined stress as a nonspecific response, indicating that all stressors evoke a similar response, although the influence of psychological factors on corticosteroid levels suggests this is not totally true (Mason, 1971; Barton, 2002). The ability of fish to assess stressful situations is debatable, although studies have revealed that the magnitude of the stress response in fish is modulated by awareness of the stressor (Barton, 2002; Eddy, 1981; Schreck, 1981; Galhardo and Oliveira, 2009). Furthermore, different husbandry stressors provoke different physiological effects in fish and can lead to a "learning-adaptation" response, while others may result in a maladaptive response with considerable costs in energy and compromised disease resistance (Barton and Iwama, 1991; EL-Khaldi, 2010; Pickering, 1998; Varsamos et al., 2006).

Short-term responses to stress are not necessarily linked to reduced welfare, but chronic stress is a major concern in aquaculture and is strongly correlated with poor health and is an indicator of disturbed welfare (Ashley, 2007). Welfare indicators in aquaculture are largely associated with the tertiary effects of the stress response when fish are confronted with chronic stressors, and clearly reveal the detrimental effects of stress on the health and welfare of farmed fish (Ashley, 2007; Martins et al., 2012). Thus, even though chronically stressed fish may compensate for the stressor by returning plasma cortisol to resting levels, the associated physiological alterations may compromise health.

8.1.2.2 Allostasis

The concept of "allostasis" was introduced by Sterling and Eyer (1988) to clarify the ambiguity of the term homeostasis. Allostasis means "achieving stability through change" and represents the active process by which the body responds to daily events to maintain an optimal range of physiological systems such as pH, body temperature, oxygen tension, and glucose levels during specific periods of the life cycle (homeostasis) (McEwen and Wingfield, 2003). In this context, activation of the HPI axis and sympathetic nervous system after sensing a stressor is an active process of allostasis, and the allostatic state is the result of an unbalanced production of glucocorticoids and other mediators such as catecholamines and cytokines for a limited period of time (Koob and Le Moal, 2001; Romero et al., 2009). The problem of chronic stress is that it prolongs the duration of the allostatic state and this can cause "wear and tear" on the regulatory systems and also the brain (McEwen, 2005; Flier et al., 1998). The cumulative effects of an allostatic state result in an "allostatic load," and "allostatic overload" occurs when there is an additional load that predisposes the individual to disease, which is largely dependent on the energy demands and availability during the allostatic state (McEwen and Wingfield, 2003; McEwen, 2005; Schreck, 2010).

Selye (1946) believed that understanding the stress response would contribute to understanding the establishment of diseases of adaptation. Indeed, endocrine responses directly affect the secondary response of stress and lead to changes in plasma metabolites and ions and the expression of heat shock proteins (Barton, 2002). An extensive literature exists about the physiological effects of stress on farmed fish (for reviews see, e.g., Barton, 2002; Barton and Iwama, 1991; Pickering, 1998; Sumpter, 1997; Wendelaar Bonga, 1997; Schreck, 1981, 2000), but the mechanisms behind stress that precipitate disease are still largely unexplored.

8.1.2.3 Stress and the Immune System

One of the best recognized detrimental effects of stress is the suppression of the immune system, which can lead to increased disease incidence and mortality rates (Section 8.1.3). In fish the effects of perturbations of the immune system by stress are variable and depend on a number of key factors such as species, stressor, intensity and persistence of the stimulus, and the time delay at which the immune indicator was measured after the onset of the stress response (Tort et al., 2004a).

Plasma glucose and liver glycogen content have been extensively studied as indicators of the effects of stress on carbohydrate metabolism (reviewed by Van Der Boon et al., 1991). Cortisol has been recognized as a major stimulator of glucose production and of gluconeogenesis enzyme activity in the liver (Pottinger and Pickering, 1997), which allows fish to cope with urgent needs of plasma glucose for energy production. Because of this evidence, plasma glucose has been extensively used as a stress indicator (Martínez-Porchas et al., 2009). However, the usefulness of plasma glucose as a biomarker of the stress response is questionable (Martínez-Porchas et al., 2009; Pottinger, 1998; Simontacchi et al., 2008). This is because glucose levels are highly dependent on species, developmental stage, and the metabolic status of the fish and can be increased, be decreased, or remain unaltered following exposure to increased cortisol levels (reviewed by Mommsen et al., 1999).

8.1.2.4 Stress and Metabolism

Studies focused on the molecular basis of the adaptive response to stress revealed that acute stress induced a rapid upregulation of genes involved in energy metabolism, immune function, and protein degradation in the liver of rainbow trout, *Oncorhynchus mykiss*. Wiseman et al. (2007) showed that acute stress increased liver metabolic capacity in response to increased energy demand through carbohydrate metabolism, including the glycogenolysis, glycolysis, and gluconeogenesis pathways. Catecholamine, cortisol, and insulin-like growth factors were the mediators of the metabolic response to acute stress (Wiseman et al., 2007). In the case of chronically stressed gilthead sea bream, *Sparus aurata*, enzymes involved in carbohydrate metabolism were downregulated and enzymes related to mobilization of lipid reserves were upregulated in the liver (Alves et al., 2010). This suggests that in gilthead sea bream carbohydrate reserves in the liver are insufficient to meet the increased energy demands during long-term exposure to stress and for this reason lipid reserves are mobilized (Alves et al., 2010). An increase in lipid utilization in response to chronic stress may be associated with a decrease in the hepatosomatic index in gilthead sea bream (Montero et al., 1999). Additionally, exhaustion of the liver capacity to maintain homeostasis during chronic stress may be linked to its reduced capacity to eliminate ammonia, since glutamine synthetase is downregulated (Alves et al., 2010). The increased energy requirements caused by chronic stress reduce the growth potential of fish (reviewed in Ellis et al., 2012; Pickering, 1993; Van Weerd and Komen, 1998). The factors contributing to reduced growth potential in chronically stressed fish are limitations in energy store mobilization from liver and muscles, suppressed appetite (Power et al., 2000; Ibarz et al., 2005, 2007; Tort et al., 2004b), inhibition of protein synthesis, and stimulated protein catabolism (Van Der Boon et al., 1991).

8.1.2.5 Stress and Ion Homeostasis

Many stressor-specific effects occur in ion homeostasis and water uptake in fish exposed to stress, and this is associated with disturbed plasma ion levels and acid–base balance (Wendelaar Bonga, 1997). It has been proposed that osmoregulatory disturbance occurs only in the case of chronic severe stress (Davis, 2006). The effects of stress on ionic balance and osmoregulation are variable in fish exposed to stress or cortisol administration, and decreased or unchanged osmolality has been reported, together with increased, unchanged, or decreased sodium and/or potassium levels, although these effects are also largely dependent on water salinity (Hwang et al., 2011; Arends et al., 1999; Laiz-Carrión et al., 2002, 2003, 2009; Mancera et al., 2002; Rotllant et al., 2001). Understanding the specific causes and consequences of ionic and acid–base disturbances induced by stress is complex, since they may result from both direct and indirect effects of stress hormones (McDonald and Milligan, 1997). For example, vigorous activity may induce anaerobiosis that leads to extracellular acid–base disturbances; suboptimal water quality (related to oxygen, hardness, pH, ammonia, and salt levels) may result in deleterious morphological changes in the gills; physical injury of the skin (scale loss and wounds) may result in general physiological disturbances.

Structural modifications in gills during acute and chronic stress compromise ion and osmotic homeostasis (Wendelaar Bonga, 1997), since they are the main organs responsible for the balance between influx and efflux of ions and water (Evans et al., 2005). Structural modifications in gills caused by stress include necrosis of branchial epithelial cells, epithelial separation from the basement membrane, and dilation of blood sinuses of the branchial lamellae (reviewed by Wendelaar Bonga, 1997). Although compensatory mechanisms exist, the maintenance of ionic/osmotic regulation during chronic stress is energy demanding (Hwang et al., 2011), and the inability of the liver to supply the necessary energy may explain the allostatic overload that may predispose to disease and compromise the survival of fish. Chronic stress affects the number of branchial chloride cells and Na^+-K^+-ATPase activity, and the overall outcome is an increase in passive ion influxes and water loss in seawater fish and increased passive ion efflux and water uptake in freshwater fish (Wendelaar Bonga, 1997).

The skin of teleosts is an important osmoregulatory organ (Marshall and Grosell, 2006) and mucous cells are important for the osmoregulatory activity of the skin (Shephard, 1994). Stress and cortisol administration have variable effects on skin mucous cells; in some studies they increase in number (Vatsos et al., 2010; Cruz et al., 2013; Iger et al., 1988, 1990; Zuchelkowski et al., 1981) and in others they decrease in number (Iger et al., 1988; Iger and Bonga, 1994; Kalogianni et al., 2011). Wounds and scale loss cause osmotic dysfunction (Black and Tredwell, 1967; Olsen et al., 2012; Smith, 1993), and a study from our group revealed that when 50% of the scales were removed from one flank of chronically stressed fish there was a decrease in osmolality (own unpublished results). More studies are required to explore the direct and indirect physiological effects of stress.

8.1.3 Endocrine Control and Mediators

The interaction between the neuroendocrine and the immune systems can easily be observed in the head kidney (Tort et al., 2004a; Nardocci et al., 2014). This organ is composed of the interrenal cells, which synthesize the stress hormone cortisol, and the chromaffin cells, which belong to the sympathetic nervous system and produce catecholamines. The head kidney in fish is also the major hematopoietic organ and produces the immune cells from the lymphoid and myeloid lines, and also the erythrocytes (Gallo and Civinini, 2003; Tort et al., 2004a; Verburg-Van Kemenade et al., 2011; Zapata et al., 1997). Although there is a need to explore this subject in fish, the notion that stress modulates immune function through mediators of a bidirectional communication is not new in vertebrates, including fish (Harris and Bird, 2000; Wendelaar Bonga, 1997; Weyts et al., 1999), and components of the immune system (levels of IgM, serum hemolytic activity, phagocytosis activity, number of circulating lymphocytes, lysozyme activity, and complement pathway) have been used as indicators of the stress response (Cnaani, 2006).

8.1.3.1 Cortisol and Glucocorticoids

The levels of cortisol in response to stress have been used in aquaculture to select genetic lines of Atlantic salmon, *Salmo salar*, and rainbow trout with improved disease resistance (Fevolden et al., 1991, 1992, 1993; Weber et al., 2008). Negative genetic correlations have also been found between cortisol response to acute stress and body weight or length in European sea bass, *Dicentrarchus labrax* (Volckaert et al., 2012), but not gilthead sea bream (Boulton et al., 2011). However, analysis of disease resistance in high or low stress-responder lines was ambiguous, since low cortisol responders did not always have enhanced immunity and lower death rates (Pottinger and Pickering, 1997). Nonetheless, administration of an acute stressor to trout elicits increased cortisol levels (range of 50–200 ng/mL) (Ellis et al., 2012; Tort et al., 2011; Pickering and Pottinger, 1989) and enhanced levels of lysozyme and complement of the innate immune response (Bowden, 2008; Demers and Bayne, 1997; Fast et al., 2008; Stolte et al., 2006; Weyts et al., 1999). Furthermore, chronic stress is associated with reduction in circulating leukocytes, antibody levels, and cytokine expression, indicative of a suppressive effect on the adaptive immune system (Bowden, 2008; Castillo et al., 2009; Fast et al., 2008; Stolte et al., 2006; Verburg-Van Kemenade et al., 2009; Weyts et al., 1999; Ellis et al., 2012; Tort et al., 2011; Pickering and Pottinger, 1989). Temperature can also act as a stressor, and European sea bass reared in stressful water temperatures have decreased serum IgM and increased susceptibility to viral challenges and chronic activation of the HPI axis (Varsamos et al., 2006).

The effects of cortisol are mediated through GR signaling and this is part of the negative feedback loop that regulates the HPI axis and release of cortisol (Alderman et al., 2012; Prunet et al., 2006). Changes in the GR are therefore proposed as a good candidate indicator for the activity of the stress axis (Terova et al., 2005). However, since the autoregulation of GR by cortisol is dependent on the type of the stressor, acute or chronic stress, the target tissue, and other factors, the relationship between the GR and stress is not clear-cut (Kiilerich and Prunet, 2011; Schreck, 2000).

The role of the GR in the modulation of the immune system gained substantial attention when it was found to be expressed in the main organs of the fish immune system, the epithelia of the gills, the skin, the gut, the head kidney, and the spleen (Acerete et al., 2007; Stolte et al., 2008) and also the leukocytes (Maule and Schreck, 1990; Weyts et al., 1998; Vizzini et al., 2007). It is well established in mammals that insufficient glucocorticoid signaling reduces glucocorticoid responsiveness and that this plays a significant role in dampening disease resistance in stress-related disorders (Engler et al., 2005; Raison and Miller, 2003). The study of the relationship between glucocorticoid resistance and the immune system in fish is even more interesting and intriguing because of the existence of duplicate GR genes in addition to splice variants of these genes, which contributes to their variable cortisol sensitivity and differential expression in tissues (Stolte et al., 2006; Schoonheim et al., 2010). However, the interplay between the expression of GR and sensitivity to cortisol in fish immune cells is not clearly understood and more studies are required.

One of the best-characterized effects of increased expression of the GR is the downregulation of proinflammatory cytokine interleukin (IL)-1β (Acerete et al., 2007; Stolte et al., 2009; Quabius et al., 2005) even when other immune-related genes are upregulated (Valenzuela et al., 2015). The downregulation of proinflammatory cytokines, enzymes, and their receptors is linked to GR-mediated inhibition of transcription factors such as nuclear factor κB (Zhang et al., 2012; Chen et al., 2015) and activator protein (reviewed by McKay and Cidlowski, 1999). GR expression can also be modulated by activation of the immune system, and injection of lipopolysaccharide (LPS) increased expression of the GR in immune-related organs of the gilthead sea bream (Acerete et al., 2007). In the common carp, *Cyprinus carpio*, increased GR1 expression occurred in leukocytes during inflammation (Stolte et al., 2009), while GR2 was downregulated (Stolte et al., 2008), and this was associated with the higher sensitivity of GR2 to cortisol and lower sensitivity of GR1 (Stolte et al., 2006, 2009; Verburg-Van Kemenade et al., 2011). Added complexity is evident from studies in Coho salmon, *Oncorhynchus kisutch*, in which high expression of the GR in the anterior kidney and splenic leukocytes was linked to their reduced affinity for cortisol (Maule et al., 1993; Maule and Schreck, 1991), indicating that increased expression of the GR is not necessarily linked to increased sensitivity to cortisol. This is also true in the Eurasian perch, *Perca fluviatilis*, in which immune parameters were negatively correlated with GR expression during an acute stress challenge, with an increase in plasma lysozyme levels and a decrease in GR expression observed in leukocytes isolated from the head kidney (Milla et al., 2010). A decrease in GR expression was observed in leukocytes isolated from the head kidney and in peripheral blood leukocytes in seawater-acclimated trout, even though without significant differences in GR expression in the head kidney and spleen (Yada et al., 2008). Upregulation of immune genes [major histocompatibility complex (MHC) I, MHC II, cytokine receptors, and tumor necrosis factor α (TNF-α)] concomitant with a significant downregulation of the GR was also observed in fine flounder, *Paralichthys adspersus*, 24 h after refeeding (Valenzuela et al., 2015).

8.1.3.2 Other Factors of the HPI Axis

Overall, it cannot be assumed that cortisol is immunosuppressive (reviewed by Tort et al., 2004a). Such observations have led to studies to identify other mediators. Attention has focused on factors of the HPI axis such as CRF- and POMC-derived peptides (α-MSH, MCH, and ACTH) and their respective receptors that are also present in fish immune organs (Harris and Bird, 2000; Tort, 2011). In mammals CRF is a proinflammatory factor while cortisol and POMC are predominantly antiinflammatory (Brazzini et al., 2003; Slominski, 2007; Slominski et al., 2006). In fish there have been relatively few studies of how other HPI axis factors affect the immune system. In carp lines with different susceptibilities to disease exposed to acute stress, CRF induced IL-1β expression in the brain, and in Mozambique tilapia, *Oreochromis mossambicus*, challenged with LPS, CRF mediated a response in tissues that are the first line of defense against external stressors (Pepels et al., 2004; Pijanowski et al., 2015; Mola et al., 2011; Mazon et al., 2006). POMC end-products stimulate phagocytosis, the respiratory burst, and mitogenesis (reviewed by Bayne and Levy, 1991; Tort, 2011) and more specifically ACTH is an antiinflammatory agent (Yarahmadi et al., 2016) that downregulates proinflammatory cytokines (TNF-α, IL-6, and IL-1β) in isolated head kidney cells (Castillo et al., 2009). The results of the studies of ACTH actions in fish as of this writing make it a strong candidate for the immunosuppressive effects of chronic stress.

8.1.3.3 The Adrenergic System

The adrenergic system and associated catecholamines are implicated in the enhanced immune response during stress (Narnaware and Baker, 1996; Nardocci et al., 2014; Tort, 2011). The immune-regulatory role of catecholamines was suspected when noradrenergic fibers and adrenergic neurons were identified in the spleen of Coho salmon (Flory, 1989), and catecholamine receptors were identified in head kidney and splenic leukocytes of channel catfish, *Ictalurus punctatus* (Finkenbine et al., 2002), and in the immune organs and leukocytes of common carp (Chadzinska et al., 2012). Modulation of immune function by catecholamines was observed using catecholamine receptor agonists, which suppressed phagocytosis but stimulated the antibody response, and had ambiguous effects on in vitro leukocyte proliferation (mainly macrophage and neutrophils) and on the respiratory burst in rainbow trout (Flory and Bayne, 1991). Reduction of phagocytosis by macrophages seems to be an established effect of catecholamines, and in spotted murrel, *Channa punctatus*, they also stimulate the respiratory burst of macrophages in vitro (Roy and Rai, 2008). In gilthead sea bream macrophages, adrenaline caused a reduction in the expression of IL-1β but did not modify the expression of the proinflammatory cytokines TNF-α, transforming growth factor β, and IL-6 (Castillo et al., 2009), and in common carp caused downregulation of proinflammatory IL-1β after LPS stimulation and enhanced arginase activity in phagocytes (Chadzinska et al., 2012). Nonetheless, the physiological significance of catecholamines during stress is unclear. We hypothesize that elevated catecholamine levels associated with severe stress can reduce the detrimental effects of stress on the immune system but that during chronic stress catecholamine levels are unchanged and so suppression of immune function is the dominant effect (Perry and Bernier, 1999). Further studies will be required to establish the interplay that occurs between catecholamines, stress, and the immune response.

8.1.3.4 The Somatotropin Family

Part of the difficulty in interpreting the exact role of cortisol and other key mediators of the stress response on the immune response in fish stems from the fact that the stress axis modulates other endocrine axes that also regulate the immune system. The ambiguous effects of the stress response on the immune response may be linked to the antagonistic effects of growth hormone (GH) on the immunosuppression induced by chronic stress (reviewed by Yada, 2007). The stimulatory effects of GH include increased phagocytosis, leukocyte proliferation, antibody production, respiratory burst, complement activity, and proinflammatory cytokine production (reviewed by Tort, 2011). Hypophysectomy, which ablates GH production, has a strong suppressive effect on the immune system (Yada and Azuma, 2002; Yada et al., 2001). The close evolutionary association between somatotropic hormones and their receptors and cytokines of the immune system is revealed by their common membership in the cytokine superfamily and this may explain the strong cross talk that occurs between family members (reviewed by Yada, 2007). The functional link between the somatotropic hormones and the immune system is further supported by their expression in tissues of the fish immune system, including the head kidney, spleen, and leukocytes (Calduch-Giner et al., 1995; Yang et al., 1997; Yada et al., 2002; Mori and Devlin, 1999; Calduch-Giner and Pérez-Sánchez, 1999).

A universal demonstration that GH mRNA expression and plasma levels are correlated with cortisol does not exist in fish. For example, in Atlantic salmon repeated acute stress is associated with increased plasma GH levels (McCormick et al., 1998), and GH mRNA abundance increases in rainbow trout fed with cortisol (Yada et al., 2005). However, other studies of rainbow trout reported decreased plasma GH levels after acute stress and increased plasma GH after chronic stress (Pickering et al., 1991; Farbridge and Leatherland, 1992). It is possible that the initial rise in cortisol level in acute stress may induce a prolonged suppression of GH secretion (Farbridge and Leatherland, 1992). Establishment of the factors and mechanisms that explain the immunoenhancer effect of acute stress and immunosuppression by chronic stress is an area that deserves further investigation.

8.1.4 Gene Network and Regulation Linking the Immune System to Disease Resistance

The interaction between key modulators of the stress axis and the immune system during acute and chronic stress in teleosts is well described. However, the connection between the HPI axis and the immune system is far from clear, particularly in relation to how it affects resistance to disease. The general characteristics of the vertebrate immune system are conserved in teleost fish, although morphological and functional specializations exist (for reviews see Uribe et al., 2011; Tort et al., 2003; Rauta et al., 2012).

An exhaustive overview of the immune response of fish against pathogens will not be given as there are several reviews (for reviews see Brown and Johnson, 2008; Langevin et al., 2013; Ellis, 2001). Instead brief summary tables (Tables 8.1–8.3) are provided for the most commercially important aquaculture species (including zebrafish), by life-cycle status and the immune-related genes modulated during viral, bacterial, and parasitic infections. Many of the genes listed in Tables 8.1–8.3 have come from the application of functional genomics (microarray technology, gene library construction, transcriptomic analysis, next-generation sequencing, and gene expression profiling) and capture multiple genes activated by an immune challenge (reviewed by Kaur, 2013). Some of the studies were conducted on fish selected for disease resistance or susceptibility to identify mechanisms underlying host resistance. The identification of genes differently expressed in resistant lines, particularly those involved in higher rates of survival and elimination of the pathogen, enables the definition of genetic markers for disease resistance. Genes linked to disease resistance are of interest as potential biomarkers of robustness but also to determine their contribution to the reduced disease resistance observed in fish exposed to stress. A brief consideration of the genes that have been detected in various fish species challenged with disease now follows.

8.1.4.1 Innate Immunity

Innate (nonspecific) immunity is a fundamental mechanism of defense against pathogens in fish since they have a more limited adaptive (specific) immune response (e.g., repertoire of antibodies, maturation, and memory). In general, innate immunity (nonspecific defense), rather than adaptive immunity (specific defense), offers the greatest protection. The role of nonspecific immune parameters on disease resistance in fish depends largely on the production of a broad spectrum of antimicrobial peptides (AMPs), acute-phase proteins, lectins, nonclassical activation of complement, inflammation, and phagocytosis, whereby mucosal epithelial surfaces (skin, gills, and intestine) prevent the entry and dissemination of pathogens in an organism (Bayne and Gerwick, 2001; Tort et al., 2003; Ellis, 2001; Magnadóttir, 2006; Aoki et al., 2008). Several innate immune parameters are expressed during viral, bacterial, and parasitic infections (Tables 8.1–8.3), but only hepcidin, mannose-binding lectin (MBL), and ceruloplasmin have been thoroughly explored because of their potential role as resistance biomarkers.

The importance of innate immunity to protect against pathogens in teleosts is evident from the upregulation of hepcidin in many immune organs of fish infected with bacteria or viruses or after in vitro stimulation with mitogens (Cho et al., 2009; Cuesta et al., 2008). Hepcidin is a potent AMP (reviewed in Katzenback, 2015), it is a liver-expressed AMP

TABLE 8.1 List of Immune-Related Genes Differentially Modulated During Bacterial Infections of Fish

Fish	Fish Tissue	Pathogen	Upregulated	Downregulated	References
Oncorhynchus mykiss	HK macrophages	*Renibacterium salmoninarum*	IL-1β, MHC II, COX-2, iNOS	TNF-α	Grayson et al. (2002)
Ictalurus punctatus	Spleen	*Edwardsiella ictaluri*	LEAP-2		Bao et al. (2006)
Dicentrarchus labrax	Liver	*Photobacterium damselae*	Hepcidin		Rodrigues et al. (2006)
Scophthalmus maximus	Liver, spleen, HK	*Vibrio anguillarum*		MHC Iiβ	Zhang and Chen (2006)
I. punctatus	Liver	*E. ictaluri*	Coagulation factors, proteinase inhibitors, transport proteins, complement components, iron homeostasis-related genes	AMP-2	Peatman et al. (2007)
Ictalurus furcatus	Liver	*E. ictaluri*	Acute-phase response, complement activation, metal ion binding/transport, immune/defense response-associated genes	Selenoprotein H, selenoprotein P	Peatman et al. (2008)
Seriola quinqueradiata	Primary cultured kidney cells	LPS	Inflammation, cytokine activity, antigen-presentation and antigen-binding genes		Darawiroj et al. (2008)
Gadus morhua	Blood	*V. anguillarum*	Apolipoprotein A-1, IL-1β and IL-8, NCCRP-1		Caipang et al. (2008)
Psetta maxima	Kidney, spleen	*Vibrio harveyi*	MHC Iα, HSP70, signaling molecules (src-family tyrosine kinase SCK, sgk-1 serine–threonine protein kinase, and amyloid precursor-like protein 2)		Wang et al. (2008)
S. maximus	Liver, spleen, and HK	*Aeromonas salmonicida*	Glutathione *S*-transferase, HSP, cytochrome P450, elastases, MHC, coagulation factors involved in innate immunity, IFN, perforin, hepcidin, nephrosin, α2-macroglobulin, iron metabolism-related genes (haptoglobin fragment 1, globin-related proteins, hepcidin precursor)	Proteolytic activity (chymotrypsin B precursor, trypsinogen 1, chymotrypsinogen 2, trypsinogen-like serine protease, and elastase precursor)	Pardo et al. (2008)
Salmo salar	Liver	*A. salmonicida*	**Low resistance**: Recruitment of immune cell genes (LECT2), regulators of immune responses, proteasome components, and extracellular proteases	**Low resistance**: Recruitment of immune cell genes (SKAP2), scavengers of reactive oxygen species, and genes for proteins of iron metabolism	Škugor et al. (2009)
			High resistance: Regulators of immune responses, protease inhibitors, components of membrane attack complex, and the complement proteins	**High resistance**: genes implicated in recruitment and motility of immune cells	

Solea senegalensis	HK	LPS	NCCRP-1, CEBPB, SQSTM1, NDUFA4, C7, HP, TF, CIRBP, TRFA, C3	TKT	Prieto-Álamo et al. (2009)
Hippoglossus hippoglossus	Liver, HK, PBL, spleen	*V. anguillarum*	TNF-α-induced protein 9; CD36; CD63; MHC Iα chain variants 1, 2, and 4; MHC IIβ variant 2; MHC II invariant chain; immunoglobulin light chain; complements/complement factors; proteasome; α2-macroglobulin; Fas (tumor necrosis factor receptor superfamily 6)-associated via death domain; mast cell proteinase-3; programmed cell death 6 and 8		Dumrongphol et al. (2009)
G. morhua	Spleen and HK	*A. salmonicida*	**Spleen**: SCYA, IRF1, hepcidin antimicrobial peptide, IL-8, basic transcription factor 3, DNA damage-inducible transcript 4, IL-1β, and serum lectin isoforms 1 and 2 **HK**: Proteasome activator subunit 2, translationally controlled tumor protein, CD84 molecule, LPS binding protein, IL-5 receptor α, and inhibitor of NF-κBα. **Both**: Cytokine signaling (IL-1, IL-1R, IL-8, SCYA, and IRF1), apoptosis (basic transcription factor 3, MCL1), iron homeostasis (FTH and HAMP), and antibacterial defense response (CAMP and HAMP)	HSP, tyrosine kinase 2, mitogen-activated protein kinase 14a, scavenger receptor class B member 2, IL-1 receptor-like protein precursor, lymphocyte antigen 75, complement receptor-like protein 1, a novel immune type receptor 4, TLR23, apoptosis regulation (caspase 8, leukocyte elastase inhibitor, cell division cycle and apoptosis regulator 1)	Feng et al. (2009)
S. salar	HK	*A. salmonicida*	Inflammatory-related genes (anterior gradient-like protein); innate humoral defense (lysozyme C, type P precursor, lysozyme II, C1q-like adipose specific protein, secernin 1, C3-1, C6); antigen recognition, processing, and presentation (MHC II, TAP2, C-type lectin receptor A, proteasome subunit α type 4, mannose-binding protein C precursor, lysozyme type II, MHC II invariant chain, MHC I region, MHC IIα, proteasome subunit β type 7 precursor); cell signaling (CC chemokine, leukocyte cell-derived chemotaxin 2 precursor, IFN regulatory factor 1); immunoglobulin-related genes	Inflammatory-related genes (B cell translocation gene 1 protein); innate humoral defense (integrin β2 precursor); antigen recognition, processing, and presentation (MHC I antigen, TAP2, proteasome subunit α type 5); cell signaling (regulator of G protein signaling domain, guanine nucleotide-binding protein G); T-cell-related genes (T cell receptor α chain, tyrosine-protein kinase ZAP-70)	Mutoloki et al. (2010)

Continued

TABLE 8.1 List of Immune-Related Genes Differentially Modulated During Bacterial Infections of Fish—cont'd

Fish	Fish Tissue	Pathogen	Upregulated	Downregulated	References
S. maximus	Spleen	*A. salmonicida*	Acute-phase response/inflammation (α2 HS glycoprotein, α2 macroglobulin, ceruloplasmin, fibrinogen α chain and β polypeptide, γ-fibrinogen); carriers and transporters (apolipoprotein, serotransferrin, transferrin); cell division/apoptosis (plasminogen); immune response (chemokine CC-like-protein, chemotaxin, γ gene loci of immunoglobulin heavy chains); protein synthesis, processing, or degradation [α1 antitrypsin, α1-microglobulin, serine (or cysteine) proteinase inhibitor, similar to inter-α-trypsin inhibitor, heavy chain 3]; stress and/or defense response (C1 inhibitor, complement binding protein, C3, C8B, C9, gelatinase, haptoglobin fragment 1, hepcidin precursor	Metabolism (xylose isomerase)	Millán et al. (2010)
Labeo rohita	HK	*Edwardsiella tarda*	IL-1β, iNOS, C3, CXCa, C-type and G-type lysozymes	β2-microglobulin and TNF-α	Mohanty and Sahoo (2010)
Paralichthys olivaceus	HK	*E. tarda*	**Susceptible**: Innate immune-response genes (IL-1β, C3, and factor B)		Yasuike et al. (2010)
			Resistant: MHC I pathway (MHC Iα and β, β2-microglobulin, proteasome activator PA28a subunit, and 20S proteasome b subunit), cytokine and cytokine receptor (TNF ligand supergroup member 13b, IFN-induced protein 44), histone proteins		
Pseudosciaena crocea		*A. hydrophila*	Prkcb1, Hspa5, Radd45a, Dusp7, Rac1, Casp1, TCR signaling (Khdrbs1, Scap2, Vasp, Pik3r2, Cebpb, Zap70, and Cbl); TLR pathway (TLR1, TLR2, TLR3, R22); cytokine genes (TNF-α, IL-1β, and IL-8); chemokine and chemokine receptor genes (CCL-4, CCL-c25v, CCR-1, CCR-12.3); apoptosis-related genes (Casp9 and Fas)	MAPK pathway (Map3k12, Crkl, Jun, and Raf1), TCR signaling (Was, Lyn, Ptpn6, Ctnnb1, Itk, Crkl, Jun, and Ripk2)	Mu et al. (2010)
I. punctatus	Whole fish	*Flavobacterium columnare*	Putative histone H3, triglyceride lipase, glyoxalase domain-containing 4, fetuin-B, hemoglobin-β, nascent polypeptide-associated complex subunit α, intelectin 2, pyrophosphatase (inorganic) 1, C1R/C1S subunit of Ca^{2+}-dependent complex, membrane-spanning 4-domains subfamily A member 8A, tyrosine aminotransferase, matrix metalloproteinase, and CD59		Pridgeon and Klesius (2010)
P. olivaceus	HK	*S. iniae* and *E. tarda*	MMP-9, MMP-13, CXC chemokine, CD20 antigen receptor and hepcidin	IFN-inducible Mx protein, MHC II-associated invariant chain, MHC IIα and β encoding genes,	Aoki et al. (2011)

G. morhua	Spleen	*A. salmonicida*	Immune, inflammatory, and bactericidal responses (cathelicidin 1, CC chemokine, CD63, hepcidin precursor, integrin β2, IFN-stimulated gene 15-2, IFN-inducible GTPase_b, IL-8 variant 5); proteolysis (cathepsin L precursor, cysteine protease, proteasome subunit α type-6, proteasome subunit β type-7 precursor)		Booman et al. (2011)
I. punctatus	Gills	*F. columnare*	Pathogen/antigen recognition (C-type mannose receptor 2-like, MHC Iα chain, MHC IIβ chain, NLR3-like, rhamnose-binding lectin), anti-apoptosis (GRINA, JunB protooncogene), proapoptosis (calpain 8, caspase 8, cyclophilin D, DNA damage-inducible transcript 4 protein-like), autophagy/lysosome/phagosome (cathepsin B preproprotein, cathepsin L precursor, lysosomal protective protein, nucleolin), antiinflammatory/immunosuppressive (leukocyte elastase inhibitor, lymphocyte activation gene 3 precursor, NF-κB inhibitor α-like proteins A and B, Toll-interacting protein, TRAF2), proinflammatory (CC chemokine, C7, complement receptor-like, immunoresponsive gene 1, IFN-induced protein 44, MBL serine protease 2, iNOS2b, SAM domain and HD domain-containing protein 1, transmembrane protease serine 9-like, VHSV-induced protein)	Pathogen/antigen recognition (NCAMP-1, NITR10), antiapoptosis (G3BP1), antiapoptosis (GRINA, JunB protooncogene), proapoptosis (caspase-1-like), proinflammatory (IFN-induced GTP-binding protein Mx)	Sun et al. (2012)
Danio rerio	Liver	*E. tarda*	Acute-phase proteins (CRP and SAP); C3 and the complement C4; C2; C7 and C9; C1 inhibitor; leukocyte cell-derived chemotaxin 2; TLR5; cytokine genes related to the JAK/STAT, MAPK, TGF-β, apoptosis and VEGF signaling pathways; MHC I antigen processing pathways; ER-resident chaperone calreticulin; calnexin; endoplasmin (grp94); TAP binding protein; proteasome activator (PA28); HSP superfamilies; and cathepsin L	Acute-phase proteins (apolipoprotein AIV and α2-HS-glycoprotein), C1q, MHC-II processing pathway (LAMP2, MHC II DAB, MHC IIβ chain, MHC II invariant chain (CD74 molecule), MHC II transactivator (CIITA), and cathepsin S	Yang et al. (2012)

Continued

TABLE 8.1 List of Immune-Related Genes Differentially Modulated During Bacterial Infections of Fish—cont'd

Fish	Fish Tissue	Pathogen	Upregulated	Downregulated	References
I. punctatus	Entire intestinal tract	*E. ictaluri*	Lysosome/phagosome (cathepsin Z, CD63, lysosomal membrane glycoprotein 2, lysosomal protective protein, MHC I ZE-like, sialidase-1), immune activation/inflammation (apolipoprotein B, chemokine CXCL12, complement C4-1-like, IgG Fc-binding protein, IL-11 receptor, macrophage MIF, prostaglandin E synthase 3, TNF receptor superfamily)	Lysosome/phagosome (hexosaminidase B, MHC I UXA2, MHC IIα chain 1, MHC IIβ, NADPH oxidase 1, NOXO1), immune activation/inflammation (C1Q subcomponent-binding protein, C1q-like 3, CC chemokine 25-like, CC chemokine, glutathione peroxidase 3, metallothionein-2, MMP13, serum amyloid P component-like 2, tumor protein p53-inducible protein 1)	Li et al. (2012)
Oryzias melastigma	Liver	*Vibrio parahaemolyticus*	C1q, C3-2, C4, BF, HF, MASP		Bo et al. (2012)
L. rohita	Liver, intestine, muscle, kidney, spleen, and brain	*A. hydrophila*	**Resistant line**: MHC, HSP30, HSP70, HSP90, serum lectin genes		Robinson et al. (2012)
S. maximus	Brain-hypophysis–gonadal axis	*A. salmonicida* and *Philasterides dicentrarchi*	Complement, Toll-like receptor signaling, B cell receptor signaling, T cell receptor signaling and apoptosis cascade, chemokine signaling-related genes		Ribas et al. (2013)
L. rohita	Skin, muscle, liver, anterior kidney, spleen, gill, eye, and brain	*A. hydrophila*	**Susceptible**: HSP70 **Resistant**: HSP30 and HSP90, NQO1		Das et al. (2014)
S. salar	HK	*Piscirickettsia salmonis*	**Susceptible**: Iron ion-binding, regulation of apoptosis, adaptive immune system, posttranslational protein modification-associated genes	**Susceptible**: Chemotaxis, cellular metal ion homeostasis, mTOR signaling pathway, natural killer cell-mediated cytotoxicity associated genes **Resistant**: β2-microglobulin, iron ion binding, regulation of apoptosis, adaptive immune system, posttranslational protein modification-related genes	Pulgar et al. (2015)
O. mykiss	HK	*A. salmonicida*	Ferritin, CD209, IL-13 receptor α-1, and VDAC2	Arachidonate 5-lipoxygenase, adenosine deaminase, ATP5J, and nucleoside diphosphate kinase	Long et al. (2015)
O. mykiss	Whole fish	*Flavobacterium psychrophilum*	**Susceptible**: MHC II, acute-phase proteins and cytokines, immune-relevant gene nlrc5-partial, Ig-like V-type domain-containing protein family 187a-like **Resistant**: IgM, TNF receptor superfamily member 14b-like isoform x1, IL-1 receptor-like 1		Marancik et al. (2014)

TABLE 8.2 List of Immune-Related Genes Differentially Modulated During Viral Infections of Fish

Fish	Fish Organ	Pathogen	Upregulated	Downregulated	References
Paralichthys olivaceus	Kidney	HRV	Apoptosis (IAP, NR-13, NGF-induced protein, HSP70 and HSP90, phospholipase C), cell-activating genes	Mx1 and β2-microglobulin	Kurobe et al. (2005)
Pseudosciaena crocea	Spleen	Poly [I:C]	Hepcidin, Mx, β2m, Cyba		Zheng et al. (2006)
P. olivaceus	HK	VHSV	**G vaccine**: Component C3, histone H1, IFN-induced protein 56kDa, TNF (ligand) superfamily, IgM, MHC II-associated invariant chain, humoral immune-related genes like CD20 receptor or IgM		Byon et al. (2006)
			DNA vaccine: MIPI-α, apoptosis-associated protein, Mx protein, complement component C3, CD8 α chain, T cell immune regulator		
Salmo salar	Lymphoid tissues	ISAV	Type I and II IFN and MHC I		Jørgensen et al. (2007)
Sparus aurata	Brain	Nodavirus strain 475-9/99	Fms-interacting protein, TNF-α-induced protein 8, ubiquitin-conjugating enzyme 7-interacting protein, IFN induced with helicase C domain protein 1, HSP70-1		Dios et al. (2007)
Gadus morhua	Spleen	Double-stranded RNA poly-riboinosinic polyribocytidylic acid	IFN-stimulated gene 15, VHSV-induced protein, MHC I, Toll-like receptor 3, small inducible cytokine SCYA104	α1-Microglobulin, α2-macroglobulin, acute-phase-related proteins, C-type lectin domain family 4 member M, complement C3	Rise et al. (2008)
S. salar	Lymphoid tissues and heart	ISAV	**Early mortalities**: IFN-related genes (src-type tyrosine kinase JAK/STAT1, IFN-induced protein 44, ADAR-1), antigen processing and presentation (β2m encoding MHC I light chain, tapasin, proteasome activator complex, ubiquitin, cathepsins), leukocyte migration and recruitment (upregulation of chemokines and receptors; SCYA110-2/CCL7-like, SCYA106/CCL21-like, CXCR4-like), IgM, CD4, and TGF-β	**Early and late mortalities**: B cell regulatory genes (MHC IIα- and invariant chain II	Jørgensen et al. (2008)
			Late mortalities: Adaptive immunity (Ig-related genes, B cells involved in signal transduction, and differentiation genes), CD8α		

Continued

TABLE 8.2 List of Immune-Related Genes Differentially Modulated During Viral Infections of Fish—cont'd

Fish	Fish Organ	Pathogen	Upregulated	Downregulated	References
Hippoglossus hippoglossus	Thymus	Nodavirus, IPVN	Receptors involved in T cell receptor complex (CD3 γ/δ, T cell receptor α and β chain, T cell surface glycoproteins), cytokines/chemokines and their receptors (chemokine receptor, CXC chemokine receptor), genes and proteins involved in cytokine cascade reaction (IFN-inducible protein 56, IFN-stimulated gene 15, TNF-α-induced 2, 6, 8), antigens (CD18, CD200, MHC Iα chain, MHC II), immunoglobulins (Fc receptor, IgE, high-affinity I, γ polypeptide), complement component C1q B chain, and proteasomes		Patel et al. (2009)
Psetta maxima	HK	Nodavirus	Mx, IFN-inducible protein 35, saxitoxin-binding protein 1, serum lectin isoform 4, serum-inducible protein kinase, complement pathway and coagulation cascade (kinnogin I, haptoglobin, thrombin, and proteinase activated receptor 3)	MHC IIα chain, F-box only protein 25, 5-aminolevulinate synthase, phosphatidylinositol 4-kinase	Park et al. (2009)
S. salar	Spleen, liver, HK, and gill	IPNV	IL-1β, IL-10, IFN-α, IFN-c, Mx, MHC-I, MHC-II, TCR-α, CD8-α, and mIgM		Ingerslev et al. (2009)
Ictalurus punctatus	42TA and CCO cell lines (macrophages and fibroblast-like cells, respectively)	Poly [I:C]	Type I IFN, ISG15, Mx1, IFN regulatory factor 1 (IAP-1), and the chemokine CXCL10		Milev-Milovanovic et al. (2009)
Danio rerio	Fins and immune organs	VHSV	Immune-related transcripts in fins (MHC I, Ig domain 4 (sid4), macrophage stimulating factor (mst1), IL-17d, IL-22); immune-related transcripts in organs (c9 and sla/lpl)	Immune-related transcripts in organs (crfb12)	Encinas et al. (2010)
S. salar	HK	IPNV	**Both**: Mx and IFN-γ, IFN-α1 **High mortality group**: Antigen presentation (β2-microglobulin, MHC I, proteasome subunit), signal transduction downstream of IFN (JAK/STAT) and editing viral RNA (double-stranded RNA-specific adenosine deaminase–ADAR)	**Both**: TLR8, TLR9, TLR22, RIG-I, MDA5, PKR, MyD88	Skjesol et al. (2011)
P. olivaceus	HK	HRV	IFN-stimulated gene 15 kDa (ISG15), ISG56, Mx		Aoki et al. (2011)
Oncorhynchus mykiss	HK	VHSV	**Resistant fish**: Response to non-self, stress and defense response, NF-κB signal transduction, antigen processing and presentation, and ATPase activity		Jørgensen et al. (2011)

S. salar	Heart	Piscine myocarditis virus	**High responder**: Apoptosis (TNF decoy receptor, TPase IMAP7 family member 7, cell death activator CIDE-3)		Timmerhaus et al. (2012)
			Low responder: Cytokine–cytokine receptor interaction genes, natural killer cell-mediated cytotoxicity, and NOD-like receptor signaling pathway genes		
Ctenopharyngodon idella	HK	Grass carp reovirus	Ig heavy chain V-III region CAM, β2-microglobulin, complement-binding mitochondrial, IFN-regulatory factor 1	Complement C3, myosin-if, adenylate kinase mitochondrial, integrin αl, β2-microglobulin, C-X-C chemokine receptor type 4, fucolectin, lysozyme C, IL enhancer-binding factor 2 homolog, MMP-9, chemokine-like factor	Chen et al. (2012)
	Epithelioma papulosum cyprini cells	Ranaviruses (family Iridoviridae)	β2M, TNF-α, IL-1β, IL-10	TGF-β	Holopainen et al. (2012)
G. morhua	Brain	Nervous necrosis virus	B and T cell receptor signaling, NOD-like receptor signaling, Toll-like receptor signaling, JAK/STAT signaling, MAPK signaling, NF-κB cascade, chemokine signaling and activity, cytokine–cytokine receptor interaction, antigen processing and presentation, lysozyme activity, phagosome, natural killer cell-mediated cytotoxicity, complement and coagulation cascades	Neuropeptide signaling, Notch signaling	Krasnov et al. (2013)
S. salar	HK	IPNV	**Resistant**: TGF-β1, NF-κB, STAT 1, IFN-α, PKR, Vig-2		Cofre et al. (2014)

HK, head kidney.

TABLE 8.3 List of Immune-Related Genes Differentially Modulated During Parasitic Infections of Fish

Fish	Tissue	Pathogen	Upregulated	Downregulated	Unchanged	References
Cyprinus carpio	HK, phagocytes	*Trypanoplasma borreli*	**Resistant**: TNF1 and 2			Saeij et al. (2003)
Oncorhynchus mykiss	Skin	*Gyrodactylus derjavini*	Mx, TNF-a1, iNOS, COX-2	MHC II	IL-8, TCR	Lindenstrøm et al. (2004)
O. mykiss	Skin, HK, spleen	*Ichthyophthirius multifiliis*	**Skin:** C3, IgM, MHC II, iNOS; HK, IgM, MHC II; spleen, C3	**HK:** C3; spleen, IgM, MHC II, iNOS		Sigh et al. (2004)
Salmo salar	HK	*Lepeophtheirus salmonis*	IL-1β, TNF-α, MHC I and II, TGF-β, COX-2			Fast et al. (2006)
C. carpio	Skin, liver	*I. multifiliis*	C7 and SCYA103	PGDS and β2-microglobulin		Gonzalez et al. (2007)
O. mykiss	Skin	*Myxobolus cerebralis*	**Susceptible**: Ubiquitin-like protein 1, IFN-regulating factor 1, PPARα-interacting complex protein 285, similar to IFN-inducible protein Gig2			Baerwald et al. (2008)
			Resistant: Ubiquitin-like protein 1, IFN-regulating factor 1, PPARα-interacting complex protein 285, similar to IFN-inducible protein Gig2, metallothionein B, IFN-regulatory factor 7, cyclin-dependent kinase 4 inhibitor B, β2-microglobulin precursor, haptoglobin precursor, VHSV-induced protein			
S. salar	Lesions on gills	*Neoparamoeba perurans*	C1q-like adipose specific protein, C-type lectin 2, CCAAT/enhancer-binding protein, chemokine CC-like protein	IFN-regulatory factor (IRF)-1, independent of IFN-α, IFN-γ, and IRF-2 expression; MHC I and II pathway-related genes, C type lectin receptors A and B		Young et al. (2008)
Dicentrarchus labrax	Gills	*Diplectanum aequans*	IL-1β, TGF-β		TCR-β	Faliex et al. (2008)
Scophthalmus maximus	Liver, spleen, and HK	*Philasterides dicentrarchi*	HSP, cytochrome P450, elastases, MHC, coagulation factors involved in innate immunity, IFN, perforin, hepcidin, nephrosin or α2-macroglobulin, antioxidant genes (glutathione peroxidase and glutathione-*S*-transferase), profilin and lysozyme, acute-phase proteins (transferrin and pentraxin)			Pardo et al. (2008)

O. mykiss	Gill	*I. multifiliis*	Immunoglobulins (IgM, IgT), complement factors (C3, BF), acute-phase proteins (SAA), cytokines (IL-8, IL-22, IFN-γ), and cellular receptors (CD4, CD8, MHC II)		IL-1β, iNOS	Olsen et al. (2011)
S. salar	Skin and spleen	*L. salmonis*	**Skin:** lectins and enzymes of eicosanoid metabolism;	MHC I and II, β2-microglobulin		Tadiso et al. (2011)
			Spleen: acute-phase proteins, matrix metalloproteinases			
Sparus aurata	Intestine, skin	*Enteromyxum leei*	**Intestine:** apoptosis (Fas apoptotic inhibitory molecule 2, apoptosis-associated speck-like protein containing a CARD, peptidyl-prolyl *cis–trans* isomerase), cell proliferation, transcription and translation (cysteine-rich protein 1, fatty acid-binding protein, brain, nuclear receptor coactivator 5, 40S ribosomal protein S3), ROS scavenging to prevent oxidative damage (glutathione *S*-transferase 3, phospholipid hydroperoxide glutathione peroxidase)	**Intestine:** complement activation, acute-phase response (fibronectin) and cell adhesion (mannose binding lectin 2, complement C3-1, complement C1s, fibronectin, adhesive plaque matrix protein), carbohydrate metabolism		Davey et al. (2011)
			Skin: cell cycle from DNA replication to cytokinesis and cell division (cell division cycle-associated protein 8, proliferating cell nuclear antigen, mitotic spindle assembly checkpoint protein MAD2A)	**Skin:** complement activation (complement C3-1, complement C2), APR (prothrombin), innate and humoral immune responses (fucolectin-1, Ig heavy chain Mem5), proteolysis (carboxypeptidase A1, carboxypeptidase B, cathepsin B), lipid metabolism and transport, serine protease inhibition		
O. mykiss	Caudal fin	*M. cerebralis*	**Resistant**: iNOS, STAT3		**Susceptible**: STAT3	Baerwald (2013)
			Both: IFN-γ and IRF-1		**Both**: KLF2 and STAT5	

HK, head kidney.

(Krause et al., 2000; Park et al., 2001), and has key functions in iron metabolism (reviewed by Shi and Camus, 2006). The potential role of hepcidin as a biomarker for resistance comes from hepcidin transgenic fish, which have upregulated hepcidin and enhanced resistance to *Vibrio vulnificus* (Hsieh et al., 2010). However, in the turbot, *Scophthalmus maximus*, viral infection provoked downregulation of AMPs in resistant fish compared to susceptible fish, although the peptides were highly expressed in the resistant fish before the immune challenge, suggesting a preexisting phenotype that caused a more robust immune response (Díaz-Rosales et al., 2012). The role of hepcidin in iron metabolism may explain its importance for resistance against this bacterial infection, since iron is an important growth factor for *V. vulnificus* (Ashrafian, 2003). The existence of polymorphism in hepcidin could explain the differential resistance of fish but this is an issue that remains to be explored (Pereiro et al., 2012; Cuesta et al., 2008). The complement pathway is also linked to disease resistance, and polymorphism of MBL, a C-type lectin that activates the lectin pathway (Nakao et al., 2006), is associated with increased survival in zebrafish, *Danio rerio*, infected with *Listonella anguillarum* (Jackson et al., 2007).

Two metal-binding proteins, ceruloplasmin and metallothionein B, that are involved in detoxification and iron metabolism (reviewed in Sevcikova et al., 2011) are proposed as biomarkers of disease resistance in fish (Tables 8.1–8.3 for references). Exposure of resistant and susceptible rohu, *Labeo rohita*, to *Aeromonas hydrophila* caused upregulation of ceruloplasmin in the liver of the former fish (Sahoo et al., 2013), and uninfected resistant fish have increased serum ceruloplasmin. Metallothionein B has a role in inflammation, apoptosis, and detoxification (reviewed in Coyle et al., 2002; Thirumoorthy et al., 2007) and in resistant rainbow trout is significantly upregulated relative to susceptible fish when they are exposed to parasites (Baerwald et al., 2008) (see Table 8.3 for more information). In the case of metallothionein in the rainbow trout resistant to the parasite *Myxobolus cerebralis*, there was consistent upregulation of the signal transducer and activator of transcription (STAT). The JAK/STAT pathway has been implicated in the modulation of the immune response in fish (Philip and Vijayan, 2015). Cortisol induces upregulation of suppressors of cytokine signaling and inhibits the JAK/STAT signaling pathway in the liver of rainbow trout, both actions being mediated by GR activation and a potential mechanism by which chronic stress causes immunosuppression.

8.1.4.2 Acquired Immunity

Acquired immunity plays a less prominent role in disease resistance in teleost fish, although it is frequently difficult to categorize innate and acquired immunity separately. One important group of immune genes that underpin the cellular response of acquired immunity and have been associated with disease resistance in many studies (Tables 8.1–8.3) comprises the polymorphic MHC genes (Lohm et al., 2002; Langefors et al., 2001; Grimholt et al., 2003; Arkush et al., 2002; Zhang et al., 2006). MHC genes encode class I MHC molecules that are present in the membrane of all nucleated cells as quaternary proteins composed of a polymorphic α chain and β2-microglobulin. MHC I presents intracellular antigens to CD8-positive lymphocytes. Class II MHC molecules are in the cell membrane of antigen-presenting cells (dendritic, B lymphocytes, and macrophage) and consist of two polymorphic α and β chains and present exogenous antigens to CD4-positive lymphocytes (reviews in Grimholt et al., 2015; Dijkstra et al., 2013). In resistant fish challenged with bacterial, viral, and parasitic infections MHC I and II pathway genes are mainly upregulated (Tables 8.1–8.3) but this does not occur in susceptible fish. In general, downregulation of MHC is associated with either reduced survival or increased pathogen load. Thus, downregulation of MHC-associated genes occurs in turbot challenged with the pathogenic bacteria *Vibrio anguillarum* (Zhang and Chen, 2006), in rainbow trout infected with *Renibacterium salmoninarum* (Grayson et al., 2002), and in the gills of Atlantic salmon infected with the parasite *Neoparamoeba perurans* (Young et al., 2008). The general response to parasitic infection in Atlantic salmon gills includes downregulation of interferon (IFN) cytokines and IFN-inducible genes (see Table 8.3). Higher resistance to parasitic infection in the three-spined stickleback, *Gasterosteus aculeatus* (lowest parasite load and highest rate of survival), and to bacterial infection in Atlantic salmon has been associated, respectively, with polymorphism of the MHC IIβ loci (Wegner et al., 2008; Eizaguirre et al., 2012) and with the presence of a high-resistance allele in the MHC IIβ locus (Lohm et al., 2002). Additionally, higher susceptibility to bacterial infection is associated with MHC IIβ homozygosity in Atlantic salmon with a bacterial infection (Turner et al., 2007).

Apoptotic genes are also modulated during viral and bacterial infections (Tables 8.1 and 8.2) and include the inhibitor of apoptosis protein (IAP), heat shock proteins (HSP) 70 and 90, phospholipase C, ubiquitin-conjugating enzyme 7, myeloid leukemia differentiation protein, programmed cell death (6, 7, and 8), tumor necrosis factor receptor superfamily-associated via death domain protein, B cell lymphoma protein, Toll-like receptor-3-like protein, cell death activator CIDE-3, and members of the caspase and BCL-2 families (references in Tables 8.1 and 8.2). The BCL-2 protein family comprises both pro- and antiapoptotic members and the proapoptotic proteins possess a death domain, BCL-2 homology 3 (BH3). BH3 is activated by oxidative cellular stress, which inactivates the antiapoptotic members (BCL-2) and activates the proapoptotic BAX and BAK members. The BH3 activation is coupled to mitochondrial cytochrome *c* release that culminates in the activation of caspase family members (Gross, 2016), which mediate apoptosis (reviewed by Degterev et al., 2003).

The regulation of apoptosis is an equilibrium between activation and suppression of caspases, in which the IAPs play an important inhibition role (Roy et al., 1997). The upregulation of apoptosis-associated genes in viral immune challenge is an antiviral defense mechanism since host cell death prevents viral replication and spreading, as apoptotic cells are eliminated by phagocytic cells (Elliott and Ravichandran, 2010; Fujimoto et al., 2000). In Atlantic salmon susceptible (high responders) to the piscine myocarditis virus (Haugland et al., 2011), upregulation of proapoptotic genes occurs in fish with increased viral levels and pathology compared with resistant fish (low responders) that have no detectable pathology (Timmerhaus et al., 2012). Some bacterial immune challenges also provoke overexpression of apoptosis-associated genes (references in Table 8.1) but it is unclear if this is a consistent defense mechanism during bacterial infections. In the absence of apoptosis, *Piscirickettsia salmonis* survives in host cells and is disseminated to neighboring cells (Rojas et al., 2009). Furthermore, proapoptotic genes are upregulated and antiapoptotic genes are downregulated in gill epithelial cells during infection with *Flavobacterium columnare* bacteria, supporting the role of apoptosis as a local defense mechanism against bacterial infections (Sun et al., 2012). Nonetheless, it remains to be clarified if induced apoptosis after viral and bacterial infections has a protective role for the host or offers the pathogen access to neighboring tissue (reviewed in Grassme et al., 2001).

Activation of apoptosis-associated genes in both viral and bacterial infections seems to be under the control of oxidative stress, with significant upregulation of HSPs and reactive oxygen species (ROS) (see Tables 8.1 and 8.2). ROS induction of apoptosis is mediated through activation of proapoptotic members of the BCL-2 protein family (Olavarría et al., 2015). In general, the induction of the oxidative stress response in both viral and bacterial infections may play a relevant role in host defense mechanisms (references in Tables 8.1 and 8.2).

Proinflammatory cytokines, TNF-α and IL-1β, have also been associated with the induction of apoptosis through stimulation of IFN-γ, a potent mediator of apoptosis (reviewed by Goetz et al., 2004), during a viral challenge (Holopainen et al., 2012). TNF-α binds to its cell surface receptor TNFR and activates the apoptotic signaling pathway (reviewed in Dos Santos et al., 2008). However, TNF-α may not be considered a marker of disease in salmonids since during the early phases of infection with *R. salmoninarum* there is decreased expression of TNF-α (Grayson et al., 2002) and increased susceptibility to mycobacterial infection (Tobin et al., 2012). However, TNF-α has been proposed as a marker of disease susceptibility in common carp since TNF-α polymorphism increases susceptibility to the parasite *Trypanoplasma borreli* in a susceptible carp line (Saeij et al., 2003).

A common feature of chronic stress and activation of the acquired immune response is increased energy demand. In the case of immune activation caused by parasitic infection the increased energy demand has been linked to poor condition and increased oxidative stress in fish (Kurtz et al., 2006). In fact the downregulation of immune genes is frequently associated with the activation of genes linked to metabolic pathways and cell respiration (see Table 8.3 for more information) (Timmerhaus et al., 2012). For this reason, alternative markers for susceptibility and disease resistance may be nonimmune and linked to the secondary consequences of infection in fish. An example of this is peroxisome proliferator-activated receptor γ, a nuclear receptor that regulates lipid and glucose homeostasis in vertebrates (reviewed in Grygiel-Górniak, 2014) and is associated with the survival of Atlantic salmon infected with *Aeromonas salmonicida* (Sundvold et al., 2010).

8.1.4.3 Stress and Immune-Associated Disease

Functional genomics has contributed to identifying the effects of stress on immune-associated genes in aquaculture species (see also Aluru and Vijayan, 2009, for early studies). In general, the majority of the candidate biomarkers (Table 8.3) for disease resistance are downregulated by cortisol treatment or exposure to stress. The exceptions are the HSPs and other genes related to ROS and apoptosis (both pro and anti). Activation of apoptotic pathways combined with negative regulation of cell proliferation (Table 8.4) may influence susceptibility to disease of stressed fish, since these genes are upregulated in susceptible Atlantic salmon suffering from viral infection (Timmerhaus et al., 2012). The general upregulation of HSPs in fish exposed to stress (Table 8.4) protects against oxidative stress at the cellular level and induces the immune response, mediating cytokine production and macrophage activation (reviewed in Iwama et al., 2004; Roberts et al., 2010). In contrast, if stress (cortisol) downregulates HSPs this may increase the susceptibility of fish to immune challenges (Gadan et al., 2012).

Elements of the innate immune response modified by stress that have the potential to be detrimental to the survival of fish include downregulation of lysozyme, hepcidin, and key proinflammatory cytokines, TNF-α and IL-1β (Table 8.4). In contrast, acute stress simulated by cortisol administration in rainbow trout or exposure to ammonia of pufferfish increased

TABLE 8.4 List of Fish Immune Genes Modulated by Stress or Cortisol Treatment From Functional Genomics Studies

Fish	Stressor	Tissue	Upregulated	Downregulated	Unchanged	References
Perca fluviatilis	1 min air exposure	Spleen		Lysozyme, hepcidin	Complement unit 3, apolipoprotein A1, and 14-kDa chemotaxin	Milla et al. (2010)
Sparus aurata	Confinement	Brain, HK, gills, and liver	Apoptosis (cell death-inducing DFFA-like effector protein C, ecto-ADP-ribosyltransferase 4) and cell proliferation (BolA-like protein 2, interferon-related developmental regulator 1), regulation of transcription and translation (zinc finger proteins 9, 330, and 479, transcription factor Sp1, threonyl-tRNA synthetase), inhibitors of protein breakdown (α2-macroglobulin, HSP70-binding protein 1), immune response (complement pathway, cyclosporine A-binding protein, hepcidin), iron metabolism (ferroxidase/ceruloplasmin)	Acute immune response (α2-HS glycoprotein), antigen binding and processing (Ig heavy chain Mem5, histocompatibility complex), lectin pathway (fucolectin-1), inflammatory and interferon signaling (macrophage migration inhibitory factor, interferon regulatory factor 8)		Calduch-Giner et al. (2010)
S. aurata	Cortisol implants	Liver	Cell death activator CIDE3, proteolytic enzymes (cdc48 and pitrilysin), metalloproteinase 1, zinc-binding protein 1	T cell receptor α and β chain, MHC II antigen β chain, and MHC II antigen-associated invariant chain, complement C4 and C9, interleukin-22, interleukin-8-like protein, CXC chemokine, interferon regulatory factor 1, acute-phase proteins		Teles et al. (2013)

Oncorhynchus mykiss	Cortisol treatment	HK	IL-1β, TNF-α, complement component C3	IL-6, TGF-β, lysozyme		Cortés et al. (2013)
Takifugu obscurus	High temperature	Blood cells	p53, BAX, caspase 9 and caspase 3, Mn-SOD, Cu/Zn-SOD, CAT, GPx, HSP90, and HSP70			Cheng et al. (2015a)
T. obscurus	Ammonia	Liver	Bax, caspase 9, caspase 8, caspase 3, BAFF, TNF-α, IL-6, IL-12, Mn-SOD, CAT, GPx, and GR	BCL-2		Cheng et al. (2015b)
O. mykiss	Cu and hypoxia	Liver/gill	Metallothionein (A/B), COX4-2	AMPKα1		Sappal et al. (2016)
S. aurata	Heavy metals	Skin, liver	Liver, Cyp1a1	Skin, metallothionein A, HSP90	Liver, metallothionein A, HSP70 and HSP90; skin, Cyp1a1	Benhamed et al. (2016)
O. mykiss	Crowding (chronic stress)	HK	HSP70	LyzII, TNF-1α, IL-1β, IL-8, and IFN-γ1		Yarahmadi et al. (2016)

HK, head kidney.

the expression of TNF-α and IL-1β (Table 8.4) and had positive effects on the immune response. Such results highlight that the outcomes of chronic and acute stress on the immune response in fish are divergent.

Stress also regulates the adaptive immune response, and genes associated with T cell function and class I and II MHC pathways are downregulated in fish exposed to stress (Table 8.4). In contrast, ceruloplasmin and metallothionein B are increased by stress, although the effect of stress on metallothionein is clearer than the effect on ceruloplasmin. The levels of metallothionein protein were increased in the liver of rainbow trout exposed either to cadmium or to handling stress (Tort et al., 1996) and in the liver and kidney of crucian carp, *Carassius cuvieri*, exposed to air-pumping stress, dexamethasone, cadmium, and zinc (Muto et al., 1999). The expression of metallothionein (A and B) was also significantly upregulated in the liver and gills of rainbow trout subjected to copper and hypoxia and in gilthead sea bream exposed to confinement stress or cortisol treatment (Table 8.4). The increase in both proteins during stress conditions, mainly after exposure to metals, is essential for preventing damage to cells due to the inhibition of ROS production (Sevcikova et al., 2011), which makes ceruloplasmin and metallothionein markers of oxidative stress and pollutants (reviewed by Mahboob, 2013). The pattern of immune gene expression contributes to understanding how chronic stress and acute stress can have divergent effects on fish immune function and compromise the resistance to disease.

8.1.5 Stress Management

Aquaculture has developed rapidly since 1965 to fulfill the global requirements of fish protein consumption. To achieve this, the systems and techniques used in fish aquaculture have developed to improve growth and survival by modifying husbandry and aquaculture systems (Bostock et al., 2010). Fish from aquaculture are confined at high density to a production system they cannot escape and are frequently under conditions normally associated with chronic stress (Rottmann et al., 1992). Chronic stress of aquaculture fish causes an imbalance in the endocrine mediators of the HPI axis (e.g., cortisol), which leads to overall suppression of the immune system and a significant decrease in resistance to disease. The best approach to minimize the susceptibility of aquaculture fish to disease is to improve husbandry procedures to minimize exposure to chronic stress, to develop noninvasive monitoring methods to assess the health condition and stress status, and to prevent the introduction of pathogens (Gabriel and Akinrotimi, 2011; Tucker and Hargreaves, 2009; Davis, 2006). A number of reviews have suggested measures to reduce stress without affecting the sustainability of aquaculture production (Hollingsworth et al., 2006; Gabriel and Akinrotimi, 2011; Conte, 2004).

Stress management refers to a wide range of strategies and therapies directed at controlling and minimizing stress, especially under conditions of chronic stress (Ong et al., 2004; Moberg, 2000). In fish aquaculture, stress management includes control of environmental quality, improvement of food quality, establishment of the ideal tank density, and avoidance of excessive manipulation of fish. Maintenance of water quality in the production system must consider the physiological needs of the species being cultured and include factors such as temperature, pH, oxygen levels, and the accumulation of nitrogenous wastes. The establishment of an optimal fish density in tanks should consider the competition for food, oxygen, and space and the incidence of physical injuries. Malnourished fish are chronically stressed and have increased disease susceptibility and so a high-quality diet should be provided that fulfills the nutritional needs of the fish. Transportation is a major stressor that can be minimized if it is performed early in the morning or late in the evening or by sedating fish with anesthetics (for more details see, e.g., Gabriel and Akinrotimi, 2011; Conte, 2004). Knowledge and recognition of specific behavioral traits are key components of stress prevention and diagnosis (Martins et al., 2012; Romero et al., 2009; Rupia et al., 2016; Abreu et al., 2016).

Selective breeding for the stress response based on cortisol levels is a part of stress management programs that aims to select fish with a reduced response to the common stressors associated with aquaculture practice (Cnaani, 2006; Davis, 2006; Yáñez et al., 2014). However, the results of previous studies (Fevolden et al., 1993, 1991, 1992; Weber et al., 2008) have been ambiguous and there is no clear conclusion about the impact of high or low cortisol response on the response to an immune challenge. The focus of future genetics research should be on the overall HPI response to chronic stress conditions, which are strongly correlated with suppression of the immune function (Weyts et al., 1999; Harris and Bird, 2000; Bowden, 2008; Tort, 2011; Aluru and Vijayan, 2009; Barton et al., 2005; Rotllant et al., 2001; Rotllant and Tort, 1997).

The way in which stress conditions during early development affect the juvenile and adult response to stress remains obscure in teleost fish. The effect of early life experience on juvenile adult performance has been explored and there is evidence that early stress/glucocorticoid exposure influences adult behavior and metabolism (Wilson et al., 2016; Tsalafouta et al., 2015). The rationale behind such studies is that stressing fish larvae by manipulation might be a strategy for stress management in aquaculture since the resulting adult fish have a reduced response to cortisol (Auperin and Geslin, 2008). Furthermore, improved growth is a general beneficial effect of early life exposure to stress (Varsamos et al., 2006; Boglione and Costa, 2011; Wilson et al., 2016). Nonetheless, species-specific variations occur and when larval European sea bass or gilthead sea bream are exposed to stress during early development and then as adults are subsequently exposed to an acute stressor, sea bass have increased and divergent plasma cortisol levels relative to gilthead sea bream (authors' unpublished results). These results highlight that approaches to stress management in aquaculture should be species specific.

Strategies for monitoring stress management programs include measurement of cortisol and other parameters indicative of animals' welfare. Cortisol is most commonly used given that it tends to increase after exposure to a wide range of stress conditions (Martínez-Porchas et al., 2009; Mommsen et al., 1999; Barton, 2002; Wendelaar Bonga, 1997). Nonetheless, there are discrepancies within the scientific literature and sometimes reported plasma cortisol levels do not always correspond to a stressful situation (reviewed in Mommsen et al., 1999; Harper and Wolf, 2009). Inconsistencies in the cortisol response to stress are frequent in chronic stress (reviewed by Harper and Wolf, 2009; Laidley and Leatherland, 1988; Molinero et al., 1997) and highlight the challenge of using cortisol as an indicator of stress and to monitor stress management strategies in aquaculture.

For monitoring stress management, cortisol has mainly been measured in plasma samples (reviewed in Ellis et al., 2012, 2013; Harper and Wolf, 2009; Ellis et al., 2013). However, alternative samples that permit noninvasive monitoring of cortisol and give reliable and informative cortisol levels have been developed. There is a strong correlation between cortisol levels found in plasma and water after acute or chronic stress (Tsalafouta et al., 2015; Scott and Ellis, 2007; Wong et al., 2008; Félix et al., 2013; Fanouraki et al., 2008, 2011; Scott et al., 2001, 2008). Although there are some technical limitations to the approach, the advantages tend to outweigh the disadvantages. Alternative matrices to plasma and water to measure cortisol levels have also been suggested, such as mucus and gut content (Bertotto et al., 2010; Simontacchi et al., 2008). Scales from common carp (*C. carpio*) appear to capture systemic cortisol during long exposure periods to stress in contrast to blood and water, which reveal the cortisol status only at the moment of sampling (Aerts et al., 2015).

While cortisol levels, especially those in plasma samples, have been used as the primary indicator of fish welfare, particularly for acute stress, several methodologies to assess fish welfare have been proposed to ameliorate stress management: assess if the animal is coping well with the environmental and husbandry conditions (related to the physiological responses to stress), evaluate changes in the normal behavior, and assess the fishes' perception of the environmental and husbandry conditions (Martins et al., 2012; Huntingford et al., 2006; Prunet et al., 2012; Poli et al., 2005; Segner et al., 2012). Behavioral welfare indicators such as feed intake, ventilator activity, swimming activity, and skin darkness seem to be plausible indicators for assessing fish welfare during chronic stressful conditions, given the facts that their measurement relies on noninvasive methods (video recordings or direct observation) and that they are modified during early stages of potential welfare problems (reviewed by Martins et al., 2012; Ellis et al., 2012).

The understanding of the husbandry practices performed in aquaculture that force the fish to live under chronic stress conditions is of utmost importance for farmers to reduce the stress levels of fish and improve their resistance to diseases, which will lead to increased production and therefore to the achievement of higher profits. In addition, the implementation of stress management strategies and the use of reliable and noninvasive biomarkers for cortisol and behavior responses during husbandry practices must be a priority for farmers, which costs will be necessarily inferior to those related to mortality and inferior flesh quality associated with increased production systems without the awareness of fish welfare. Accordingly, Berrill et al. (2012) have reviewed this conflict of interest between the ethics of achieving fish welfare during aquaculture production and the economic interests of stakeholders. Several priorities came out from this discussion, including "better understanding of what good fish welfare is," "the role of genetic selection in producing fish suited to the farming environment," and "a need for integration and application of behavioral and physiological measures." Implementation of these priorities will lead to improved fish welfare in the fish farm and increased product quality, productivity, and consumer acceptance of farmed fish.

8.1.6 Conclusions

Intensive fish farming favors chronic stress as a result of high densities, suboptimal physical environment, and muted social interactions. The HPI axis response toward regaining homeostasis compromises the immune system, increases vulnerability to disease, and reduces growth. However, early conditioning to stress offers the potential to reprogram behavior and metabolism to minimize the impact of stress later in life. The use of nonintrusive monitoring methods based on behavior and physiology offers a practical way to minimize or abolish sources of stress. Although some markers of stress-responsive immune genes are known, more studies are required to understand their role and mechanisms of action and to extend this information to other potential markers. A key area for future research is the understanding of the interactions between the HPI and growth axes and the immune system.

ACKNOWLEDGMENT

The authors acknowledge support by national funds from FCT—Foundation for Science and Technology—through Project UID/Multi/04326/2013.

REFERENCES

Abreu, M.S., Giacomini, A., Kalueff, A.V., Barcellos, L.J.G., 2016. The smell of "anxiety": behavioral modulation by experimental anosmia in zebrafish. Physiology & Behavior 157, 67–71.

Acerete, L., Balasch, J.C., Castellana, B., Redruello, B., Roher, N., Canario, A.V., Planas, J.V., Mackenzie, S., Tort, L., 2007. Cloning of the glucocorticoid receptor (GR) in gilthead seabream (*Sparus aurata*). Differential expression of GR and immune genes in gilthead seabream after an immune challenge. Comparative Biochemistry and Physiology – Part B: Biochemistry and Molecular Biology 148, 32–43.

Aerts, J., Metz, J.R., Ampe, B., Decostere, A., Flik, G., De Saeger, S., 2015. Scales tell a story on the stress history of fish. PLoS One 10, e0123411.

Alderman, S.L., Mcguire, A., Bernier, N.J., Vijayan, M.M., 2012. Central and peripheral glucocorticoid receptors are involved in the plasma cortisol response to an acute stressor in rainbow trout. General and Comparative Endocrinology 176, 79–85.

Aluru, N., Vijayan, M.M., 2009. Stress transcriptomics in fish: a role for genomic cortisol signaling. General and Comparative Endocrinology 164, 142–150.

Alves, R.N., Cordeiro, O., Silva, T.S., Richard, N., De Vareilles, M., Marino, G., Di Marco, P., Rodrigues, P.M., Conceição, L.E., 2010. Metabolic molecular indicators of chronic stress in gilthead seabream (*Sparus aurata*) using comparative proteomics. Aquaculture 299, 57–66.

Aoki, T., Hirono, I., Kondo, H., Hikima, J.-I., Jung, T.S., 2011. Microarray technology is an effective tool for identifying genes related to the aquacultural improvement of Japanese flounder, *Paralichthys olivaceus*. Comparative Biochemistry and Physiology Part D: Genomics and Proteomics 6, 39–43.

Aoki, T., Takano, T., Santos, M.D., Kondo, H., Hirono, I., 2008. Molecular innate immunity in teleost fish: review and future perspectives. In: Tsukamoto, K., Kawamura, T., Takeuchi, T., Beard JR, T., Kaiser, M. (Eds.), Fisheries for Global Welfare and Environment, 5th World Fisheries Congress. Terrapub, Tokyo, pp. 263–276.

Arends, R., Mancera, J., Munoz, J., Bonga, S.W., Flik, G., 1999. The stress response of the gilthead sea bream (*Sparus aurata* L.) to air exposure and confinement. Journal of Endocrinology 163, 149–157.

Arkush, K.D., Giese, A.R., Mendonca, H.L., Mcbride, A.M., Marty, G.D., Hedrick, P.W., 2002. Resistance to three pathogens in the endangered winter-run chinook salmon (*Oncorhynchus tshawytscha*): effects of inbreeding and major histocompatibility complex genotypes. Canadian Journal of Fisheries and Aquatic Sciences 59, 966–975.

Ashley, P.J., 2007. Fish welfare: current issues in aquaculture. Applied Animal Behaviour Science 104, 199–235.

Ashrafian, H., 2003. Hepcidin: the missing link between hemochromatosis and infections. Infection and Immunity 71, 6693–6700.

Auperin, B., Geslin, M., 2008. Plasma cortisol response to stress in juvenile rainbow trout is influenced by their life history during early development and by egg cortisol content. General and Comparative Endocrinology 158, 234–239.

Avella, M., Schreck, C.B., Prunet, P., 1991. Plasma prolactin and cortisol concentrations of stressed coho salmon, *Oncorhynchus kisutch*, in fresh water or salt water. General and Comparative Endocrinology 81, 21–27.

Baerwald, M.R., 2013. Temporal expression patterns of rainbow trout immune-related genes in response to *Myxobolus cerebralis* exposure. Fish and Shellfish Immunolog 35, 965–971.

Baerwald, M.R., Welsh, A.B., Hedrick, R.P., May, B., 2008. Discovery of genes implicated in whirling disease infection and resistance in rainbow trout using genome-wide expression profiling. BMC Genomics 9, 37.

Bao, B., Peatman, E., Xu, P., Li, P., Zeng, H., He, C., Liu, Z., 2006. The catfish liver-expressed antimicrobial peptide 2 (LEAP-2) gene is expressed in a wide range of tissues and developmentally regulated. Molecular Immunology 43, 367–377.

Barton, B.A., 2002. Stress in fishes: a diversity of responses with particular reference to changes in circulating corticosteroids. Integrative and Comparative Biology 42, 517–525.

Barton, B.A., 2011. Stress in finfish: past, present and future – a historical perspective. In: Iwama, G.K., Pickering, A.D., Sumpter, J.P., Schreck, C.B. (Eds.), Fish Stress and Health in Aquaculture, Society for Experimental Biology Seminar Series 62. Cambridge University Press, Cambridge, pp. 1–34.

Barton, B.A., Iwama, G.K., 1991. Physiological changes in fish from stress in aquaculture with emphasis on the response and effects of corticosteroids. Annual Review of Fish Diseases 1, 3–26.

Barton, B.A., Ribas, L., Acerete, L., Tort, L., 2005. Effects of chronic confinement on physiological responses of juvenile gilthead sea bream, *Sparus aurata* L., to acute handling. Aquaculture Research 36, 172–179.

Bayne, C.J., Gerwick, L., 2001. The acute phase response and innate immunity of fish. Developmental and Comparative Immunology 25, 725–743.

Bayne, C.J., Levy, S., 1991. Modulation of the oxidative burst in trout myeloid cells by adrenocorticotropic hormone and catecholamines: mechanisms of action. Journal of Leukocyte Biology 50, 554–560.

Benhamed, S., Guardiola, F.A., Martínez, S., Martínez-Sánchez, M., Pérez-Sirvent, C., Mars, M., Esteban, M.A., 2016. Exposure of the gilthead seabream (*Sparus aurata*) to sediments contaminated with heavy metals down-regulates the gene expression of stress biomarkers. Toxicology Reports 3, 364–372.

Berrill, I., Cooper, T., Macintyre, C., Ellis, T., Knowles, T.G., Jones, E.K., Turnbull, J., 2012. Achieving consensus on current and future priorities for farmed fish welfare: a case study from the UK. Fish Physiology and Biochemistry 38, 219–229.

Bertotto, D., Poltronieri, C., Negrato, E., Majolini, D., Radaelli, G., Simontacchi, C., 2010. Alternative matrices for cortisol measurement in fish. Aquaculture Research 41, 1261–1267.

Black, E.C., Tredwell, S., 1967. Effect of a partial loss of scales and mucous on carbohydrate metabolism in rainbow trout (*Salmo gairdneri*). Journal of the Fisheries Board of Canada 24, 939–953.

Bo, J., Giesy, J.P., Ye, R., Wang, K.-J., Lee, J.-S., Au, D.W., 2012. Identification of differentially expressed genes and quantitative expression of complement genes in the liver of marine medaka *Oryzias melastigma* challenged with *Vibrio parahaemolyticus*. Comparative Biochemistry and Physiology Part D: Genomics and Proteomics 7, 191–200.

Boglione, C., Costa, C., 2011. Skeletal deformities and juvenile quality. In: Pavlidis, M.A., Mylonas, C.C. (Eds.), Sparidae: Biology and Aquaculture of Gilthead Sea Bream and Other Species. Wiley-Blackwell, Chichester, pp. 233–294.

Booman, M., Borza, T., Feng, C.Y., Hori, T.S., Higgins, B., Culf, A., Léger, D., Chute, I.C., Belkaid, A., Rise, M., 2011. Development and experimental validation of a 20K Atlantic cod (*Gadus morhua*) oligonucleotide microarray based on a collection of over 150,000 ESTs. Marine Biotechnology 13, 733–750.

Bostock, J., Mcandrew, B., Richards, R., Jauncey, K., Telfer, T., Lorenzen, K., Little, D., Ross, L., Handisyde, N., Gatward, I., Corner, R., 2010. Aquaculture: global status and trends. Philosophical Transactions of the Royal Society B: Biological Sciences 365, 2897–2912.

Boulton, K., Massault, C., Houston, R.D., De Koning, D.J., Haley, C.S., Bovenhuis, H., Batargias, C., Canario, A.V.M., Kotoulas, G., Tsigenopoulos, C.S., 2011. QTL affecting morphometric traits and stress response in the gilthead seabream (*Sparus aurata*). Aquaculture 319, 58–66.

Bowden, T.J., 2008. Modulation of the immune system of fish by their environment. Fish and Shellfish Immunology 25, 373–383.

Braithwaite, V., Ebbesson, L., 2014. Pain and stress responses in farmed fish. Revue Scientifique et Technique: Office International des Epizooties 33, 245–253.

Brazzini, B., Ghersetich, I., Hercogova, J., Lotti, T., 2003. The neuro-immuno-cutaneous-endocrine network: relationship between mind and skin. Dermatologic Therapy 16, 123–131.

Brett, J., 1958. Implications and assessments of environmental stress. In: Larkin, P.A. (Ed.), The Investigation of Fish-Power Problems, H.R. MacMillan Lectures in Fisheries. University of British Columbia, Vancouver, pp. 69–83.

Brown, L., Johnson, S., 2008. Molecular interaction between fish pathogens and host aquatic animals. In: Tsukamoto, K., Kawamura, T., Takeuchi, T., Beard JR, T., Kaiser, M. (Eds.), Fisheries for Global Welfare and Environment. Terrapub, Tokyo, pp. 277–288.

Brown, S., Fedoruk, K., Eales, J., 1978. Physical injury due to injection or blood removal causes transitory elevations of plasma thyroxine in rainbow trout, *Salmo gairdneri*. Canadian Journal of Zoology 56, 1998–2003.

Byon, J.Y., Ohira, T., Hirono, I., Aoki, T., 2006. Comparative immune responses in Japanese flounder, *Paralichthys olivaceus* after vaccination with viral hemorrhagic septicemia virus (VHSV) recombinant glycoprotein and DNA vaccine using a microarray analysis. Vaccine 24, 921–930.

Caipang, C.M.A., Hynes, N., Puangkaew, J., Brinchmann, M.F., Kiron, V., 2008. Intraperitoneal vaccination of Atlantic cod, *Gadus morhua* with heat-killed *Listonella anguillarum* enhances serum antibacterial activity and expression of immune response genes. Fish and Shellfish Immunology 24, 314–322.

Calduch-Giner, J., Sitja-Bobadilla, A., Alvarez-Pellitero, P., Perez-Sanchez, J., 1995. Evidence for a direct action of GH on haemopoietic cells of a marine fish, the gilthead sea bream (*Sparus aurata*). Journal of Endocrinology 146, 459–467.

Calduch-Giner, J.A., Davey, G., Saera-Vila, A., Houeix, B., Talbot, A., Prunet, P., Cairns, M.T., Pérez-Sánchez, J., 2010. Use of microarray technology to assess the time course of liver stress response after confinement exposure in gilthead sea bream (*Sparus aurata* L.). BMC Genomics 11, 193.

Calduch-Giner, J.A., Pérez-Sánchez, J., 1999. Expression of growth hormone gene in the head kidney of gilthead sea bream (*Sparus aurata*). Journal of Experimental Zoology 283, 326–330.

Castillo, J., Teles, M., Mackenzie, S., Tort, L., 2009. Stress-related hormones modulate cytokine expression in the head kidney of gilthead seabream (*Sparus aurata*). Fish and Shellfish Immunology 27, 493–499.

Chadzinska, M., Tertil, E., Kepka, M., Hermsen, T., Scheer, M., Verburg-Van Kemenade, B.L., 2012. Adrenergic regulation of the innate immune response in common carp (*Cyprinus carpio* L.). Developmental and Comparative Immunology 36, 306–316.

Chen, J., Li, C., Huang, R., Du, F., Liao, L., Zhu, Z., Wang, Y., 2012. Transcriptome analysis of head kidney in grass carp and discovery of immune-related genes. BMC Veterinary Research 8, 1.

Chen, Y.-P., Jiang, W.-D., Liu, Y., Jiang, J., Wu, P., Zhao, J., Kuang, S.-Y., Tang, L., Tang, W.-N., Zhang, Y.-A., 2015. Exogenous phospholipids supplementation improves growth and modulates immune response and physical barrier referring to NF-κB, TOR, MLCK and Nrf2 signaling factors in the intestine of juvenile grass carp (*Ctenopharyngodon idella*). Fish and Shellfish Immunology 47, 46–62.

Cheng, C.-H., Yang, F.-F., Liao, S.-A., Miao, Y.-T., Ye, C.-X., Wang, A.-L., Tan, J.-W., Chen, X.-Y., 2015a. High temperature induces apoptosis and oxidative stress in pufferfish (*Takifugu obscurus*) blood cells. Journal of Thermal Biology 53, 172–179.

Cheng, C.-H., Yang, F.-F., Ling, R.-Z., Liao, S.-A., Miao, Y.-T., Ye, C.-X., Wang, A.-L., 2015b. Effects of ammonia exposure on apoptosis, oxidative stress and immune response in pufferfish (*Takifugu obscurus*). Aquatic Toxicology 164, 61–71.

Cho, Y.S., Lee, S.Y., Kim, K.H., Kim, S.K., Kim, D.S., Nam, Y.K., 2009. Gene structure and differential modulation of multiple rockbream (*Oplegnathus fasciatus*) hepcidin isoforms resulting from different biological stimulations. Developmental and Comparative Immunology 33, 46–58.

Chrousos, G.P., 1998. Stressors, stress, and neuroendocrine integration of the adaptive response: the 1997 Hans Selye Memorial Lecture. Annals of the New York Academy of Sciences 851, 311–335.

Cnaani, A., 2006. Genetic perspective on stress response and disease resistance in aquaculture. Israeli Journal of Aquaculture-Bamidgeh 58, 375–383.

Cofre, C., Gonzalez, R., Moya, J., Vidal, R., 2014. Phenotype gene expression differences between resistant and susceptible salmon families to IPNV. Fish Physiology and Biochemistry 40, 887–896.

Conte, F., 2004. Stress and the welfare of cultured fish. Applied Animal Behaviour Science 86, 205–223.

Cortés, R., Teles, M., Trídico, R., Acerete, L., Tort, L., 2013. Effects of cortisol administered through slow-release implants on innate immune responses in rainbow trout (*Oncorhynchus mykiss*). International Journal of Genomics 2013.

Coyle, P., Philcox, J., Carey, L., Rofe, A., 2002. Metallothionein: the multipurpose protein. Cellular and Molecular Life Sciences 59, 627–647.

Cruz, S.A., Lin, C.-H., Chao, P.-L., Hwang, P.-P., 2013. Glucocorticoid receptor, but not mineralocorticoid receptor, mediates cortisol regulation of epidermal ionocyte development and ion transport in zebrafish (*Danio rerio*). PLoS One 8, e77997.

Cuesta, A., Meseguer, J., Esteban, M.A., 2008. The antimicrobial peptide hepcidin exerts an important role in the innate immunity against bacteria in the bony fish gilthead seabream. Molecular Immunology 45, 2333–2342.

Darawiroj, D., Kondo, H., Hirono, I., Aoki, T., 2008. Immune-related gene expression profiling of yellowtail (*Seriola quinqueradiata*) kidney cells stimulated with ConA and LPS using microarray analysis. Fish and Shellfish Immunology 24, 260–266.

Das, S., Chhottaray, C., Mahapatra, K.D., Saha, J.N., Baranski, M., Robinson, N., Sahoo, P., 2014. Analysis of immune-related ESTs and differential expression analysis of few important genes in lines of rohu (*Labeo rohita*) selected for resistance and susceptibility to Aeromonas hydrophila infection. Molecular Biology Reports 41, 7361–7371.

Davey, G.C., Calduch-Giner, J.A., Houeix, B., Talbot, A., Sitjà-Bobadilla, A., Prunet, P., Pérez-Sánchez, J., Cairns, M.T., 2011. Molecular profiling of the gilthead sea bream (*Sparus aurata* L.) response to chronic exposure to the myxosporean parasite *Enteromyxum leei*. Molecular Immunology 48, 2102–2112.

Davis, K.B., 2006. Management of physiological stress in finfish aquaculture. North American Journal of Aquaculture 68, 116–121.

Degterev, A., Boyce, M., Yuan, J., 2003. A decade of caspases. Oncogene 22, 8543–8567.

Demers, N.E., Bayne, C.J., 1997. The immediate effects of stress on hormones and plasma lysozyme in rainbow trout. Developmental and Comparative Immunology 21, 363–373.

Díaz-Rosales, P., Romero, A., Balseiro, P., Dios, S., Novoa, B., Figueras, A., 2012. Microarray-based identification of differentially expressed genes in families of turbot (*Scophthalmus maximus*) after infection with viral haemorrhagic septicaemia virus (VHSV). Marine Biotechnology 14, 515–529.

Dijkstra, J.M., Grimholt, U., Leong, J., Koop, B.F., Hashimoto, K., 2013. Comprehensive analysis of MHC class II genes in teleost fish genomes reveals dispensability of the peptide-loading DM system in a large part of vertebrates. BMC Evolutionary Biology 13, 260.

Dios, S., Poisa-Beiro, L., Figueras, A., Novoa, B., 2007. Suppression subtraction hybridization (SSH) and macroarray techniques reveal differential gene expression profiles in brain of sea bream infected with nodavirus. Molecular Immunology 44, 2195–2204.

Dos Santos, N., Vale, A.D., Reis, M., Silva, M., 2008. Fish and apoptosis: molecules and pathways. Current Pharmaceutical Design 14, 148–169.

Dumrongphol, Y., Hirota, T., Kondo, H., Aoki, T., Hirono, I., 2009. Identification of novel genes in Japanese flounder (*Paralichthys olivaceus*) head kidney up-regulated after vaccination with *Streptococcus iniae* formalin-killed cells. Fish and Shellfish Immunology 26, 197–200.

Eddy, F.B., 1981. Effects of stress on osmotic and ionic regulation in fish. In: Pickering, A.D. (Ed.), Stress and Fish. Academic Press, London, pp. 77–102.

Eizaguirre, C., Lenz, T.L., Kalbe, M., Milinski, M., 2012. Rapid and adaptive evolution of MHC genes under parasite selection in experimental vertebrate populations. Nature Communications 3, 621.

El-Khaldi, A.T., 2010. Effect of different stress factors on some physiological parameters of Nile tilapia (*Oreochromis niloticus*). Saudi Journal of Biological Sciences 17, 241–246.

Elliott, M.R., Ravichandran, K.S., 2010. Clearance of apoptotic cells: implications in health and disease. The Journal of Cell Biology 189, 1059–1070.

Ellis, A., 2001. Innate host defense mechanisms of fish against viruses and bacteria. Developmental and Comparative Immunology 25, 827–839.

Ellis, T., Sanders, M., Scott, A., 2013. Non-invasive monitoring of steroids in fishes. Veterinary Medicine Austria 100, 255–269.

Ellis, T., Yildiz, H.Y., López-Olmeda, J., Spedicato, M.T., Tort, L., Øverli, Ø., Martins, C.I., 2012. Cortisol and finfish welfare. Fish Physiology and Biochemistry 38, 163–188.

Encinas, P., Rodriguez-Milla, M.A., Novoa, B., Estepa, A., Figueras, A., Coll, J., 2010. Zebrafish fin immune responses during high mortality infections with viral haemorrhagic septicemia rhabdovirus. A proteomic and transcriptomic approach. BMC Genomics 11, 1.

Engler, H., Engler, A., Bailey, M.T., Sheridan, J.F., 2005. Tissue-specific alterations in the glucocorticoid sensitivity of immune cells following repeated social defeat in mice. Journal of Neuroimmunology 163, 110–119.

Evans, D.H., Piermarini, P.M., Choe, K.P., 2005. The multifunctional fish gill: dominant site of gas exchange, osmoregulation, acid-base regulation, and excretion of nitrogenous waste. Physiological Reviews 85, 97–177.

Faliex, E., Da Silva, C., Simon, G., Sasal, P., 2008. Dynamic expression of immune response genes in the sea bass, *Dicentrarchus labrax*, experimentally infected with the monogenean *Diplectanum aequans*. Fish and Shellfish Immunology 24, 759–767.

Fanouraki, E., Mylonas, C.C., Papandroulakis, N., Pavlidis, M., 2011. Species specificity in the magnitude and duration of the acute stress response in Mediterranean marine fish in culture. General and Comparative Endocrinology 173, 313–322.

Fanouraki, E., Papandroulakis, N., Ellis, T., Mylonas, C., Scott, A., Pavlidis, M., 2008. Water cortisol is a reliable indicator of stress in European sea bass, *Dicentrarchus labrax*. Behaviour 145, 1267–1281.

Farbridge, K., Leatherland, J., 1992. Plasma growth hormone levels in fed and fasted rainbow trout (*Oncorhynchus mykiss*) are decreased following handling stress. Fish Physiology and Biochemistry 10, 67–73.

Fast, M., Muise, D., Easy, R., Ross, N., Johnson, S., 2006. The effects of *Lepeophtheirus salmonis* infections on the stress response and immunological status of Atlantic salmon (*Salmo salar*). Fish and Shellfish Immunology 21, 228–241.

Fast, M.D., Hosoya, S., Johnson, S.C., Afonso, L.O., 2008. Cortisol response and immune-related effects of Atlantic salmon (*Salmo salar* Linnaeus) subjected to short-and long-term stress. Fish and Shellfish Immunology 24, 194–204.

Félix, A.S., Faustino, A.I., Cabral, E.M., Oliveira, R.F., 2013. Noninvasive measurement of steroid hormones in zebrafish holding-water. Zebrafish 10, 110–115.

Feng, C.Y., Johnson, S.C., Hori, T.S., Rise, M., Hall, J.R., Gamperl, A.K., Hubert, S., Kimball, J., Bowman, S., Rise, M.L., 2009. Identification and analysis of differentially expressed genes in immune tissues of Atlantic cod stimulated with formalin-killed, atypical *Aeromonas salmonicida*. Physiological Genomics 37, 149–163.

Fevolden, S.E., Nordmo, R., Refstie, T., Rø, K.H., 1993. Disease resistance in Atlantic salmon (*Salmo salar*) selected for high or low responses to stress. Aquaculture 109, 215–224.

Fevolden, S.E., Refstie, T., Rø, K.H., 1991. Selection for high and low cortisol stress response in Atlantic salmon (*Salmo salar*) and rainbow trout (*Oncorhynchus mykiss*). Aquaculture 95, 53–65.

Fevolden, S.E., Refstie, T., Rø, K.H., 1992. Disease resistance in rainbow trout (*Oncorhynchus mykiss*) selected for stress response. Aquaculture 104, 19–29.

Finkenbine, S.S., Gettys, T.W., Burnett, K.G., 2002. Beta-adrenergic receptors on leukocytes of the channel catfish, *Ictalurus punctatus*. Comparative Biochemistry and Physiology Part C: Toxicology and Pharmacology 131, 27–37.

Flier, J.S., Underhill, L.H., Mcewen, B.S., 1998. Protective and damaging effects of stress mediators. New England Journal of Medicine 338, 171–179.

Flory, C.M., 1989. Autonomic innervation of the spleen of the coho salmon, *Oncorhynchus kisutch*: a histochemical demonstration and preliminary assessment of its immunoregulatory role. Brain, Behavior and Immunity 3, 331–344.

Flory, C.M., Bayne, C.J., 1991. The influence of adrenergic and cholinergic agents on the chemiluminescent and mitogenic responses of leukocytes from the rainbow trout, *Oncorhyncus mykiss*. Developmental and Comparative Immunology 15, 135–142.

Fujimoto, I., Pan, J., Takizawa, T., Nakanishi, Y., 2000. Virus clearance through apoptosis-dependent phagocytosis of influenza A virus-infected cells by macrophages. Journal of Virology 74, 3399–3403.

Gabriel, U.U., Akinrotimi, O.A., 2011. Management of stress in fish for sustainable aquaculture development. Researcher 3, 28–38.

Gadan, K., Marjara, I.S., Sundh, H., Sundell, K., Evensen, Ø., 2012. Slow release cortisol implants result in impaired innate immune responses and higher infection prevalence following experimental challenge with infectious pancreatic necrosis virus in Atlantic salmon (*Salmo salar*) parr. Fish and Shellfish Immunology 32, 637–644.

Galhardo, L., Oliveira, R.F., 2009. Psychological stress and welfare in fish. Annual Review of Biomedical Sciences 1–20.

Gallo, V.P., Civinini, A., 2003. Survey of the adrenal homolog in teleosts. International Review of Cytology 230, 89–187.

Goetz, F.W., Planas, J.V., Mackenzie, S., 2004. Tumor necrosis factors. Developmental and Comparative Immunology 28, 487–497.

Gonzalez, S.F., Chatziandreou, N., Nielsen, M.E., Li, W., Rogers, J., Taylor, R., Santos, Y., Cossins, A., 2007. Cutaneous immune responses in the common carp detected using transcript analysis. Molecular Immunology 44, 1664–1679.

Grassme, H., Jendrossek, V., Gulbins, E., 2001. Molecular mechanisms of bacteria induced apoptosis. Apoptosis 6, 441–445.

Grayson, T.H., Cooper, L.F., Wrathmell, A.B., Evenden, J.R., Andrew, J., Gilpin, M.L., 2002. Host responses to *Renibacterium salmoninarum* and specific components of the pathogen reveal the mechanisms of immune suppression and activation. Immunology 106, 273–283.

Grimholt, U., Larsen, S., Nordmo, R., Midtlyng, P., Kjoeglum, S., Storset, A., Saebø, S., Stet, R.J., 2003. MHC polymorphism and disease resistance in Atlantic salmon (*Salmo salar*); facing pathogens with single expressed major histocompatibility class I and class II loci. Immunogenetics 55, 210–219.

Grimholt, U., Tsukamoto, K., Azuma, T., Leong, J., Koop, B.F., Dijkstra, J.M., 2015. A comprehensive analysis of teleost MHC class I sequences. BMC Evolutionary Biology 15, 1.

Gross, A., 2016. BCL-2 family proteins as regulators of mitochondria metabolism. Biochimica et Biophysica Acta – Bioenergetics.

Grygiel-Górniak, B., 2014. Peroxisome proliferator-activated receptors and their ligands: nutritional and clinical implications – a review. Nutrition Journal 13, 1.

Harper, C., Wolf, J.C., 2009. Morphologic effects of the stress response in fish. Ilar Journal 50, 387–396.

Harris, J., Bird, D.J., 2000. Modulation of the fish immune system by hormones. Veterinary Immunology and Immunopathology 77, 163–176.

Haugland, Ø., Mikalsen, A.B., Nilsen, P., Lindmo, K., Thu, B.J., Eliassen, T.M., Roos, N., Rode, M., Evensen, Ø., 2011. Cardiomyopathy syndrome of Atlantic salmon (*Salmo salar* L.) is caused by a double-stranded RNA virus of the *Totiviridae* family. Journal of Virology 85, 5275–5286.

Hollingsworth, C.S., Baldwin, R., Wilda, K., Ellis, R., Soares, S., 2006. Best Management Practices for Finfish Aquaculture in Massachusetts. Western Massachusetts Center for Sustainable Aquaculture.

Holopainen, R., Tapiovaara, H., Honkanen, J., 2012. Expression analysis of immune response genes in fish epithelial cells following ranavirus infection. Fish and Shellfish Immunology 32, 1095–1105.

Hsieh, J.-C., Pan, C.-Y., Chen, J.-Y., 2010. Tilapia hepcidin (TH)2-3 as a transgene in transgenic fish enhances resistance to *Vibrio vulnificus* infection and causes variations in immune-related genes after infection by different bacterial species. Fish and Shellfish Immunology 29, 430–439.

Huntingford, F., Kadri, S., 2014. Defining, assessing and promoting the welfare of farmed fish. Revue Scientifique et Technique (International Office of Epizootics) 33, 233–244.

Huntingford, F.A., Adams, C., Braithwaite, V., Kadri, S., Pottinger, T., Sandøe, P., Turnbull, J., 2006. Current issues in fish welfare. Journal of Fish Biology 68, 332–372.

Hwang, P.-P., Lee, T.-H., Lin, L.-Y., 2011. Ion regulation in fish gills: recent progress in the cellular and molecular mechanisms. American Journal of Physiology – Regulatory, Integrative and Comparative Physiology 301, R28–R47.

Ibarz, A., Beltrán, M., Fernández-Borràs, J., Gallardo, M., Sánchez, J., Blasco, J., 2007. Alterations in lipid metabolism and use of energy depots of gilthead sea bream (*Sparus aurata*) at low temperatures. Aquaculture 262, 470–480.

Ibarz, A., Blasco, J., Beltrán, M., Gallardo, M., Sánchez, J., Sala, R., Fernández-Borràs, J., 2005. Cold-induced alterations on proximate composition and fatty acid profiles of several tissues in gilthead sea bream (*Sparus aurata*). Aquaculture 249, 477–486.

Iger, Y., Abraham, M., Dotan, A., Fattal, B., Rahamim, E., 1988. Cellular responses in the skin of carp maintained in organically fertilized water. Journal of Fish Biology 33, 711–720.

Iger, Y., Bonga, S.W., 1994. Cellular responses of the skin of carp (*Cyprinus carpio*) exposed to acidified water. Cell and Tissue Research 275, 481–492.

Iger, Y., Hilge, V., Abraham, M., 1992. The ultrastructure of fish skin during stress in aquaculture. In: Progress in Aquaculture Research: Proceedings of the 4th German-Isreali Status Seminar Held on October 30–31, 1990. European Aquaculture Society, Oostende, pp. 205–214.

Ingerslev, H.C., Rønneseth, A., Pettersen, E., Wergeland, H., 2009. Differential expression of immune genes in Atlantic salmon (*Salmo salar* L.) challenged intraperitoneally or by cohabitation with IPNV. Scandinavian Journal of Immunology 69, 90–98.

Iwama, G.K., Afonso, L.O., Todgham, A., Ackerman, P., Nakano, K., 2004. Are hsps suitable for indicating stressed states in fish? Journal of Experimental Biology 207, 15–19.

Iwama, G.K., Afonso, L.O.B., Vijayan, M.M., 2005. Stress in fishes. In: Evans, D.H., Claiborne, J.B. (Eds.), The Physiology of Fishes, third ed. Taylor and Francis, Boca Raton, FL, pp. 319–342.

Iwama, G.K., Pickering, A., Sumpter, J., Schreck, C., 2011. Fish Stress and Health in Aquaculture. Cambridge University Press.

Jackson, A.N., Mclure, C.A., Dawkins, R.L., Keating, P.J., 2007. Mannose binding lectin (MBL) copy number polymorphism in Zebrafish (*D. rerio*) and identification of haplotypes resistant to *L. anguillarum*. Immunogenetics 59, 861–872.

Jørgensen, H.B., Sørensen, P., Cooper, G., Lorenzen, E., Lorenzen, N., Hansen, M.H., Koop, B., Henryon, M., 2011. General and family-specific gene expression responses to viral hemorrhagic septicaemia virus infection in rainbow trout (*Oncorhynchus mykiss*). Molecular Immunology 48, 1046–1058.

Jørgensen, S.M., Afanasyev, S., Krasnov, A., 2008. Gene expression analyses in Atlantic salmon challenged with infectious salmon anemia virus reveal differences between individuals with early, intermediate and late mortality. BMC Genomics 9, 1.

Jørgensen, S.M., Hetland, D.L., Press, C.M., Grimholt, U., Gjøen, T., 2007. Effect of early infectious salmon anaemia virus (ISAV) infection on expression of MHC pathway genes and type I and II interferon in Atlantic salmon (*Salmo salar* L.) tissues. Fish and Shellfish Immunology 23, 576–588.

Kalogianni, E., Alexis, M., Tsangaris, C., Abraham, M., Wendelaar Bonga, S., Iger, Y., Van Ham, E., Stoumboudi, M.T., 2011. Cellular responses in the skin of the gilthead sea bream *Sparus aurata* L. and the sea bass *Dicentrarchus labrax* (L.) exposed to high ammonia. Journal of Fish Biology 78, 1152–1169.

Katzenback, B.A., 2015. Antimicrobial peptides as mediators of innate immunity in teleosts. Biology 4, 607–639.

Kaur, S., 2013. Genomics. In: Maloy, S., Hughes, K. (Eds.), Brenner's Encyclopedia of Genetics. Academic Press, San Diego, pp. 310–312.

Kiilerich, P., Prunet, P., 2011. Hormonal control of metabolism and ionic regulation – corticosteroids. In: Farrell, A.P. (Ed.), Encyclopedia of Fish Physiology: From Genome to Environment. Academic Press, pp. 1474–1482.

Koob, G.F., Le Moal, M., 2001. Drug addiction, dysregulation of reward, and allostasis. Neuropsychopharmacology 24, 97–129.

Krasnov, A., Kileng, Ø., Skugor, S., Jørgensen, S.M., Afanasyev, S., Timmerhaus, G., Sommer, A.-I., Jensen, I., 2013. Genomic analysis of the host response to nervous necrosis virus in Atlantic cod (*Gadus morhua*) brain. Molecular Immunology 54, 443–452.

Krause, A., Neitz, S., Mägert, H.-J., Schulz, A., Forssmann, W.-G., Schulz-Knappe, P., Adermann, K., 2000. LEAP-1, a novel highly disulfide-bonded human peptide, exhibits antimicrobial activity. FEBS Letters 480, 147–150.

Kurobe, T., Yasuike, M., Kimura, T., Hirono, I., Aoki, T., 2005. Expression profiling of immune-related genes from Japanese flounder *Paralichthys olivaceus* kidney cells using cDNA microarrays. Developmental and Comparative Immunology 29, 515–523.

Kurtz, J., Wegner, K.M., Kalbe, M., Reusch, T.B., Schaschl, H., Hasselquist, D., Milinski, M., 2006. MHC genes and oxidative stress in sticklebacks: an immuno-ecological approach. Proceedings of the Royal Society of London B: Biological Sciences 273, 1407–1414.

Laidley, C., Leatherland, J., 1988. Cohort sampling, anaesthesia and stocking-density effects on plasma cortisol, thyroid hormone, metabolite and ion levels in rainbow trout, *Salmo gairdneri* Richardson. Journal of Fish Biology 33, 73–88.

Laiz-Carrión, R., Fuentes, J., Redruello, B., Guzmán, J.M., Del Río, M.P.M., Power, D., Mancera, J.M., 2009. Expression of pituitary prolactin, growth hormone and somatolactin is modified in response to different stressors (salinity, crowding and food-deprivation) in gilthead sea bream *Sparus auratus*. General and Comparative Endocrinology 162, 293–300.

Laiz-Carrión, R., Sangiao-Alvarellos, S., Guzmán, J.M., Del Río, M.P.M., Míguez, J.M., Soengas, J.L., Mancera, J.M., 2002. Energy metabolism in fish tissues related to osmoregulation and cortisol action. Fish Physiology and Biochemistry 27, 179–188.

Laiz-Carrión, R., Martín Del Río, M.P., Miguez, J.M., Mancera, J.M., Soengas, J.L., 2003. Influence of cortisol on osmoregulation and energy metabolism in gilthead seabream *Sparus aurata*. Journal of Experimental Zoology Part A: Comparative Experimental Biology 298, 105–118.

Langefors, Å., Lohm, J., Grahn, M., Andersen, Ø., Von Schantz, T., 2001. Association between major histocompatibility complex class IIB alleles and resistance to *Aeromonas salmonicida* in Atlantic salmon. Proceedings of the Royal Society of London B: Biological Sciences 268, 479–485.

Langevin, C., Aleksejeva, E., Passoni, G., Palha, N., Levraud, J.-P., Boudinot, P., 2013. The antiviral innate immune response in fish: evolution and conservation of the IFN system. Journal of Molecular Biology 425, 4904–4920.

Li, C., Zhang, Y., Wang, R., Lu, J., Nandi, S., Mohanty, S., Terhune, J., Liu, Z., Peatman, E., 2012. RNA-seq analysis of mucosal immune responses reveals signatures of intestinal barrier disruption and pathogen entry following *Edwardsiella ictaluri* infection in channel catfish, *Ictalurus punctatus*. Fish and Shellfish Immunology 32, 816–827.

Lindenstrøm, T., Secombes, C.J., Buchmann, K., 2004. Expression of immune response genes in rainbow trout skin induced by *Gyrodactylus derjavini* infections. Veterinary Immunology and Immunopathology 97, 137–148.

Lohm, J., Grahn, M., Langefors, Å., Andersen, Ø., Storset, A., Von Schantz, T., 2002. Experimental evidence for major histocompatibility complex–allele–specific resistance to a bacterial infection. Proceedings of the Royal Society of London B: Biological Sciences 269, 2029–2033.

Long, M., Zhao, J., Li, T., Tafalla, C., Zhang, Q., Wang, X., Gong, X., Shen, Z., Li, A., 2015. Transcriptomic and proteomic analyses of splenic immune mechanisms of rainbow trout (*Oncorhynchus mykiss*) infected by *Aeromonas salmonicida* subsp. *salmonicida*. Journal of Proteomics 122, 41–54.

Magnadóttir, B., 2006. Innate immunity of fish (overview). Fish and Shellfish Immunology 20, 137–151.

Mahboob, S., 2013. Environmental pollution of heavy metals as a cause of oxidative stress in fish: a review. Life Science Journal 10, 336–347.

Mancera, J.M., Carrión, R.L., del Río, M.D.P.M., 2002. Osmoregulatory action of PRL, GH, and cortisol in the gilthead seabream (*Sparus aurata* L.). General and Comparative Endocrinology 129, 95–103.

Marancik, D., Gao, G., Paneru, B., Ma, H., Hernandez, A.G., Salem, M., Yao, J., Palti, Y., Wiens, G.D., 2014. Whole-body transcriptome of selectively bred, resistant-, control-, and susceptible-line rainbow trout following experimental challenge with *Flavobacterium psychrophilum*. Frontiers in Genetics 5, 453.

Marshall, W., Grosell, M., 2006. Ion transport, osmoregulation, and acid-base balance. In: Evans, D.H., Claiborne, J.B. (Eds.), The Physiology of Fishes. CRC Press, Boca Raton, pp. 177–230.

Martínez-Porchas, M., Martínez-Córdova, L.R., Ramos-Enriquez, R., 2009. Cortisol and glucose: reliable indicators of fish stress. Pan-American Journal of Aquatic Sciences 4, 158–178.

Martins, C.I., Galhardo, L., Noble, C., Damsgård, B., Spedicato, M.T., Zupa, W., Beauchaud, M., Kulczykowska, E., Massabuau, J.-C., Carter, T., 2012. Behavioural indicators of welfare in farmed fish. Fish Physiology and Biochemistry 38, 17–41.

Mason, J.W., 1971. A re-evaluation of the concept of "non-specificity" in stress theory. Journal of Psychiatric Research 8, 323–333.

Maule, A.G., Schreck, C.B., 1990. Glucocorticoid receptors in leukocytes and gill of juvenile coho salmon (*Oncorhynchus kisutch*). General and Comparative Endocrinology 77, 448–455.

Maule, A.G., Schreck, C.B., 1991. Stress and cortisol treatment changed affinity and number of glucocorticoid receptors in leukocytes and gill of coho salmon. General and Comparative Endocrinology 84, 83–93.

Maule, A.G., Schreck, C.B., Sharpe, C., 1993. Seasonal changes in cortisol sensitivity and glucocorticoid receptor affinity and number in leukocytes of coho salmon. Fish Physiology and Biochemistry 10, 497–506.

Mazon, A., Verburg-Van Kemenade, B., Flik, G., Huising, M., 2006. Corticotropin-releasing hormone-receptor 1 (CRH-R1) and CRH-binding protein (CRH-BP) are expressed in the gills and skin of common carp *Cyprinus carpio* L. and respond to acute stress and infection. Journal of Experimental Biology 209, 510–517.

McCormick, S., Shrimpton, J., Carey, J., O'dea, M., Sloan, K., Moriyama, S., Björnsson, B.T., 1998. Repeated acute stress reduces growth rate of Atlantic salmon parr and alters plasma levels of growth hormone, insulin-like growth factor I and cortisol. Aquaculture 168, 221–235.

McDonald, G., Milligan, L., 1997. Ionic, osmotic and acid-base regulation in stress. In: Iwama, G.K., Pickering, A.D., Sumpter, J.P., Schreck, C.B. (Eds.), Fish Stress and Health in Aquaculture, Society for Experimental Biology Seminar Series. Cambridge University Press, Cambridge, pp. 119–145.

McEwen, B.S., 2000. The neurobiology of stress: from serendipity to clinical relevance. Brain Research 886, 172–189.

McEwen, B.S., 2005. Stressed or stressed out: what is the difference? Journal of Psychiatry and Neuroscience 30, 315.

McEwen, B.S., Wingfield, J.C., 2003. The concept of allostasis in biology and biomedicine. Hormones and Behavior 43, 2–15.

McKay, L.I., Cidlowski, J.A., 1999. Molecular control of immune/inflammatory responses: interactions between nuclear factor-κB and steroid receptor-signaling pathways. Endocrine Reviews 20, 435–459.

Milev-Milovanovic, I., Majji, S., Thodima, V., Deng, Y., Hanson, L., Arnizaut, A., Waldbieser, G., Chinchar, V.G., 2009. Identification and expression analyses of poly [I:C]-stimulated genes in channel catfish (*Ictalurus punctatus*). Fish and Shellfish Immunology 26, 811–820.

Milla, S., Mathieu, C., Wang, N., Lambert, S., Nadzialek, S., Massart, S., Henrotte, E., Douxfils, J., Mélard, C., Mandiki, S., 2010. Spleen immune status is affected after acute handling stress but not regulated by cortisol in Eurasian perch, *Perca fluviatilis*. Fish and Shellfish Immunology 28, 931–941.

Milla, S., Wang, N., Mandiki, S.N.M., Kestemont, P., 2009. Corticosteroids: friends or foes of teleost fish reproduction? Comparative Biochemistry and Physiology A: Molecular and Integrative Physiology 153, 242–251.

Millán, A., Gómez-Tato, A., Fernández, C., Pardo, B.G., Álvarez-Dios, J.A., Calaza, M., Bouza, C., Vázquez, M., Cabaleiro, S., Martínez, P., 2010. Design and performance of a turbot (*Scophthalmus maximus*) oligo-microarray based on ESTs from immune tissues. Marine Biotechnology 12, 452–465.

Miller, D.B., O'callaghan, J.P., 2002. Neuroendocrine aspects of the response to stress. Metabolism 51, 5–10.

Moberg, G., 2000. Biological response to stress: implications for animal welfare. In: Moberg, G., Mench, J. (Eds.), The Biology of Animal Stress: Basic Principles and Implications for Animal Welfare. CABI Publishing, London, pp. 1–21.

Mohanty, B., Sahoo, P., 2010. Immune responses and expression profiles of some immune-related genes in Indian major carp, *Labeo rohita* to *Edwardsiella tarda* infection. Fish and Shellfish Immunology 28, 613–621.

Mola, L., Gambarelli, A., Pederzoli, A., 2011. Immunolocalization of corticotropin-releasing factor (CRF) and corticotropin-releasing factor receptor 2 (CRF-R2) in the developing gut of the sea bass (*Dicentrarchus labrax* L.). Acta Histochemica 113, 290–293.

Molinero, A., Gómez, E., Balasch, J., Tort, L., 1997. Stress by fish removal in the gilthead sea bream, *Sparus aurata*: a time course study on the remaining fish in the same tank. Journal of Applied Aquaculture 7, 1–12.

Mommsen, T.P., Vijayan, M.M., Moon, T.W., 1999. Cortisol in teleosts: dynamics, mechanisms of action, and metabolic regulation. Reviews in Fish Biology and Fisheries 9, 211–268.

Montero, D., Izquierdo, M., Tort, L., Robaina, L., Vergara, J., 1999. High stocking density produces crowding stress altering some physiological and biochemical parameters in gilthead seabream, *Sparus aurata*, juveniles. Fish Physiology and Biochemistry 20, 53–60.

Mori, T., Devlin, R.H., 1999. Transgene and host growth hormone gene expression in pituitary and nonpituitary tissues of normal and growth hormone transgenic salmon. Molecular and Cellular Endocrinology 149, 129–139.

Mu, Y., Ding, F., Cui, P., Ao, J., Hu, S., Chen, X., 2010. Transcriptome and expression profiling analysis revealed changes of multiple signaling pathways involved in immunity in the large yellow croaker during *Aeromonas hydrophila* infection. BMC Genomics 11, 1.

Muto, N., Ren, H.-W., Hwang, G.-S., Tominaga, S., Itoh, N., Tanaka, K., 1999. Induction of two major isoforms of metallothionein in crucian carp (*Carassius cuvieri*) by air-pumping stress, dexamethasone, and metals. Comparative Biochemistry and Physiology Part C: Pharmacology, Toxicology and Endocrinology 122, 75–82.

Mutoloki, S., Cooper, G.A., Marjara, I.S., Koop, B.F., Evensen, Ø., 2010. High gene expression of inflammatory markers and IL-17A correlates with severity of injection site reactions of Atlantic salmon vaccinated with oil-adjuvanted vaccines. BMC Genomics 11, 1.

Nakao, M., Kajiya, T., Sato, Y., Somamoto, T., Kato-Unoki, Y., Matsushita, M., Nakata, M., Fujita, T., Yano, T., 2006. Lectin pathway of bony fish complement: identification of two homologs of the mannose-binding lectin associated with MASP2 in the common carp (*Cyprinus carpio*). The Journal of Immunology 177, 5471–5479.

Nardocci, G., Navarro, C., Cortés, P.P., Imarai, M., Montoya, M., Valenzuela, B., Jara, P., Acuña-Castillo, C., Fernández, R., 2014. Neuroendocrine mechanisms for immune system regulation during stress in fish. Fish and Shellfish Immunology 40, 531–538.

Narnaware, Y., Baker, B., 1996. Evidence that cortisol may protect against the immediate effects of stress on circulating leukocytes in the trout. General and Comparative Endocrinology 103, 359–366.

Olavarría, V.H., Recabarren, P., Fredericksen, F., Villalba, M., Yáñez, A., 2015. ISAV infection promotes apoptosis of SHK-1 cells through a ROS/p38 MAPK/Bad signaling pathway. Molecular Immunology 64, 1–8.

Olsen, M.M., Kania, P.W., Heinecke, R.D., Skjoedt, K., Rasmussen, K.J., Buchmann, K., 2011. Cellular and humoral factors involved in the response of rainbow trout gills to *Ichthyophthirius multifiliis* infections: molecular and immunohistochemical studies. Fish and Shellfish Immunology 30, 859–869.

Olsen, R.E., Oppedal, F., Tenningen, M., Vold, A., 2012. Physiological response and mortality caused by scale loss in Atlantic herring. Fisheries Research 129, 21–27.

Ong, L., Linden, W., Young, S., 2004. Stress management: what is it? Journal of Psychosomatic Research 56, 133–137.

Pardo, B.G., Fernández, C., Millán, A., Bouza, C., Vázquez-López, A., Vera, M., Alvarez-Dios, J.A., Calaza, M., Gómez-Tato, A., Vázquez, M., 2008. Expressed sequence tags (ESTs) from immune tissues of turbot (*Scophthalmus maximus*) challenged with pathogens. BMC Veterinary Research 4, 37.

Park, C.H., Valore, E.V., Waring, A.J., Ganz, T., 2001. Hepcidin, a urinary antimicrobial peptide synthesized in the liver. Journal of Biological Chemistry 276, 7806–7810.

Park, K.C., Osborne, J.A., Montes, A., Dios, S., Nerland, A.H., Novoa, B., Figueras, A., Brown, L.L., Johnson, S.C., 2009. Immunological responses of turbot (*Psetta maxima*) to nodavirus infection or polyriboinosinic polyribocytidylic acid (pIC) stimulation, using expressed sequence tags (ESTs) analysis and cDNA microarrays. Fish and Shellfish Immunology 26, 91–108.

Patel, S., Malde, K., Lanzen, A., Olsen, R.H., Nerland, A.H., 2009. Identification of immune related genes in Atlantic halibut (*Hippoglossus hippoglossus* L.) following in vivo antigenic and in vitro mitogenic stimulation. Fish and Shellfish Immunology 27, 729–738.

Pavlidis, M., Mylonas, C. (Eds.), 2011. Sparidae: Biology and Aquaculture of Gilthead Sea Bream and Other Species. John Wiley & Sons.

Peatman, E., Baoprasertkul, P., Terhune, J., Xu, P., Nandi, S., Kucuktas, H., Li, P., Wang, S., Somridhivej, B., Dunham, R., 2007. Expression analysis of the acute phase response in channel catfish (*Ictalurus punctatus*) after infection with a Gram-negative bacterium. Developmental and Comparative Immunology 31, 1183–1196.

Peatman, E., Terhune, J., Baoprasertkul, P., Xu, P., Nandi, S., Wang, S., Somridhivej, B., Kucuktas, H., Li, P., Dunham, R., 2008. Microarray analysis of gene expression in the blue catfish liver reveals early activation of the MHC class I pathway after infection with *Edwardsiella ictaluri*. Molecular Immunology 45, 553–566.

Pepels, P., Bonga, S.W., Balm, P., 2004. Bacterial lipopolysaccharide (LPS) modulates corticotropin-releasing hormone (CRH) content and release in the brain of juvenile and adult tilapia (*Oreochromis mossambicus*; Teleostei). Journal of Experimental Biology 207, 4479–4488.

Pereiro, P., Figueras, A., Novoa, B., 2012. A novel hepcidin-like in turbot (*Scophthalmus maximus* L.) highly expressed after pathogen challenge but not after iron overload. Fish and Shellfish Immunology 32, 879–889.

Perry, S.F., Bernier, N.J., 1999. The acute humoral adrenergic stress response in fish: facts and fiction. Aquaculture 177, 285–295.

Philip, A.M., Vijayan, M.M., 2015. Stress-immune-growth interactions: cortisol modulates suppressors of cytokine signaling and JAK/STAT pathway in rainbow trout liver. PLoS One 10, e0129299.

Pickering, A., 1993. Growth and stress in fish production. Aquaculture 111, 51–63.

Pickering, A., Pottinger, T., 1989. Stress responses and disease resistance in salmonid fish: effects of chronic elevation of plasma cortisol. Fish Physiology and Biochemistry 7, 253–258.

Pickering, A., Pottinger, T., Sumpter, J., Carragher, J., Le Bail, P., 1991. Effects of acute and chronic stress on the levels of circulating growth hormone in the rainbow trout, *Oncorhynchus mykiss*. General and Comparative Endocrinology 83, 86–93.

Pickering, A.D., 1998. Stress responses of farmed fish. In: Black, K.D., Pickering, A.D. (Eds.), Biology of Farmed Fish. CRC Press, Boca Raton, FL, pp. 222–255.

Pijanowski, L., Jurecka, P., Irnazarow, I., Kepka, M., Szwejser, E., Verburg-Van Kemenade, B., Chadzinska, M., 2015. Activity of the hypothalamus–pituitary–interrenal axis (HPI axis) and immune response in carp lines with different susceptibility to disease. Fish Physiology and Biochemistry 41, 1261–1278.

Poli, B., Parisi, G., Scappini, F., Zampacavallo, G., 2005. Fish welfare and quality as affected by pre-slaughter and slaughter management. Aquaculture International 13, 29–49.

Pottinger, T., 1998. Changes in blood cortisol, glucose and lactate in carp retained in anglers' keepnets. Journal of Fish Biology 53, 728–742.

Pottinger, T., Pickering, A., 1997. Genetic basis to the stress response: selective breeding for stress-tolerant fish. In: Iwama, G., Pickering, A.D., Sumpter, J., Schreck, C. (Eds.), Fish Stress and Health in Aquaculture. Cambridge University Press, Cambridge, pp. 171–193.

Pottinger, T.G., 2008. The stress response in fish-mechanisms, effects and measurement. In: Branson, E.J. (Ed.), Fish Welfare. Wiley-Blackwell, Oxford, UK, pp. 32–48.

Power, D., Melo, J., Santos, C., 2000. The effect of food deprivation and refeeding on the liver, thyroid hormones and transthyretin in sea bream. Journal of Fish Biology 56, 374–387.

Pridgeon, J.W., Klesius, P.H., 2010. Identification and expression profile of multiple genes in channel catfish fry 10 min after modified live *Flavobacterium columnare* vaccination. Veterinary Immunology and Immunopathology 138, 25–33.

Prieto-Álamo, M.-J., Abril, N., Osuna-Jiménez, I., Pueyo, C., 2009. *Solea senegalensis* genes responding to lipopolysaccharide and copper sulphate challenges: large-scale identification by suppression subtractive hybridization and absolute quantification of transcriptional profiles by real-time RT-PCR. Aquatic Toxicology 91, 312–319.

Prunet, P., Øverli, Ø., Douxfils, J., Bernardini, G., Kestemont, P., Baron, D., 2012. Fish welfare and genomics. Fish Physiology and Biochemistry 38, 43–60.

Prunet, P., Sturm, A., Milla, S., 2006. Multiple corticosteroid receptors in fish: from old ideas to new concepts. General and Comparative Endocrinology 147, 17–23.

Pulgar, R., Hödar, C., Travisany, D., Zuñiga, A., Domínguez, C., Maass, A., González, M., Cambiazo, V., 2015. Transcriptional response of Atlantic salmon families to *Piscirickettsia salmonis* infection highlights the relevance of the iron-deprivation defence system. BMC Genomics 16, 1.

Quabius, E.S., Krupp, G., Secombes, C.J., 2005. Polychlorinated biphenyl 126 affects expression of genes involved in stress-immune interaction in primary cultures of rainbow trout anterior kidney cells. Environmental Toxicology and Chemistry 24, 3053–3060.

Raison, C.L., Miller, A.H., 2003. When not enough is too much: the role of insufficient glucocorticoid signaling in the pathophysiology of stress-related disorders. American Journal of Psychiatry.

Rand-Weaver, M., Pottinger, T., Sumpter, J., 1993. Plasma somatolactin concentrations in salmonid fish are elevated by stress. Journal of Endocrinology 138, 509–515.

Rauta, P.R., Nayak, B., Das, S., 2012. Immune system and immune responses in fish and their role in comparative immunity study: a model for higher organisms. Immunology Letters 148, 23–33.

Reid, S.G., Bernier, N.J., Perry, S.F., 1998. The adrenergic stress response in fish: control of catecholamine storage and release. Comparative Biochemistry and Physiology Part C: Pharmacology, Toxicology and Endocrinology 120, 1–27.

Ribas, L., Pardo, B.G., Fernández, C., Álvarez-Diós, J.A., Gómez-Tato, A., Quiroga, M.I., Planas, J.V., Sitjà-Bobadilla, A., Martínez, P., Piferrer, F., 2013. A combined strategy involving Sanger and 454 pyrosequencing increases genomic resources to aid in the management of reproduction, disease control and genetic selection in the turbot (*Scophthalmus maximus*). BMC Genomics 14, 1.

Rise, M.L., Hall, J., Rise, M., Hori, T., Gamperl, A.K., Kimball, J., Hubert, S., Bowman, S., Johnson, S.C., 2008. Functional genomic analysis of the response of Atlantic cod (*Gadus morhua*) spleen to the viral mimic polyriboinosinic polyribocytidylic acid (pIC). Developmental and Comparative Immunology 32, 916–931.

Roberts, R., Agius, C., Saliba, C., Bossier, P., Sung, Y., 2010. Heat shock proteins (chaperones) in fish and shellfish and their potential role in relation to fish health: a review. Journal of Fish Diseases 33, 789–801.

Robinson, N., Sahoo, P.K., Baranski, M., Mahapatra, K.D., Saha, J.N., Das, S., Mishra, Y., Das, P., Barman, H.K., Eknath, A.E., 2012. Expressed sequences and polymorphisms in rohu carp (*Labeo rohita*, Hamilton) revealed by mRNA-seq. Marine Biotechnology 14, 620–633.

Rodrigues, P.N., Vázquez-Dorado, S., Neves, J.V., Wilson, J.M., 2006. Dual function of fish hepcidin: response to experimental iron overload and bacterial infection in sea bass (*Dicentrarchus labrax*). Developmental and Comparative Immunology 30, 1156–1167.

Rojas, V., Galanti, N., Bols, N.C., Marshall, S.H., 2009. Productive infection of *Piscirickettsia salmonis* in macrophages and monocyte-like cells from rainbow trout, a possible survival strategy. Journal of Cellular Biochemistry 108, 631–637.

Romero, L.M., Dickens, M.J., Cyr, N.E., 2009. The reactive scope model—a new model integrating homeostasis, allostasis, and stress. Hormones and Behavior 55, 375–389.

Rotllant, J., Balm, P., Perez-Sanchez, J., Wendelaar-Bonga, S., Tort, L., 2001. Pituitary and interrenal function in gilthead sea bream (*Sparus aurata* L., Teleostei) after handling and confinement stress. General and Comparative Endocrinology 121, 333–342.

Rotllant, J., Tort, L., 1997. Cortisol and glucose responses after acute stress by net handling in the sparid red porgy previously subjected to crowding stress. Journal of Fish Biology 51, 21–28.

Rottmann, R., Francis-Floyd, R., Durborow, R., 1992. The Role of Stress in Fish Disease. Southern Regional Aquaculture Center Publication, no. 474.

Roy, B., Rai, U., 2008. Role of adrenoceptor-coupled second messenger system in sympatho-adrenomedullary modulation of splenic macrophage functions in live fish *Channa punctatus*. General and Comparative Endocrinology 155, 298–306.

Roy, N., Deveraux, Q.L., Takahashi, R., Salvesen, G.S., Reed, J.C., 1997. The c-IAP-1 and c-IAP-2 proteins are direct inhibitors of specific caspases. The EMBO Journal 16, 6914–6925.

Rupia, E.J., Binning, S.A., Roche, D.G., Lu, W.Q., 2016. Fight-flight or freeze-hide? Personality and metabolic phenotype mediate physiological defence responses in flatfish. Journal of Animal Ecology 85, 927–937.

Saeij, J.P., Stet, R.J., De Vries, B.J., Van Muiswinkel, W.B., Wiegertjes, G.F., 2003. Molecular and functional characterization of carp TNF: a link between TNF polymorphism and trypanotolerance? Developmental and Comparative Immunology 27, 29–41.

Sahoo, P., Das, S., Mahapatra, K.D., Saha, J.N., Baranski, M., Ødegård, J., Robinson, N., 2013. Characterization of the ceruloplasmin gene and its potential role as an indirect marker for selection to *Aeromonas hydrophila* resistance in rohu, *Labeo rohita*. Fish and Shellfish Immunology 34, 1325–1334.

Sapolsky, R.M., 2002. Endocrinology of the stress-response. In: Becker, J.B., Breedlove, S., Crews, D., Mccarthy, M. (Eds.), Behavioral Endocrinology. MIT Press, Cambridge, MA, pp. 409–450.

Sappal, R., Fast, M., Purcell, S., Macdonald, N., Stevens, D., Kibenge, F., Siah, A., Kamunde, C., 2016. Copper and hypoxia modulate transcriptional and mitochondrial functional-biochemical responses in warm acclimated rainbow trout (*Oncorhynchus mykiss*). Environmental Pollution 211, 291–306.

Schaaf, M.J.M., Chatzopoulou, A., Spaink, H.P., 2009. The zebrafish as a model system for glucocorticoid receptor research. Comparative Biochemistry and Physiology A: Molecular and Integrative Physiology 153, 75–82.

Schoonheim, P.J., Chatzopoulou, A., Schaaf, M.J., 2010. The zebrafish as an in vivo model system for glucocorticoid resistance. Steroids 75, 918–925.

Schreck, C., 1981. Stress and compensation in teleostean fishes: response to social and physical factors. In: Pickering, A.D. (Ed.), Stress and Fish. Academic Press, New York, pp. 295–321.

Schreck, C., 2000. Accumulation and long-term effects of stress in fish. In: Moberg, G.P., Mench, J.A. (Eds.), The Biology of Animal Stress. CABI Publishing, Wallingford, UK, pp. 147–158.

Schreck, C.B., 2010. Stress and fish reproduction: the roles of allostasis and hormesis. General and Comparative Endocrinology 165, 549–556.

Scott, A.P., Ellis, T., 2007. Measurement of fish steroids in water—a review. General and Comparative Endocrinology 153, 392–400.

Scott, A., Pinillos, M., Ellis, T., 2001. Why measure steroids in fish plasma when you can measure them in water. In: Perspectives in Comparative Endocrinology: Unity and Diversity: The Proceedings of the 14th International Congress of Comparative Endocrinology, Sorrento, Italy, pp. 26–30.

Scott, A.P., Hirschenhauser, K., Bender, N., Oliveira, R., Earley, R.L., Sebire, M., Ellis, T., Pavlidis, M., Hubbard, P.C., Huertas, M., Canario, A., 2008. Non-invasive measurement of steroids in fish-holding water: important considerations when applying the procedure to behaviour studies. Behaviour 145, 1307–1328.

Segner, H., Sundh, H., Buchmann, K., Douxfils, J., Sundell, K.S., Mathieu, C., Ruane, N., Jutfelt, F., Toften, H., Vaughan, L., 2012. Health of farmed fish: its relation to fish welfare and its utility as welfare indicator. Fish Physiology and Biochemistry 38, 85–105.

Selye, H., 1936. A syndrome produced by diverse nocuous agents. Nature 138, 32.

Selye, H., 1946. The general adaptation syndrome and the diseases of adaptation. The Journal of Clinical Endocrinology and Metabolism 6, 117–230.

Selye, H., 1973. The evolution of the stress concept: the originator of the concept traces its development from the discovery in 1936 of the alarm reaction to modern therapeutic applications of syntoxic and catatoxic hormones. American Scientist 61, 692–699.

Sevcikova, M., Modra, H., Slaninova, A., Svobodova, Z., 2011. Metals as a cause of oxidative stress in fish: a review. Veterinarni Medicina 56, 537–546.

Shephard, K.L., 1994. Functions for fish mucus. Reviews in Fish Biology and Fisheries 4, 401–429.

Shi, J., Camus, A.C., 2006. Hepcidins in amphibians and fishes: antimicrobial peptides or iron-regulatory hormones? Developmental and Comparative Immunology 30, 746–755.

Sigh, J., Lindenstrøm, T., Buchmann, K., 2004. The parasitic ciliate *Ichthyophthirius multifiliis* induces expression of immune relevant genes in rainbow trout, *Oncorhynchus mykiss* (Walbaum). Journal of Fish Diseases 27, 409–417.

Simontacchi, C., Poltronieri, C., Carraro, C., Bertotto, D., Xiccato, G., Trocino, A., Radaelli, G., 2008. Alternative stress indicators in sea bass *Dicentrarchus labrax*, L. Journal of Fish Biology 72, 747–752.

Skjesol, A., Skjæveland, I., Elnæs, M., Timmerhaus, G., Fredriksen, B.N., Jørgensen, S.M., Krasnov, A., Jørgensen, J.B., 2011. IPNV with high and low virulence: host immune responses and viral mutations during infection. Virology Journal 8, 1.

Škugor, S., Jørgensen, S.M., Gjerde, B., Krasnov, A., 2009. Hepatic gene expression profiling reveals protective responses in Atlantic salmon vaccinated against furunculosis. BMC Genomics 10, 503.

Slominski, A., 2007. A nervous breakdown in the skin: stress and the epidermal barrier. The Journal of Clinical Investigation 117, 3166–3169.

Slominski, A., Zbytek, B., Zmijewski, M., Slominski, R.M., Kauser, S., Wortsman, J., Tobin, D.J., 2006. Corticotropin releasing hormone and the skin. Frontiers in Bioscience 11, 2230–2248.

Smith, L.S., 1993. Trying to explain scale loss mortality: a continuing puzzle. Reviews in Fisheries Science 1, 337–355.

Sterling, P., Eyer, J., 1988. Allostasis: a new paradigm to explain arousal pathology. In: Fisher, S., Reason, J. (Eds.), Handbook of Life Stress, Cognition and Health. Wiley, New York, pp. 629–649.

Stolte, E.H., Van Kemenade, B.L.V., Savelkoul, H.F., Flik, G., 2006. Evolution of glucocorticoid receptors with different glucocorticoid sensitivity. Journal of Endocrinology 190, 17–28.

Stolte, E.H., Chadzinska, M., Przybylska, D., Flik, G., Savelkoul, H.F., Verburg-Van Kemenade, B.L., 2009. The immune response differentially regulates Hsp70 and glucocorticoid receptor expression in vitro and in vivo in common carp (*Cyprinus carpio* L.). Fish and Shellfish Immunology 27, 9–16.

Stolte, E.H., Nabuurs, S.B., Bury, N.R., Sturm, A., Flik, G., Savelkoul, H.F., Verburg-Van Kemenade, B.L., 2008. Stress and innate immunity in carp: corticosteroid receptors and pro-inflammatory cytokines. Molecular Immunology 46, 70–79.

Sumpter, J., 1997. The endocrinology of stress. In: Iwama, G.K., Pickering, A.D., Sumpter, J.P., Schreck, C.B. (Eds.), Fish Stress and Health in Aquaculture. Cambridge University Press, Cambridge, pp. 95–118.

Sun, F., Peatman, E., Li, C., Liu, S., Jiang, Y., Zhou, Z., Liu, Z., 2012. Transcriptomic signatures of attachment, NF-κB suppression and IFN stimulation in the catfish gill following columnaris bacterial infection. Developmental and Comparative Immunology 38, 169–180.

Sundvold, H., Ruyter, B., Østbye, T.-K., Moen, T., 2010. Identification of a novel allele of peroxisome proliferator-activated receptor gamma (PPARG) and its association with resistance to *Aeromonas salmonicida* in Atlantic salmon (*Salmo salar*). Fish and Shellfish Immunology 28, 394–400.

Tadiso, T.M., Krasnov, A., Skugor, S., Afanasyev, S., Hordvik, I., Nilsen, F., 2011. Gene expression analyses of immune responses in Atlantic salmon during early stages of infection by salmon louse (*Lepeophtheirus salmonis*) revealed bi-phasic responses coinciding with the copepod-chalimus transition. BMC Genomics 12, 1.

Teles, M., Boltaña, S., Reyes-López, F., Santos, M.A., Mackenzie, S., Tort, L., 2013. Effects of chronic cortisol administration on global expression of GR and the liver transcriptome in *Sparus aurata*. Marine Biotechnology 15, 104–114.

Terova, G., Gornati, R., Rimoldi, S., Bernardini, G., Saroglia, M., 2005. Quantification of a glucocorticoid receptor in sea bass (*Dicentrarchus labrax*, L.) reared at high stocking density. Gene 357, 144–151.

Thirumoorthy, N., Manisenthil Kumar, K., Shyam Sundar, A., Panayappan, L., Chatterjee, M., 2007. Metallothionein: an overview. World Journal of Gastroenterology 13, 993.

Timmerhaus, G., Krasnov, A., Takle, H., Afanasyev, S., Nilsen, P., Rode, M., Jørgensen, S.M., 2012. Comparison of Atlantic salmon individuals with different outcomes of cardiomyopathy syndrome (CMS). BMC Genomics 13, 205.

Tobin, D.M., Roca, F.J., Oh, S.F., Mcfarland, R., Vickery, T.W., Ray, J.P., Ko, D.C., Zou, Y., Bang, N.D., Chau, T.T., 2012. Host genotype-specific therapies can optimize the inflammatory response to mycobacterial infections. Cell 148, 434–446.

Tort, L., 2011. Stress and immune modulation in fish. Developmental and Comparative Immunology 35, 1366–1375.

Tort, L., Balasch, J., Mackenzie, S., 2003. Fish immune system. A crossroads between innate and adaptive responses. Inmunologia 22, 277–286.

Tort, L., Balasch, J., Mackenzie, S., 2004a. Fish health challenge after stress. Indicators of immunocompetence. Contributions to Science 2, 443–454.

Tort, L., Kargacin, B., Torres, P., Giralt, M., Hidalgo, J., 1996. The effect of cadmium exposure and stress on plasma cortisol, metallothionein levels and oxidative status in rainbow trout (*Oncorhynchus mykiss*) liver. Comparative Biochemistry and Physiology Part C: Pharmacology, Toxicology and Endocrinology 114, 29–34.

Tort, L., Pavlidis, M., Woo, N.Y., 2011. Stress and welfare in sparid fishes. In: Pavlidis, M., Mylonas, C. (Eds.), Sparidae. Biology and Aquaculture. Wiley-Blackwell, Oxford, UK, pp. 75–94.

Tort, L., Rotllant, J., Liarte, C., Acerete, L., Hernandez, A., Ceulemans, S., Coutteau, P., Padros, F., 2004b. Effects of temperature decrease on feeding rates, immune indicators and histopathological changes of gilthead sea bream *Sparus aurata* fed with an experimental diet. Aquaculture 229, 55–65.

Tort, L., Teles, M., 2011. The endocrine response to stress – a comparative view. In: Akin, F. (Ed.), Basic and Clinical Endocrinology Up-to-date, the Endocrine Response to Stress – a Comparative View. InTech. Available: http://www.intechopen.com/books/basic-and-clinical-endocrinology-up-to-date/the-endocrine-response-to-stress-a-comparative-view.

Tsalafouta, A., Papandroulakis, N., Pavlidis, M., 2015. Early life stress and effects at subsequent stages of development in European sea bass (*D. labrax*). Aquaculture 436, 27–33.

Tucker, C.S., Hargreaves, J.A., 2009. Environmental Best Management Practices for Aquaculture. John Wiley & Sons.

Turner, S., Faisal, M., Dewoody, J., 2007. Zygosity at the major histocompatibility class IIB locus predicts susceptibility to *Renibacterium salmoninarum* in Atlantic salmon (*Salmo salar* L.). Animal Genetics 38, 517–519.

Uribe, C., Folch, H., Enriquez, R., Moran, G., 2011. Innate and adaptive immunity in teleost fish: a review. Veterinarni Medicina 56, 486–503.

Valenzuela, C.A., Escobar, D., Perez, L., Zuloaga, R., Estrada, J.M., Mercado, L., Valdés, J.A., Molina, A., 2015. Transcriptional dynamics of immune, growth and stress related genes in skeletal muscle of the fine flounder (*Paralichthys adpersus*) during different nutritional statuses. Developmental and Comparative Immunology 53, 145–157.

Van Der Boon, J., Van Den Thillart, G.E., Addink, A.D., 1991. The effects of cortisol administration on intermediary metabolism in teleost fish. Comparative Biochemistry and Physiology Part A: Physiology 100, 47–53.

Van Weerd, J., Komen, J., 1998. The effects of chronic stress on growth in fish: a critical appraisal. Comparative Biochemistry and Physiology Part A: Molecular and Integrative Physiology 120, 107–112.

Varsamos, S., Flik, G., Pepin, J.F., Bonga, S.E., Breuil, G., 2006. Husbandry stress during early life stages affects the stress response and health status of juvenile sea bass, *Dicentrarchus labrax*. Fish and Shellfish Immunology 20, 83–96.

Vatsos, I., Kotzamanis, Y., Henry, M., Angelidis, P., Alexis, M., 2010. Monitoring stress in fish by applying image analysis to their skin mucous cells. European Journal of Histochemistry 54.

Verburg-Van Kemenade, B., Ribeiro, C., Chadzinska, M., 2011. Neuroendocrine–immune interaction in fish: differential regulation of phagocyte activity by neuroendocrine factors. General and Comparative Endocrinology 172, 31–38.

Verburg-Van Kemenade, B.L., Stolte, E.H., Metz, J.R., Chadzinska, M., 2009. Neuroendocrine–immune interactions in teleost fish. In: Bernier, N.J., Van Der Kraak, G., Farrel, A.P., Brauner, C.J. (Eds.), Fish Neuroendocrinology, Fish Physiology. Elsevier, pp. 313–364.

Vijayan, M.M., Aluru, N., Leatherland, J.F., 2010. Stress response and the role of cortisol. In: Leatherland, J.F., Woo, P.T.F. (Eds.), Fish Diseases and Disorders, vol. 2. CABI Publishing, pp. 181–201.

Vizzini, A., Vazzana, M., Cammarata, M., Parrinello, N., 2007. Peritoneal cavity phagocytes from the teleost sea bass express a glucocorticoid receptor (cloned and sequenced) involved in genomic modulation of the in vitro chemiluminescence response to zymosan. General and Comparative Endocrinology 150, 114–123.

Volckaert, F.A., Hellemans, B., Batargias, C., Louro, B., Massault, C., Van Houdt, J.K., Haley, C., De Koning, D.J., Canario, A.V., 2012. Heritability of cortisol response to confinement stress in European sea bass *Dicentrarchus labrax*. Genetics Selection Evolution 44, 15.

Wang, C., Zhang, X.H., Jia, A., Chen, J., Austin, B., 2008. Identification of immune-related genes from kidney and spleen of turbot, *Psetta maxima* (L.), by suppression subtractive hybridization following challenge with *Vibrio harveyi*. Journal of Fish Diseases 31, 505–514.

Weber, G.M., Vallejo, R.L., Lankford, S.E., Silverstein, J.T., Welch, T.J., 2008. Cortisol response to a crowding stress: heritability and association with disease resistance to *Yersinia ruckeri* in rainbow trout. North American Journal of Aquaculture 70, 425–433.

Wegner, K.M., Kalbe, M., Milinski, M., Reusch, T.B., 2008. Mortality selection during the 2003 European heat wave in three-spined sticklebacks: effects of parasites and MHC genotype. BMC Evolutionary Biology 8, 1.

Wendelaar Bonga, S.E., 1997. The stress response in fish. Physiological Reviews 77, 591–625.

Weyts, F., Cohen, N., Flik, G., Verburg-Van Kemenade, B., 1999. Interactions between the immune system and the hypothalamo-pituitary-interrenal axis in fish. Fish and Shellfish Immunology 9, 1–20.

Weyts, F., Verburg-Van Kemenade, B., Flik, G., 1998. Characterisation of glucocorticoid receptors in peripheral blood leukocytes of carp, *Cyprinus carpio* L. General and Comparative Endocrinology 111, 1–8.

Wilson, K.S., Tucker, C.S., Al-Dujaili, E.A.S., Holmes, M.C., Hadoke, P.W.F., Kenyon, C.J., Denvir, M.A., 2016. Early-life glucocorticoids programme behaviour and metabolism in adulthood in zebrafish. Journal of Endocrinology 230, 125–142.

Wiseman, S., Osachoff, H., Bassett, E., Malhotra, J., Bruno, J., Vanaggelen, G., Mommsen, T.P., Vijayan, M.M., 2007. Gene expression pattern in the liver during recovery from an acute stressor in rainbow trout. Comparative Biochemistry and Physiology Part D: Genomics and Proteomics 2, 234–244.

Wong, S.C., Dykstra, M., Campbell, J.M., Earley, R.L., 2008. Measuring water-borne cortisol in convict cichlids (*Amatitlania nigrofasciata*): is the procedure a stressor? Behaviour 145, 1283–1305.

Yada, T., 2007. Growth hormone and fish immune system. General and Comparative Endocrinology 152, 353–358.
Yada, T., Azuma, T., 2002. Hypophysectomy depresses immune functions in rainbow trout. Comparative Biochemistry and Physiology Part C: Toxicology and Pharmacology 131, 93–100.
Yada, T., Azuma, T., Hirano, T., Grau, E., 2001. Effects of hypophysectomy on immune functions in channel catfish. In: Perspective in Comparative Endocrinology: Unity and Diversity. Monduzzi Editore, Bologna, pp. 369–376.
Yada, T., Hyodo, S., Schreck, C.B., 2008. Effects of seawater acclimation on mRNA levels of corticosteroid receptor genes in osmoregulatory and immune systems in trout. General and Comparative Endocrinology 156, 622–627.
Yada, T., Muto, K., Azuma, T., Hyodo, S., Schreck, C.B., 2005. Cortisol stimulates growth hormone gene expression in rainbow trout leucocytes in vitro. General and Comparative Endocrinology 142, 248–255.
Yada, T., Uchida, K., Kajimura, S., Azuma, T., Hirano, T., Grau, E., 2002. Immunomodulatory effects of prolactin and growth hormone in the tilapia, *Oreochromis mossambicus*. Journal of Endocrinology 173, 483–492.
Yáñez, J.M., Houston, R.D., Newman, S., 2014. Genetics and genomics of disease resistance in salmonid species. Frontiers in Genetics 5, 415.
Yang, B.-Y., Arab, M., Chen, T., 1997. Cloning and characterization of rainbow trout (*Oncorhynchus mykiss*) somatolactin cDNA and its expression in pituitary and nonpituitary tissues. General and Comparative Endocrinology 106, 271–280.
Yang, D., Liu, Q., Yang, M., Wu, H., Wang, Q., Xiao, J., Zhang, Y., 2012. RNA-seq liver transcriptome analysis reveals an activated MHC-I pathway and an inhibited MHC-II pathway at the early stage of vaccine immunization in zebrafish. BMC Genomics 13, 1.
Yarahmadi, P., Miandare, H.K., Fayaz, S., Caipang, C.M.A., 2016. Increased stocking density causes changes in expression of selected stress-and immune-related genes, humoral innate immune parameters and stress responses of rainbow trout (*Oncorhynchus mykiss*). Fish and Shellfish Immunology 48, 43–53.
Yasuike, M., Takano, T., Kondo, H., Hirono, I., Aoki, T., 2010. Differential gene expression profiles in Japanese flounder (*Paralichthys olivaceus*) with different susceptibilities to edwardsiellosis. Fish and Shellfish Immunology 29, 747–752.
Young, N., Cooper, G., Nowak, B., Koop, B., Morrison, R., 2008. Coordinated down-regulation of the antigen processing machinery in the gills of amoebic gill disease-affected Atlantic salmon (*Salmo salar* L.). Molecular Immunology 45, 2581–2597.
Zapata, A.G., Chibá, A., Varas, A., 1997. Cells and tissues of the immune system of fish. In: Iwama, G.K., Nakanishi, T. (Eds.), Organism, Pathogen, and Environment, Fish Physiology. Elsevier, pp. 1–62.
Zhang, Y.X., Chen, S.L., 2006. Molecular identification, polymorphism, and expression analysis of major histocompatibility complex class IIA and B genes of turbot (*Scophthalmus maximus*). Marine Biotechnology 8, 611–623.
Zhang, A., Chen, D., Wei, H., Du, L., Zhao, T., Wang, X., Zhou, H., 2012. Functional characterization of TNF-α in grass carp head kidney leukocytes: induction and involvement in the regulation of NF-κB signaling. Fish and Shellfish Immunology 33, 1123–1132.
Zhang, Y., Chen, S., Liu, Y., Sha, Z., Liu, Z., 2006. Major histocompatibility complex class IIB allele polymorphism and its association with resistance/susceptibility to *Vibrio anguillarum* in Japanese flounder (*Paralichthys olivaceus*). Marine Biotechnology 8, 600–610.
Zheng, W., Liu, G., Ao, J., Chen, X., 2006. Expression analysis of immune-relevant genes in the spleen of large yellow croaker (*Pseudosciaena crocea*) stimulated with poly I:C. Fish and Shellfish Immunology 21, 414–430.
Zuchelkowski, E.M., Lantz, R.C., Hinton, D.E., 1981. Effects of acid-stress on epidermal mucous cells of the brown bullhead *Ictalurus nebulosus* (LeSeur): a morphometric study. The Anatomical Record 200, 33–39.

Chapter 9

Planning a Fish-Health Program

Mohamed Faisal[1], Hamed Samaha[2], Thomas P. Loch[1]
[1]*Michigan State University, East Lansing, MI, United States;* [2]*Alexandria University, Edfina, Egypt*

9.1 INTRODUCTION

In 1973, Snieszko adopted a simple diagram (Fig. 9.1) made of three intersecting circles to show the interactions of three components that are commonly necessary for an infectious disease outbreak to occur; unfavorable environmental conditions, a virulent pathogen, and a susceptible host. An effective fish-health management plan (FHMP) aims to disentangle the three circles using common sense, knowledge of pathogen–host interactions, and strict application of biosecurity measures. The details of an FHMP often vary from one facility and geographic region to another depending on several variables, including fish species, endemic pathogens, and prevailing environmental conditions. In general, there are three levels of fish-health management planning: facility/farm, regional/national, and international. Ideally, an FHMP should start at the international/global level, with the aim of preventing the introduction and spread of fish pathogens with live fish trade into the country. Unfortunately, the international fish-health standards that regulate the trade of live fish and their products are not equally implemented throughout the world and legislation varies from one country to another. This may also be the case at the regional/national level. As a result, strict implementation of disease-prevention and control practices at the facility level remains the cornerstone of any effective FHMP.

9.1.1 Formulating a Fish-Health Management Plan at the Farm/Facility Level

Disease prevention and control are intertwined with each and every aspect of the facility layout and design. Indeed, disease-prevention planning should start long before the aquaculture facility is built, which should then be followed by strict biosecurity, as well as preventive and control measures, during the operation.

9.2 FISH DISEASE-PREVENTION PLAN DURING THE FACILITY SITE SELECTION, DESIGN, AND CONSTRUCTION

In most extensive aquaculture operations, it is well known that success is dependent on choice of site, water source, and the design and layout of the facility. Unsurprisingly, these same factors have significant effects on the health and welfare of farmed fish. There are a number of published books/chapters and review articles that have addressed aquaculture facility design with an eye on disease prevention (Westers and Wedemeyer, 2002; Pillay and Kutty, 2005).

9.2.1 Choosing an Optimal Site for an Aquaculture Facility Can Minimize Disease Outbreaks

Despite the vast technological advances in aquaculture, including the expansion of intensive rearing and recirculating systems, fish raised in earthen ponds still comprise a major source of farmed fish worldwide. From the disease point of view, ponds that cannot be drained or dried (i.e., due to high level of surrounding groundwater) can be a constant source of fish pathogens that can also be spread to other parts of the farm. Therefore, a thorough survey of the site topology and hydrobiology should be undertaken to ensure that proper disinfection of ponds between production periods is feasible.

It is well recognized that an ample supply of water at the optimal temperature and water chemistry (dissolved oxygen, water hardness, pH, etc.) for the fish species of interest is imperative for health and efficient growth. For extensive farming, water can be drawn from rivers, creeks, impoundments, small dams, lakes, irrigation canals, and underground sources.

Fish Diseases. http://dx.doi.org/10.1016/B978-0-12-804564-0.00009-0

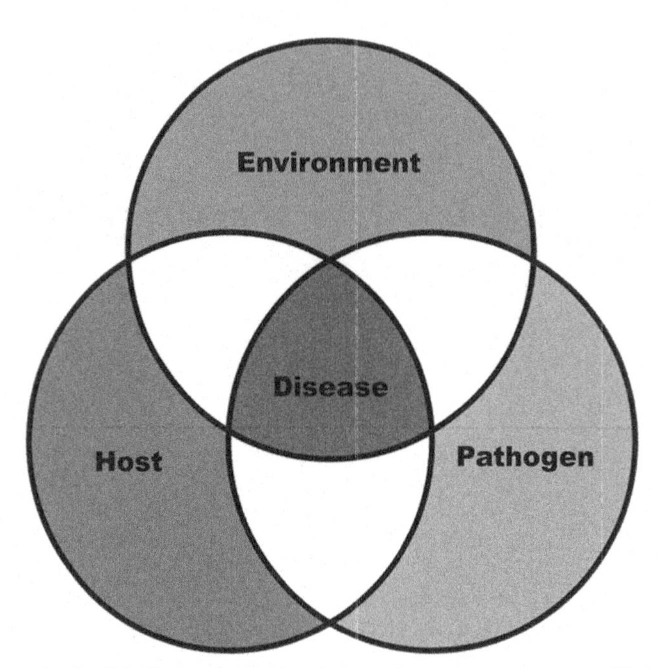

FIGURE 9.1 Diagrammatic representation for fish disease ecology. In order for a disease to ensue, a susceptible host, under a set of potentially adverse environmental conditions, becomes exposed to a virulent pathogen.

Box 9.1 Criteria to Be Considered When Selecting the Water Source for an Aquaculture Facility

- Water abundance
- Optimal temperature for the farmed fish species
- Preferably stable temperature and water chemistry year-round
- Free of known fish pathogens
- Should not contain excess nitrogen
- Not contaminated with sewage and other dissolved wastes
- Not contaminated with toxic chemicals, such as heavy metals (e.g., lead, zinc, and iron), oils, pesticides, and herbicides
- Free of harmful gases, such as chlorine, methane, hydrogen sulfide
- Low turbidity due to suspended particles (e.g., silt and clay colloids)
- Ideally free of fish and other aquatic organisms
- Water from domestic supplies should be avoided

The type, size, location, and topography of a farm, as well as types of water available, will determine the best or most practical water source. Groundwater (i.e., well water) is often optimal for intensively cultured fish because fish pathogens are likely not present.

9.2.2 Water Quality

Not only is the abundance of available water at the facility site important, but also its quality. A number of factors should be considered when selecting the source for the farm water supply (Box 9.1). From the disease point of view, aquaculture facilities receiving untreated surface water from streams or reservoirs that contain wild fish have an elevated risk of virulent pathogen introduction. Therefore, it is imperative that incoming surface water be filtered to prevent the introduction of wild fish/eggs and aquatic invertebrates, as well as other unwanted items (e.g., sediment), and then disinfected with ultraviolet radiation or ozone to minimize the possibility of pathogen introduction. The details of surface water treatment have been reviewed previously (Yanong, 2003; Pillay and Kutty, 2005; Noga, 2010; Liltved and Landfald, 2000). It is important to note, however, that microbial fish pathogens vary in their susceptibility to various water treatments; thus, specific treatment regimens should be designed to ensure that any endemic pathogens of concern will be reduced to acceptable levels or eliminated altogether.

9.2.3 Temperature

Temperature in particular is very important and a determinant of fish well-being since they are poikilotherms (cold-blooded animals). Temperatures that deviate beyond the upper and lower ranges tolerated by the species may lead to health deterioration and eventually death. It is therefore important to select the site of the facility in an area where the prevailing temperature fluctuations are within the tolerance range of the species and that enable maximal growth, efficient food conversion, and a healthy immune response.

9.2.4 Facility Design

From the construction point of view, the design of an aquaculture facility must consider the site topography, nature of the soil, water source, and funding. Many commercial scale hatcheries are designed to include a brood stock holding and spawning area, an area for food storage, an egg incubation area, a larval or juvenile culture area, and pump facilities. The design should denote areas where disinfection of tools and equipment can and should be performed. Disinfectant baths and places to change clothes should also be included in the design. A laboratory equipped with a microscope and dissection/necropsy capabilities is also important to have in a facility. A separate building with an independent, isolated culture system should be assigned as the facility's quarantine space. New fish should be kept in the quarantine facility for observation, testing, and potential treatment of pathogens. Upon design of the aquaculture facility, strict biosecurity planning must be seriously considered. Examples of such considerations are listed in Box 9.2.

9.3 FISH DISEASE-PREVENTION PLAN PRIOR TO OPERATION

Ideally, an effective FHMP should be in place before any fish or eggs are introduced into the facility. The FHMP at this stage is multifaceted and involves the following:

9.3.1 Training of Personnel

All key personnel should be familiarized with the following aspects:

- The physiology of the fish species to be raised, including the optimal temperature range, oxygen threshold, proper nutrition, food quantity and quality, and tolerance to chemical and physical deviations of the water.
- The basic concepts of fish-health maintenance and a working knowledge on the susceptibility of the target fish species to pathogens of concern. For example, although *Renibacterium salmoninarum* infects a number of fish families, it causes disease (i.e., bacterial kidney disease) only in salmonid species. Similarly, fish species/strains vary in their susceptibility to infection, which also has implications as to where fish should be placed within the fish farm.

Box 9.2 Points to Be Considered for Enhanced Biosecurity During Aquaculture Facility Design and Construction

- Maximize the use of nonporous construction materials, such as plastic, PVC, and stainless steel, as they are easier to disinfect, and keep the use of wood to a minimum.
- Design a quarantine and isolation facility in each farm.
- Avoid bringing any equipment or fish from other operating farms.
- Discourage the presence of unwanted animals (e.g., invertebrates, rodents, birds, amphibians, reptiles) through the use of enclosures and screens, as they can be a source for fish pathogens.
- Include areas to accommodate disinfection baths for incoming vehicles, equipment/tools, and boots/shoes.
- Use circular tanks or other self-cleaning designs (e.g., raceways with baffles) when possible.
- Include partitions between individual or groups of tanks/raceways to minimize pathogen spread.
- Design plumbing that enables individual rearing units to be taken offline for cleaning, chemotherapy/vaccine administration, or isolation.
- Design plumbing with access points that allow for mechanical cleaning (i.e., brushing/scraping) and that reroute away from any fish during periodic disinfection.
- Include features to facilitate the removal of solids from systems.
- If visitors will be allowed, design viewing windows and a tour flow that will not put brood stock or the facility at risk for pathogen introduction.
- Incorporate separate equipment/tools for each fish-rearing unit (can be labeled or color-coded) and ensure that they are spatially separated from neighboring tools.
- Ensure that equipment is kept clean to ensure maximal disinfection efficacy.

- The signs of disease in fish, as early diagnosis is critical for minimizing losses. In a similar vein, personnel should be aware of who to contact when signs of disease emerge, as well as how to collect, submit, and ship fish for clinical evaluation and diagnostic testing. An example of a submission form for diagnostic testing and some of the information that may be requested can be found in Fig. 9.2. For more on this, please see Section 9.5.
- The concepts of biosecurity, along with the best ways for their implementation. There are a number of publications dealing with fish health that contain such information (Yanong, 2003; Scarfe et al., 2008; Yanong and Erlacher-Reid, 2012), and more on biosecurity can be found in Section 9.8.
- The monitoring of water quality parameters (e.g., temperature, dissolved oxygen, pH, salinity, ammonia, nitrite, alkalinity, general hardness).

9.3.2 Fish Eggs

Unless impossible, all fish coming into the facility should be brought in as eggs. The introduction of live fish, particularly brood stock, should be avoided, as they can be carriers of pathogens without exhibiting any disease signs. However, eggs can also be a major pathogen source for fish farms, especially for those pathogens that are vertically transmitted

Date: ___________ **Hatchery:** ___________

Submitted by: ___________ **Phone:** ___________

Facility Address: ___________

Billing name/address: ___________

Reporting name/address: ___________

Reason for fish submission:

_____ *Increasing Mortalities* _____ *Morbidity* _____ *Planned Sample*

_____ *Other (Explain)* ___________

What treatments have been applied to this lot in the past 6 months? ___________

____

FISH IDENTIFICATION

Species: ___________ **Strain:** ___________ **Lot ID:** ___________

Age: ___________ **Sex:** ___________ **Number of fish submitted** ___________

Anticipated stocking date: ___________

ORIGIN OF EGGS

Wild/feral caught ☐ ***Hatchery*** ☐

Location/State ___________

Treated with: Erythromycin ☐ **Iodine** ☐ **Other** ☐ ***Explain*** ___________

HISTORY OF MORTALITIES & MORBIDITY *(if applicable)*

Date mortalities began _____/_____/_____

Progression and current status ___________

____

____

Describe fish behavior: *(Normal, gasping at the surface, lethargic, flashing etc.)* ___________

____

____

Describe recent eating pattern *(Be specific)* ___________

____

Describe clinical signs:

_____ Color change   _____ Fuzzy gills/body   _____ Dropsy

_____ White spots on skin   _____ Fin/tail erosion   _____ Reddened/Hemorrhagic fins

_____ Red spots covering body _____ Exophthalmia/Bulging eyes _____ Furuncles

_____ Other *(write in)* ____________________

Gill scraping results: ____________________

Skin scraping results: ____________________

HATCHERY CONDITIONS

Number of fish in each culture system of concern: ____________________

Density (Kg/m^3) of fish in each culture system of concern: ____________________

Type of water supply: ____________________ **Flow Rate (L/min):** ____________________

Tank volume turnover time (minutes): _____ **Feeding Interval:** ____________________

Water conditions: *Temperature* __________ *D.O.* _______ *pH* __________ *NH_3* ____________________

NO_2^- __________ NO_3^- __________ **Other abnormal conditions** *(describe)* ____________________

Other culture systems that share the same water supply:

Upstream ____________________

Downstream ____________________

Parallel ____________________

Describe other raceways and/or species with mortality/morbidity? ____________________

Describe weather preceding mortalities ____________________

Describe weather during mortalities (for outside raceways/ponds) ____________________

HISTORY

Any recent history of disease at the hatchery? ____________________

Any historical disease problems with this species at the hatchery? ____________________

____

Any historical disease problems with this culture system? ____________________

FIGURE 9.2 Example of fish-health submission form.

(e.g., *R. salmoninarum*, *Flavobacterium psychrophilum*, and infectious pancreatic necrosis virus). Other non-vertically transmitted pathogens can also adhere to the egg surface and be carried into the hatchery system. Indeed, there are numerous documented cases where devastating disease outbreaks have occurred in new aquaculture facilities because pathogens were carried on or in eggs from unreliable sources. Eggs originating from outside the facility should be accompanied by a health certificate based upon tests performed in reputable laboratories and signed by a certified fish health expert. It is preferred that a subsample of the eggs be retested for pathogens that are endemic to the area of egg origination. As detailed in Section 9.4.1, eggs must also be surface-disinfected upon arrival and maintained under quarantine conditions.

9.3.3 Food Source and Storage

The quantity and quality of food is integral to the success of any aquaculture project. In addition to its direct effect on growth and weight gain, imbalanced or nutritionally deficient food can lead to serious negative health effects. Therefore, food must be obtained from a reputable and inspected feed mill, and accompanied by guaranteed ingredient/nutritional information. Upon receipt, the date of expiration should be inspected, and the feed should be stored in a cool, dry, pest-proof building that is designated for feed storage. Ideally, dry feeds should be stored at temperatures <20°C and humidity <75%.

9.3.4 Quarantine

Prior to operation, protocols for the facility's quarantine should be in place (also see Section 9.10.7). Most quarantine standard operations call for the following:

- A subset of all new arrivals should be subjected to a thorough internal and external clinical examination, followed by appropriate diagnostic testing.

- Fish should be kept in quarantine for a period that exceeds the incubation time of all pathogens of concern.
- Optimal water parameters for the species should be maintained and fish kept at densities similar to those of the production tanks.
- Strict biosecurity measures should be implemented in the quarantine area, including disinfection of footwear, use of separate equipment/tools for every rearing unit, change of clothing, exclusion of pests/unwanted animals (see also Box 9.1), and, preferably, disinfection of water effluent via UV/ozone treatment.

9.3.5 Medicine Cabinet

A multitude of compounds should be available for use at any aquaculture facility, a portion of which is listed in Table 9.1. It must be noted that regulations for some of their uses/withdrawal time vary according to region and should be checked before use in food fish. For aquaculture operations in the United States, the reader is referred to the US Fish & Wildlife Service—Aquatic Animal Drug Approval Partnership Program Website for materials on approved/investigational treatments for use in food fish (https://www.fws.gov/fisheries/aadap/home.htm).

9.4 FISH-HEALTH MAINTENANCE PLAN DURING OPERATION

As mentioned earlier, an FHMP is multifaceted, targeting not only infectious pathogen prevention and control, but also emphasizing maintenance of optimal environmental conditions and culture practices. In other words, successful health maintenance does not depend upon a single procedure, but rather is the culmination of applying multiple integrated concepts and management with biosecurity occupying a central position.

Despite numerous technological advances, many aspects of fish diseases and their causation remain unsolved. Due to the complex nature of aquaculture operations, it will likely be impossible to fully eliminate pathogen entry into aquaculture facilities and, in fact, if an infectious agent becomes endemic in a facility, it may be nearly impossible to totally eradicate

TABLE 9.1 List of Some Potential Compounds and Items That Should Be Kept on Hand at an Aquaculture Facility

Purpose in Aquaculture Operations	Examples of Relevant Compounds
Sedation/anesthesia/euthanasia	Tricaine methanesulfonate (MS-222), eugenol, benzocaine
Egg disinfection and treatment	Povidone iodine, formalin, hydrogen peroxide
Surface disinfectants for equipment, footwear, etc.	Calcium hypochlorite, sodium hydroxide, perooxygenated compounds (e.g., Virkon), povidone iodine; chlorhexidine diacetate (e.g., Nolvasan); sodium hypochlorite (i.e., bleach)
Stress reduction/maintenance of osmoregulation	Calcium chloride, sodium chloride, potassium chloride
External parasite control[a]	Formalin, hydrogen peroxide, acetic acid, calcium oxide, magnesium sulfate, sodium chloride
External flavobacteriosis control[a]	Chloramine-T, hydrogen peroxide
General	Sodium thiosulfate (treatment of chlorine toxicity in fish, removal of chlorine from municipal water, and/or neutralization of chloramine-T), sodium bicarbonate (to buffer alkalinity of water, especially when MS-222 is used)
Monitoring water parameters	Water test kit
Rapid response to arising fish health issues	List of important contacts: • Fish-health professional/aquatic veterinarian • Local/regional water authority • Competent authority • Local/regional natural resource agency • Agency responsible for investigational new animal drugs

Note that regulations for their use/withdrawal time vary according to region and should be checked before use in food fish. Vendor instructions should be followed in conjunction with the procedures recommended by your fish-health professional/aquatic veterinarian. For aquaculture operations in the United States, the reader is referred to the US Fish & Wildlife Service–Aquatic Animal Drug Approval Partnership Program website for materials on approved/investigational treatments for use in food fish (https://www.fws.gov/fisheries/aadap/home.htm).
[a]*Fish-health professionals and/or aquatic veterinarians should be consulted prior to use.*

without depopulation. This leads to two major approaches; prevention, i.e., preventing disease from occurring; and control, i.e., minimizing the losses associated with disease and limiting its spread.

Prophylactic measures can prevent disease occurrence in an aquaculture facility and constitute an effective strategy for preventing diseases from ever beginning. Indeed, the ancient quote that an ounce of prevention is worth more than a million pounds of cure is resoundingly true in aquaculture. Prevention is a holistic approach that encompasses every step through the fish-rearing process. The following are examples of prophylactic measures that are commonly used in successful aquaculture projects.

9.4.1 Prophylactic Measures to Minimize Pathogen Introduction Through Gametes

Effective prophylaxis starts with the selection of the fish egg source. In addition to purchasing fertilized eggs from dependable commercial sources, many fish farmers raise their own brood stock. Most recently, eggs from brood stock that are resistant to a specific disease have become commercially available; however, the susceptibility of these stocks to other diseases frequently goes unevaluated and such strains may also suffer from reduced growth rates. In governmental hatcheries where stock enhancement, rehabilitation, and conservation of genetic diversity are the main goals, wild fish frequently serve as gamete donors. Such stocks may carry pathogens that are vertically transmitted (Brown et al., 1997; Loch et al., 2012; Loch and Faisal, 2016a, 2016b).

In order to reduce vertical transmission of pathogens along with eggs/reproductive fluids, a number of strategies can be employed. For example, to reduce transmission of *R. salmoninarum* in salmonids, several studies examined the effect of administering the macrolide antibiotic erythromycin to brood stock via injection several weeks prior to spawning to allow for antibiotic accumulation within the egg and found this practice to be effective (Bullock and Leek, 1986; Moffitt, 1991). Another promising strategy for reducing *R. salmoninarum* vertical transmission is by testing reproductive fluids (e.g., ovarian and seminal fluid) for the presence of the bacterium and subsequently culling any eggs that resulted from positive male/female pairings (Pascho et al., 1991; Faisal et al., 2012).

Regardless of source, each individual brood fish should be externally examined and those that exhibit gross signs of disease should not be spawned and, ideally, culled. Examples of gross disease signs are displayed in Table 9.2. In the case of semelparous (i.e., terminal) spawning fish, an examination of internal organs should also be performed and gametes from fish with gross lesions (Table 9.3) should be excluded. When feasible, and prior to the spawning season, a subsample of the brood stock should be subjected to laboratory examination to determine if they harbor any of the serious fish pathogens.

The prevention of gamete contamination by pathogens, vertically transmitted or otherwise, is also imperative. During egg collections, strict hygienic measures should be followed to prevent cross-contamination of egg groups. Many hatchery personnel prefer to disinfect surfaces, buckets, beakers, egg-take tools, and other relevant egg collection equipment with a warm solution of sodium hypochlorite, followed by thorough rinsing and waste-water neutralization with sodium thiosulfate. However, other disinfectants are also effective (see Table 9.1 for examples). Prior to artificial stripping of eggs or milt, the fish should be rinsed with clean water to reduce debris, feces, and microbes that may be attached to the mucus and fins. Terminal spawners should be dipped in an iodine solution, the skin/mucus around the vent wiped with disposable paper towel, and then rinsed with clean water. The eggs/milt should then be collected in individual containers (i.e., not reused for multiple fish unless disinfected first) and labeled to denote the female/male that was used for the pairing process if gamete testing/culling will be performed. Any groups of eggs with an unusual appearance (e.g., opaque, coagulated ovarian fluid) should be discarded. Blood and fecal material should be avoided and either removed from eggs/milt or those samples discarded.

Egg-surface disinfection is one of the most effective methods for minimizing pathogen transfer from brood stock to offspring and for reducing pathogen spread among fish groups. Disinfection can be safely performed following fertilization during or after water hardening. For over three decades, iodine-containing compounds have been safely used as egg-surface disinfectants. One advantage of the iodine-containing compounds is that their color lightens when disinfection conditions are suboptimal and/or disinfection efficacy is reduced, thereby allowing technicians to monitor effectiveness throughout the disinfection process and add fresh iodine solution as needed. These same compounds can also be used for equipment/surface disinfection. Trucks, tanks, and containers used to transport eggs/fish to and from aquaculture facilities should be cleaned by power-hose, and disinfected before and after each delivery. For more on disinfection of equipment and eggs, the reader is referred to Danner and Merrill (2006) and Chapter 1.1.3 on methods for disinfection in aquaculture establishments (OIE, 2009).

During incubation, fish eggs are susceptible to infections with saprophytic fungi and water molds, which can spread rapidly within and among incubators, and lead to poor eye-up and hatching rates. There are a number of compounds that are effective fungicides and safe to use on embryonated eggs, such as formalin, hydrogen peroxide, and iodine-containing

TABLE 9.2 Examples of Common External Lesions Indicative of Disease Conditions in Cultured Fish

Abnormal Signs	Possible Disease Causes
	Multifocal to coalescing hemorrhage suggestive of a systemic viral and/or bacterial infection, or heavy external parasitism
	Diffuse hemorrhages, along with a hemorrhagic vent, suggestive of a systemic subacute viral and/or bacterial infection
	Furuncle suggestive of infection with *Aeromonas salmonicida*
	Deep hemorrhagic ulcer suggestive of bacterial infection
	Fin erosin and deep ulcer on the caudal peduncle suggestive of flavobacterial infection
	Exophthalmia with an ocular hemorrhage suggestive of a systemic viral and/or bacterial infection
	Severely pale gills suggestive of anemia, possibly induced by viral or bacterial infection
	Gills showing extensive tissue loss suggestive of flavobacterial infection and some viral infections (e.g., Koi herpervirus)

disinfectants. Being careful to adhere to local/regional effluent restrictions, regular use of formalin for the prevention and treatment of fungi/water molds is a widespread method employed by aquaculture facilities. More information on preventing and treating egg infections is detailed in Stoskopf (1993), Bruno et al. (2011), and Bowker et al. (2015).

9.4.2 Food, Nutrition, Immunostimulants, and Probiotics

Proper nutrition impacts both the health of farmed fish and the profitability of the project. Therefore, feeding programs must consider the quality and freshness of the feed, the nutritional requirements needed for growth and health maintenance of the target fish species, fish age, water temperature, and a proper feed size that is commensurate with the age/size of the fish. In fish

TABLE 9.3 Examples of Common Internal Lesions Indicative of Disease Conditions in Cultured Fish

Abnormal Signs	Possible Disease Causes
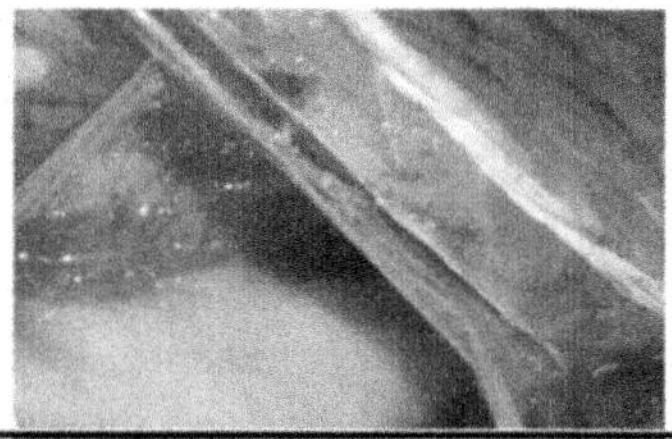	Presence of fluids in the abdominal cavity suggestive of a systemic bacterial and/or viral disease
	Hemorrhage in visceral fat suggestive of nutritional deficiency or systemic viral/bacterial disease
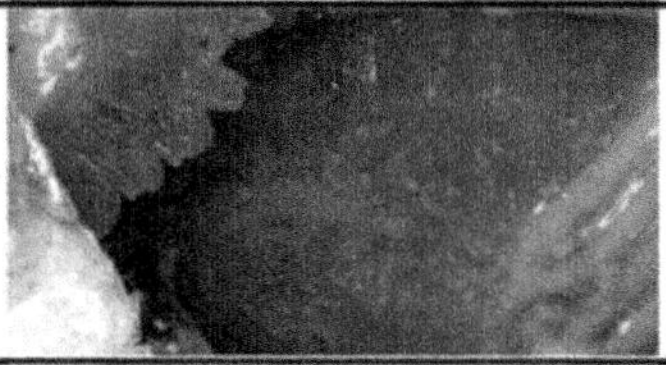	Multiple whitish nodules in the liver suggestive of granulomatous diseases, such as mycobacteriosis, bacterial kidney disease, or piscirickettsiosis. The same lesions can be caused by encysted metacercariae of larval trematodes
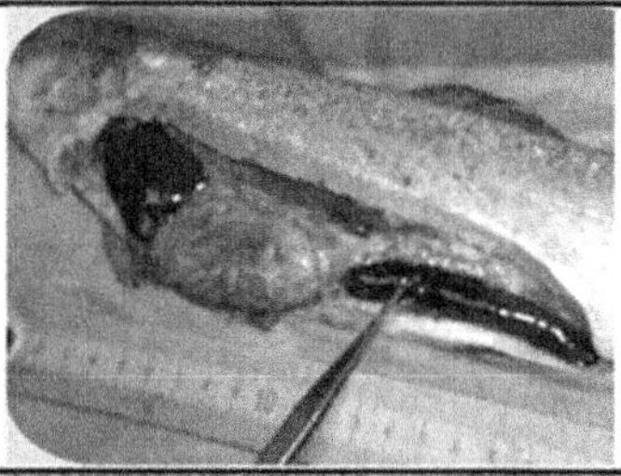	Hemorrhagic inflammation of the intestine suggestive of toxicosis or enteric redmouth disease caused by *Yersinia ruckeri*
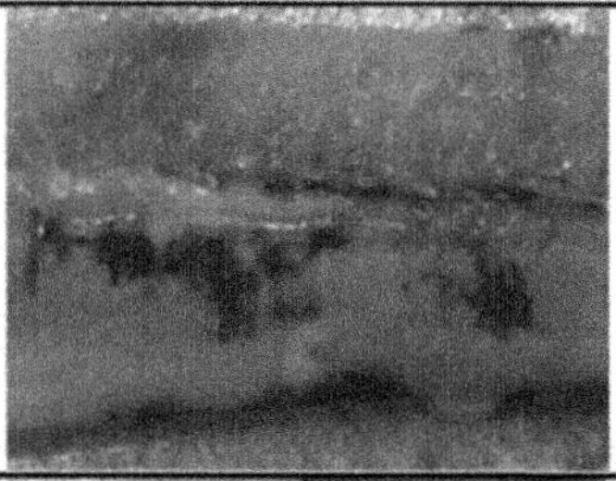	Hemorrhages in the swimbladder suggestive of systemic viral and/or bacterial disease
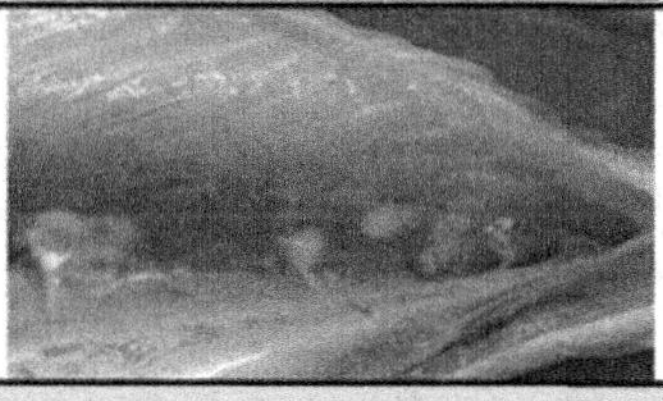	Whitish nodules in the kidney parenchyma suggestive of bacterial kidney disease

species with a physostomous swim bladder, Faisal et al. (2007) described large scale mortality in farmed chinook salmon due to a generalized systemic mycosis caused by *Phoma herbarum*. The underlying cause of this condition was attributed to the blockage of the pneumatic duct by small feed pellets fed to fish of a relatively large size. When the pellet size was corrected, the fungal infection was resolved. Needless to say, proper storage of dry feeds in a cool (<20°C) and low humidity (<75%) environment (i.e., a dedicated food storage area) ensures the maintenance of its nutritive value and attractive taste, as well as reduces the likelihood of rancidity and/or fungal contamination. The quantity and frequency of feeding also impacts the health of raised fish. For example, overfeeding can lead to poor water quality, excess organics, and suspended solids that can predispose fish to bacterial and viral infections, whereas underfeeding leads to poor immunocompetence and growth. To ensure that fish are fed at the optimal frequency and quantity, the feed conversion ratio and daily feeding rates should be calculated using a number of available formulas, including those outlined by Westers and Wedemeyer (2002). It is important to note that actively feeding fish is an excellent biomarker for their health; when a group of fish begins to feed less or cease feeding altogether (and is unrelated to a decrease in temperature), it often indicates that the environmental conditions are suboptimal and/or a disease process has begun. Thus, observing and recording rates of wasted feed on a daily basis is not only an invaluable tool for tracking fish health, but also ensures that profitability can be maximized through avoidance of feed waste.

Over the last few decades, immunostimulants have been used to heighten the host defense mechanisms of farmed fish. Immunostimulants interact with and activate cells of the immune system, thereby increasing their capacity for engaging pathogens and/or secreting germicides. Immunostimulants can be administered by injection, immersion, or orally along with feed; however, injection was found to be the most effective for some immunostimulants, particularly in larger fish (Chen and Ainsworth, 1992). Immersion of younger fish in immunostimulant solutions proved effective during acclimation of juveniles to ponds under field conditions (Gannam and Schrock, 1999). Immunostimulants mixed with fish oil and coated onto feed pellets is a cost-effective method of administration to extensively cultured fish. As recommended by Anderson (1992), immunostimulants should be applied before disease outbreaks occur (i.e., at times of year prior to when disease outbreaks cyclically occur, prior to stress-associated events, such as grading, moving, stocking) to reduce disease-related losses. Effective doses, exposure times, and routes of administration depend upon the fish species, age, culture system, and endemic diseases.

Immunostimulants that have been used successfully in aquaculture vary in their nature and mode of action (reviewed in Barman et al., 2013). One such immunostimulant is levamisole, which protected rainbow trout (*Oncorhynchus mykiss*) bathed in a 5–25 μg/mL solution from *Yersinia ruckeri*, the causative agent of enteric redmouth disease (Ispir, 2009). Even at very low doses, it inactivated whole bacteria and bacterial wall components, such as lipopolysaccharide and *N*-acetylmuramyl-L-alanyl-D-isoglutamine, and increased fish resistance and survival upon subsequent challenge with a number of fish pathogens (e.g., *Aeromonas salmonicida*, the causative agent of furunculosis, and *A. hydrophilla*, one of the causative agents of motile aeromonas septicemia). Similarly, glucans, which are long-chain polysaccharides extracted from different yeast species, proved to be excellent immunostimulators when added to the feed of a number of fish species or injected intraperitoneally. For example, glucan-treated fish showed increased resistance against a range of fish-pathogenic bacteria, including *Vibrio anguilllarum*, *Vibrio salmonicida*, and *Y. ruckeri* (reviewed in Barman et al., 2013).

In recent years, probiotics have taken center stage as feed additives for cultured fish and shellfish. Probiotics are live microorganisms that, when administered in adequate amounts, confer health benefits to the host. Extensive in vivo studies have demonstrated that probiotics heighten disease resistance, increase growth performance, improve feed utilization, lessen negative-stress impacts, and harmonize the microbial balance between the host and the surrounding microbial communities (reviewed in Verschuere et al., 2000; Merrifield et al., 2010). Probiotics can be of bacterial or nonbacterial origin, but it should be noted that their use is prohibited in some regions. Thus, local restrictions should be investigated prior to use.

A wide range of primarily Gram-positive and, to a lesser extent, Gram-negative bacteria, has been employed as probiotics in aquaculture with success. *Bacillus subtilis*, an endospore-forming bacterium, has been widely used in finfish aquaculture (Hong and Cutting, 2005). Other Gram-positive bacteria have also been used as probiotics, albeit with varying success depending upon the fish species, age, and geographic location, including *Arthrobacter* sp., *Brevibacillus* sp., *Brochothrix* sp., *Clostridium* sp., *Carnobacterium* spp., *Enterococcus* spp., *Kocuria* sp., *Lactobacillus* spp., *Microbacterium* sp., *Micrococcus* spp., *Pediococcus* spp., *Streptococcus* sp., *Vagococcus* sp., and *Weissella* sp. (reviewed in Hai, 2015). More specifically, there are strong data on the effectiveness of certain bacterial probiotics protecting cultured fish from pathogenic infections, including *Bacillus* spp. against *Streptococcus iniae* (Cha et al., 2013), *Brevibacillus brevis* against *Vibrio* spp. (Mahdhi et al., 2012), and *Pseudomonas* M162/M174 against *F. psychrophilum* (Korkea-Aho et al., 2012).

Other nonbacterial organisms have also been utilized as beneficial probiotics in aquaculture, including bacteriophages, microalgae, and yeasts. Bacteriophages of the families of Myoviridae and Podoviridae protected ayu (*Plecoglossus altivelis*) from a highly pathogenic strain of the Gram-negative bacterium, *Pseudomonas plecoglossicida* (Park et al., 2000). However, the effects of bacteriophage administration on beneficial bacteria in the host have yet to be adequately studied. For microalgae, species of the genera *Chaetoceros* spp., *Tetraselmis* sp., and *Phaeodactylum* sp. prevented fish-pathogenic *Vibrio* spp. from causing disease in fish (Naviner et al., 1999). Yeasts of several genera have also had profound probiotic effects on cultured fish. For example, *Saccharomyces cerevisiae* improved resistance of Nile tilapia (*Oreochromis niloticus*; Lara-Flores et al., 2003; Meurer et al., 2006) and common carp (*Cyprinus carpio*; Faramarzi et al., 2011) to a number of bacterial pathogens. In the same context, the yeast *Debaryomyces hansenii* enhanced the growth performance of sea bass (*Dicentrarchus labra*) larvae and improved its health status (Tovar-Ramírez et al., 2010).

In finfish aquaculture, most probiotics are designed to be mixed with feed. Although most studies on probiotics have focused on the use of single bacterial cultures, some multi-strain and multispecies probiotics enhanced protection against pathogenic infection when compared to single-strain preparations (Timmerman et al., 2004). For example, a mixture of *Lactococcus lactis* and *Lactobacillus plantarum* protected Japanese flounder (*Paralichthys olivaceus*) against *Streptococcus iniae* (Beck et al., 2015), and similar effects were reported in rohu (*Labeo rohita*) during fry stages (Jha et al., 2015).

Adjusting the dose of probiotics is integral to their success, since responses to different dietary probiotic levels varied greatly (Panigrahi et al., 2004; Bagheri et al., 2008). For example, a diet supplemented with *Lactobacillus brevis* at 10^9 cells/g protected hybrid tilapia (*Oreochromis niloticus* × *Oreochromis aureus*) against *Aeromonas hydrophila* (Liu et al., 2013). However, rainbow trout that were fed the probiotic diet at the same concentration did not exhibit a similar protection. In general, a high dose of probiotic does not mean a greater level of protection; rather, other factors, such as the probiont species, fish species and their physiological status, rearing conditions, and the specific goal of the applications determine the outcome (Merrifield et al., 2010). The period of administration is also considered an important factor, as fish fed with probiotics over a long period of time tend to develop less resistance to pathogens. For example, grass carp (*Ctenopharyngodon idellus*) fed with the probiotic mixture of *Shewanella xiamenensis* and *Aeromonas veronii* for 28 days exhibited higher cumulative mortality upon challenge with *A. hydrophila* compared to those that received the probiotic for 14 days (Wu et al., 2015). It is therefore strongly recommended that the exact dose of probiotic and length and frequency of administration be carefully evaluated.

The mode of action for probiotics is only partially understood. It is believed that competitive exclusion is one of the underlying mechanisms for probiotic protective action against invading pathogens. In the digestive tract, probiotics interfere with the action of potential pathogens by the production of inhibitory molecules and/or direct competition for space, nutrients, or oxygen (Fuller, 1989). Some probiotics were also found to produce organic acid, which causes a drop in pH, thereby antagonizing some pathogenic bacteria (Ma et al., 2009). Another group of probiotics was found to exert its effects by decomposing organic matter, reducing nitrogen and phosphorus concentrations, and controlling ammonia, nitrite, and hydrogen sulfide concentrations, thereby improving water quality, which is an integral element for keeping fish healthy (Cha et al., 2013).

9.4.3 Maintain an Optimum Environment

Maintaining optimal water quality for the target species is paramount to fish health and for minimizing their susceptibility to infection. In any aquaculture facility, regardless of the type of rearing unit (pond, raceway, tank, etc.), water quality should be continuously monitored, recorded, and managed in a way that takes fish age and density, feeding rate, volume, filter/turn-over time, and water supply into account. Removing excess organic material and metabolic waste is essential for maintaining optimal water conditions and reducing proliferation of unwanted, facultatively pathogenic microbes. Each facility should have alarms and a contingency plan in place that protects fish from catastrophic water quality failure events (e.g., pump failure, burst pipes, filter malfunction). Equally integral is the presence of backup power generators that come online in the event of power outages. Fish behavior (e.g., gasping) and distribution in the water column (e.g., aggregation at the water inlet or outlet) mirror the water quality; therefore, frequent monitoring of fish behavior is also essential. Detailed information on water quality in aquaculture facilities can be found in several chapters of this book, and is also reviewed in Stoskopf (1993), Westers and Wedemeyer (2002), and Noga (2010).

9.4.4 Disease-Prevention by Segregation and Logistical Planning

An important principle of disease-prevention is segregation of fish according to species, strain, and age, as well as their strategic placement within the aquaculture facility relative to one another. Indeed, fish species vary widely in their susceptibility to fish-pathogenic microbes. For example, the salmonid lake herring (*Coregonus artedi*) is highly susceptible to viral hemorrhagic septicemia virus genotype IVb (Weeks et al., 2011), while rainbow trout, another salmonid species, is far less susceptible to the disease caused by the same virus sublineage (Kim and Faisal, 2010) but has the ability to shed the virus for extended periods. Similarly, different genetic strains within the same susceptible fish species are not equally vulnerable to microbes, as evidenced by the substantial variation in susceptibility to *F. psychrophilum*-induced bacterial coldwater disease that occurs among different rainbow trout strains (Wiens et al., 2013). Likewise, fish of different ages vary widely in disease susceptibility. Special attention should be paid to newly hatched fish, as their immune system is not well developed, leading to far less disease resistance than their older counterparts. In addition, some tissues in young fish are more susceptible to destruction and/or invasion. For instance, *Myxobolus cerebralis*, the causative agent of whirling disease, can destroy swaths of the skeleton and cartilage in young trout, where large portions have yet to be fully ossified. The multiplying parasite continues to digest infected cartilage, often leading to collapse of the vertebral column. However, once fish are older and skeletal ossification is nearly or fully complete, most become nearly refractile to infection. Thus, young, whirling disease-susceptible fish must not be kept on surface water that contains this parasite unless it is first UV- or ozone-treated, but preferably should be maintained in groundwater (i.e., well water). Additionally, young salmonids are very susceptible to infectious pancreatic necrosis virus (IPNV), whereby infections lead to substantial losses. On the contrary, older fish can contract the infection, but become asymptomatic carriers, shedding the virus in the surrounding environment without any signs of disease. For all of these reasons, it is prudent to implement strict segregation among fish stocks reared in the same facility and strategically place the fish with differential susceptibility in mind. It is also crucial to allow for adequate fallowing time (i.e., temporal separation) between different groups/age classes of fish in a facility.

9.4.5 Fish Handling

Aquacultured fish are exposed to continuous stress that frequently makes them more prone to infectious disease. One of the most stressful events for farmed fish is handling, not only because of the expended energy and stress response associated with the flight response and eventual capture, but also because handling leads to disturbance of the protective mucus coat and sometimes further damage to the fish's external barriers, such as the skin and gills. These areas then go on to serve as portals for pathogen invasion. Therefore, fish handling should be done in a way that minimizes stress and integument damage, and also eliminates the risk of escape. Box 9.3 lists a number of precautions that should be followed to reduce handling stress.

9.4.6 Monitoring Fish for Disease Signs

All groups of fish in every culture unit should be thoroughly monitored for signs of health and disease. All personnel in an aquaculture facility should be familiar with such signs and report any unusual behavior or gross abnormalities to their fish-health professional, especially if the prevalence begins to increase. Changes in fish behavior frequently begin prior to the development of physical disease signs and thus serve as an early warning signal of a disease process. While observing fish behavior, it is important that the fish are disturbed as little as possible. Table 9.4 denotes some of the behavioral changes associated with the early stages of disease. Fish should also be monitored for the appearance of external abnormalities that are characteristic of ongoing disease processes (Table 9.2). Any moribund fish that are culled/euthanized, as well as dead fish, should also be examined for the presence of internal lesions that are associated with the disease in cultured fish (Table 9.3).

9.4.7 Sedation and Anesthesia in Fish

Extensively handled and/or transported fish are frequently sedated to reduce the associated stress. Additionally, some gamete-collection procedures for artificial spawning, along with many fish-health procedures, require that fish be anesthetized. For anesthesia, a number of chemicals have been used, including tricaine methanesulfonate (MS-222), eugenol, and benzocaine, among others. Such compounds vary in their mode of action, duration of anesthesia, withdrawal time, and postanesthesia treatment. The duration should be minimized while still ensuring that the anesthetic level is adequate for the procedure. Anesthetized fish should be carefully monitored all times, as should the water quality of the anesthetic bath (in particular, the dissolved oxygen concentration) to ensure fish survival. In all the cases, regional restrictions and approved

Box 9.3 Measures to Be Followed for Reducing Stress During Fish Handling

- Prior to handling, feed should be withheld for a minimum of 12h.
- Tools used for handling fish should have smooth surfaces as appropriate.
- Keep time that fish are out of water to a minimum.
- If fish are to be handled for extended periods, they should be sedated or anesthetized.
- Surfaces that will contact the fish should be kept clean and wet.
- Specially formulated water conditioners, collectively known as mucus replacements/protectants, can be used on handling equipment/surfaces and in baths or transport water to protect the fish mucus coat.
- Dip net loads should not contain excessive numbers of fish.
- Mesh size of fish nets should commensurate with fish size to prevent injury.
- Handled fish should be monitored closely and dead or injured fish removed.

TABLE 9.4 Examples of Behavioral Changes Indicative of Disease Conditions in Cultured Fish

Signs of Abnormal Behavior	Possible Disease Causes
Abnormal swimming	Lethargy due to systemic disease, severe nutritional deficiency, suboptimal water quality
Gathering at the water supply inlet	Hypoxia, anemia, gill disease, elevated suspended solids, high density, suboptimal water quality, external parasites
Gathering at the water outlet	Lethargy due to diseases, weakness, severe fin erosions, bloating, swimbladder dysfunction
Unpropelled motion and elevated in the water column	Lethargy from systemic disease, blindness or cataract, poor intestinal motility, general weakness, hypoxia, neurologic disorders
Sluggish swimming at the bottom	Disturbed buoyancy, severe digestive problems, presence of worms or tumors in the swimbladder, systemic disease
Gulping air at the surface (piping)	Elevated temperature, hypoxia, ammonia poisoning, nitrite poisoning, elevated suspended solids, gill disease
Abnormal body axis posture	Subordination, neurologic disorder, swimbladder disorder, following long transportation
Rubbing against objects	Ectoparasites, presence of toxic chemicals, skin irritation
Erratic swimming	Ectoparasites, neurologic disorders, systemic disease
Swimming in circles (whirling)	Blindness, infection with *Myxobolus cerebralis*, flavobacterial infection
Decreased appetite	Disease, anxiety, high temperature, too much food, pellets are of improper size
Food spitting	Stress, infection, incorrect pellet size
Increased opercular movements	Stress, high temperature, high density, infection, gill disease/parasite, hypoxia, anemia, poor water quality, incorrect light cycle
Decreased opercular movements	Starvation, infection, lethargy due to diseases
Cannibalism	Starvation, stress, social conflicts, increased light cycle, fish vary in size too much

uses must be investigated prior to use of sedatives/anesthetics, particularly in food fish. For more on sedation and anesthesia in fish, the reader is referred to Stoskopf (1993), Harms (2005), and Noga (2010). A list of approved and conditionally approved compounds for sedation/anesthesia in food fish within the United States can be found in Bowker et al. (2015).

9.5 PROPER SPECIMEN COLLECTION FOR DIAGNOSTIC PURPOSES

Despite efforts to the contrary, disease outbreaks in farmed fish may still occur, which require timely diagnosis and management. Live specimens are ideal for diagnostic purposes, but are not necessarily required for all diagnostic tests. However, it is imperative that specimens for diagnostic purposes be properly collected and optimal sample integrity maintained so that accurate and useful laboratory results can be achieved. In order to know which samples to submit, how many, and how to submit them, the following must be determined in consultation with your fish-health professional prior to collection/submission:

9.5.1 Purpose of Investigation

What is the purpose for sending samples to the laboratory? If a health certificate is required, what is to be done with these fish? What are the requirements of the receiving jurisdiction? If disease diagnostics are required, is anything in particular suspected? What are the clinical signs? Is there anything else that must be ruled out? For pathogen screening, what are the pathogens of interest/concern? What else does laboratory testing need to reveal?

9.5.2 Samples Needed

Once the purpose of investigation is solidified, the types of required samples can be determined by answering a few additional questions. What is the size of the fish population and what is the assumed pathogen-prevalence level, both of which dictate the number of specimens that should be analyzed? Taking the age of the fish into account, what is the ideal tissue to submit for the required assay(s)? What is the required tissue quantity and is more than one tissue type required? Do samples need to be randomly collected, or should moribund fish be targeted?

9.5.3 Sample Integrity

Submitting live fish provides the most flexibility for the selection of analyses that can be performed, but may not be necessary in all situations (Fig. 9.3). Fresh dead fish on ice, frozen, or formalin-fixed whole fish or tissues can be acceptable alternatives for some tests. Thus, a series of questions should be answered to most efficiently achieve the desired results. For example, what is the ideal condition of the sample for this assay? Can other conditions provide adequate results? Do whole fish need to be submitted or can select tissues be collected and subsequently submitted?

9.5.4 Available Resources

Assess what is required and available in order to determine what types of specimens can realistically be submitted. For example, proximity to a fish-health laboratory may determine how fish are delivered. Are hauling vehicles, transport tanks, live wells, air pumps, compressed oxygen, and/or dry ice available for use? What are the sizes and species of fish, and will they tolerate the required transit time? What collection methods are practical? Are the fish to be submitted in such poor health that they are unlikely to survive transport? Are the necessary tools, supplies, equipment, and expertise present at the aquaculture facility? What other resources are available?

9.5.5 Consultation

Prior to collection of specimens/samples, discuss all of the above with a veterinarian, fish-health professional, and/or diagnostic laboratory. Using all of the provided information, the laboratory will help determine which specimens/tissues are required, how many samples are needed, and how they should be submitted. The laboratory will also provide specific instructions, forms (Figs. 9.2 and 9.4), and a timeline. It is also crucial that specimens/tissues be clearly and accurately labeled with relevant information, including the collection date, location/source (including hatchery rearing unit designation if appropriate), fish species/strain, specimen number, sample type, and type of preservative (when applicable). If any hazardous chemicals/fixatives are included in the shipment, they must be clearly denoted and shipping requirements met.

9.5.6 Case History

Provide the fish-health professional/laboratory with a complete case history and facility observations, which should include, but is not limited to, species, size/age, water source, location description, method of collection, water temperature and other water quality parameters, environmental perturbations preceding events, other recent potential stressors, health concerns, mortality/morbidity trends, gross disease signs and behavioral changes, history of disease and treatment at the facility, vaccination history, and recent feeding patterns.

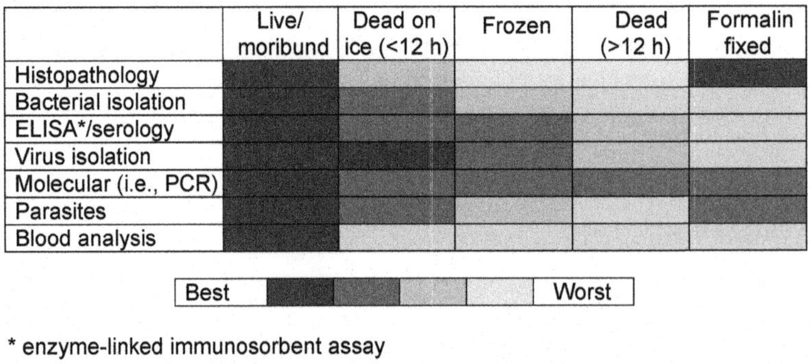

	Live/ moribund	Dead on ice (<12 h)	Frozen	Dead (>12 h)	Formalin fixed
Histopathology					
Bacterial isolation					
ELISA*/serology					
Virus isolation					
Molecular (i.e., PCR)					
Parasites					
Blood analysis					

Best — Worst

* enzyme-linked immunosorbent assay

FIGURE 9.3 General guide depicting fish/tissue status and their respective suitability for diagnostic testing.

<table>
<tr><th>Fish Species</th><th>Strain</th><th>Total # Fish Submitted</th><th>Total # in Lot</th><th>Length</th><th>Age</th><th>Sex</th><th>Lot ID</th><th>Source</th><th>Sample Identification</th><th>Tests Required</th><th># Submitted as Whole Fish</th><th># Submitted as Tissues Only: (type:______)</th><th># Submitted as Tissues Only: (type:______)</th><th># Submitted as Tissues Only: (type:______)</th></tr>
<tr><td></td><td></td><td></td><td></td><td></td><td></td><td></td><td></td><td></td><td></td><td></td><td></td><td></td><td></td><td></td></tr>
<tr><td></td><td></td><td></td><td></td><td></td><td></td><td></td><td></td><td></td><td></td><td></td><td></td><td></td><td></td><td></td></tr>
<tr><td></td><td></td><td></td><td></td><td></td><td></td><td></td><td></td><td></td><td></td><td></td><td></td><td></td><td></td><td></td></tr>
<tr><td></td><td></td><td></td><td></td><td></td><td></td><td></td><td></td><td></td><td></td><td></td><td></td><td></td><td></td><td></td></tr>
<tr><td></td><td></td><td></td><td></td><td></td><td></td><td></td><td></td><td></td><td></td><td></td><td></td><td></td><td></td><td></td></tr>
<tr><td></td><td></td><td></td><td></td><td></td><td></td><td></td><td></td><td></td><td></td><td></td><td></td><td></td><td></td><td></td></tr>
<tr><td>Total</td><td></td><td></td><td></td><td></td><td></td><td></td><td></td><td></td><td></td><td></td><td></td><td></td><td></td><td></td></tr>
<tr><td colspan="15">I certify that all fish were collected and inspected as listed and according to the current version of the Inspection Section of the AFS-FHS Blue Book or the OIE Manual and Code.</td></tr>
<tr><td colspan="4">Collected by</td><td colspan="6"></td><td colspan="2">Collector's signature</td><td colspan="3"></td></tr>
<tr><td colspan="4">Collector's phone/fax/email</td><td colspan="6"></td><td>Date of collection</td><td></td><td>Date of submission</td><td colspan="2"></td></tr>
</table>

FIGURE 9.4 Example of a form that can accompany a submission of fish samples.

9.6 FISH TRANSPORTATION

9.6.1 Shipping Live Specimens

Prior to shipping live fish, all required permits should be obtained and current. Fish should also be taken off feed 12–24h prior to transportation. Live fish should be submitted immediately upon capture whenever possible, and crowding during the capture process should be avoided. For shipping, live fish should be placed in a strong plastic bag with no more than one-third water, and the remaining two-third with compressed oxygen. The use of mucus protectants, salt, or sedatives may be useful for minimizing stress during transit, however, their use may affect laboratory test results and thus should be discussed with your fish-health professional first. Twist and fold the top of the bag, then seal tightly with a strong rubber band. Pack the bag in another bag, seal with a rubber band in the same manner, and label appropriately (see above). Place bags in a strong, styrofoam container with cardboard exterior or in an insulated cooler. For coldwater fish species, surround bags with ice packs or crushed ice in sealed plastic bags. Fish too large to ship should be transported to the lab in a vehicle equipped with live wells, transport tanks and oxygen, or in a stocking trailer. Whenever possible, the transported fish should be visually checked for signs of significant distress, and functionality of oxygen tanks and air stones confirmed. After transport, all vehicles and equipment used should be cleaned and disinfected.

9.6.2 Shipping Fresh Fish Specimens on Ice

Freshly dead fish or tissues should be collected, placed immediately on ice, and submitted to the laboratory to arrive <24h after death. Collected tissues must be placed in sterile sample containers and, depending on the purpose of submission (i.e., health certification), be harvested by an aquatic veterinarian and/or certified fish-health professional. Whole fish should be placed in strong, airtight plastic bags (five fish per bag, or as appropriate for size) without water, and labeled as described above. Sealed bags should be placed above and below crushed ice bags or ice packs. If multiple fish are collected and shipped in the same container, fish should be arranged so that only one layer of fish is in each bag and that ice surrounds each bag of fish. Fish samples and ice should be contained within another sealed plastic bag and placed in a strong insulated shipping container to prevent water from leaking as the ice melts. Fresh specimens must be submitted to the laboratory as soon as possible.

9.6.3 Frozen Specimens

Whole fish or tissues should be sealed in strong, airtight plastic bags (up to five fish per bag, or as appropriate for size) and labeled appropriately. Samples must be frozen immediately after collection and remain frozen until testing. Frozen samples should be shipped with an appropriate amount of dry ice, whereby samples should be packed at the bottom of the container with dry ice on top of them, and newspaper or packing peanuts used to fill the air space above.

Whether sending live, freshly dead, or frozen fish specimens, all arrangements for the submission of samples with the laboratory should be coordinated in advance. Shipments should be sent overnight with a tracking number and accompanied by the appropriate submission form in a waterproof bag that also is provided to the laboratory ahead of time.

9.7 VACCINATION

Vaccination of fish against major fish pathogens is one of the most effective tools for disease-prevention in aquaculture. Unlike aquatic invertebrates (e.g., shrimp and mollusks), fish possess a robust immune system that can produce a potent response when exposed to a foreign antigen (e.g., bacteria or virus). When the fish are reexposed to the same antigen, a stronger immune response develops in a much shorter time as compared to the first exposure. This immunological memory encouraged fish-health professionals to administer dead bacteria, inactivated virus, or recombinant proteins of fish pathogens (collectively known as vaccines) to fish as a method of preventing disease following subsequent pathogen exposure. If pathogens invade an effectively vaccinated fish, the secondary immune response will, in many cases, abolish the infection in its early stages.

Rather than directly antagonizing a target pathogen, vaccines harness the power of the fish's immune system to prevent disease, a strategy that has distinct advantages over the use of drugs. First, many vaccines induce long-term protection. Second, they increase aquaculture profitability by providing fixed cost disease-prevention measures that reduce and/or eliminate mortality associated with the target pathogen. Third, unlike antibiotics, most vaccines do not leave residues in the edible portions of the fish. Fourth, vaccines reduce the dependency upon antibiotics to control mortality, which not only reduces losses, but also avoids contributing to the development of antibiotic-resistant microbes. Fifth, vaccines can be developed against specific strains of a pathogen (i.e., autogenous vaccines, Section 9.7) that are perpetually problematic at a particular aquaculture facility, thereby providing targeted disease-prevention specificity. Like other strategies in aquaculture, however, vaccines alone cannot prevent all fish disease problems. But when effective vaccines are combined with strict biosecurity measures and early disease diagnosis, many health issues can be prevented or controlled. Box 9.4 lists the criteria for acceptable vaccines for use in aquacultured fish. Because the nature of vaccine preparations varies, they must be administered in accordance with the manufacturer's instructions and in consultation with a fish-health professional/aquatic veterinarian. It is also imperative that facility personnel be appropriately trained prior to undertaking the vaccination procedure so that effectiveness can be ensured.

Currently, there are over 30 licensed fish vaccines available in Europe and the USA. Vaccines made from bacteria isolated from a disease condition in the same farm, called autogenous vaccines, can be prepared by some biotechnology companies and have proven effective against common endemic infections. Vaccines can be monovalent, meaning they are effective against a single pathogen, or multivalent, meaning they contain antigens of two or more pathogens. Although most vaccines are made of dead bacteria (i.e., bacterins), inactivated virus, or recombinant proteins, in some infections of fish, the vaccine preparation must contain live/active attenuated bacteria/virus to be effective, as is the case with *Edwardsiella ictaluri*, a pathogen of the channel catfish. Table 9.5 lists some examples of the vaccines that are approved for use in US aquaculture operations.

Box 9.4 Characteristics of an Acceptable Vaccine for Use in Aquaculture

- Safe for vaccinated fish
- Effective in conferring protection against a specific disease
- Provides protection when fish are in the most susceptible stage and under intensive rearing conditions
- Provides protection of a long duration
- Protects against all indigenous strains of the same pathogen
- Easily administered
- Cost-effective

TABLE 9.5 Examples of Vaccine Preparations That Are Approved for Use in Farmed Fish in the United States

Vaccine Preparation	For Prevention of	Comments
Aeromonas salmonicida bacterin	Furunculosis in salmonids	Killed preparation; immersion delivery; ≥2 g salmonids
A. salmonicida, Vibrio anguillarum, Vibrio salmonicida, Vibrio ordalii bacterin	Furunculosis, vibriosis, and coldwater vibriosis in salmonids	Killed preparation; injection delivery; ≥10 g salmonids
Arthrobacter vaccine	Bacterial kidney disease caused by *Renibacterium salmoninarum*	Live culture preparation; injection administration; ≥10 g salmonids
Cyprinid herpesvirus 3 vaccine	Koi herpesvirus disease	Modified active virus preparation; immersion delivery
Edwardsiella ictaluri vaccine	Enteric septicemia of catfish	Avirulent live culture preparation; immersion delivery
Flavobacterium columnare bacterin	Columnaris disease in salmonids	Killed preparation; immersion delivery; ≥3 g salmonids
F. columnare vaccine	Columnaris disease in catfish and large-mouth bass	Avirulent live culture preparation; immersion delivery
Infectious salmon anemia, *A. salmonicida, V. anguillarum, V. salmonicida, V. ordalii* vaccine	Infectious salmon anemia, furunculosis, vibriosis, and coldwater vibriosis in salmonids	Inactivated/killed preparation; injection administration; ≥30 g salmonids
V. anguillarum, V. ordalii bacterin	Vibriosis in salmonids	Killed preparation; immersion delivery; ≥2 g salmonids; against *V. anguillarum* serotypes I & II
Yersinia ruckeri bacterin	Enteric redmouth disease in salmonids	Killed preparation; immersion delivery; ≥2 g salmonids; against *Y. ruckeri* serotype I

Because the nature of vaccine preparations varies, they must be administered in accordance with the manufacturer's instructions and in consultation with a fish-health professional/aquatic veterinarian. It is also imperative that facility personnel be appropriately trained prior to undertaking the vaccination procedure so that effectiveness can be ensured. For more information on approved vaccines in the United States, please see Bowker et al. (2015).

Fish vaccines are applied via a number of routes, including injection, immersion, spraying, or orally with food. Injectable vaccines are primarily for larger fish and are injected either intraperitoneally, intramuscularly, or subcutaneously. Injectable vaccines tend to elicit the highest level of protection and require relatively small amounts of vaccine. For smaller fish, vaccination by dipping fish in a solution of diluted vaccine, for seconds to minutes, elicits acceptable levels of protection, and fish are less stressed than when injected. Spraying or showering is a modification of immersion vaccination that provides a high level of protection and can accommodate a higher weight of fish/vaccine unit. Oral vaccines elicit variable degrees of protection and may require large vaccine quantities.

Prior to vaccination, fish should be taken off feed for 24–72 h (age-, species-, vaccine-, and environment-dependent). Water quality should be maintained at optimal levels during the procedure, particularly dissolved oxygen and temperature. Since the concentration of the vaccine preparation is calculated based upon fish weight, fish should be captured in a dip net or bucket and weighed. Following vaccination and regardless of the method of vaccine administration, fish should be monitored closely for signs of injury from handling. The typical feeding regimen should resume on the day following vaccination if their behavior and appearance are normal. It is recommended that fish not be placed in a situation where they will be exposed to the target pathogen(s) before maximal vaccine immunostimulation/memory has commenced, which is temperature-, fish species/age-, and time-dependent. For more on vaccination in fish, the reader is referred to a number of excellent reviews (Woo and Bruno, 2011; Gudding and Van Muiswinkel, 2013; Tafalla et al., 2013).

9.8 BIOSECURITY

Biosecurity is a cornerstone of disease prevention. The term has multiple meanings; however, in the context of aquaculture, it refers to a set of preventive measures designed to reduce the risk of introducing infectious agents into an aquaculture facility, prevent their transmission in farmed fish within the facility, and avert dissemination to neighboring facilities/the environment. Biosecurity measures should also aim to reduce conditions that increase susceptibility to infection and disease (e.g., reduce stress). The employed biosecurity measures should reflect the risk involved. For example, when there is a dangerous, endemic disease in the region that is not yet present in the facility, it does not mean the pathogen presents no risk. Disease agents and vectors may persist even when fish have been removed, thus requiring year-round biosecurity measures. In general, there are two key biosecurity principles: (1) minimizing the introduction of pathogens into the aquaculture facility, and (2) implementing the best hygienic practices, which minimize pathogen loads in the premises and interrupts the routes through which a pathogen can be transmitted within the facility itself. Although several aspects of biosecurity were covered in previous sections of this chapter, the following are integral to any biosecurity plan at the facility level.

9.8.1 Minimizing Pathogen Introductions

Precautions for minimizing pathogen introductions via water, fish/eggs, and feed have been described above. Personnel and equipment can also introduce pathogens into the facility, as well as spread them among different fish lots. It is recommended that personnel do not travel between hatcheries, but if such travel is unavoidable, personnel must not return to a clean facility after visiting a disease-suspect one. Staff members should always disinfect footwear between sites, and change into clean, dry clothing as appropriate. During hatchery cleaning/maintenance, personnel should be sure to service sections with sick fish last. Washing of hands and arms with antibacterial soap should be repeated several times a day, and some facilities also employ hand disinfectant stations throughout different hatchery sections.

For vehicles, parking lots should be on the periphery of facility grounds and access to the facility should be restricted to a minimum number of people. However, in public hatcheries, visitors are common for outreach purposes. In this case, visitors should be instructed not to touch or lean on anything in culture rooms. Areas holding critical life stages (i.e., incubation rooms, brood stock rooms) should be off-limits, as should any areas holding fish amid a disease outbreak. The importance of reducing stress, which includes that associated with increased amounts of human traffic/activity, cannot be overstated.

Disinfection and equipment allocation/separation is also integral to any biosecurity plan. Each section in the facility should have designated equipment (dip nets, buckets, tools, and feeding equipment) that should not be used in other sections. Fish-rearing units should be cleaned, disinfected, and dried between outgoing and incoming fish groups. Tanks should be regularly cleaned to prevent the buildup of residual feed, organic matter, and algal growth, all of which can lead to the proliferation of pathogenic microbes. High pressure water may be insufficient in removing all biological matter and thus may require manual scrubbing. On the rare occasion when equipment must be shared with other sites, it must be cleaned, disinfected, and dried prior to leaving the site and prior to returning to the originating site. Vehicles and vessels used between multiple sites should be cleaned and disinfected prior to use at another site.

Effective methods of disinfection include "fallowing," drying, ultraviolet exposure, and the use of chemical disinfectants, but in general, most items to be disinfected should first be thoroughly cleaned, as organic buildup decreases the efficacy of many, but not all, compounds. There are many different compounds that can be used as aquaculture disinfectants and are available commercially, a portion of which are listed in Table 9.1. It must be noted, however, that some disinfectants are effective for some applications, but not others. For instance, chlorine derivatives are very effective for disinfecting rearing unit equipment and footwear, but are highly toxic to fish and thus cannot be used as an egg disinfectant. On the contrary, povidone iodine is a widely used egg-surface disinfectant, but can stain clothing and equipment. Thus, in general, the choice of disinfectant for each application should be based upon the desired degree of microbial killing, the nature of the surfaces involved (i.e., rubber, stainless steel, concrete), required contact time/feasibility, cost, safety, and ease of use. Disinfectants should always be used according to the manufacturer's instructions. For more on specific disinfectants that can be used in aquaculture situations, the reader is referred to Chapter 1.1.3 on disinfection in the *OIE Diagnostic Manual* (OIE, 2009), as well as the review on aquaculture disinfection in general (Danner and Merrill, 2006) and for specific pathogens of relevance to aquaculture (Woo and Bruno, 2011).

During disinfection, the disinfectant concentration should be monitored and maintained either via testing (e.g., using test strips), by visual checks for color change that indicates decreased disinfection efficacy (e.g., a change from brown to yellow or clear for iodine, loss of original color for Virkon), or through regular and timely renewal of the product. Previously cleaned equipment can be disinfected by either immersion in a disinfectant bath or sprayed down if too large, but attention should be paid to the minimum contact time for the utilized compound. Disinfection of footwear by foot bath and/or foot mat at the entrance(s) of the facility sections should be strictly implemented and disinfectant solution regularly maintained according to manufacturer instructions.

9.8.2 Hygienic Practices to Minimize Pathogen Loads and Interrupt Their Transmission

There are a number of practices for minimizing pathogen loads within a facility. Among these, regular removal of dead fish minimizes pathogen shedding into water, as does the culling of overtly moribund fish. The removal of dead and dying fish as frequently as possible becomes even more imperative during disease events, which disrupts horizontal transmission among cohorts. Mortality curves (Fig. 9.5) should be recorded daily, as they are important indicators of imminent disease events and aid in disease diagnosis. If daily mortality/morbidity exceeds 0.05% for a particular rearing unit or fish group, a fish-health professional/aquatic veterinarian should be contacted and diagnostic submission arranged. All fish-health data should be recorded (Box 9.5) and kept in the facility archives for future use. The regular removal of uneaten feed, feces, and organic films is also an effective method of reducing microbial proliferation/loads. Likewise, ultraviolet treatment of incoming water (see above), as well as water that is to be reused among rearing units (i.e., between first and second pass water) is also effective.

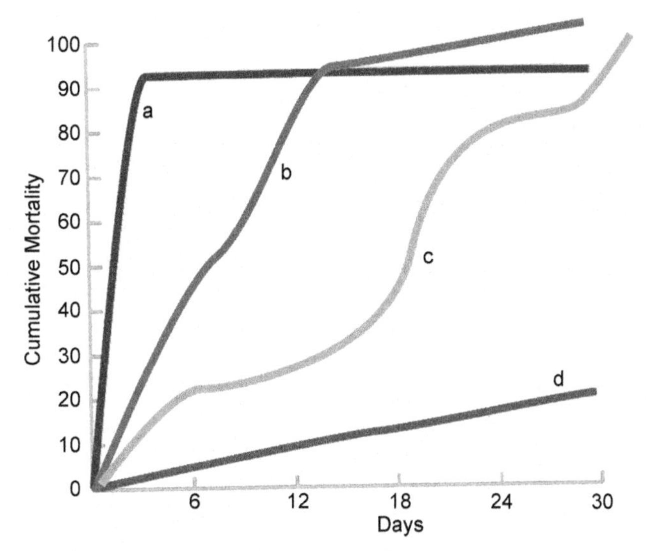

FIGURE 9.5 Examples of mortality curves in cultured finfish. (a) Peracute mortality, such as what would be experienced when fish are exposed to toxic chemicals or hypoxic conditions; (b) an acute disease course, such as those caused by some virulent viruses and bacteria; (c) recurrent curve indicating a subacute viral or bacterial infection and the pathogen is being shed by survivors, causing new infections; and (d) lingering mortality due to a chronic infection.

Box 9.5 Fish Health Records Should Contain, But Not Be Limited to, the Following:

- An inventory of all fish lots in the facility
- Egg source
- Water quality data
- Growth curves
- Daily feed consumption, growth rate, and feeding behavior
- Records of vaccination
- Records of previous treatment
- Response to treatment
- Mortality curves
- Clinical signs observed
- Number of culled fish
- Previous disease history

Daily fish culture activities can also minimize pathogen spread. For example, reasonable rearing densities should be maintained, as overcrowding not only directly stresses most fish, but also leads to suboptimal water quality, more efficient pathogen transmission, and a weakened immune system. Fish in their early life stages should be cared for before older fish, which reduces the likelihood of pathogen transfer from older fish to their younger, more susceptible counterparts (see segregation, Section 9.4.4 above). Fish that jump from tanks to the floor should be humanely euthanized and not returned to the tank.

9.9 CONTROL MEASURES UPON ERUPTION OF A DISEASE OUTBREAK

A primary constraint on aquaculture expansion is infectious disease. Many aquaculturists have experienced fish disease outbreaks despite the concerted implementation of prophylactic measures. Although observation of early disease signs and rapid detection are keys for managing this risk, there are additional steps to follow so that disease outbreaks can be effectively managed. Efforts should be directed in at least two primary directions; first, confine the pathogen to the currently affected ponds/tanks/sections; and second, minimize the pathogen load to levels that will allow as many fish as possible to survive. Means to these ends are listed in Box 9.6.

The options for managing a disease outbreak can indeed be limited. For instance, if the disease is caused by a virulent pathogen that causes wide-scale mortality and/or is of reportable significance, total depopulation of the facility may be the only option. This may also be required in the event that surviving fish suffer from indefinite growth rate impairment. Another option may be chemotherapeutic treatment with drugs, such as antibiotics. Although some antibiotics and drugs have historically been used for prophylactic purposes and/or enhanced production, there is a general trend in aquaculture (and especially the United States) to limit antibiotic use for treatment purposes only. Although drugs can be administered in several ways and under a variety of conditions, it is imperative to note that they are only approved for the uses under which they are labeled, unless prescribed under extra-label use via a licensed veterinarian or under an Investigational New Animal Drug (INAD) study protocol. Nevertheless, many of these drugs have the advantage of being able to be used under multiple scenarios/conditions. In addition, if drugs are administered properly and given in the right dose, most are effective

Box 9.6 Key Points to Consider When Managing a Disease Outbreak in Aquaculture

- If a reportable agent is suspected, notify proper authorities.
- Affected fish should be isolated/quarantined to the best degree possible until the disease has been diagnosed and managed.
- Remove dead and dying fish as frequently as possible.
- Minimize stress to affected and neighboring groups of fish.
- Movement of all fish within the facility should be suspended.
- Equipment should not move on or off site without special arrangements.
- Biosecurity measures should be strictly enforced, including disinfection of potentially contaminated materials/equipment.
- Appropriate samples should be collected/submitted to a diagnostic laboratory.
- Fish-health records, including mortality curves, should be provided to the investigating fish-health professional.
- At least until disease resolution, special efforts to maintain hygiene and peak water quality for the affected species should be made.
- Feed amount should be adjusted to the fish appetite to avoid accumulation.
- Clean affected rearing units last.

in curing the disease, minimizing mortality, and preventing further infection in the short term. Unfortunately, there is a reticence among many pharmaceutical companies to invest in drug development for aquaculture because the approval process is strenuous, time-consuming, expensive, and accompanied by an uncertain outcome. Since drug approval status varies among countries, regulatory agencies also limit a number of drugs from being used.

On the contrary, there are multiple disadvantages for the use of drugs, such as their short-term effect, meaning reinfection can happen following the cessation of drug application. This is especially true if pathogen loads are not reduced and any predisposing environmental conditions not rectified. In the case of antibiotics, another potential disadvantage is the effect that treatments may have on the beneficial fish-associated microbes, whose role in the development of healthy host immune systems and overall health/growth are just becoming better understood. Another disadvantage is that bacteria may develop resistance to antibiotics over time, especially if they are over-utilized, administered at an incorrect dose, or for an incomplete treatment duration. It is therefore of paramount importance that an in vitro antibiotic sensitivity test be performed to determine the most effective antibiotic to use prior to treatment recommendation and application. In the United States, the number of drugs that are approved/conditionally approved for use in farmed finfish is limited (Table 9.6), but nevertheless

TABLE 9.6 Examples of Compounds That Are Approved/Conditionally Approved to Treat Parasitic Infestations and/or External Bacterial Infections in Aquaculture

	Compound	Indication(s)[a]	Comments
Approved and conditionally approved compounds for use in aquaculture	Florfenicol (50% active ingredient)	Furunculosis caused by *Aeromonas salmonicida* subsp. *salmonicida*, in freshwater-reared salmonids; bacterial coldwater disease, caused by *Flavobacterium psychrophilum*, in freshwater-reared salmonids; enteric septicemia, due to *Edwardsiella ictaluri*, in catfish; columnaris disease, caused by *F. columnare*, in freshwater-reared finfish; streptococcal septicemia, caused by *Streptococcus iniae*, in warmwater finfish	Oral administration via food; requires a veterinary feed directive (VFD); 15 days withdrawal period; also use under INAD for other indications[b]
	Sulfadimethoxine and ormetoprim	Furunculosis in salmonids; enteric septicemia in catfish	Oral administration via food; requires a VFD as of January 1, 2017; 3 (catfish) to 42 (salmonids) day withdrawal time
	Oxytetracycline dihydrate	Furunculosis, ulcer disease, bacterial hemorrhagic septicemia, and *pseudomonas* disease in salmonids; bacterial coldwater disease in freshwater-reared salmonids; bacterial hemorrhagic septicemia and *pseudomonas* disease in catfish; columnaris disease in freshwater-reared rainbow trout	Oral administration via food; requires a VFD as of January 1, 2017. 21 days withdrawal time; also use under INAD for other indications[b]
	Chloramine-T	Bacterial gill disease in freshwater-reared salmonids; external columnaris disease in walleye and warmwater finfish	Bath immersion administration; 0 day withdrawal time; also use under INAD for other indications[b]
	Formalin (10%)	External protozoa and monogeneans in all finfish; saprolegniasis in finfish eggs	Bath immersion (fish) or flow-through immersion (eggs); 0 day withdrawal time; dose and duration species and temperature- dependent
	Hydrogen peroxide (35%)	Bacterial gill disease in salmonids; columnaris disease in channel catfish and coolwater finfish; saprolegniasis in all freshwater-reared finfish eggs	Bath immersion (fish) or flow-through immersion (eggs); 0 day withdrawal time; also use under INAD for other indications[b]

Continued

TABLE 9.6 Examples of Compounds That Are Approved/Conditionally Approved to Treat Parasitic Infestations and/or External Bacterial Infections in Aquaculture—cont'd

	Compound	Indication(s)[a]	Comments
Investigational new animal drug (INAD) status[b]	Emamectin benzoate	External parasites (copepods) in freshwater fish	Oral via food; 60 days withdrawal time
	Oxytetracycline hydrochloride	Furunculosis, bacterial hemorrhagic septicemia, enteric redmouth caused by *Yersinia ruckeri*, flexibacteriosis, and vibriosis in salmonids; enteric septicemia in catfish; bacterial hemorrhagic septicemia, *pseudomonas* disease, and flexibacteriosis in cool and warmwater fish	Bath immersion; 21–60 days withdrawal time (species dependent)
	Diquat dibromide (37.3%)	Bacterial gill disease and external columnaris, and external flavobacteriosis in freshwater fish	Bath immersion; 5–30 days withdrawal time (species dependent)
	Potassium permanganate	External protozoa and metazoan, bacterial and fungal infections in warmwater fish	Bath immersion
	Copper sulfate	External protozoa and metazoan, bacterial and fungal infections in warmwater fish	Bath immersion; 7 days withdrawal time; water quality parameters should be checked prior to use[b]; kills aquatic invertebrates, plants, and algae
	Erythromycin	Bacterial kidney disease caused by *Renibacterium salmoninarum*; prevention of *R. salmoninarum* transmission in eggs	Injection (brood stock for prevention of vertical transmission) and oral via feed; 60 days withdrawal time
LRP	Acetic acid	External parasites in fish	Dip or bath immersion; see relevant regulations for approved use and withdrawal times
	Calcium oxide		
	Magnesium sulfate		
	Sodium chloride		
	Povidone iodine	Egg-surface disinfectant	Bath immersion

Those that are under an investigational new animal drug (INAD) status, or those that are of low regulatory priority (LRP) in the United States are current as of the time this chapter was written. It is incumbent upon the administering individual to ensure that such treatments are in compliance with local, regional, and/or national regulations and that they are conducted according to the label instructions. Note that it is also advisable to first perform a bioassay on a small number of fish prior to wide-scale application.
[a]*All treatments must be conducted in consultation with an aquatic veterinarian/fish-health professional.*
[b]*See Bowker et al. (2015) and the US Food and Drug Administration Center for Veterinary Medicine (FDA-CVM) Website (http://www.fda.gov/AnimalVeterinary/DevelopmentApprovalProcess/Aquaculture/ucm132954.htm) for more information, restrictions, and protocols for uses under INAD.*

has grown slightly over the years. There are also multiple compounds that can be used in the United States under an INAD status, or are of low regulatory priority according to the US Food and Drug Administration Center for Veterinary Medicine (FDA-CVM; Table 9.6). In all cases, a fish-health professional/aquatic veterinarian should be consulted prior to any treatment application. It is incumbent upon the administering individual to ensure that such treatments are in compliance with local, regional, and/or national regulations and that they are conducted according to the label instructions. For more information on approved treatments, restrictions, and INAD protocols, the reader is referred to the FDA-CVM Website (http://www.fda.gov/AnimalVeterinary/DevelopmentApprovalProcess/Aquaculture/ucm132954.htm), and Bowker et al. (2015).

Before a treatment is applied, some important aspects must be evaluated, some of which relate to water quality and others relate to the species to be treated, the chemical nature of the drug, and the causative agent. For example, total alkalinity and hardness, pH, organic load, and temperature of the water should be known, as these factors influence the efficacy and toxicity of some drugs. Another uncertainty is that chemical toxicity of a drug varies among fish species, strains within species, and between fish age groups. Therefore, it is recommended to conduct a bioassay on a small group of fish prior to widespread application. In many cases, the accuracy of diagnosis can determine the outcome of treatment. For example, infections involving more than one pathogen often occur in fish (Loch et al., 2012) and the drug to be used may not be effective against all pathogens. Similarly, the pathogen to be treated must be susceptible to the drug of choice (see above).

Once a treatment has begun, fish should be monitored closely for any adverse effects. Gathering at the inflow, gasping at the surface, attempts to jump out of the water, etc. may be indicative of treatment toxicity, especially if such signs persist or worsen. Thus, it may become necessary to immediately stop the treatment and, in flow-through systems, increase the water flow to dilute and flush out the offending chemical (or perform water changes in recirculation systems). After completion of a treatment course, fish are not suitable for human consumption until after the withdrawal time of the drug has passed (see Table 9.6 for examples), which ensures that no drug residues remain in the edible fish tissues.

9.10 FORMULATING A FISH-HEALTH MANAGEMENT PLAN AT THE REGIONAL/NATIONAL AND INTERNATIONAL LEVELS

A fish-health management plan at the facility level is invaluable, but indeed may not be adequate for preventing the spread of pathogens to or from wider geographic locations. Therefore, policies and regulations at the regional/national and international levels must be implemented. Such wide-scale plans require a proper understanding of the fish disease ecology but unfortunately, many aspects remain partially or fully unknown. For example, baseline data on the distribution of nearly all fish pathogens, both nation- and worldwide, are either nonexistent or relatively primitive. This lack of knowledge impedes accurate disease-risk analysis, increases the difficulty of differentiating between exotic and endemic infections, and hinders the selection of disease-management options. This problem is further complicated by the lack of sensitive diagnostic tools that can detect subclinical carriers for many infectious agents of fish. Although the field of fish medicine is somewhat in its infancy, it is faced with challenges of unprecedented magnitude.

One such challenge is the increased trade and translocation of fish and their products all around the world. With the continuous expansion of aquaculture and species translocations (Coward and Little, 2001; Moffitt, 2005), many harmful fish parasites and pathogens have spread, not only to farmed fish in other facilities (Bartley and Subasinghe, 1996; Naylor et al., 2000; Murray and Peeler, 2005), but also to wild fish populations through water or escapees (Gagen and Pitts, 1990; Blanc, 1997), thereby causing losses in neighboring wild stocks. For example, the vertical transmission of *R. salmoninarum* within salmonid eggs, coupled with the ease of egg transport, contributed to the global spread of BKD among salmonid populations, farmed and wild alike. Likewise, the transmission of *F. psychrophilum*, causative agent of bacterial coldwater disease, along with eggs is believed to be one of the primary drivers of its transcontinental dissemination to almost all areas where salmonids are intensively aquacultured (Van Vliet et al., 2016, and references therein). Similarly, the importation of fish into the Asian-Pacific region for aquaculture purposes from multiple sources resulted in the introduction of multiple fish pathogens/diseases, including the copepod *Lernaea cyprinacea*, myxosporeans of the genus *Myxobolus*, and the epizootic ulcerative syndrome, which have now spread throughout much of South and Southeast Asia and resulted in tremendous financial losses (Tonguthai, 1985; Arthur and Shariff, 1991; Lilley et al., 1992). Even the trade of frozen fish or fish products has resulted in fish pathogen translocations, such as the suspected introduction of *M. cerebralis* into the United States along with the importation of fish fillets from Denmark (Marnell, 1986). Live bait (e.g., minnows, frogs, worms, and squid) is routinely used by recreational and commercial anglers to catch fish, but in some regions, no health certification is required for its transfer from one area to another. This practice has the potential to spread pathogens into noninfected geographical areas (Hoole et al., 2001; Gaughan, 2001; Miller et al., 2011).

9.10.1 National, Regional, and International Efforts to Control Pathogen Transfer

In response to the aforementioned challenges, fishery organizations at all levels joined forces to significantly lower the risk of pathogen introduction without impeding the booming internal trade and aquaculture expansion. Four primary principles were identified to aid in resolving this issue:

- enforcing appropriate and effective national, regional, and interregional policies and regulatory frameworks that focus on the introduction and movement of aquatic animals and their products;
- implementing national disease reporting systems;
- developing sensitive diagnostic methods and safe therapeutants; and
- promoting a holistic system approach to fish-health management, emphasizing preventative measures and the maintenance a healthy ecosystem.

These principles have guided several nations in the development of national health plans. The Diseases of Fish Act, issued in 1937 in Great Britain, is the longest-standing example of national legislation specifically devised to control fish diseases (Hill, 1996). The Act prohibited importation of live salmonids into Great Britain, and made it illegal to import salmonid ova and all live freshwater fish species without a license. Moreover, the Act enabled any disease to be designated as "notifiable," meaning that even the suspicion of its presence in any waters must be reported to the official services. Although these principles remain at the core of most national policies, other countries amended and expanded their policies as needed to face emerging challenges (Brückner, 1996; Campos and Valenzuela, 1996; Carey, 1996; Doyle et al., 1996; Hill, 1996; Schlotfeldt, 1996).

At the regional level, two health plans emerged and exhibited positive impacts on the health of fish in their respective regions. One such plan was developed by the European Union (Daelman, 1996), whereas the second was developed by regulatory fishery agencies of the United States and Canadian provinces bordering the Laurentian Great Lakes of North America (Hnath, 1993). This regional approach significantly extended the scope of aquatic animal health legislation to include recommended directives and decisions that do not impede the movement of live fish and their products, provided that a high level of animal health is guaranteed. Both national and regional plans now include lists of notifiable diseases, a system of certification, and descriptions of reliable protocols for use in laboratory diagnostic testing. An excellent example of such coordination at the national level includes the National Animal Health Laboratory Network (NAHLN), which is a cooperative effort between the United States Department of Agriculture—Animal Plant Health Inspection Service (USDA—APHIS), the USDA National Institute of Food and Agriculture (USDA—NIFA), the American Association of Veterinary Laboratory Diagnosticians (AAVLD), the USDA National Veterinary Services Laboratories (NVSL), and State/University laboratories. This partnership between State and Federal agencies allows for an efficient means of coping with and tracking emerging animal disease outbreaks (https://www.aphis.usda.gov/aphis/ourfocus/animalhealth/lab-info-services/nahln/).

At the international level, a number of international organizations and agreements have also been developed. For example, standards for the prevention of pathogen introductions through live fish trade have been developed, such as the Code of Practice for Introduction and Transfer of Marine Organisms by the International Council for the Exploration of the Sea (ICES) and the North American Free Trade Agreement (NAFTA). The most comprehensive international fish-health plan has been developed and implemented by the United Nations World Organization of Animal Health (OIE). The OIE-Aquatic Animal Health Standards Commission (AAHSC) crafted two seminal documents that are regularly updated: the *International Aquatic Animal Health Code* (OIE, 2016a) and the *Diagnostic Manual for Aquatic Animal Diseases* (OIE, 2016b). Information within these documents serves as a guideline for the preparation of regulations on importing and exporting live animals and products, including protocols for appropriate and standardized assays for pathogen detection.

The *OIE Aquatic Code* (OIE, 2016a) currently lists the following pathogens as notifiable:

- Epizootic hematopoietic necrosis virus
- *Aphanomyces invadans* (Epizootic ulcerative syndrome)
- *Gyrodactylus salaris*
- Infectious salmon anemia virus
- Salmonid alpha virus
- Infectious hematopoietic necrosis
- Koi herpes virus
- Red sea bream irido virus
- Spring viremia of carp
- Viral hemorrhagic septicemia virus
- *Oncorhynchus masou* virus
- Nodaviruses causing viral encephalopathy and retinopathy

The OIE obligates its member countries to report if one of these pathogens is suspected or confirmed within its territory within 24h. However, notification is also required for other significant diseases of aquatic animals of which the international community should be made aware. Moreover, OIE has designated a number of laboratories and experts as reference laboratories for notifiable diseases, with the goal of standardizing diagnostic techniques for the designated diseases and providing technical assistance and advice as needed. Reporting is required not only when an infectious disease occurs, but also when the disease is eradicated.

9.10.2 Components of a Fish-Health Plan at the National and International Levels

The cornerstone of national and international fish-health plans is to establish an accurate disease surveillance system, create a system of certification, establish disease-free zones, carry out risk assessments for pathogen introductions associated with the trade, and establish quarantine plans for introduced species before permitting access into the importing country.

9.10.3 Disease Surveillance

The OIE code defines "surveillance" as "a systematic series of investigations of a given population of aquatic animals to detect the occurrence of disease for control purposes, and which may involve testing samples of a population." "Monitoring" is defined as "an on-going program directed at the detection of changes in the prevalence of disease in a given population and in its environment." The term "surveillance program" encompasses both surveillance and monitoring tasks. Surveillance and monitoring serve multiple purposes, such as providing an early warning of serious and emerging disease outbreaks; providing evidence of freedom from diseases; revealing the potential source and subsequent spread of pathogens; and most importantly, assessing the efficacy of control or eradication measures for a particular disease or diseases.

9.10.4 Certification

An official health certificate must accompany each shipment of live fish. Diagnosis and official certification should be based upon the standards, guidelines, and recommendations of the *OIE Manual of Diagnostic Tests for Aquatic Animals* (OIE, 2016b).

9.10.5 Zoning

Disease zoning aims to divide a geographical location into two main zones for a specific disease: infected and uninfected (Murray, 2002). An uninfected zone can be established for a specific disease within a particular geographic or hydrogeographic area within a country. In the aquatic environment, catchment areas and rivers may be used to define continental zones. The simplest freshwater zonation system is to designate an entire river system or water catchment area as a zone. Ideally, a free zone should be separated from a neighboring infected zone by a so-called buffer zone, where surveillance is constantly performed. In an infected zone, no live aquatic animals should be permitted to leave the zone unless for immediate slaughter.

9.10.6 Risk Assessment

When a proposal to introduce an exotic fish species is made, a risk assessment for the possibility of pathogen introduction into the importing country must be conducted prior to the arrival of any fish. The analysis should be conducted by the importing country. A sequence of events has been proposed by OIE and ICES that involves the following:

- Development of a proposal by the entity moving an exotic species. The proposal should include the location of the facility, planned use, and source of the exotic species.
- Review and evaluate the risk for pathogen introductions, along with the potential for the pathogen(s) to become established within the importing location should they be introduced. The review should be conducted by an independent panel of experts.
- The independent experts then advise whether the proposal should be accepted, refined, or rejected.
- If approval to introduce a species is granted, then quarantine, containment, monitoring, and reporting systems should be implemented.

9.10.7 Quarantine

Quarantine denotes retaining animals in facilities designed specifically to prevent the release of animals, or the pathogens they may carry, into regions where these animals or their pathogens do not exist (Doyle et al., 1996). The ICES developed codes linking exotic species introduction to the availability of a sound quarantine system in the importing country (Bartley et al., 1997; ICES, 1998). In these codes, the following has been suggested:

- Live fish should not be imported, but rather eggs.
- The eggs should be delivered directly to the quarantine facility.
- A brood stock should be developed from these eggs.
- Fish produced from the eggs, and the brood stock, should be examined for the presence of pathogens following the protocols detailed in the *OIE Diagnostic Manual* (OIE, 2016b).
- If no pathogens become evident, then the first generation of progeny (F1), but not the original imports, can be released to culture sites.
- No more fish of this species should be imported.
- Routine disease testing should be conducted.

9.11 CONCLUSION

The inhospitable environments that are associated with some aquatic habitats, along with the artificial and potentially stressful conditions that accompany some aquaculture operations, place the health of farmed fish in a precarious position, especially if an adequate fish-health management plan is not developed, implemented, and maintained. Likewise, if such a plan is improperly designed or incompletely implemented, the risk for infectious and noninfectious disease outbreaks increases and frequently also translates into poor growth, suboptimal yields, and diminishing economic returns—a matter that can "make or break" the viability of an aquaculture operation. However, with foresight, adequate planning, and collaboration with fish-health professionals/aquatic veterinarians, facility personnel can provide optimal fish-rearing conditions, avoid pathogen introductions, prevent disease outbreaks, quickly and efficiently address and manage disease issues should they arise, and prevent pathogen spread within and among the facility and its surrounding environment, all of which contribute significantly to successful fish farming and help ensure that wild aquatic species and their habitats thrive alongside them.

REFERENCES

Anderson, D.P., 1992. Immunostimulants, adjuvants, and vaccine carriers in fish: applications to aquaculture. Annual Review of Fish Diseases 2, 281–307.

Arthur, J., Shariff, M., 1991. Towards International Fish Disease Control in Southeast Asia. Infofish Intern, Kuala Lumpur.

Bagheri, T., Hedayati, S.A., Yavari, V., Alizade, M., Farzanfar, A., 2008. Growth, survival and gut microbial load of rainbow trout (*Onchorhynchus mykiss*) fry given diet supplemented with probiotic during the two months of first feeding. Turkish Journal of Fisheries and Aquatic Sciences 8, 43–48.

Barman, D., Nen, P., Mandal, S.C., Kumar, V., 2013. Immunostimulants for aquaculture health management. Journal of Marine Science: Research and Development.

Bartley, D., Subasinghe, R., Coates, D., 1997. Draft Framework for the Responsible Use of Introduced Species. FAO. Fisheries Report (FAO).

Bartley, D.M., Subasinghe, R.P., 1996. Historical aspects of international movement of living aquatic species. Revue scientifique et technique (International Office of Epizootics) 15, 387–400.

Beck, B.R., Kim, D., Jeon, J., Lee, S.-M., Kim, H.K., et al., 2015. The effects of combined dietary probiotics *Lactococcus lactis* BFE920 and *Lactobacillus plantarum* FGL0001 on innate immunity and disease resistance in olive flounder (*Paralichthys olivaceus*). Fish and Shellfish Immunology 42, 177–183.

Blanc, G., 1997. L'introduction des agents pathogènes dans les écosystèmes aquatiques: aspects théoriques et réalités. Bulletin Français de la Pêche et de la Piscicultur 489–513.

Bowker, J., Trushenski, J., Gaikowski, M., Straus, D., 2015. Guide to using drugs, biologics, and other chemicals in aquaculture. American Fisheries Society Fish Culture Section.

Brown, L., Cox, W., Levine, R., 1997. Evidence that the causal agent of bacterial cold-water disease *Flavobacterium psychrophilum* is transmitted within salmonid eggs. Diseases of Aquatic Organisms 29, 213–218.

Brückner, G., 1996. Review of disease control in aquaculture in the Republic of South Africa. Revue scientifique et technique (International Office of Epizootics) 15, 703–710.

Bruno, D.W., Van West, P., Beakes, G.W., 2011. *Saprolegnia* and other oomycetes. In: Woo, P.T.K., Bruno, D.W. (Eds.), Fish Diseases and Disorders. CABI Publishing.

Bullock, G.L., Leek, S.L., 1986. Use of erythromycin in reducing vertical transmission of bacterial kidney disease. Veterinary and Human Toxicology 28 (Suppl. 1), 18–20.

Campos, L.M., Valenzuela, A.M., 1996. Chilean legislation for the control of diseases of aquatic species. Revue scientifique et technique (International Office of Epizootics) 15, 675–686.

Carey, T., 1996. Finfish health protection regulations in Canada. Revue scientifique et technique (International Office of Epizootics) 15, 647–658.

Cha, J.-H., Rahimnejad, S., Yang, S.-Y., Kim, K.-W., Lee, K.-J., 2013. Evaluations of *Bacillus* spp. as dietary additives on growth performance, innate immunity and disease resistance of olive flounder (*Paralichthys olivaceus*) against *Streptococcus iniae* and as water additives. Aquaculture 402, 50–57.

Chen, D., Ainsworth, A.J., 1992. Glucan administration potentiates immune defence mechanisms of channel catfish, *Ictalurus punctatus* Rafinesque. Journal of Fish Diseases 15, 295–304.

Coward, K., Little, D., 2001. Culture of the aquatic chicken: present concerns and future prospects. Biologist (London, England) 48, 12–16.

Daelman, W., 1996. Animal health and the trade in aquatic animals within and to the European Union. Revue scientifique et technique (International Office of Epizootics) 15, 711–722.

Danner, G.R., Merrill, P., 2006. Chapter 8: Disinfectants, disinfection, and biosecurity in aquaculture. In: Scarfe, A.D., Lee, C., O'bryen, P.J. (Eds.), Aquaculture Biosecurity: Prevention, Control, and Eradication of Aquatic Animal Disease. Blackwell Publishing, Aimes, Iowa.

Doyle, K., Beers, P., Wilson, D., 1996. Quarantine of aquatic animals in Australia. Revue scientifique et technique (International Office of Epizootics) 15, 659–673.

Faisal, M., Elsayed, E., Fitzgerald, S.D., Silva, V., Mendoza, L., 2007. Outbreaks of phaeohyphomycosis in the chinook salmon (*Oncorhynchus tshawytscha*) caused by *Phoma herbarum*. Mycopathologia 163, 41–48.

Faisal, M., Schulz, C., Eissa, A., Brenden, T., Winters, A., et al., 2012. Epidemiological investigation of *Renibacterium salmoninarum* in three *Oncorhynchus* spp. in Michigan from 2001 to 2010. Preventive Veterinary Medicine 107, 260–274.

Faramarzi, M., Kiaalvandi, S., Iranshahi, F., 2011. The effect of probiotics on growth performance and body composition of common carp (*Cyprinus carpio*). Journal of Animal and Veterinary Advances 10, 2408–2413.

Fuller, R., 1989. A review of probiotics in man and animals. Journal of Applied Bacteriology 66, 365–378.

Gagen, C.J., Pitts, J.L., 1990. Fish, money, and science in puget sound. Science 248, 290–291.

Gannam, A.L., Schrock, R.M., 1999. Immunostimulants in fish diets. Journal of Applied Aquaculture 9, 53–89.

Gaughan, D.J., 2001. Disease-translocation across geographic boundaries must be recognized as a risk even in the absence of disease identification: the case with Australian *Sardinops*. Reviews in Fish Biology and Fisheries 11, 113–123.

Gudding, R., Van Muiswinkel, W.B., 2013. A history of fish vaccination: science-based disease prevention in aquaculture. Fish and Shellfish Immunology 35, 1683–1688.

Hai, N., 2015. The use of probiotics in aquaculture. Journal of Applied Microbiology 119, 917–935.

Harms, C.A., 2005. Surgery in fish research: common procedures and postoperative care. Lab Animal 34, 28–34.

Hill, B., 1996. National legislation in Great Britain for the control of fish diseases. Revue scientifique et technique (International Office of Epizootics) 15, 633–645.

Hnath, J.G., 1993. Great Lakes Fish Disease Control Policy and Model Program. Great Lakes Fishery Commission, Ann Arbor, MI.

Hong, H.A., Cutting, S.M., 2005. The use of bacterial spore formers as probiotics. FEMS Microbiology Reviews 29, 813–835.

Hoole, D., Bucke, D., Burgess, P., Wellby, I., 2001. Diseases of Carp and Other Cyprinid Fishes. Wiley Online Library, London.

ICES, 1998. ICES Code of Practice on the Introductions and Transfers of Marine Organisms. International Council for the Exploration of the Sea, Copenhagen, Denmark.

Ispir, U., 2009. Prophylactic effect of levamisole on rainbow trout (*Oncorhynchus mykiss*) against *Yersinia ruckeri*. Pesquisa Veterinária Brasileira 29, 700–702.

Jha, D.K., Bhujel, R.C., Anal, A.K., 2015. Dietary supplementation of probiotics improves survival and growth of Rohu (*Labeo rohita* Ham.) hatchlings and fry in outdoor tanks. Aquaculture 435, 475–479.

Kim, R., Faisal, M., 2010. Comparative susceptibility of representative Great Lakes fish species to the North American viral hemorrhagic septicemia virus sublineage IVb. Diseases of Aquatic Organisms 91, 23–34.

Korkea-Aho, T.L., Papadopoulou, A., Heikkinen, J., Von Wright, A., Adams, A., et al., 2012. *Pseudomonas* M162 confers protection against rainbow trout fry syndrome by stimulating immunity. Journal of Applied Microbiology 113, 24–35.

Lara-Flores, M., Olvera-Novoa, M.A., Guzmán-Méndez, B.Z.E., López-Madrid, W., 2003. Use of the bacteria *Streptococcus faecium* and *Lactobacillus acidophilus*, and the yeast *Saccharomyces cerevisiae* as growth promoters in Nile tilapia (*Oreochromis niloticus*). Aquaculture 216, 193–201.

Lilley, J.H., Phillips, M.J., Tonguthai, K., 1992. A Review of Epizootic Ulcerative Syndrome (EUS) in Asia. Aquatic Animal Health Research Institute, Bangkok, Thailand.

Liltved, H., Landfald, B., 2000. Effects of high intensity light on ultraviolet-irradiated and non-irradiated fish pathogenic bacteria. Water Research 34, 481–486.

Liu, W., Ren, P., He, S., Xu, L., Yang, Y., et al., 2013. Comparison of adhesive gut bacteria composition, immunity, and disease resistance in juvenile hybrid tilapia fed two different *Lactobacillus* strains. Fish and Shellfish Immunology 35, 54–62.

Loch, T.P., Faisal, M., 2016a. Flavobacteria isolated from the milt of feral Chinook salmon of the Great Lakes. North American Journal of Aquaculture 78, 25–33.

Loch, T.P., Faisal, M., 2016b. Gamete-associated flavobacteria of the oviparous Chinook salmon (*Oncorhynchus tshawytscha*) in lakes Michigan and Huron, North America. Journal of Microbiology 54, 477–486.

Loch, T.P., Scribner, K., Tempelman, R., Whelan, G., Faisal, M., 2012. Bacterial infections of Chinook salmon, *Oncorhynchus tshawytscha* (Walbaum), returning to gamete collecting weirs in Michigan. Journal of Fish Diseases 35, 39–50.

Ma, C.-W., Cho, Y.-S., Oh, K.-H., 2009. Removal of pathogenic bacteria and nitrogens by *Lactobacillus* spp. JK-8 and JK-11. Aquaculture 287, 266–270.

Mahdhi, A., Kamoun, F., Messina, C., Bakhrouf, A., 2012. Probiotic properties of *Brevibacillus brevis* and its influence on sea bass (*Dicentrarchus labrax*) larval rearing. African Journal of Microbiology Research 6, 6487–6495.

Marnell, L.F., 1986. Impacts of Hatchery Stocks on Wild Fish Populations. Fish Culture in Fisheries Management. American Fisheries Society, Fish Culture Section and Fisheries Management Section, Bethesda, Maryland.

Merrifield, D.L., Dimitroglou, A., Foey, A., Davies, S.J., Baker, R.T., et al., 2010. The current status and future focus of probiotic and prebiotic applications for salmonids. Aquaculture 302, 1–18.

Meurer, F., Hayashi, C., Costa, M.M.D., Mauerwerk, V.L., Freccia, A., 2006. *Saccharomyces cerevisiae* as probiotic for Nile tilapia during the sexual reversion phase under a sanitary challenge. Revista Brasileira de Zootecnia 35, 1881–1886.

Miller, D., Gray, M., Storfer, A., 2011. Ecopathology of ranaviruses infecting amphibians. Viruses 3, 2351–2373.

Moffitt, C.M., 1991. Oral and injectable applications of erythromycin in salmonid fish culture. Veterinary and Human Toxicology 33 (Suppl. 1), 49–53.

Moffitt, C.M., 2005. Environmental, economic and social aspects of animal protein production and the opportunities for aquaculture. Fisheries 13, 10–12.

Murray, A.G., 2002. Making the case for zoning. Australian Veterinary Journal 80, 458.

Murray, A.G., Peeler, E.J., 2005. A framework for understanding the potential for emerging diseases in aquaculture. Preventive Veterinary Medicine 67, 223–235.

Naviner, M., Bergé, J.-P., Durand, P., Le Bris, H., 1999. Antibacterial activity of the marine diatom *Skeletonema costatum* against aquacultural pathogens. Aquaculture 174, 15–24.

Naylor, R.L., Goldburg, R.J., Primavera, J.H., Kautsky, N., Beveridge, M.C., et al., 2000. Effect of aquaculture on world fish supplies. Nature 405, 1017–1024.

Noga, E.J., 2010. Fish Disease: Diagnosis and Treatment. Wiley-Blackwell.

OIE, 2009. Manual of Diagnostic Tests for Aquatic Animals. OIE, Paris, France.

OIE, 2016a. Aquatic Animal Health Code, sixth ed. Office International des Epizooties, Paris, France.

OIE, 2016b. Manual of Diagnostic Tests for Aquatic Animals, fourth ed. Office International des Epizooties, Paris, France.

Panigrahi, A., Kiron, V., Kobayashi, T., Puangkaew, J., Satoh, S., et al., 2004. Immune responses in rainbow trout *Oncorhynchus mykiss* induced by a potential probiotic bacteria *Lactobacillus rhamnosus* JCM 1136. Veterinary Immunology and Immunopathology 102, 379–388.

Park, S.C., Shimamura, I., Fukunaga, M., Mori, K.-I., Nakai, T., 2000. Isolation of bacteriophages specific to a fish pathogen, *Pseudomonas plecoglossicida*, as a candidate for disease control. Applied and Environmental Microbiology 66, 1416–1422.

Pascho, R.J., Elliott, D.G., Streufert, J.M., 1991. Brood stock segregation of spring chinook salmon *Oncorhynchus tshawytscha* by use of the enzyme-linked immunosorbent assay (ELISA) and the fluorescent antibody technique (FAT) affects the prevalence and levels of *Renibacterium salmoninarum* infection in progeny. Diseases of Aquatic Organisms 12, 25–40.

Pillay, T., Kutty, M., 2005. Aquaculture: Principles and Practices. Blackwell Publishing Ltd.

Scarfe, A.D., Lee, C.-S., O'bryen, P.J., 2008. Aquaculture Biosecurity: Prevention, Control, and Eradication of Aquatic Animal Disease. Willey-Blackwell.

Schlotfeldt, H., 1996. Synopsis of freshwater aquaculture legislation in Germany since national reunification. Revue scientifique et technique (International Office of Epizootics) 15, 687–701.

Snieszko, S., 1973. Recent advances in scientific knowledge and developments pertaining to diseases of fishes. Advances in Veterinary Science and Comparative Medicine 17, 291–314.

Stoskopf, M., 1993. Fish Medicine. W.B. Saunders Company, Philadelphia, Pennsylvania.

Tafalla, C., Bogwald, J., Dalmo, R.A., 2013. Adjuvants and immunostimulants in fish vaccines: current knowledge and future perspectives. Fish and Shellfish Immunology 35, 1740–1750.

Timmerman, H., Koning, C., Mulder, L., Rombouts, F., Beynen, A., 2004. Monostrain, multistrain and multispecies probiotics—a comparison of functionality and efficacy. International Journal of Food Microbiology 96, 219–233.

Tonguthai, K., 1985. A Preliminary Account of Ulcerative Fish Diseases in the Indo-Pacific Region (A Comprehensive Study Based on Thai Experiences). Department of Fisheries, Ministry of Agriculture and Cooperation, Bangkok, Thailand.

Tovar-Ramírez, D., Mazurais, D., Gatesoupe, J., Quazuguel, P., Cahu, C., et al., 2010. Dietary probiotic live yeast modulates antioxidant enzyme activities and gene expression of sea bass (*Dicentrarchus labrax*) larvae. Aquaculture 300, 142–147.

Van Vliet, D., Wiens, G., Loch, T., Nicolas, P., Faisal, M., 2016. Genetic diversity of *Flavobacterium psychrophilum* isolated from three *Oncorhynchus* spp. in the U.S.A. revealed by multilocus sequence typing. Applied and Environmental Microbiology 82, 3246–3255.

Verschuere, L., Rombaut, G., Sorgeloos, P., Verstraete, W., 2000. Probiotic bacteria as biological control agents in aquaculture. Microbiology and Molecular Biology Reviews 64, 655–671.

Weeks, C., Kim, R., Wolgamod, M., Whelan, G., Faisal, M., 2011. Experimental infection studies demonstrate the high susceptibility of the salmonid, lake herring, *Coregonus artedi* (Le Sueur), to the Great Lakes strain of viral haemorrhagic septicaemia virus (genotype IVb). Journal of Fish Diseases 34, 887–891.

Westers, H., Wedemeyer, G., 2002. Fish Hatchery Management. American Fisheries Society.

Wiens, G.D., Lapatra, S.E., Welch, T.J., Evenhuis, J.P., Rexroad III, C.E., et al., 2013. On-farm performance of rainbow trout (*Oncorhynchus mykiss*) selectively bred for resistance to bacterial cold water disease: effect of rearing environment on survival phenotype. Aquaculture 388, 128–136.

Woo, P., Bruno, D., 2011. Fish Diseases and Disorders: Viral, Bacterial and Fungal Infections. CAB International, Oxfordshire, UK.

Wu, Z.-Q., Jiang, C., Ling, F., Wang, G.-X., 2015. Effects of dietary supplementation of intestinal autochthonous bacteria on the innate immunity and disease resistance of grass carp (*Ctenopharyngodon idellus*). Aquaculture 438, 105–114.

Yanong, R.P., 2003. Fish health management considerations in recirculating aquaculture systems – part 2: pathogens. UF/IFAS Circular 121.

Yanong, R.P., Erlacher-Reid, C., 2012. Biosecurity in Aquaculture, Part 1: An Overview. USDA Southern Regional Aquaculture Center Special Publication. United States Department of Agriculture.

Chapter 10

Aquatic Animal Health and the Environmental Impacts

Aaron A. McNevin
World Wildlife Fund, Washington, DC, United States

10.1 INTRODUCTION

The implications and effects that a particular culture environment and the broader environment have on species being reared are the factors that many aquaculturists consider and give ample attention to. These considerations are often related to the potential for aquatic species to be stressed, such as low dissolved oxygen content in a pond or ample tidal flushing for siting an entire farm in a particular estuarine environment. However, there is less consideration for how aquatic animal–health issues relate to or affect the natural environment.

Negative environmental impacts occur in any form of animal husbandry, and aquaculture is no different. Some of these impacts are simply a matter of energy transfer and waste in the production of animal proteins. Other impacts can be attributed to farm management. Successful aquaculturists are better at managing aquatic animal health-related problems. Avoidance of disease is clearly the optimal situation and many producers take measures to ensure animals are not stressed, and they recognize the limitations of their particular culture system. Of course, even the most isolated and biosecure facilities can be affected by pathogenic organisms, and the manner in which they attempt to remedy these issues can have an impact on the natural environment.

There is an inherent challenge in determining how management of aquatic animal-health issues relates to the surrounding environment of a farm. This is because aquaculture seldom exists as the sole "user" of a particular landscape. Thus, determining the impacts of one or more aquaculture operations in a specific geography that contains numerous "users" of the environment requires the ability to parse out aquaculture's impacts from other users. For example, estuaries are considered to be the bottom of the watershed where all of the effects of upstream users are realized. If an aquaculture operation is located near this estuary, the challenge of determining the magnitude of impact that aquaculture has on the local environmental condition is great. Most of the negative impacts in the estuarine setting are a result of the cumulative use and abuse of the environment within the entire watershed. That being stated, there are examples that will be discussed where aquaculture and the aquatic animal health of the culture stock is directly related to the condition of other species in the environment. Further, there are examples where aquaculture operations take specific actions that can be attributed to an outcome in the natural environment.

The environmental nongovernmental organizations (eNGOs) and others have exerted substantial effort to raise awareness of the environmental impacts of aquaculture; however, they differ in their approach to addressing environmental issues. Fundamentally, natural resources and their appropriate and efficient use should be the basis of those organizations concerned with the environment. With regard to aquaculture, these natural resources are land, water, fossil fuels (and the effects of their use), and wild fish. If focus could remain on these key natural resources and their use, much progress in the debate on responsible aquaculture could be achieved.

The most efficient use of natural resources is ultimately a benefit for the producer. Less energy, lower feed conversion ratios (FCRs), better survival, and more production per unit land all have financial implications for the producer. If these variables of production and farm management are not optimized, more waste and more natural resources will be used to produce the culture species. Thus, a producer that is not efficient with the use of natural resources will suffer financially. In a globalized market, poorer performing producers will be at a disadvantage because the competition for sale of aquaculture species is fierce.

This chapter is intended to invert the conventional thinking of how a particular environment affects aquatic animal health to consider how aquatic animal health and interventions affect the environment in turn. Emphasis will be on feed-based aquaculture systems and their natural resource use, but issues such as disease transfer and chemical use also will be addressed.

Fish Diseases. http://dx.doi.org/10.1016/B978-0-12-804564-0.00010-7

10.2 NATURAL RESOURCE USE

10.2.1 Land

The siting of land-based aquaculture requires a large portion of the site to be converted to an aquatic system. In so doing, the aquatic system will not have the same ecological function as would a terrestrial ecosystem. If an undisturbed habitat is converted to a largely aquatic habitat, some portion of the original biodiversity will be lost.

Direct land use by aquaculture includes the farm and support facilities to maintain operations. Indirect land or more appropriately, embodied land use, is the land that is utilized off-site of the farming and operations activities. Embodied land is largely in the form of the land area necessary to produce the plant-based feed ingredients and will be addressed in a subsequent section with other embodied resource uses.

Direct land use in aquaculture can span from nearly no land used (cage-based systems) to large tracts of land being used (extensive pond culture). The effects of aquatic animal health on direct land use is not great. However, the magnitude of economic impact of disease does have a relation to the direct use of land for aquaculture. When a land is abandoned, because the activity is no longer economically viable, it may not be able to recover its previous ecosystem function. This effect is most notable when the land used has a high ecosystem value—whether for assimilation purposes (wetlands) or high biodiversity areas (tropical rain forests). Conversion of terrestrial areas to ponds can often recover their terrestrial ecosystem value over time. A simple example of a shallow pond that has been reclaimed by plant agriculture is provided (Fig. 10.1).

The ability for an ecosystem to recover through natural processes requires conditions similar to those in the former environment. The magnitude of habitat conversion is proportional to the time necessary for recovery to the previous condition (Dobson et al., 1997). As the natural process of eutrophication in lakes is essentially the transition from aquatic to wetland to land, aquaculture ponds will tend to follow this same succession. However, the rate of succession can be increased if water is restricted from accumulating in the former pond. As vegetation regrowth occurs, so too will the accumulation of organic matter in the former pond bottom, and over time the pond will become less shallow and begin to take the shape of the surrounding terrestrial environment. It is important to understand that the hydrology of the environment will be an important component to reestablish such that the former environmental condition can be realized.

The land dedicated to the culturing of the organism is seldom affected by the aquatic animal health and health management interventions. In rare cases where disease results in the inability of the company or individual to maintain ownership of the land, abandonment may render the land unfit to recover to natural conditions. The most prominent examples of this situation are from the shrimp aquaculture sector where disease left aquaculture operations financially insolvent requiring abandonment of the activity. In some cases, these operations were sited in the intertidal zones previously occupied by mangrove forests. The main reason that former shrimp ponds are difficult to be reclaimed by natural processes in mangrove forests is that the construction of ponds disrupts the hydrology of the site. When tidal action is restricted, natural mangrove

FIGURE 10.1 Reclaimed aquaculture ponds for terrestrial agriculture.

regrowth is hindered. The result is the remnants of ponds remaining and lack of ecosystem services provided by mangroves (Fig. 10.2).

The starkest example of aquatic animal health impacting the direct use of land is likely the early development of the shrimp aquaculture industry. In the late 1980–90s, it was thought that the ideal place to locate ponds was where shrimp were found. Intertidal mangrove forests presented a site that had access to habitat where shrimp were naturally found, and tidal forces which could fill and empty ponds. There are numerous reasons why mangrove forests are inappropriate sites for shrimp aquaculture (Lebel et al., 2002) and the shrimp aquaculture sector's early prospectors made serious mistakes in siting in these wetland ecosystems that ultimately led to considerable environmental degradation.

Unlike the scenario of the reclamation of a simple pond in a terrestrial environment, shrimp ponds located in former mangrove forests caused a dramatic change in the hydrology of the coastal wetland ecosystems (Boyd and Clay, 1998). Levee ponds were constructed which restricted the contact of the natural tidal fluxes with the area converted to ponds. Thus, to reclaim these ponds, embankments must be leveled.

The widespread outbreaks of white spot syndrome virus and the Taura syndrome virus reduced production (Brock et al., 1997; Lightner, 1999) to the point where many shrimp farmers could not recover and left their land to fallow. Some of these ponds were used for other purposes, but some remain abandoned with no restoration of hydrology, thus no recovery of mangrove forest and loss of ecosystem function (Dierberg and Kiattisimkul, 1996; Primavera, 1997; Sathirathai and Barbier, 2001).

Reclamation of aquaculture sites that are forced to close because of disease would not likely consider reclamation of the site as a priority, thus allowance for siting of aquaculture operations should be highly scrutinized to consider the fundamental environmental changes that will take place. Improper siting of aquaculture facilities accounts for most of the negative environmental impacts of the activity.

10.2.2 Water

Water is of critical importance to aquaculture producers and is also required for human well-being and the rest of the life on earth. Many regions of the world have scarce water resources and the availability of water for aquaculture is considered more commonly now with the potential human well-being interactions.

As water is the culture medium for aquaculture, the amount of water available and the quality of that water are key factors for the ability to sustain an aquaculture activity. Additionally, water quantity and quality will be major factors in the siting of aquaculture facilities, the type of farming system, and the species raised. Aquatic animal health does impact water resources in quantity more so than quality. However, chemicals used to treat health issues at aquaculture sites do impact water quality but those issues will be discussed in the appropriate context later in this chapter.

Depending on species, aquatic animals require certain attributes of water quality, which is one of the main factors affecting animal stress. Further, because aquaculture typically involves the feeding of aquatic organisms, nutrient

FIGURE 10.2 Abandoned shrimp pond constructed in mangrove forest.

FIGURE 10.3 Pump station for Vietnamese catfish farm.

accumulation and water-quality deterioration are common occurrences (Boyd and Tucker, 1998). The combined water-quality requirements of species cultured and the loading of nutrients in the water will have an impact on aquatic animal health, particularly in the form of stress. Thus, water management on farms is related to the desired level of stress reduction.

A common means to reduce the stress caused by impaired water quality is the flushing of water out of the culture system and replacing with less impaired water (water exchange). Water exchange is typically reported in the percent volume of water in a culture enclosure replaced per day (Yoo and Boyd, 1994). The use of water exchange is dependent on species and system type. The use of flowing water for trout culture requires near-constant water exchange (Soderberg, 1994). Brune and Drapcho (1991) described how water exchange can be used to supplement aeration and reduce the effects of eutrophication of pond water used for aquaculture. Hargreaves and Tucker (2004) quantified the amount of water exchange necessary for ponds to dilute ammonia concentrations that can stress fish. The Pangasius catfish industry in Vietnam has used water exchange (Fig. 10.3) to lower stress and prevent discoloration and off-flavor in fish. The rates of water exchange in Vietnam Pangasisus aquaculture can range from 30 to 100% of pond volume per day (Phu, 2015).

The culture of shrimp also uses water, but there is a tendency to not consider water use in the same light as water used for freshwater species because brackish water is not potable and cannot be used to irrigate crops or be consumed by humans. Nevertheless, shrimp aquaculture will require a range of water exchange rates depending on the system, and the use of this water has potential implications for the environmental health of receiving water bodies. Thus, although water use in shrimp farming does not typically compete with other water users, the presence of disease-causing organisms or outbreaks of disease can trigger shrimp producers to exchange more water or expel all water and leave ponds to dry.

Shrimp aquaculture has been plagued by numerous disease epidemics and while shrimp farming used substantial amounts of water exchange in its early stages of expansion (AQUACOP, 1984), there is a growing trend to reduce the water exchange—explicitly to combat the spread of early mortality syndrome in Asia.

High-volume, high-quality freshwater resources are scarce in the main areas of aquaculture production. Water resource availability is not expected to increase over time and aquaculture will be required to produce more fish with less available water. This may foster greater development of cage culture in open water systems where water is not pumped, but other environmental implications for the expansion of cage culture will be realized.

10.2.3 Energy

Fossil fuels are a result of deposition of organic carbon in dead biomass from centuries ago and their gradual transformation to oil, coal/peat, and natural gas. When these materials are combusted, concentrations of carbon dioxide, methane, and several other gases result and these gases can enhance the ability of the earth's atmosphere to trap and hold heat. Thus, these gases can contribute to climate change. There is a large effort to reduce the amount of greenhouse gases emitted into the atmosphere and every industrial sector in existence has been compelled to take steps to reduce these emissions.

Energy generated for aquaculture can come from a variety of sources depending on geography. Coal fire, hydroelectric, natural gas, and geothermal are all used to varying extent in aquaculture production. While there is a natural tendency to optimize energy use on a farm to reduce costs, there are occasions where greater energy is used to maintain the culture stock on the farm or to conduct specific tasks (such as harvesting). There are aspects of compromised aquatic animal health where greater energy is utilized than would be under normal operating conditions.

It is clear that flushing water through culture enclosures is one strategy discussed previously to reduce stress of culture species, but in many cases the flushing is carried out using pumps powered by gasoline or diesel fuel, and to a lesser extent, by electricity.

Persistent low DO concentrations can significantly affect production because cultured species may consume less feed, grow more slowly, convert feed less efficiently, be more susceptible to infections and diseases, and even suffocate and die (Tucker, 2005).

10.2.4 Embodied Natural Resources

The natural resource use burden for aquaculture is composed of the natural resources used directly on the farm for the culture activity (direct natural resource use) and the natural resources that are embodied in feed (indirect natural resource use). The latter is less concise than the former because the resources embodied in feeds are not constant and the ingredients in feed change frequently. The feed ingredients purchased by major aquaculture feed companies are commodities and follow trends of market prices. The lowest cost formulation for feeds is often used and this approach will require substitution based on commodity pricing. Thus, the specific ingredients in feed are seldom known outside of the feed manufacturer and the manufacturer does not share formulations publically. Additionally, productivity and resources used for plant crops change with variables such as soil health, precipitation, temperature. Many of these factors are specific to a geographic region, thus some regions of the world have a greater potential for producing plant crops with less natural resources. Nevertheless, estimates based on average yields of crops and the resources they require can allow for an estimation of embodied natural resource use in aquaculture feeds.

The main source of nonfarm natural resource use is embodied in the production of feed ingredients. A variety of feed ingredients are utilized in the production of aquaculture feeds and typically includes fish meal, animal by-product meals, fish oil, plant meals and other plant products and by-products, vegetable oil, vitamins, minerals, and other additives. Thus, when producing an aquaculture species with feed, there are external factors in the production of specific feed ingredients that result in an embodied environmental burden. The production of all ingredients in feed requires natural resources. Land, water, energy, and wild fish are the critical natural resources that are utilized in the production of aquaculture feeds. The relationship between natural resources dedicated for terrestrial feed ingredients and aquatic animal health rests in the amount of feed utilized for culture organisms that do not enter the global food system. Essentially, the feed used to produce aquatic organisms that die before being harvested represents a quantity of wasted resources (Table 10.1). For a detailed description of natural resource use in aquaculture, the reader is referred to Boyd and McNevin (2015).

TABLE 10.1 Change in Natural Resources Used per Metric Ton of Production of Commonly Traded Aquaculture Species at Two Different Survival Rates

	Energy (GJ/t)		Land (ha/t)		Water (m^3/t)		Wild Fish (kg/t)	
	95%	75%	95%	75%	95%	75%	95%	75%
Atlantic salmon	13.1	16.6	0.15	0.19	894	1132	3032	3840
Channel catfish	5.2	6.5	0.19	0.24	1292	1636	–	–
Pangasius	5.6	7.1	0.23	0.30	1263	1600	635	804
Tilapia	6.1	7.8	0.33	0.42	1774	2247	211	267
White leg shrimp	9.5	12.1	0.26	0.33	1175	1488	861	1091

An estimate of natural resources per unit of production was provided by Chatvijitkul et al. (2016).

10.3 DISEASE TRANSFER

The transfer of disease from aquacultured species to natural fauna is one of the most obvious interactions between aquatic animal health on farms and the environment. There are impacts from the introduction of novel species which can be accompanied by other disease-causing organisms, but there are also impacts that result from the intensification of aquaculture. The intensification of aquaculture may foster the cultivation of organisms that can be detrimental to flora and fauna in the natural environment.

Mechanisms, such as risk assessments, are not uniform and every country (and even more local jurisdictions) will have a range of steps that are necessary to be taken to impede the negative impacts of the introduction of the culture and carrier species. Even with highly conservative efforts to mitigate risk, it is possible for oversights and mistakes to be made leading to considerable harm of the surrounding environment.

Aquaculture operations in open systems can act as a sink for potential pathogens. Wallace et al. (2008) demonstrated a higher prevalence for infection of infectious pancreatic necrosis virus (IPNV) in wild marine fish from Scottish waters in the vicinity of Atlantic salmon pens. Gregory et al. (2007) found repositories of IPNV in a variety of other organisms besides fish. Krkosek et al. (2007) suggested that there is strong empirical evidence that salmon farm-induced *Lepeophtheirus salmonis* infestations of juvenile pink salmon have depressed wild pink salmon populations and may lead to localized extinctions. Further, Nylund et al. (1999) found that *L. salmonis* can act as a vector of infectious salmon anemia (ISA). Nowak and LaPatra (2006) showed the increased bacterial loads of bacterial kidney disease (BKD) in cultured fish versus wild. Jones et al. (1999) noted that dense culture systems tend to have increased risk of disease and also shed disease agents to the surrounding environment.

The salmon aquaculture industry has been one of the most scrutinized sector of the global aquaculture industry. It is not surprising that the impacts of disease transmission in the salmon industry have drawn considerable attention. However, salmon is not alone in the transmission of disease to wild stocks. Even those species thought of as benign can have considerable effects on their wild counter parts. The koi herpes virus (KHV) has been transferred throughout the world as koi are considered a valuable ornamental fish and carp are a staple fish for many Asian cultures. In Indonesia, introductions of koi seed for aquaculture resulted in the transmission of KHV to common carp, and in 2002 the virus began to spread from East Java to Sumatra with mortality of wild and farmed carp reaching 80–90% (Sunarto et al., 2005).

Bivalve mollusks are cultured widely throughout the world, and there are numerous species and strains that are desirable for their specific taste and their aquaculture production attributes. Further, it is important to note that often these mollusks are being cultured near or with native and naturally occurring species. Thus, introduction of a new species into the same environment where similar or the same species inhabit can lead to the transfer of pathogenic organisms.

The oyster aquaculture industry in the United States managed several introductions of novel disease agents. Multinucleated unknown (MSX) disease first appeared in 1957 in Delaware Bay where it caused massive mortalities of eastern oysters (*Crassostrea virginica*) (Barber, 1999). Juvenile oyster disease was first noted in 1988 in hatchery-produced eastern oyster seed held at nursery sites in Maine, New York, and Massachusetts (Elston, 1990). It has been responsible for mortalities of over 90% in certain locations. First noticed in oysters brought into the United States from Europe, *Bonamia ostreae* became a prominent protozoan disease of oysters in California. This disease was subsequently transported with seed to Brittany, France, where it led to the demise of the flat oyster industry throughout Europe. Dermo disease was first documented in the 1940s in the Gulf of Mexico, where it caused extensive oyster mortalities. The disease was found in Chesapeake Bay in 1949, and since then it had been consistently present there. Dermo disease has spread in the United States and can be found in other places including Delaware Bay, Cape Cod, New Jersey, and Maine. The range extension was attributed to introductions by many means over many years in conjunction with recent increases in sea-surface temperatures during the winter (Ford, 1996; Cook et al., 1998; Ford et al., 2000).

The high value of various univalve species of abalone has fostered the cultivation of numerous species across Asia, Australia, South Africa, and the United States. Its relatively slow growth increases the exposure to various stressors and disease-causing organisms. Examples of disease transfer from abalone farms also shows the risks associated with importation and translocation of species. The parasitic sabellid polychaete worm, native to South Africa, colonizes the shell of abalone altering its appearance and making it less marketable. The polychaete infects *Haliotis rufescens*, *H. fulgens*, *H. corrugate*, *H. midae*, and other species of abalone as well as some species of marine gastropods, and it was first introduced to California by importation of South African abalone for commercial aquaculture research (Culver et al., 1997). California abalone aquaculturists have been the most severely affected by this parasite. The abalone viral ganglioneuritis (AVG) virus first appeared in abalone farms in southern Victoria in late 2005, and subsequently spread to the wild abalone fishery along 280 km of coastline at a rate of 5–10 km/month (Hills, 2007), causing a reduction of total allowable catch (TAC) in the fishery from 280 to 16 mt (Mayfield et al., 2011).

It is clear that disease transmission occurs and in several instances the resulting outbreaks can be severe for both the farmed and wild species. Governmental regulations and international normative guidelines boast of numerous measures to prevent negative impacts of the introductions of aquatic organisms, yet there are continual cases in which these measures have failed. It is notable that considerable research and risk-based assessments have occurred for the introduction of exotic species, yet the negative effects are still realized at present. Obviously, the banning of exotic species would be the best measure to prevent the various disease organisms from affecting wild species. Unfortunately, even the cultivation of native species has been shown to have implications on the aquatic health of wild species.

10.4 CHEMICAL USE

Chemicals used in aquaculture are varied with many different purposes. Many of the chemicals, such as agricultural limestone used to buffer pH fluctuations in ponds are benign, and others are quite toxic, such as cytochalasin B which is used to induce polyploidy in various aquaculture species. Consequently, the effects of chemical use to control aquatic animal health-related problems on the environment can be attributed to the toxicity or effect of the chemical in the environment, the amount of chemical released into the environment, and the frequency of its use. The chemicals most widely used in aquaculture are fertilizers and liming materials, but a wide array of other chemicals are also used including oxidants, coagulants, osmoregulators, algicides, herbicides, fish toxicants, antifoulants, disinfectants, anesthetics, agricultural pesticides, and hormones (Boyd and McNevin, 2015). It is beyond the scope of this chapter to describe all of the interactions of these chemicals with the natural environment. However, there are categorical statements that can be made regarding the use of certain types of chemicals that can have a major impact on the environment.

10.4.1 Fertilizers

In much of the feed-based aquaculture, fertilizers are seldom used. There are instances in the early life stages of certain aquatic organisms which require planktonic food sources and in such cases fertilizers are used to stimulate primary production. The stimulation of primary production is the main impact of fertilizer use on the natural environment. Although it is wasteful to discharge fertilizers in effluent, some extensive forms of production require both fertilization and water exchange to maintain water quality. In these circumstances, fertilizer discharge in the form of nutrients can foster a greater amount of primary productivity in a receiving water body. This can lead to greater dissolved oxygen fluctuation and potential dead zones if enough organic matter accumulates. Fertilizers also require energy for mining and manufacturing, thus contributing to the emission of greenhouse gases.

10.4.2 Antibiotics

Antibiotics are designed to be utilized to treat bacterial infections, but they are also misused for viral or protozoan infestations and in some cases used prophylactically in an attempt to prevent aquatic animal health issues. Dosage rates and specific antibiotics used are often difficult to summarize for global aquaculture production, but Table 10.2 depicts the common antimicrobials observed to be used in aquaculture facilities. Some antibiotics that are utilized in aquaculture have been tested and approved for use for specific species under specific circumstances. Some producers will follow the guidelines for their use and others will not. A large portion of the antibiotics administered in fish farms (70–80%) has been reported to reach the environment (Schneider, 1994; Hektoen et al., 1995).

Approximately 70–80% of the antibiotics in medicated feed enter the water via urinary or fecal excretion and uneaten feed (Boyd and McNevin, 2015). The antibiotic remains in the water or soil until degraded by natural processes. Heuer et al. (2009) noted a wide usage of tetracyclines, sulfonamides, and fluoroquinolones in aquaculture and Gräslund et al. (2003) surveyed shrimp producers in Thailand and found that a large proportion used antimicrobial agents in the production process. Xiong et al. (2015) found three classes of antibiotics in water and soil in aquaculture zones of Guangdong Province, China, including nine antibiotics (sulfametoxydiazine, sulfamethazine, sulfamethoxazole, oxytetracycline, chlorotetracycline, doxycycline, ciprofloxacin, norfloxacin, and enrofloxacin) were detected in sediment and water samples. While most of the concern over antibiotic use is the development of microbial resistance that, by horizontal transmission, could lead to human health concerns, these chemicals are toxic for the bacterial flora of the aquatic environment. Moreover, there is a large portion of photosynthetic bacteria that are also negatively affected by the release of antibiotics.

TABLE 10.2 Antibiotics Commonly Used in Aquaculture

Class	Compound
Aminoglycosides	Neomycin*
	Gentamycin*
	Streptomycin (obsolete)
β-Lactams; penicillin	Amoxicillin*
	Ampicillin*
Fenicoles	Chloramphenicol
	Florfenicol
Fluoroquinolones	Ciprofloxacin*
	Enrofloxacin*
	Flumequin*
Macrolides	Erythromycin*
Non-fluorinated guinolones	Oxolinic acid
	Sarafloxin
Sulfonamides	Sulfamethazine
	Sulfamerazine
	Sulfadimethoxine
Trimethoprim	Trimethoprim*
Tetracyclines	Chlortetracycline
	Oxytetracycline*
	Tetracycline*

*Commonly used in human medicine.
Boyd, C., McNevin, A., 2015. Aquaculture, Resource Use, and the Environment. John Wiley & Sons.

10.4.3 Insecticides

Agricultural insecticides have been used to control unwanted insects and also as a piscicide in some aquaculture sectors. These chemicals are used to remove predators to juvenile fish or postlarval shrimp (Fig. 10.4). Insecticides have been developed for row crop agriculture and not for aquatic environments. The use of these chemicals is inappropriate for aquaculture and much of the result sought from their use can be achieved via other treatments or proper barrier placement.

10.4.4 Herbicides

Herbicides are used to control aquatic macrophytes or to reduce the magnitude of algal blooms. Copper-based herbicides are most common and applied primarily to ponds. Copper has also been used in cage construction for its antifouling attributes. Copper is toxic to aquatic life as it inhibits cell division in plant life. If copper is released into natural waterways, it will cause algal kills. However, copper rapidly dissipates from water by either binding to organic matter or adsorption to soil particles. Within 1–2 days of application, copper tends to revert to background concentration in water (McNevin and Boyd, 2004). Large doses of copper released from aquaculture sites could kill large quantities of algae, fish, and other organisms, and may reduce dissolved oxygen concentration because of the large organic matter load that would result from a large kill. However, the likelihood of this situation occurring is low as a large application of copper would also kill the culture organisms.

10.4.5 Other Chemicals

Oxidizing agents and hormones are also used in the aquaculture sector. Oxidizing agents are used to sterilize and disinfect equipment. Hormones (17-methyl testosterone) are typically used to induce masculinization of tilapia fry. All of these

FIGURE 10.4 Insecticide containers found on the banks of a shrimp pond.

chemicals degrade relatively quickly with exposure to air or sunlight, and if used appropriately should not pose a significant threat to the environment.

10.5 CONCLUSION

Appropriate sites for new aquaculture operations are difficult to locate because much of the optimal land is occupied. Other areas where aquaculture could exist may be too polluted to support a viable culture operation. The environment that supports aquaculture must be maintained. There are numerous notions of what constitutes responsible aquaculture, but the biodiversity surrounding the aquaculture operation and the natural resources used to produce a product are the most important indications of impact.

Aquatic animal–health does influence the natural resources that are used and the surrounding environment. Land converted to aquaculture facilities can be abandoned because disease renders the operation insolvent. Greater the magnitude of change that has occurred through the conversion of former habitat, the longer the environment will take to recover and restore ecosystem function. Water and aeration are the two methods to improve culture conditions to reduce stress on animals. More energy is required to run aerators and water pumps, and if water exchange is employed, inefficient use of water resources is possible. There is a considerable amount of natural resource embodied in feed and that feed is considered wasted because the culture species does not survive until harvest, and those resources will also have been inefficiently utilized. The introduction of novel species can also bring novel disease-causing agents. Aquaculture can also be a vector of diseases for native or similar species that inhabit the environment surrounding the farm. If conditions are optimal, these hitchhiker organisms cultivated or introduced by a farm could harm or kill the natural fauna. Chemicals are used for a variety of aquatic animal health-related issues. If these chemicals are to kill organisms on a farm—bacteria, plants, predatory fish—they will also kill those in the natural environment. Aquaculture operations would be most efficient with chemical use and least impactful to the environment if there was no release of these chemicals before being taken up by the target organisms on the farm or until they are adequately degraded.

The global environment has been compromised in many ways and we continue to experience challenges with availability of natural resources. It is critical that for aquaculture to grow that it maintains a continuously improving efficiency with natural resource use. The efficient use of natural resources, appropriate use of chemicals, and maintenance of the surrounding environmental condition will not only reduce aquaculture ecological footprint but also increase farm efficiency.

REFERENCES

AQUACOP, 1984. Review of ten year of experimental penaeid shrimp culture in Tahiti and New Caledonia (South Pacific). Journal of the World Mariculture Society 15, 73–91.

Barber, B.J., 1999. A Guide to Bivalve Diseases for Aquaculturists in the Northeastern US. Sea Grant Maine, New Hampshire.

Boyd, C., McNevin, A., 2015. Aquaculture, Resource Use, and the Environment. John Wiley & Sons.

Boyd, C.E., Clay, J.W., 1998. Shrimp aquaculture and the environment. Scientific American 278 (6), 58–65.

Boyd, C.E., Tucker, C.S., 1998. Pond Aquaculture Water Quality Management. Kluwer Academic Publishers.

Brock, J.A., Gose, R.B., Lightner, D.V., Hasson, K.W., 1997. Recent developments and an overview of Taura syndrome of farmed shrimp in the Americas. Diseases in Asian Aquaculture 3, 267–283.

Brune, D.E., Drapcho, C.M., 1991. Fed Pond Aquaculture. Aquaculture Systems Engineering. American Society of Agriculture Engineers, ASAE Publication, pp. 15–28.

Chatvijitkul, S., Boyd, C.E., Davis, D.A., McNevin, A.A., 2016. Embodied Resources in Aquafeeds. http://dx.doi.org/10.1111/jwas.12360.

Cook, T., Folli, M., Klinck, J., Ford, S., Miller, J., 1998. The relationship between increasing sea-surface temperature and the northward spread of *Perkinsus marinus* (Dermo) disease epizootics in oysters. Estuarine, Coastal and Shelf Science 46 (4), 587–597.

Culver, C.S., Kuris, A., Beede, B., 1997. Identification and Management of the Exotic Sabellid Pest in California Cultured Abalone. California Sea Grant College System, University of California.

Dierberg, F.E., Kiattisimkul, W., 1996. Issues, impacts, and implications of shrimp aquaculture in Thailand. Environmental Management 20 (5), 649–666.

Dobson, A.P., Bradshaw, A.D., Baker, A.Á., 1997. Hopes for the future: restoration ecology and conservation biology. Science 277 (5325), 515–522.

Elston, R.A., 1990. Mollusc Diseases: Guide for the Shellfish Farmer. Washington Sea Grant.

Ford, S., Smolowitz, R., Chintala, M., 2000. The question of temperature and *Perkinsus marinus* (Dermo) activity in the northeastern United States. Journal of Shellfish Research 19, 571.

Ford, S.E., 1996. Range extension by the oyster parasite *Perkinsus marinus* into the northeastern United States: response to climate change? Oceanographic Literature Review 12 (43), 1265.

Gräslund, S., Holmström, K., Wahlström, A., 2003. A field survey of chemicals and biological products used in shrimp farming. Marine Pollution Bulletin 46 (1), 81–90.

Gregory, A., Munro, L.A., Wallace, I.S., Bain, N., Raynard, R.S., 2007. Detection of infectious pancreatic necrosis virus (IPNV) from the environment in the vicinity of IPNV-infected Atlantic salmon farms in Scotland. Journal of Fish Diseases 30 (10), 621–630.

Hargreaves, J.A., Tucker, C.S., 2004. Managing Ammonia in Fish Ponds, vol. 4603. Southern Regional Aquaculture Center, Stoneville.

Hektoen, H., Berge, J.A., Hormazabal, V., Yndestad, M., 1995. Persistence of antibacterial agents in marine sediments. Aquaculture 133 (3), 175–184.

Heuer, O.E., Kruse, H., Grave, K., Collignon, P., Karunasagar, I., Angulo, F.J., 2009. Human health consequences of use of antimicrobial agents in aquaculture. Clinical Infectious Diseases 49 (8), 1248–1253.

Hills, J., 2007. A Review of the Abalone Virus Ganglioneuritis (AVG). Ministry of Fisheries, New Zealand. Available from: http://www.biosecurity.govt.nz/...virus/avg-virusreview-julie-hills-nov07.pdf.

Jones, S.R., MacKinnon, A.M., Groman, D.B., 1999. Virulence and pathogenicity of infectious salmon anemia virus isolated from farmed salmon in Atlantic Canada. Journal of Aquatic Animal Health 11 (4), 400–405.

Krkosek, M., Ford, J.S., Morton, A., Lele, S., Myers, R.A., Lewis, M.A., 2007. Declining wild salmon populations in relation to parasites from farm salmon. Science 318 (5857), 1772–1775.

Lebel, L., Tri, N.H., Saengnoree, A., Pasong, S., Buatama, U., Thoa, L.K., 2002. Industrial transformation and shrimp aquaculture in Thailand and Vietnam: pathways to ecological, social, and economic sustainability? AMBIO: A Journal of the Human Environment 31 (4), 311–323.

Lightner, D.V., 1999. The penaeid shrimp viruses TSV, IHHNV, WSSV, and YHV: current status in the Americas, available diagnostic methods, and management strategies. Journal of Applied Aquaculture 9 (2), 27–52.

Mayfield, S., McGarvey, R., Gorfine, H.K., Peeters, H., Burch, P., Sharma, S., 2011. Survey estimates of fishable biomass following a mass mortality in an Australian molluscan fishery. Journal of Fish Diseases 34 (4), 287–302.

McNevin, A.A., Boyd, C.E., 2004. Copper concentrations in channel catfish *Ictalurus punctatus* ponds treated with copper sulfate. Journal of the World Aquaculture Society 35 (1), 16–24.

Nowak, B.F., LaPatra, S.E., 2006. Epitheliocystis in fish. Journal of Fish Diseases 29 (10), 573–588.

Nylund, A., Krossoy, B., Devold, M., Aspehaug, V., Steine, N.O., Hovlund, T., 1999. Outbreak of ISA during first feeding of salmon fry (*Salmo salar*). Bulletin of the European Association of Fish Pathologists 19, 70–74.

Phu, T.Q., 2015. Water Quality Characteristics in Striped Catfish (*Pangasianodon hypophthalmus*) Aquaculture Ponds. Striped Catfish (*Pangasianodon hypophthalmus*) Farming in Mekong Delta, Vietnam: Success and Challenges Toward Sustainable Development. Can Tho University Publishing house.

Primavera, J.H., 1997. Socio-economic impacts of shrimp culture. Aquaculture Research 28 (10), 815–827.

Sathirathai, S., Barbier, E.B., 2001. Valuing mangrove conservation in southern Thailand. Contemporary Economic Policy 19 (2), 109–122.

Schneider, J., 1994. Problems related to the usage of veterinary drugs in aquaculture–a review. QUIMICA ANALITICA-BELLATERRA 13, S34.

Soderberg, R.W., 1994. Flowing Water Fish Culture. CRC Press.

Sunarto, A., Rukyani, A., Itami, T., 2005. Indonesian experience on the outbreak of koi herpesvirus in koi and carp (*Cyprinus carpio*). Bulletin of Fisheries Research Agency, Yokohama, Japan 86, 15–21.

Tucker, C., 2005. Pond Aeration. Southern Regional Aquaculture Center, Publication, p. 3700.

Wallace, I.S., Gregory, A., Murray, A.G., Munro, E.S., Raynard, R.S., 2008. Distribution of infectious pancreatic necrosis virus (IPNV) in wild marine fish from Scottish waters with respect to clinically infected aquaculture sites producing Atlantic salmon, *Salmo salar* L. Journal of Fish Diseases 31 (3), 177–186.

Xiong, W., Sun, Y., Zhang, T., Ding, X., Li, Y., Wang, M., Zeng, Z., 2015. Antibiotics, antibiotic resistance genes, and bacterial community composition in fresh water aquaculture environment in China. Microbial Ecology 70 (2), 425–432.

Yoo, K.H., Boyd, C.E., 1994. Hydrology and Water Supply for Pond Aquaculture. Chapman & Hall.

Index

'*Note*: Page numbers followed by "f" indicate figures, "t" indicate tables.'

Printed in the United States
By Bookmasters